P9-CCB-739

PRINCIPLES OF
GENETICS

Robert H. Tamarin

Boston University

Willard Grant Press, Boston, Massachusetts

Principles of Genetics was prepared for publication by the following people:

Production Editor: *Robine Storm van Leeuwen*
Technical Editor: *Mary Lewis*
Copy Editor: *Jean Peck*
Interior Designer: *Nancy McJennett*
Cover Designer: *Trisha Hanlon*
Typesetting by *University Graphics, Inc.*; artwork by *F.W. Taylor Associates*; covers printed by *Lehigh Lithographers*; text printed and bound by *R.R. Donnelley & Sons Company*

PWS PUBLISHERS

Prindle, Weber & Schmidt · 🦫 · Willard Grant Press · **wg** · Duxbury Press · ♠
Statler Office Building · 20 Providence Street · Boston, Massachusetts 02116

© Copyright 1982 by PWS Publishers
All rights reserved. No part of this book may be reproduced or transmitted in any form or by any means, electronic or mechanical, including photocopying, recording, or any information storage or retrieval system, without permission, in writing, from the publisher.

PWS Publishers is a division of Wadsworth, Inc.

Library of Congress Cataloging in Publication Data

Tamarin, Robert H.
   Principles of genetics.
   Includes bibliographies and index.
   1. Genetics.   I. Title.
QH430.T34   575.1   81-7134
ISBN 0-87150-756-0   AACR2

About the cover: A computer-generated end view of the DNA molecule; the color representation is red = oxygen, blue = nitrogen, green = carbon, yellow = phosphorus, and hydrogen not shown. Reproduced courtesy of Robert Langridge, Ph.D., Computer Graphics Laboratory, University of California, San Francisco © the Regents of the University of California.

Printed in the United States of America
85   84   83   82   81   —   1   2   3   4   5   6   7   8   9   10

For Ginger, David, and Bonnie

Contents

PART II   MOLECULAR GENETICS

PART III EVOLUTIONARY GENETICS

Preface

Genetics, the science of heredity, deals with the rules of inheritance in cells, individuals, and populations and with the molecular mechanisms by which genes control the growth, development, and appearance of an organism. No area of biology can truly be appreciated or understood without an understanding of genetics because genes not only control cellular processes but they also determine the course of evolution. Genetics is an exciting and basic science whose concepts provide the framework for the study of modern biology.

My purpose in this text is to provide a balanced treatment of the major areas of genetics in order to prepare students adequately for upper-level courses in the curriculum and to help them share in the excitement of the research being undertaken. Most readers of this text will have taken a general biology course and will have had some background in cell biology and organic chemistry. However, for an understanding of the concepts in this text, the motivated student will need to have completed only an introductory biology course and have had some chemistry and algebra in high school.

Genetics is commonly divided into three general areas: classical, molecular, and population. Many of those teaching genetics feel that a historical approach provides a sound introduction to the field and that a thorough grounding in Mendelian genetics is necessary for an understanding of molecular and population genetics—an approach that this book follows. Others, however, may prefer to begin with molecular genetics; for this reason the chapters have been written as relatively discrete units, which allows for flexibility in the order of their use. A thorough glossary and separate subject and author indexes will help to maintain the continuity if the order of the chapters is changed.

Genetics research is a new and mushrooming industry. An understanding of genetics is crucial to advancements in medicine; and genetic controversies—such as the potential harm of recombinant DNA and the nature of the heritability of IQ—have captured the interest of the general public. Throughout the book, the implications for human health and welfare of the research conducted in laboratories and universities around the world have been pointed out. Interesting digressions, in the form of boxed material, give insights into techniques, controversies, and recent breakthroughs in genetics.

For many students, genetics is their first analytical biology course, and they may have difficulty with the quantitative aspects of the field. There is no substitute for work with pad and pencil. This text provides a large number of problems to help the reader learn and retain the material. All problems, whether within the body of the text or at the end of the chapters, should be worked through by the student as they are encountered. After students have worked out the problems, they may want to refer to the answer section appearing at the back of the book.

References to review articles, more advanced volumes, and articles in the original literature are provided for those who wish to pursue particular topics. And photographs of selected geneticists are included to add the human touch to this science. Geneticists are people who occasionally make mistakes and often disagree with each other, and their words (like the words in this book) are not chiseled in stone, correct for all eternity. Perhaps the glimpse of a face from time to time will underscore that fact.

There are many people whom I would like to thank for their encouragement and scientific and technical help during the writing of this book. This book was first conceived of with the encouragement of John Moroney and Jerry Lyons. Its completion as a readable, accurate, and, I hope, useful text rests with the constant input of Willard Grant Press's Jean-François Vilain and Robine Storm van Leeuwen; with Mary Lewis, whose editing abilities and knowledge of biology contributed greatly to the improvement of the book; and with the following reviewers: Harold Rauch—University of Massachusetts, Amherst; August P. Mueller—SUNY Binghamton; A. Thomas Weber—University of Nebraska at Omaha; and James R. Wild—Texas A & M University. Mark Goldberg, Jayne Schack, and Anna Gillis helped me with permissions. Laurel Kologe Sutton, Erika Goldberg, and Bertha Bleiwas typed various drafts of the manuscript. Wayne Rosenkrans, Tom Broker, Tom Brewster, and Gary Jacobson were especially helpful in providing me with photographs. My colleague, with whom I teach genetics at Boston University, Sonia Guterman, has been a constant source of information and encouragement. Thanks also to my wife and family for putting up with much neglect during the time of the writing of this book.

Robert H. Tamarin
Boston

PART I

MENDELISM AND THE CHROMOSOMAL THEORY

1 Mendel's Principles

Turnips produce turnip offspring and humans produce human offspring. This book and the subject of genetics are concerned with understanding the mechanism and controls of the processes by which an offspring comes to look like its parent. In this, the first section of the book, we are concerned with the rules of transmission of genes, the units that control and determine the processes of development and the ultimate appearance of individuals. Gregor Mendel discovered these rules of inheritance—we derive and expand upon his rules in this chapter.

In 1900 three botanists, Carl Correns of Germany, Erich von Tschermak of Austria, and Hugo DeVries of Holland, simultaneously and independently rediscovered the rules governing the transmission of traits from parent to offspring. These rules had been previously published in 1866 by an obscure Austrian monk named Gregor Mendel. Although his work was widely available after 1866, it was not until the turn of the century that the scientific community was ready to appreciate Mendel's great contribution. There were at least four reasons for this lapse of 34 years.

First, prior to Mendel's experiments, biologists were primarily concerned with explaining the transmission of characteristics that could be measured on a continuum from low to high such as height, cranium size, and longevity. They were looking for rules of inheritance that would explain such characteristics, or **continuous variations,** especially after Darwin's theory of evolution was put forward in 1859. Mendel, however, suggested that inherited characteristics were discrete and constant **(discontinuous):** Peas were either yellow or green. Evolutionists were looking for small changes in traits with continuous variation, whereas Mendel presented them with rules for discontinuous variation. His principles did not seem to cover the type of variation that biologists thought prevailed. Second, there was no physical element to which Mendel's inherited particles could be anchored. One could not say, upon reading Mendel's work, that yes a certain subunit of the cell

3

followed Mendel's rules. Third, Mendel worked with large numbers of offspring and converted these numbers to ratios of observed classes. Biologists, practitioners of a very descriptive science at the time, were not well versed in mathematical tools. And last, Mendel was not well known and did not persevere in his attempts to convince the academic community.

Between 1865 and 1900, two major changes took place in biological science. First, by the turn of the century, not only had chromosomes been discovered, but also their movement during cell division had become understood. Second, biologists were more prepared to handle simple mathematics by the turn of this century than they were during Mendel's time.

MENDEL'S EXPERIMENTS

Gregor Mendel (Figure 1–1) was an Austrian monk (of Brünn, Austria, which is now Brno, Czechoslovakia). The essence of his experiments was to **crossbreed** plants that had discrete, nonoverlapping characteristics and then to observe the distribution of these characteristics in the next several generations. Mendel worked with the common garden pea plant, *Pisum sativum*. (See the translation of part of Mendel's original paper.) He chose the pea plant for several reasons. (1) The garden pea was an ideal plant with which to work because of its ease in cultivation and relatively short life cycle. (2) The plant had discontinuous characteristics such as flower color and pea

Figure 1–1.
Gregor Johann
Mendel

Source: Reproduced by permission of the Moravske Museum Mendelianum.

Erich von
Tschermak-
Seysenegg (1871–
1962)
Genetics, 37 (1952):
frontispiece.

texture. (3) In part because of its anatomy, pollination of the plant could be easily controlled—foreign pollen could be kept out, and **cross-fertilization** could be artificially accomplished.

Figure 1–2 shows a cross section of the pea flower and indicates the keel in which the anthers and stigma develop. Normally, **self-fertilization** occurs when pollen falls onto the stigma before the bud opens. Mendel cross-fertilized the plants by opening the keel of a flower before the anthers matured and by placing pollen from another plant on the stigma. In more than 10,000 plants examined by Mendel, only a very few were fertilized other than the way he had intended them to be (either self- or cross-pollinated.)

Mendel used plants that had been obtained from suppliers and grew them for two years to ascertain that they were homogeneous, or true-breeding, for the particular characteristic under study (Figure 1–3). He chose for study the seven characteristics shown in Figure 1–4. Let us take as an example the characteristic of plant height. While height is often continuously distributed, Mendel used plants that showed only two alternatives: very tall or dwarf. He made the crosses shown in Figure 1–5. In the parental, or P_1, generation, tall plants were pollinated by dwarf plants, and in a **reciprocal cross** dwarf plants were pollinated by tall ones to determine whether the results were independent of the parents' sex. As we will see later on, some traits have inheritance patterns related to the sex of the parents carrying the traits. In these cases reciprocal crosses give different results; with Mendel's tall and dwarf pea plants, the results were the same.

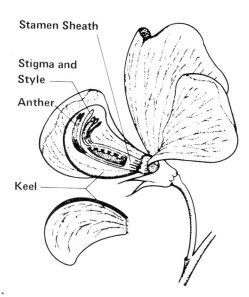

Stamen Sheath

Stigma and
Style

Anther

Keel

Figure 1–2.
Anatomy of the
Garden Pea Plant
Flower

Figure 1 – 3.   Mendel's Garden in Brno
To demonstrate Mendel's work, the curators of the Mendel Museum
originally planted tall and dwarf pea plants. Many visitors mistakenly
interpreted the genetic differences as being the result of shade; the
curators, therefore, switced to planting red and white begonias.

Source: Reproduced by permission of the Moravske Museum
Mendelianum.

Offspring of the cross of P_1 individuals are referred to as the first **filial generation,** F_1. Mendel also referred to them as **hybrids** because they were the offspring of unlike parents (tall and dwarf). We will specifically refer to the offspring of tall and dwarf peas as **monohybrids** because they are hybrid for only one characteristic (height). Since all the F_1 offspring plants were tall, Mendel referred to tallness as the **dominant** trait. The alternate, dwarfness, he referred to as **recessive.** When the F_1 offspring were self-fertilized to form the F_2 generation, both tall and dwarf offspring occurred; the dwarf characteristic reappeared. Among the F_2 offspring Mendel observed 787 tall and 277 dwarf plants for a ratio of 2.84 to 1. Many who might have done this type of experiment might not have had Mendel's insight to see an approximation to a 3:1 ratio in the numbers 787 and 277. This ratio suggested the mechanism of the inheritance of height to Mendel.

RULE OF SEGREGATION

Mendel assumed that each plant contained two determinants (which we now call **genes**) for the characteristic of height. Different forms of a gene exist within a population and are termed **alleles.** For example, a heterozygous pea plant possesses the *dominant tall allele* and the *recessive dwarf allele* for the gene that determines plant height. A homozygous pair of dwarf alleles

ALTERNATE FORMS

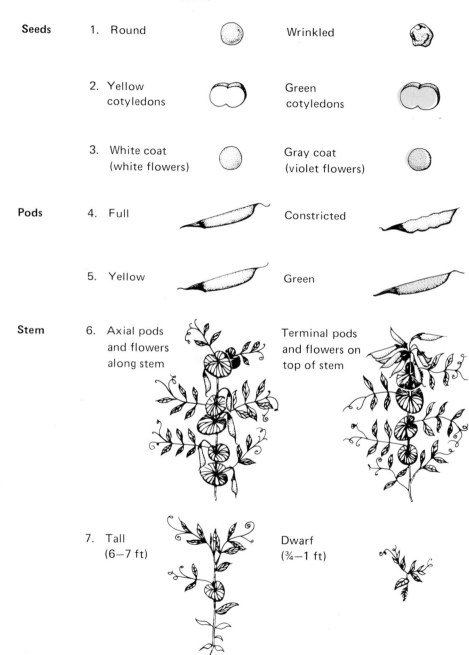

Seeds

1. Round   Wrinkled

2. Yellow cotyledons   Green cotyledons

3. White coat (white flowers)   Gray coat (violet flowers)

Pods

4. Full   Constricted

5. Yellow   Green

Stem

6. Axial pods and flowers along stem   Terminal pods and flowers on top of stem

7. Tall (6–7 ft)   Dwarf (¾–1 ft)

Figure 1–4.
Seven
Characteristics
Observed by Mendel
in Peas

7

P₁

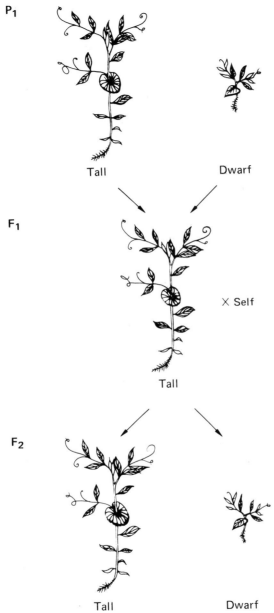

Tall   Dwarf

F₁

× Self

Tall

F₂

Tall   Dwarf

Figure 1–5.
First Two Offspring
Generations from the
Cross of Tall Plants
with Dwarf Plants

3 : 1

is required to develop the recessive phenotype. Only one of these determinants (alleles) is passed into a single gamete, and the union of two gametes to form a zygote restores the double complement of alleles. The fact that the recessive trait reappears in the F_2 generation shows that the allele controlling it was unaffected by being hidden in the F_1 individual. This explanation of the passage of these discrete trait determinants, the genes, is referred to as Mendel's first principle, the **rule of segregation.** The rule of segregation can be summarized as follows: A gamete receives only one allele from the pair of alleles possessed by an organism; fertilization (the union of two gametes) reestablishes the double number. We can visualize this process by redrawing Figure 1–5 using letters to denote the alleles. Mendel used capital letters to denote alleles that controlled dominant traits and lowercase letters for alleles that controlled recessive traits. Following this notation, T will be used for tall and t for short (dwarf). From Figure 1–6 we can see that Mendel's rule of segregation explains the homogeneity of the F_1 (all tall) generation and the 3:1 ratio of tall to dwarf in the F_2 generation.

Let us define some terms. The **genotype** of an organism is the genes it possesses. In Figure 1–6 the genotype of the parental tall plant is TT; that of the F_1 tall plant is Tt. **Phenotype** refers to the observable attributes of an organism. Both of the above two genotypes, TT and Tt, are phenotypically tall. Genotypes come in two general classes: **homozygotes,** in which both alleles are the same, as in TT or tt, and **heterozygotes,** in which the two alleles are different, as in Tt. These last two terms were coined by William Bateson in 1901. The word *gene* was first used by the Danish botanist Wilhelm Johannsen in 1909.

If we look at Figure 1–6, we can see that the TT homozygote can produce only one type of gamete, the T-bearing kind, and likewise, the tt homozygote can produce only t-bearing gametes. Thus, the F_1 individuals are uniformly heterozygous Tt, and each F_1 individual can produce two kinds of gametes in equal frequencies, T- or t-bearing. In producing the F_2 generation, these two types of gametes randomly pair during the process of fertilization. Figure 1–7 shows three ways of picturing this process.

Testing the Rule of Segregation

We can see from Figure 1–7 that the F_2 generation has a phenotypic ratio of 3:1, the classic Mendelian ratio. However, there is also a genotypic ratio of 1:2:1 for homozygote-dominant:heterozygote:homozygote-recessive. Demonstrating this genotypic ratio provides a good test of Mendel's hypothesis of segregation.

The simplest way to test the hypothesis is by **progeny testing**—that is, by self-fertilizing F_2 individuals to produce an F_3 generation, which is what Mendel did. From his hypothesis it is possible to predict the frequencies of

SOME EXCERPTS FROM MENDEL'S ORIGINAL PAPER

In February and March of 1865, Mendel delivered two lectures to the Natural History Society of Brünn. These were published as a single, 48-page article handwritten in German. The article appeared in the 1865 Proceedings of the Society, which came out in 1866. The paper was entitled "Versuche über Pflanzen-Hybriden," which means "experiments in plant hybridization." Following are some paragraphs from the English translation of the article to give some of the sense of the original.

In his introductory remarks Mendel writes:

> . . . That, so far, no generally applicable law governing the formation and development of hybrids has been successfully formulated can hardly be wondered at by anyone who is acquainted with the extent of the task, and can appreciate the difficulties with which experiments of this class have to contend. A final decision can only be arrived at when we shall have before us the results of detailed experiments made on plants belonging to the most diverse orders.
>
> Those who survey the work done in this department will arrive at the conviction that among all the numerous experiments made, not one has been carried out to such an extent and in such a way as to make it possible to determine the number of different forms under which the offspring of hybrids appear, or to arrange these forms with certainty according to their separate generations, or definitely to ascertain their statistical relations.
>
> . . .
>
> The paper now presented records the results of such a detailed experiment. This experiment was practically confined to a small plant group, and is now, after eight years' pursuit, concluded in all essentials. Whether the plan upon which the separate experiments were conducted and carried out was the best suited to attain the desired end is left to the friendly decision of the reader.

After discussing the origin of his seeds and the nature of the experiments, Mendel then proceeds to discuss the F_1, or hybrids, resulting:

> This is precisely the case with the Pea hybrids. In the case of each of the seven crosses the hybrid-character resembles that of one of the parental forms so closely that the other either escapes observation completely or cannot be detected with certainty. This circumstance is of great importance in the determination and classification of the forms under which the offspring of the hybrids appear. Henceforth in this paper those characters which are transmitted entire, or almost unchanged in the hybridisation, and therefore in themselves constitute the characters of the hybrid, are termed the *dominant,* and those which become latent in the process *recessive.* The expression "recessive" has been chosen because the characters thereby designated withdraw or entirely disappear in the hybrids, but nevertheless reappear unchanged in their progeny, as will be demonstrated later on.

He then writes about the F_2:

> In this generation there reappear, together with the dominant characters, also the recessive ones with their peculiarities fully developed, and this occurs in the defi-

nitely expressed average proportion of three to one, so that among each four plants of this generation three display the dominant character and one the recessive. This relates without exception to all the characters which were investigated in the experiments. The angular wrinkled form of the seed, the green colour of the albumen, the white colour of the seed-coats and the flowers, the constrictions of the pods, the yellow colour of the unripe pod, of the stalk, of the calyx, and of the leaf venation, the umbel-like form of the inflorescence, and the dwarfed stem, all reappear in the numerical proportion given, without any essential alteration. *Transitional forms were not observed in any experiment.*

. . .

Expt. 1. Form of seed.—From 253 hybrids 7,324 seeds were obtained in the second trial year. Among them were 5,474 round or roundish ones and 1,850 angular wrinkled ones. Therefrom the ratio 2.96 to 1 is deduced.

. . .

If A be taken as denoting one of the two constant characters, for instance the dominant, a the recessive, and Aa the hybrid form in which both are conjoined, the expression

$$A + 2Aa + a$$

shows the terms in the series for the progeny of the hybrids of two differentiating characters.

Mendel also discusses dihybrids. He mentions the genotypic ratio of 1:2:1:2:4:2:1:2:1 and the principle of independent assortment.

The fertilised seeds appeared round and yellow like those of the seed parents. The plants raised therefrom yielded seeds of four sorts, which frequently presented themselves in one pod. In all, 556 seeds were yielded by 15 plants, and of these there were

- 315 round and yellow,
- 101 wrinkled and yellow,
- 108 round and green,
- 32 wrinkled and green

. . .

Consequently the offspring of the hybrids, if two kinds of differentiating characters are combined therein, are represented by the expression

$$AB + Ab + aB + ab + 2ABb \\ + 2aBb + 2AaB + 2Aab + 4AaBb$$

This expression is indisputably a combination series in which the two expressions for the characters A and a, B and b are combined. We arrive at the full number of the classes of the series by the combination of the expressions

$$A + 2Aa + a \\ B + 2Bb + b$$

. . .

There is therefore no doubt that for the whole of the characters involved in the experiments the principle applies that *the offspring of the hybrids in which several essentially different characters are combined exhibit the terms of a series of combinations, in which the developmental series for each pair of differentiating characters are united.* It is demonstrated at the same time that *the relation of each pair of different characters in hybrid union is independent of the other differences in the two original parental stocks.*

Source: Gregor Mendel, "Versuche über Pflanzen-Hybriden," paper presented to the Natural History Society of Brünn, 1866. Translation reprinted courtesy of the Royal Horticultural Society, London.

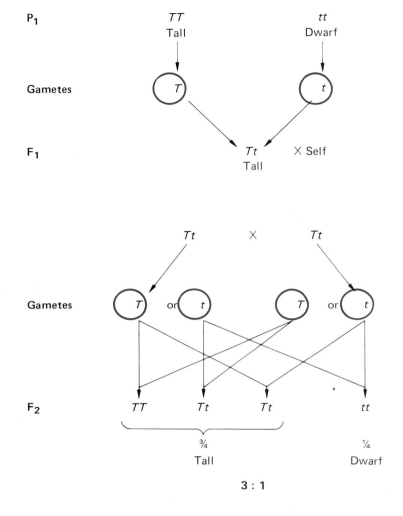

Figure 1–6.
Assigning of
Genotypes to the
Cross in Figure 1–5

the phenotypic classes that would result. The dwarf F_2 plants were homozygous recessive, and so, when **selfed** (self-fertilized), they should have produced only t-bearing gametes and had only dwarf F_3 offspring. The tall F_2 plants, however, were a heterogeneous group of which 1/3 should have been homozygous TT and 2/3 should have been heterozygous Tt. The tall homozygotes, when selfed, should have produced only tall F_3 offspring (genotypically TT). However, the F_2 heterozygotes when selfed should have produced tall and dwarf offspring in a ratio identical to that produced by the selfed F_1 plants: 3 tall : 1 dwarf. Mendel's data are presented in Figure 1–8. As you can see, all the dwarf (homozygous) F_2 bred true as predicted. Among the tall, 28% (28/100) bred true and 72% (72/100) segregated tall and dwarf

Schematic Tt × Tt

(as in Figure 1–6)

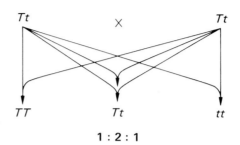

Tt × Tt

TT Tt tt

1 : 2 : 1

Diagrammatic   (Punnett Square)

Pollen

	T	t
T	TT	Tt
t	Tt	tt

Ovules

$TT : Tt : tt$
1 : 2 : 1

Probabilistic   (Multiply; see rule 2, Chapter 3.)

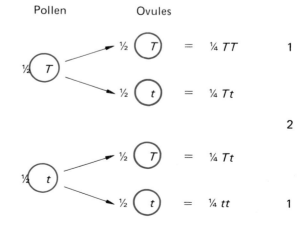

Pollen   Ovules

½ T ⟶ ½ T =   ¼ TT 1

⟶ ½ t =   ¼ Tt

2

½ t ⟶ ½ T =   ¼ Tt

⟶ ½ t =   ¼ tt 1

Figure 1-7.
Methods of
Determining F_2
Genotypic
Combinations in a
Self-Fertilized
Monohybrid
  The Punnett square
  diagram is named
  after the geneticist
  Reginald C. Punnett.

13

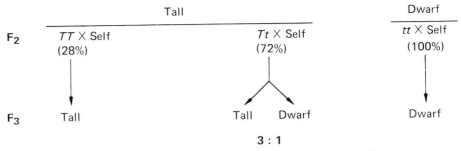

Figure 1-8. Self-Fertilized F₂ from the Monohybrid Cross in Figure 1-6
All the dwarf plants produce only dwarf progeny. However, among the
tall plants 28% produce both tall and dwarf progeny in a 3:1 ratio.

phenotypes. The prediction was 1/3 (33.3%) and 2/3 (66.7%) respectively. Mendel's observed values are well within the experimental error expected from this type of experiment (see Chapter 3). We thus conclude that our progeny-testing experiment confirms Mendel's hypothesis.

Another way to test the segregation hypothesis is with the extremely useful method of the **testcross**—that is, a cross of any organism with a homozygous recessive. (Another type of cross, a **backcross,** is the cross of a progeny with an individual that is a parental genotype. Hence, a testcross can often be a backcross.) Since the gametes of the homozygous recessive contain only recessive alleles, the alleles carried by the gametes of the other parent will determine the phenotypes of the offspring. If a gamete from the organism being tested contains a recessive, the resulting F_1 phenotype will be recessive; if it contains a dominant, the F_1 phenotype will be dominant. Thus, in using a testcross, the genotypes of the gametes from the organism tested determine the phenotypes of the offspring (Figure 1-9). A testcross of the F_2 in Figure 1-6 would produce the results shown in Figure 1-10. These results are a further confirmation of Mendel's rule of segregation.

DOMINANCE IS NOT UNIVERSAL

If dominance were universal, then we would always see the 3:1 ratio when a monohybrid is self-fertilized. However, in cases where dominance were absent, we would get a 1:2:1 phenotypic ratio when a monohybrid is selfed. As it happens, dominance is not universal. As technology has improved, more and more cases have been found in which we can differentiate the heterozygote. It is now clear that dominance and recessiveness are phenomena dependent on which alleles are interacting and what phenotypic level we are looking at. For example, in Tay-Sachs disease homozygous recessive children usually die before the age of five after suffering severe nervous degeneration; heterozygotes are seemingly normal. With the discovery of the way in which the disease works, detection of the heterozygotes has now become possible.

As with many genetic diseases, the culprit is a defective enzyme (protein

Genotype to be tested   X   Gamete of *aa*   =   Offspring

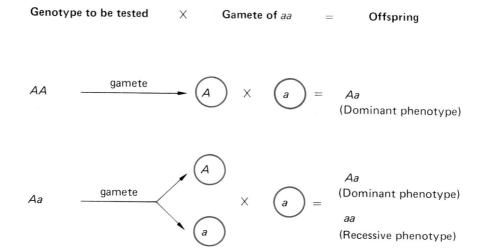

Figure 1-9.
Testcross

Tall (2 classes)

$TT \times tt$ = all Tt

$Tt \times tt$ = Tt : tt

1 : 1

Figure 1-10.
Testcrossing the
Dominant Phenotype
of F_2 from Figure
1-6

catalyst). Afflicted homozygotes have no enzyme activity, heterozygotes have about half the normal level of activity, and, of course, homozygous normals have the full level. In the case of Tay-Sachs disease, the defective enzyme is hexoseaminidase-A, needed for the proper metabolism of lipids. With modern techniques the blood of "normal" persons can be assayed for this enzyme, and heterozygotes can be identified by their intermediate level of enzyme activity. If two heterozygotes marry, they will know that there is a 25% chance that any child they bear will have the disease. They can then make an educated decision as to whether or not to marry or to have children.

Incomplete, or **partial, dominance** and **codominance** are the major classes of systems that lack simple dominance. Tay-Sachs disease is a case where an allele shows dominance at the clinical level (the heterozygote is clinically normal) and incomplete dominance at the enzymatic level. In snapdragons, if red-flowered plants are crossed with white-flowered plants, the F_1 are pink. If these are self-pollinated, the F_2 are red, pink, and white in a ratio of 1:2:1 (Figure 1-11). To the naked eye the alleles show partial dominance. At the molecular level they are codominant with molecules of red and white pigment intermixed. Another example of codominance occurs in the ABO red cell antigen system, which we discuss in the section on multiple alleles.

NOMENCLATURE

Throughout the last century, botanists, zoologists, and microbiologists have adopted different methods of naming alleles. Botanists tend to prefer the capital/lowercase letter scheme. Zoologists and microbiologists have adopted schemes that relate to the **wild type.** The wild type is the phenotype of the organism in nature and is, of course, not a predictable phenotype. Different

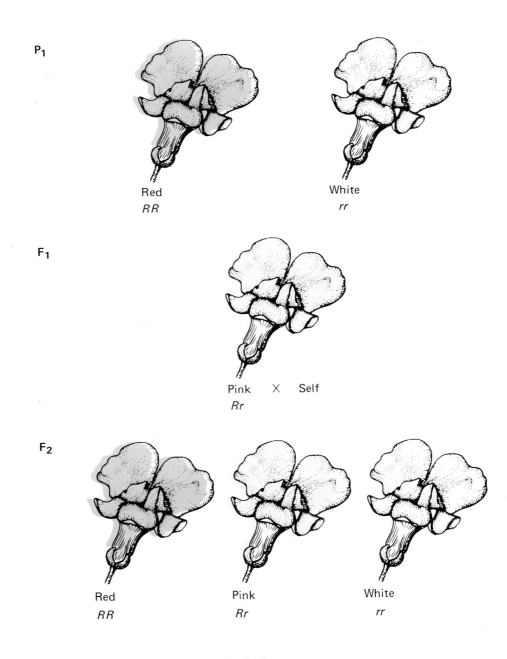

P₁ — Red *RR* / White *rr*

F₁ — Pink × Self *Rr*

F₂ — Red *RR* / Pink *Rr* / White *rr*

1 : 2 : 1

Figure 1-11.
Flower Color
Inheritance in
Snapdragons

people collecting the same organism may come up with different phenotypes from the wild. However, there is usually an agreed-upon phenotype that is referred to as the wild type. For fruit flies *(Drosophila)* it would be a red-eyed, normal-winged, etc. fly (Figure 1–12). Alternates to the wild type are

16

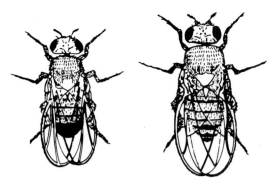

Figure 1–12.
Wild-Type Fruit Fly,
*Drosophila
melanogaster*

Adult Male   **Adult Female**

referred to as **mutants.** Thus, red eye is wild and white eye is mutant. Fruit fly genes are named after the mutant, with a capital letter if dominant and a lowercase letter if recessive. Table 1–1 gives some examples. The wild-type allele is given one of two symbols: a simple + sign or the + as a superscript to the mutant symbol. For example, w stands for white eye. The wild type (red eye) can be given the symbol + or w^+. The symbol w^+ is preferred if a + is ambiguous, as when several genes are being discussed simultaneously. By seeing the symbol w^+, we not only know that the mutant is white eyed, but we also know that it is recessive. The recessive wild-type allele of a dominant mutant might be, for example, H^+. Because H is capitalized, you know the mutant produces a dominant phenotype.

MULTIPLE ALLELES

A given gene can have more than two alleles. Although any particular diploid individual can have only two, there may be many alleles of a given gene in

TABLE 1–1. Some Mutants of *Drosophila*

Mutant Designation	Description	Dominance Relationship to Wild Type	Wild-Type Symbol (+ or . . .)	Dominance Relationship to Mutant
abrupt (*ab*)	Shortened, longitudinal, median wing vein	Recessive	ab^+	Dominant
amber (*amb*)	Pale yellow body	Recessive	amb^+	Dominant
black (*b*)	Black body	Recessive	b^+	Dominant
Bar (*B*)	Narrow, vertical eye	Dominant	B^+	Recessive
dumpy (*dp*)	Reduced wings	Recessive	dp^+	Dominant
Hairless (*H*)	Various bristles absent	Dominant	H^+	Recessive
white (*w*)	White eye	Recessive	w^+	Dominant
white-apricot (*w^a*)	Apricot-colored allele of white	Recessive	w^{a+}	Dominant

a population. The classic example of multiple alleles in humans is in the ABO blood system, discovered by Landsteiner in 1900. This is the best known of all the red cell antigen systems primarily because of its importance in blood transfusions. There are four phenotypes produced by three alleles (Table 1–2). The I^A and I^B alleles are responsible for production of the A and B antigens found on the surface of the erythrocytes (red blood cells). Antigens are substances, normally foreign to the body, that induce the immune system of the body to produce antibodies (proteins that bind to antigens). The ABO system is unusual to the extent that antibodies can be present (anti-B in a type A person, for example) without prior exposure to the antigen. Thus people with a particular ABO antigen on their red cells will have the antibody against the *other* antigen present in their serum: Type A persons have A antigen on their red cells and anti-B antibody in their serum, type O persons do not form any antigens, and type AB persons do not form any antibodies in their serum.

Adverse reactions to blood transfusions are primarily a result of the antibody in the serum of the recipient reacting with the antigen on the red cells of the donors. Thus, type A persons cannot donate blood to type B persons. Type B persons have anti-A antibody, which reacts with the A antigen on the donor red cells to cause clumping of the cells.

As can be seen from Table 1–2, the I^A and I^B alleles are codominant; a heterozygote has both the A and B antigens. Both produce phenotypes dominant to that produced by the I^O allele. This system not only shows multiple allelism; it also shows both codominance and simple dominance. According to the American Red Cross, of 100 blood donors in the United States, 46 are type O, 40 are type A, 10 are type B, and 4 are type AB. A second I^A allele has been found. Now there are I_1^A and I_2^A alleles: The I_1^A allele is dominant to the I_2^A allele, resulting in six recognizable phenotypes. As with any system, intense study yields more information, and subgroups of type B are now also known.

TABLE 1–2. ABO Blood Types with Immunity Reactions

Blood Type Corresponding to Antigens on Red Cells	Antibodies in Serum	Genotype	Reaction to Anti-A	Reaction to Anti-B
O	Anti-A and anti-B	$I^O I^O$	−	−
A	Anti-B	$I^A I^A$ or $I^A I^O$	+	−
B	Anti-A	$I^B I^B$ or $I^B I^O$	−	+
AB	None	$I^A I^B$	+	+

Many other multiple allelic systems are known. As a matter of fact, multiple alleles are the rule rather than the exception.

INDEPENDENT ASSORTMENT

Mendel also analyzed the inheritance pattern of two traits at the same time. He looked, for instance, at plants that differed in the *form* and *color* of the peas: He crossed true-breeding (homozygous) plants that had round *and* yellow seeds with plants that had wrinkled *and* green seeds. His results are shown in Figure 1–13. The F_1 all had round, yellow seeds, which demonstrated that round was dominant to wrinkled and yellow was dominant to green. When these F_1 plants were self-fertilized, they produced an F_2 that had all four possible combinations of the two seed characters: round, yellow; round, green; wrinkled, yellow; and wrinkled, green. The numbers Mendel reported in these categories were 315, 108, 101, and 32. Dividing each class by 32 will show a 9.84:3.38:3.16:1.00 ratio, which is very close to the 9:3:3:1 ratio we expect if the genes governing these two traits were behaving independently of each other.

In Figure 1–13 the letter R has been assigned to the dominant round trait and r has been assigned to the recessive wrinkled trait; Y and y have been used for yellow and green. We can rediagram the cross in Figure 1–13 using these letters to represent the alleles (Figure 1–14). The P_1 in this cross produce only one type of gamete each, RY for the dominant and ry for the recessive parent. The resulting F_1 is heterozygous for both genes **(dihybrid).** Self-fertilizing the dihybrid produces the F_2.

In the construction of the **Punnett square** to form the F_2, a critical assumption is being made: The four types of gametes will be produced in equal numbers, and, hence, every offspring category or "box" of the square is equally likely. Thus, because there are 16 boxes in the Punnett square, the ratio of F_2 offspring should be a function of sixteenths. Grouping the F_2 offspring by phenotype, we find there are 9/16 that have round, yellow seeds; 3/16 that have round, green seeds; 3/16 that have wrinkled, yellow seeds; and 1/16 that have wrinkled, green seeds. This is the origin of the expected 9:3:3:1 F_2 ratio.

Rule of Independent Assortment

This ratio comes about because the two characteristics behave independently of each other. There are four types of gametes in the F_1 plants (check Figure 1–14): RY, Ry, rY, and ry. These gametes occur in equal frequencies. Regardless of which allele for seed shape a gamete ends up with, it has a 50:50 chance of getting either of the color alleles—the two genes are segregating, or assorting, independently. This is the essence of Mendel's second

Reginald C. Punnett (1875–1967)
Genetics, 58 (1968): frontispiece.

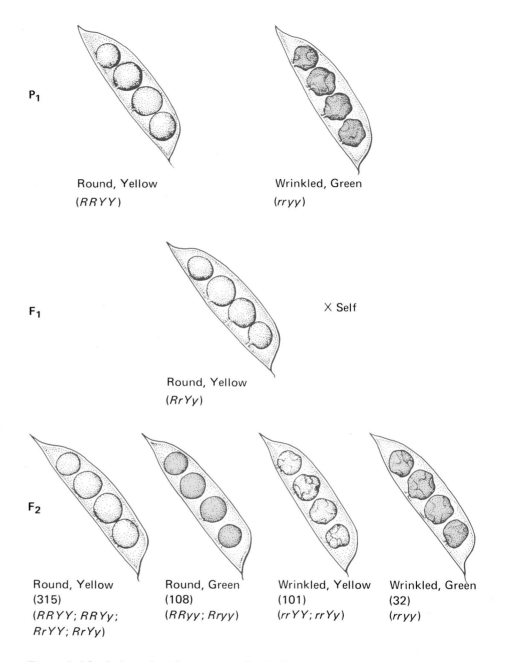

Figure 1-13. Independent Assortment in Garden Peas

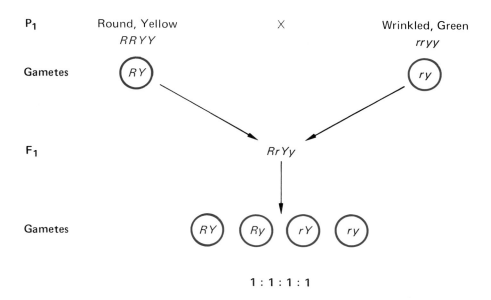

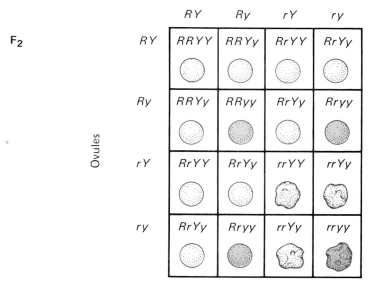

Figure 1-14.
Assigning
Genotypes in the
Cross of Figure 1-13

DID MENDEL CHEAT?

Overwhelming evidence gathered this century has proven the correctness of Mendel's conclusions. However, close scrutiny of Mendel's paper has led to some suggestions that (a) Mendel failed to report the inheritance of traits that did not show independent assortment and (b) Mendel fabricated numbers. Both these claims are on the surface difficult to ignore; both have been reasonably countered.

The first claim—that Mendel failed to report on crosses involving traits not showing independent assortment—arises from the observation that all seven traits that Mendel studied do show independent assortment and the pea plant has precisely seven pairs of chromosomes. For Mendel to have chosen seven genes, one located on each of the seven chromosomes, by chance alone seems extremely unlikely. In fact, the probability would be

$$7/7 \times 6/7 \times 5/7 \times 4/7 \times 3/7 \times 2/7 \times 1/7 = .006$$

That is, Mendel had less than 1 chance in 100 of randomly picking 7 traits on the 7 different chromosomes. However, in a recent study Douglas and Novitski (1977) have analyzed Mendel's data in a different way. To understand their analysis you have to know that two genes sufficiently far apart on the same chromosome will appear to assort independently. Thus Mendel's choice of characters showing independent assortment has to be viewed in light of the length of all the chromosomes. That is, Mendel could have chosen two genes on the same chromosome that would still show independent assortment. In fact, he did! Stem length and pod texture (wrinkled or smooth) are linked to each other on the fourth chromosome pair in peas. In their analysis Douglas and Novitski report that the probability of randomly choosing 7 characteristics that appear to assort indendently is actually between 1 in 4 and 1 in 3. So it seems that Mendel did not have to manipulate his choice of characters in order to hide the failure of independent assortment. He had a 1 in 3 chance of naively choosing the 7 characters that he did so that no deviation from independent assortment would be uncovered.

The second claim—that Mendel fabricated data—comes from a careful analysis of Mendel's paper by R. A. Fisher, a brilliant English statistician and population geneticist. In a paper in 1936, Fisher pointed out two problems of Mendel's work. First, all of Mendel's published data taken together fit their expected ratios better than predicted by rule, the **rule of independent assortment,** which states that alleles of one gene can segregate independently of alleles of other genes. Are the alleles for the two characteristics of color and form segregating properly according to Mendel's first principle?

If we look only at seed shape, we find that a *round* homozygote was crossed with a *wrinkled* homozygote in the P_1 generation ($RR \times rr$) to give only round heterozygous (Rr) offspring in the F_1. When these are self-fertilized, the result is $315 + 108$ round (RR or Rr) F_2 and $101 + 32$ wrinkled (rr) F_2. This is 423:133 or a 3.18:1.00 phenotypic ratio—very close to the expected 3:1 ratio. So, yes, the gene for seed shape is segregating normally.

chance alone. Second, and more disturbing, were cases where Mendel's data were a better fit to his expected ratios, which turn out to be the wrong expecteds. This error on Mendel's part came about as follows.

Mendel determined whether a dominant phenotype in the F_2 was a homozygote or a heterozygote by self-fertilizing it and examining ten offspring. In an F_2 composed of $1AA:2Aa:1aa$, he expected a 2:1 ratio of heterozygotes to homozygotes within the dominant phenotypic class. In fact, this ratio is not precisely correct because of the problem of misclassification of heterozygotes. There is a probability that some heterozygotes will be classified as homozygotes because all their offspring will be of the dominant phenotype. The probability that one offspring from a selfed Aa has the dominant phenotype is 3/4, or 0.75: The probability that ten offspring will be of the dominant phenotype is $(0.75)^{10}$ or 0.056. Thus Mendel misclassified heterozygotes as dominant homozygotes about 6% of the time. He should have expected a 1.89:1.11 ratio instead of a 2:1 to demonstrate segregation. Mendel classified 600 plants this way in one cross and got 201 homozygous : 399 heterozygous offspring. This is an almost perfect fit to a 2:1 ratio and thus a poorer fit to a 1.89:1.11 ratio. This bias is consistent and repeated in Mendel's trihybrid analysis.

$\chi^2 \leq 3.1$

Fisher, believing in Mendel's basic honesty, suggested that Mendel's data do not represent an experiment exactly, but more of a demonstration. In 1971 Weiling published a more convincing case in Mendel's defense. Pointing out that the data of Mendel's rediscoverers are also suspect for the same reason, he suggested that the problem lies with the process of pollen formation in plants, not with the experimenters. In an Aa heterozygote, two A and two a cells develop from a pollen mother cell. These cells tend to stay together on the anther. Thus, pollen cells do not fertilize in a strictly random fashion. A bee is more likely to take equal numbers of a and A pollen than would be expected by chance alone. The result is that the statistics used by Fisher are not applicable. By using a different statistic, Weiling showed that, in fact, no manipulations need to have been done by Mendel or his rediscoverers in order to get data that fit the expected ratios so well. By the same reasoning there would have been very little misclassification of heterozygotes.

We thus conclude that there is no compelling evidence to suggest that Mendel in any way manipulated his data to demonstrate his rules. In fact, taking into account what is known about him personally, it is much more logical that he did not "cheat."

Likewise, if we look only at the gene for color, we see that the F_2 ratio of yellow to green is 416:140 or 2.97:1.00—again, very close to a 3:1 ratio. Thus, when two genes are segregating normally according to the rule of segregation, their independent behavior will give us the rule of independent assortment.

From the Punnett square of Figure 1–14, you can see that because of dominance, all phenotypic classes except the homozygous recessive one, wrinkled green, are actually genetically heterogeneous with phenotypes that are made up of several genotypes. For example, the dominant class, the round yellow, is made up of four genotypes: $RRYY$, $RRYy$, $RrYY$, and $RrYy$. Grouping all the genotypes by phenotype shows the ratio in Figure 1–15.

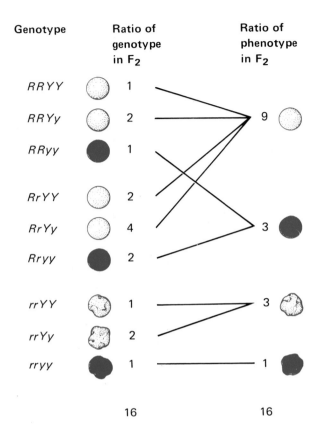

Genotype	Ratio of genotype in F$_2$	Ratio of phenotype in F$_2$

RRYY — 1

RRYy — 2 — 9

RRyy — 1

RrYY — 2

RrYy — 4 — 3

Rryy — 2

rrYY — 1 — 3

rrYy — 2

rryy — 1 — 1

16   16

Figure 1–15.
Dihybrid F$_2$
Genotypic Ratio

Thus, with complete dominance a self-fertilized dihybrid gives a 9:3:3:1 phenotypic ratio in the F$_2$. With the absence of dominance, a 1:2:1:2:4:2:1:2:1 phenotypic ratio is the F$_2$ result. What ratio would be obtained if one gene exhibited dominance and the other did not? An example of this case is given in Figure 1–16.

Testcrossing Multihybrids

A simple test of Mendel's rule of independent assortment is the testcrossing of the dihybrid. We would predict, for example, that if an *RrYy* individual were crossed with an *rryy* individual, the results would be as shown in Figure 1–17. Mendel's data verified this prediction. We will proceed to look at a **trihybrid** cross in order to develop general rules for **multihybrids.**

   The trihybrid Punnett square is shown in Figure 1–18. From this we can see that from a P$_1$ cross of a homozygous dominant and a homozygous reces-

P

P_1 $I^A I^A DD$ × $I^B I^B dd$

F_1 $I^A I^B Dd$
$(F_1 \times F_1)$

Male

F_2

	$I^A D$	$I^A d$	$I^B D$	$I^B d$
$I^A D$	$I^A I^A DD$	$I^A I^A Dd$	$I^A I^B DD$	$I^A I^B Dd$
$I^A d$	$I^A I^A Dd$	$I^A I^A dd$	$I^A I^B Dd$	$I^A I^B dd$
$I^B D$	$I^A I^B DD$	$I^A I^B Dd$	$I^B I^B DD$	$I^B I^B Dd$
$I^B d$	$I^A I^B Dd$	$I^A I^B dd$	$I^B I^B Dd$	$I^B I^B dd$

(Female, down the left side)

F_2 Summary

Phenotype		Frequency
A	Rh+	3
A	Rh−	1
B	Rh+	3
B	Rh−	1
AB	Rh+	6
AB	Rh−	2

Figure 1-16. Independent Assortment of Two Blood Systems in Humans
In the ABO system only the I^A and I^B alleles are segregating here; they are codominant. In a simplified view of the Rhesus system, the Rh^+ phenotype (D allele) is dominant to the Rh^- phenotype (d allele).

sive, the resulting generation produces 8 gamete types which, when selfed, in turn produce 27 different genotypes in a ratio of sixty-fourths in the F_2. By extrapolating from the monohybrid through the trihybrid, or simply by the rules of probability, we can construct Table 1–3 which contains the rules for F_1 gamete production and F_2 zygote formation in a multihybrid cross. For example, from this table we can figure out the F_2 of a dodecahybrid (12 segregating genes: $AABBCC \ldots LL \times aabbcc \ldots ll$). The F_1 in that cross will produce gametes with 2^{12} or 4096 different genotypes. The proportion of homozygous recessive offspring among the F^2 is $1/(2^n)^2$ where $n = 12$, or 1 in 16,777,216. There will be 4096 different F_2 phenotypes with complete dominance. If dominance is absent, there will be 3^{12}, or 531,441 different phenotypes.

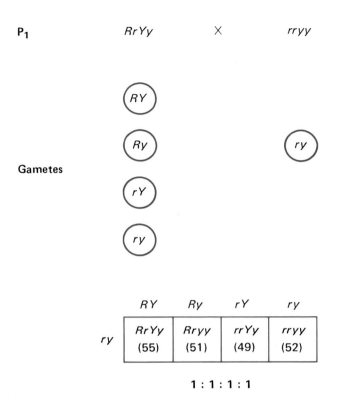

Figure 1-17.
Testcross of a
Dihybrid

GENOTYPIC INTERACTIONS

Often, several genes contribute to the same phenotype. An example is the combs of fowl (Figure 1–19). If we cross a rose-combed hen with a pea-combed cock (or vice versa), all the F_1 offspring are walnut combed. If we cross together the hens and cocks of this heterozygous F_1 group, we will get, in the F_2, walnut-, rose-, pea-, and single-combed fowl in a ratio of 9:3:3:1. Can you figure out the genotypes of this F_2 population before reading further? An immediate indication that two allelic pairs are involved is the fact that the 9:3:3:1 ratio appeared in the F_2. As we have seen, this ratio comes about when we self-fertilize dihybrids in which both genes exhibit dominance.

The analysis of this cross is given in Figure 1–20. When dominant alleles of both genes are present, the walnut comb appears. A dominant allele at the rose gene (R—) with recessives at the pea gene (pp) gives a rose comb. A dominant at the pea gene (P—) with recessives at the rose gene (rr) gives pea-combed fowl. Homozygous recessive at both genes gives the single comb. Thus a 9:3:3:1 F_2 ratio arises from selfing the dihybrid even though we are

P₁   AABBCC × aabbcc

F₁   AaBbCc × Self

	ABC	ABc	AbC	Abc	aBC	aBc	abC	abc
ABC	AABBCC	AABBCc	AABbCC	AABbCc	AaBBCC	AaBBCc	AaBbCC	AaBbCc
ABc	AABBCc	AABBcc	AABbCc	AABbcc	AaBBCc	AaBBcc	AaBbCc	AaBbcc
AbC	AABbCC	AABbCc	AAbbCC	AAbbCc	AaBbCC	AaBbCc	AabbCc	AabbCc
Abc	AABbCc	AABbcc	AAbbCc	AAbbcc	AaBbCc	AaBbcc	AabbCc	Aabbcc
aBC	AaBBCC	AaBBCc	AaBbCC	AaBbCc	aaBBCC	aaBBCc	aaBbCC	aaBbCc
aBc	AaBBCc	AaBBcc	AaBbCc	AaBbcc	aaBBCc	aaBBcc	aaBbCc	aaBbcc
abC	AaBbCC	AaBbCc	AabbCC	AabbCc	aaBbCC	aaBbCc	aabbCC	aabbCc
abc	AaBbCc	AaBbcc	AabbCc	Aabbcc	aaBbCc	aaBbcc	aabbCc	aabbcc

Figure 1-18. Trihybrid Cross

TABLE 1-3. Multihybrid Self-Fertilization, Where n Equals Number of Genes Segregating Two Alleles Each

	Monohybrid $n = 1$	Dihybrid $n = 2$	Trihybrid $n = 3$	General Rule
F₁ gamete genotypes	2	4	8	2^n
Proportion of homozygous recessives among the F₂	1/4	1/16	1/64	$1/(2^n)^2$
Number of different F₂ phenotypes given complete dominance	2	4	8	2^n
Number of different genotypes (or phenotypes if there is no dominance)	3	9	27	3^n

27

Rose

Pea

Walnut

Single

Figure 1-19.
Four Types of
Combs in Fowl

dealing with different expressions (traits) of the same phenotypic characteristic, the comb. In our previous 9:3:3:1 example (Figure 1-14), we dealt with two separate characteristics of peas: shape and color.

In corn several different field varieties produce white kernels on the ears. In certain crosses two white varieties will result in an F_1 with all the corn kernels being purple. If these purple plants are selfed, the F_2 are both purple and white in a ratio of 9:7. How can this be explained? We must be dealing with two genes, each segregating two alleles, because the ratio is in sixteenths. Therefore, the F_1 must have been a dihybrid. Furthermore, we can see that the F_2 9:7 ratio is a simple variation of the 9:3:3:1 ratio. The 3, 3, and 1 categories here are producing the same phenotype and thus make up 7/16 of the F_2 offspring. The cross is outlined in Figure 1-21. We can see from this figure that the purple color appears only when dominant alleles of both genes are present. When one or both genes have only recessives, the kernels will be white.

	Rose comb	Pea comb
P_1	$RRpp$	X $rrPP$

	Walnut comb
F_1	$RrPp$
	F_1 X F_1

F_2

	RP	Rp	rP	rp
RP	$RRPP$ Walnut	$RRPp$ Walnut	$RrPP$ Walnut	$RrPp$ Walnut
Rp	$RRPp$ Walnut	$RRpp$ Rose	$RrPp$ Walnut	$Rrpp$ Rose
rP	$RrPP$ Walnut	$RrPp$ Walnut	$rrPP$ Pea	$rrPp$ Pea
rp	$RrPp$ Walnut	$Rrpp$ Rose	$rrPp$ Pea	$rrpp$ Single

Figure 1–20.
Independent
Assortment in
Determination of
Comb Type in Fowl

Epistasis

This gets us to the concept of **epistasis,** the masking of the action of alleles of one gene by allele combinations of another gene. For example, in the preceding corn case, an aa genotype produces white kernels regardless of the B genotype. Likewise, the bb genotype masks whatever A alleles are present. We say that the *epistatic* gene masks the expression of the **hypostatic** gene. For example, apterous (wingless) in fruit flies is epistatic to any gene that controls wing characteristics: Hairy wings is hypostatic to apterous. Another example to further illustrate this principle is control of the coat color of mice.

If a pure-breeding black mouse is crossed with a pure-breeding albino mouse (pure white because all pigment is lacking), all of the offspring are agouti, (the typical brownish-grey mouse color). When the F_1 agouti are crossed to each other, agouti, black, and albino offspring appear in the F_2 in a ratio of 9:3:4. What are the genotypes of this cross? The answer is given in Figure 1–22. By now it should be apparent that this F_2 ratio is a variant of the 9:3:3:1 ratio; it indicates epistasis in a simple dihybrid cross. What is the mechanism producing this 9:3:4 ratio? Of the population comprising the

P₁ First white variety   X   Second white variety
               AAbb   aaBB

F₁   Purple
                                  AaBb X Self

F₂

	AB	Ab	aB	ab
AB	AABB	AABb	AaBB	AaBb
Ab	AABb	AAbb	AaBb	Aabb
aB	AaBB	AaBb	aaBB	aaBb
ab	AaBb	Aabb	aaBb	aabb

Figure 1-21.
Color production in
Corn

F₂

Summary

Purple : White

9 : 7

9:3:3:1 ratio, one of the 3/16 classes and the 1/16 class are combined to give a 4/16 class. Any genotype that includes $c^a c^a$ will be albino, but as long as at least one dominant C allele is present, the A gene can express itself. Dominant alleles of both genes ($A—C—$) will give the agouti color, while recessives at the A gene ($aaC—$) will give the black color. So, at the A gene, A for agouti is dominant to a for black. The albinism gene (c^a) is epistatic to black (a) and agouti (A). Black and agouti are hypostatic to the albino gene.

Mechanism of Epistasis

The physiological mechanism of epistasis is known. The pigment melanin is involved in both the black and agouti phenotypes. The agouti is a modified black hair in which yellowish stripes have been added. Thus, with melanin present agouti is dominant. Without melanin we get albino because both agouti and black depend on melanin. Without melanin the phenotype is

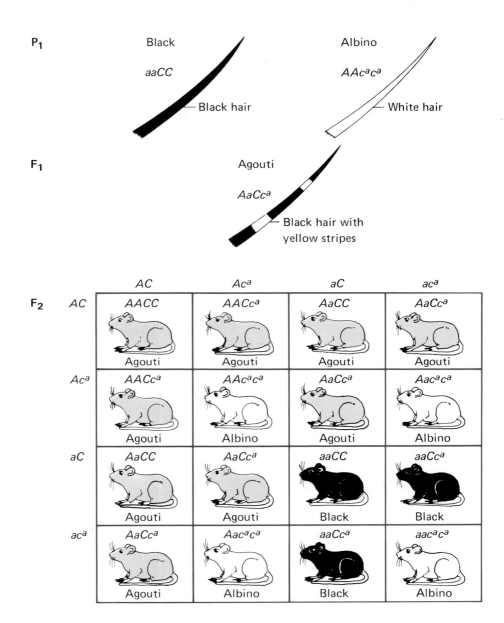

Figure 1-22.
Epistasis in Coat
Color in Rodents

Agouti : Albino : Black

9 : 4 : 3

albino regardless of the genotype of the black/agouti gene. With melanin
present we get a simple dominance effect at the black/agouti gene. Albinism
is the result of one of several defects in the enzymatic pathway for the synthesis of melanin (see Figure 1–23).

31

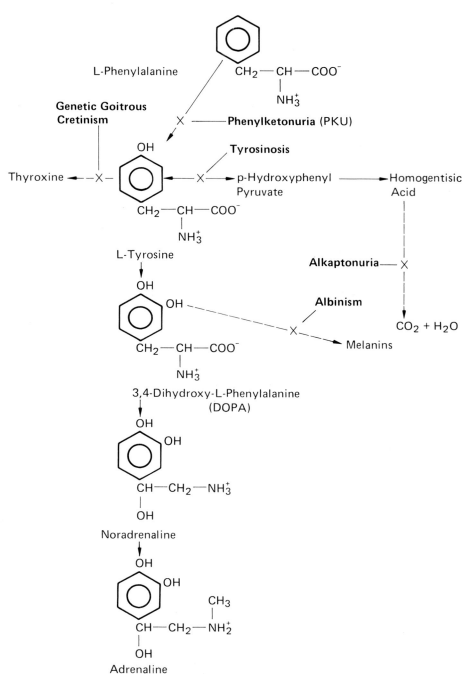

Figure 1-23.
Part of the Tyrosine
Metabolism Pathway
in Humans and
Associated Diseases
Caused by
Homozygous
Recessive
Conditions
The broken arrow
indicates that there is
more than one step
in the pathway.

With the realization that epistatic modifications of the 9:3:3:1 ratio come about through interaction at the biochemical level, we can look for a biochemical explanation for the 9:7 ratio in corn kernel color (Figure 1–21). Two possible mechanisms by which a 9:7 ratio can come about are shown in Figure 1–24. Either there is a two-step process that takes a precursor molecule and turns it into purple pigment or there are two precursors that need to be converted to final products that then combine to produce purple pigment. The dominants from the two genes control the two steps in the process. Recessive alleles are ineffective. Thus, dominants for both steps are necessary to successfully complete the pathways for an end result of purple pigment. Stopping the process at any point will prevent the production of purple color.

Other examples of epistatic interactions are known. In Table 1–4 several are listed. We do not know the exact metabolic sequence in many cases, especially when developmental processes are involved (for example, size and shape). However, from an analysis of crosses, we can still know the number of genes involved and the general nature of their interactions.

The examples of mouse coat color and corn kernel color demonstrate that genes control the formation of enzymes, which control steps in biochemical pathways. For the most part, dominant alleles control functioning enzymes that catalyze biochemical steps. Recessive alleles often produce nonfunctional enzymes that cannot catalyze specific steps. Often a heterozygote is relatively normal because one allele is producing a functional enzyme. A. E. Garrod, in *Inborn Errors of Metabolism,* published in 1909, pointed out this general concept of gene action in humans. Only nine years after Mendel was rediscovered, Garrod described many human conditions such as albinism and alkaptonuria, that are the result of homozygosity of recessive alleles.

Pathway 1

Pathway 2

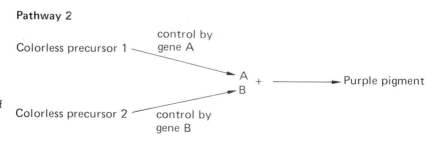

Figure 1–24. Possible Metabolic Pathways of Color Production Yielding 9:7 Ratios in the F_2 of a Self-Fertilized Dihybrid

TABLE 1-4. Some Examples of Epistatic Interactions between Alleles of Two Genes

Characteristic	Phenotype of F_1 Dihybrid (AaBb)	Phenotypic F_2 Ratio
Corn and sweet pea color	Purple	Purple:White 9:7
Mouse coat color	Agouti	Agouti:Black:Albino 9:3:4
Shepherd's purse seed capsule shape	Triangular	Triangular:Oval 15:1
Summer squash shape	Disk	Disk:Sphere:Elongate 9:6:1
Fowl color	White	White:Colored 13:3

CHAPTER SUMMARY

Genes control the phenotype. They are inherited as discrete units. Higher organisms contain two alleles of each gene, but only one allele enters each gamete. Alleles of different genes assort independently of each other. Zygote formation restores the double number of alleles in the cell. This is Mendel's rule of segregation. Mendel was the first to discern the 3:1 phenotype ratio as a pattern of inheritance in hybridization. Mendel was successful in his endeavor because he performed careful experiments using discrete characteristics, large numbers of offspring, and an organism (the pea plant) amenable to controlled fertilizations.

There can be many alleles for any one gene, although each individual organism has only two. Dominance depends on which alleles are interacting and what level of the phenotype one looks at. Genes usually control the production of enzymes, which control steps in metabolic pathways. Many metabolic diseases in humans are due to the homozygous recessive genotype of an allele that produces a nonfunctioning enzyme.

Nonallelic genes can interact in the formation of a phenotype so that alleles of one gene can alter or mask the expression of alleles of another gene. This process is termed *epistasis* and results in alterations of expected ratios.

EXERCISES AND PROBLEMS

1. Mendel crossed tall pea plants with dwarf ones. The F_1 were all tall. When these F_1 were selfed to produce F_2, he got a 3/4 tall to 1/4 dwarf ratio. If the F_2 had been selfed, give the genotypes and phenotypes and relative proportions of the F_3 generation produced.

2. Mendel self-fertilized dihybrid plants ($RrYy$) with round and yellow seeds and got a 9:3:3:1 ratio in the F_2. As a test of Mendel's hypothesis of independent assortment, predict the kind and number of progeny produced in testcrosses of these F_2 offspring.

3. In onions three bulb colors segregate: red, yellow, and white. A red parent is crossed to a white parent and all the offspring are red. When these are selfed, the following data are obtained:

   Red   119
   Yellow   32
   White   9

   What is the mode of inheritance and how do you account for the ratio?

4. Snapdragons have a color gene and a height gene with the following phenotypes:

   RR: red flower TT: tall plant
   Rr: pink flower Tt: medium height plant
   rr: white flower tt: dwarf plant

   If a dihybrid is self-fertilized, give the resulting proportions of genotypes and phenotypes produced.

5. Corn has a color gene and height gene with the following phenotypes:

   CC,Cc: purple TT: tall
   cc: white Tt: medium height
                          tt: dwarf

   If a dihybrid is selfed, give the resulting proportions of genotypes and phenotypes produced.

6. A geneticist, while studying an inherited phenomenon, discovers a phenotypic ratio of 9/16 : 6/16 : 1/16 among offspring of a given mating. Give a simple, plausible explanation for this result. How would you test your hypothesis?

7. In the ABO blood group systems in humans, I^A and I^B are codominant and both are dominant to I^O. In a paternity dispute a type AB woman claimed that one of four men was the father of her type A child. Which of the following men could be the father of the child on the basis of the evidence given?

   a. type A
   b. type B
   c. type O
   d. type AB

8. Under what circumstances can the phenotypes of the ABO system be used to refute paternity?

9. Assume that Mendel looked at four traits of his pea plants at the same time and that each trait showed dominance. If he crossed a homozygous dominant to a homozygous recessive, all the F_1 would have been of the dominant phenotype. If he then selfed the F_1, how many different genotypes of gametes would these F_1 have produced? How many different phenotypes would have resulted in the F_2? How many different genotypes would have resulted? What proportion of the F_2 would have been of the four-fold recessive phenotype?

10. In fruit flies a new dominant trait, wingless, was discovered. Describe different ways of naming the wingless gene and its mutant form.

11. State precisely the rules of segregation and independent assortment.

12. Can you think of a biochemical pathway to explain how I^A and I^B are codominant yet both are dominant to I^O?

13. The following is a list of ten genes with an allele each. These are of traits in fruit flies. Are the alleles dominant or recessive? Are they mutant or wild type?

Name of gene	Allele
yellow	y^+
Hairy wing	Hw
Abruptex	Ax^+
Confluens	Co
raven	rv^+
downy	dow
Minute(2)e	$M(2)e^+$
Jammed	J
tufted	tuf^+
burgundy	bur

14. Genes that seem to have the greatest number of alleles are involved with self-incompatibility in plants. In some cases hundreds of alleles are known. What types of constraints might exist to set a limit on the number of alleles that a gene can have?

15. Suggest possible mechanisms for the epistatic ratios given in Table 1–4. Can you add any further ratios?

SUGGESTIONS FOR FURTHER READING

Brink, R. E., ed. 1967. *Heritage from Mendel.* Madison: University of Wisconsin Press.

Douglas, L., and E. Novitski. 1977. What chance did Mendel's experiments give him of noticing linkage? *Heredity* 38:253–257.

Fisher, R. A. 1936. Has Mendel's work been rediscovered? *Ann. Sci.* 1:115–137.

Kříženecký, J., ed. 1965. *Fundamenta Genetica.* Prague: Publishing House of the Czechoslovak Academy of Sciences. (A collection of papers including Mendel's original and 27 papers published during the rediscovery era.)

Stern, C., and E. R. Sherwood, eds. 1966. *The Origin of Genetics, A Mendel Source Book.* San Francisco: Freeman. (Includes a translation of Mendel's original paper.)

Sturtevant, A. H. 1965. *A History of Genetics.* New York: Harper & Row.

Strickberger, M. 1962. *Experiments in Genetics with Drosophila.* New York: Wiley.

Stubbe, H. 1972. *History of Genetics.* Cambridge, Mass.: MIT Press.

Weiling, F. 1971. Mendel's "too good" data in *Pisum* experiments. *Folia Mendeliana* 6:75–77.

2 Mitosis and Meiosis

Most higher organisms begin life as a fertilized egg. The zygote then divides many times to produce an adult organism, which then produces gametes to start the cycle again. Both fertilization and cell division depend on exact mechanisms for gene distribution in order to be successful. Cell division requires that each daughter cell have exactly the same genetic material as the parent cell. Otherwise, cell division would produce chaos. In fertilization the gametes must have precisely half the genetic material of the adult or else the genetic constituency of an organism would be different from that of its parent. In this chapter we examine the processes of mitosis and meiosis, which allow chromosomes, the gene vehicles, to be properly apportioned among daughter cells. Keep in mind the engineering difficulties posed by these processes and the relationship of meiosis to Mendel's rules.

Mendel's work was rediscovered at the turn of the century after being ignored for 34 years. One of the major reasons that it could be appreciated in 1900 was that the behavior of chromosomes had been discovered. With a knowledge of the behavior of chromosomes, a physical basis existed for genes. That is, the way in which chromosomes behave during gamete formation is precisely the way in which Mendel predicted that genes would behave during gamete formation. In this chapter we look at the morphology of chromosomes and their behavior during somatic-cell division and gamete formation.

Modern biologists classify organisms into two major categories: **eukaryotes,** organisms that have true nuclei, and **prokaryotes,** organisms that lack true nuclei (Table 2–1). Bacteria and blue-green algae are prokaryotes (Figure 2–1). All higher organisms are eukaryotes (Figure 2–2). In the prokaryotes the genetic material is usually a simple circle of DNA (deoxyribonucleic acid), whereas in eukaryotes the genetic material, located in the nucleus, is DNA complexed with protein (nucleoprotein). In this chapter we concentrate on the nuclear division processes of eukaryotes.

TABLE 2-1. Differences between Prokaryotic and Eukaryotic Cells

	Prokaryotic Cells	*Eukaryotic Cells*
Taxonomic groups	Bacteria and blue-green algae (monera)	All plants, fungi, animals, protists (according to Whittaker's five-kingdom classification scheme)
Size	Less than 5 μm in greatest dimension	Greater than 5 μm in smallest dimension
Nucleus	No true nucleus, no nuclear membrane	Nuclear membrane
Genetic material	One circular molecule of DNA, no protein	Linear histone-containing nucleoproteins
Mitosis and meiosis	Absent	Present

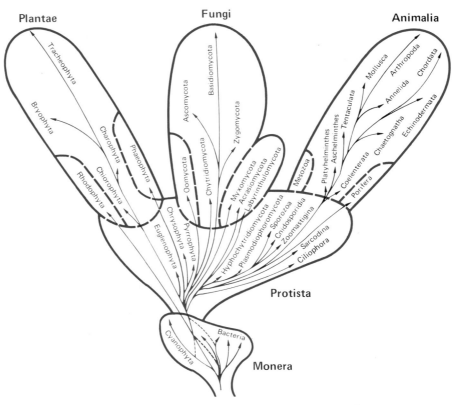

**Figure 2-1.
Five-Kingdom
Classification
Scheme of Whittaker**

Source: R. H. Whittaker, "New Concepts of Kingdoms of Organisms," *Science* 163 (1969):150–160. Copyright 1969 by the American Association for the Advancement of Science. Reproduced by permission.

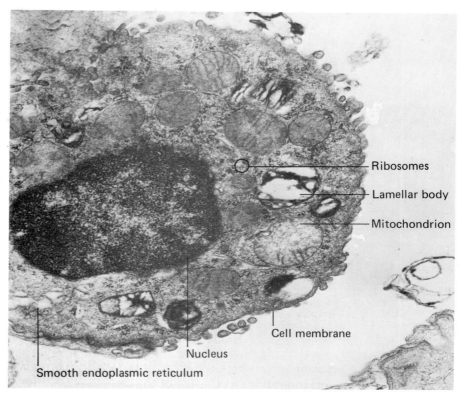

Ribosomes

Lamellar body

Mitochondrion

Cell membrane

Nucleus

Smooth endoplasmic reticulum

Figure 2-2.
Mouse Lung Cell
magnification 7000X

Source: Courtesy of Wayne Rosenkrans.

CHROMOSOMES

Chromosomes were discovered by Carl Nägeli in 1842 and named by W. Waldeyer in 1888. The term **chromosome** means "colored body" and is used because chromosomes stain a bright color with certain techniques of histology. The nucleoprotein material of the chromosomes is referred to as **chromatin.** Since we now know that the genes are located in chromosomes, we need to understand the processes through which chromosomes go in order to understand the physical basis of Mendel's rules.

Although all eukaryotes have chromosomes, between divisions, in the **interphase,** they are spread out or diffused throughout the nucleus and are usually not identifiable (Figure 2–3). Each chromosome, with very few exceptions, has a distinct attachment point for fibers that make up the mitotic and meiotic spindle apparatus. (The exceptions are in organisms with a diffuse arrangement of attachment points all along the chromosomes.) The attachment points are called **kinetochores,** and the constrictions in the chromosomes where these occur are called **centromeres** (sometimes these terms are used interchangeably). Chromosomes can be classified

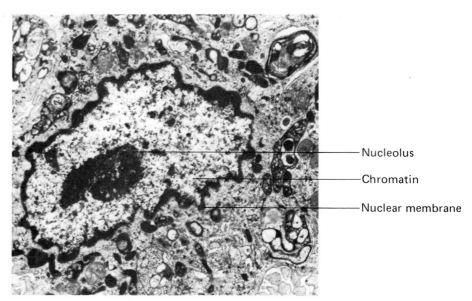

Figure 2-3. Nucleus of Rat Lung Cell
magnification 4000X

Source: Courtesy of Wayne Rosenkrans.

according to whether the centromere is in the middle of the chromosome **(metacentric)**, at the end of the chromosome **(telocentric)**, very near the end of the chromosome **(acrocentric)**, or somewhere in between **(subtelocentric** or **submetacentric).** (See Figures 2–4 and 2–5.) For any particular chromosome the position of the centromere is fixed.

Most eukaryotic cells prior to nuclear division are diploid—that is, all their chromosomes occur in pairs. We receive one member of each pair from each of our parents. Reproductive cells (gametes), however, are **haploid**— they have only one unpaired copy of each chromosome. In the diploid state, members of the same chromosome pair are referred to as **homologous chromosomes;** the two make up a homologous pair.

The total chromosome complement of a cell, the **karyotype,** can be photographed during mitosis and rearranged in pairs to make a picture referred to as an **idiogram** (Figure 2–6. The picture itself is often referred to as a karyotype.) From the karyotype it is possible to see whether there are abnormal numbers of chromosomes and to identify the sex of the organism. As you can see from Figure 2–6, all of the homologous pairs are made up of identical chromosomes. These identical pairs are **homomorphic.** A potential exception is the sex chromosome pair, which in some species are of unequal size and are therefore called **heteromorphic.**

The number of chromosomes possessed by a particular species does not

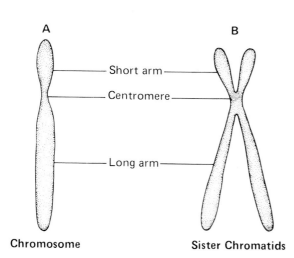

Figure 2-4. (a) Submetacentric Chromosome (b) Submetacentric Chromosome in Early Mitosis
The chromosome is best seen after it has been duplicated but prior to separation of the identical halves (sister chromatids).

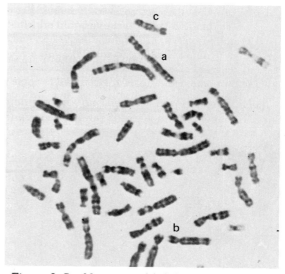

Figure 2-5. Metacentric (a), Submetacentric (b), and Acrocentric (c) Chromosomes in humans.
The centromere divides the chromosome into two arms.

Source: Reproduced courtesy of Dr. Thomas G. Brewster, Foundation for Blood Research, Scarborough, Maine.

change. Some species exist for the most part in the haploid state or have long intervals in their life cycle that are haploid. These species usually have their chromosome numbers stated in haploid numbers. For example, the pink bread mold *(Neurospora crassa)* in the haploid state has a chromosome number of 7 ($n = 7$). The diploid number is, of course, 14 ($2n = 14$). The diploid chromosome number of several species is shown in Table 2–2.

In eukaryotes there are two processes whereby the genetic material is partitioned into offspring, or daughter, cells. One is the simple division of one cell into two. In this process the two daughter cells must each receive an exact copy of the genetic material of the parent cell. The cellular process is simple cell division and the nuclear process accompanying it is **mitosis.** In the formation of gametes, or spores in plants, the genetic material must be precisely divided in half so that the diploid complement is reformed when gametes fuse to form the zygote. The cellular process is gamete formation (spore formation in plants) and the nuclear process is **meiosis.** Mitosis is from the Greek word for "a thread," referring to the chromosome. It was coined by Flemming in the 1880s. Meiosis is from the Greek word "to lessen."

From an engineering point of view, both processes of nuclear division must carefully separate chromosomes. The division of the cytoplasm of the

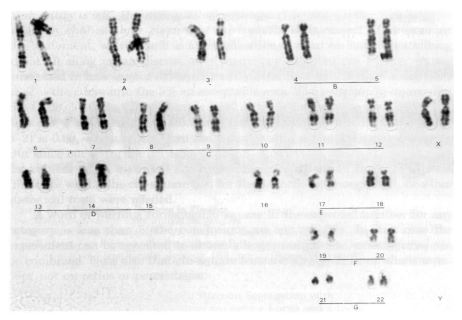

Figure 2-6.
Human Idiogram
(Karyotype)

Source: Reproduced courtesy of Dr. Thomas G. Brewster, Foundation for
Blood Research, Scarborough, Maine.

cell, **cytokinesis,** is much less organized. In animals there is a constriction
of the cell membrane that more or less randomly distributes the cytoplasm.
In plants the growth of a cell wall accomplishes the same purpose. Let us
examine the process of mitosis, where the genetic material will be equally
divided between daughter cells. Throughout this process our emphasis will
be on the behavior of the chromosomes. It is their behavior that has the
genetic implications. In this book we are less concerned with the cellular
details.

MITOSIS

Now, consider the engineering problem that mitosis must solve. It must separate the daughter chromatids in such a way that each goes into a different
daughter cell. The separation must occur for each chromosome. The two
daughter cells must end up with the identical chromosome complement that
the parent cell had. Mitosis is nature's elegant process to achieve that end—
surely an engineering marvel.

The stages of mitosis are **prophase, metaphase, anaphase,** and **telophase** (pro-, before; meta-, mid; ana-, back; telo-, end). The process is itself
a continuous one. However, for the purposes of description it is considered

TABLE 2-2. Chromosome Number for Selected Species. (2n is the diploid complement.)

Species	2n
Humans *(Homo sapiens)*	46
Garden peas *(Pisum sativum)*	14
Drosophila melanogaster	8
House mouse *(Mus musculus)*	40
Roundworm *(Ascaris spp.)*	2
Pigeon *(Columba livia)*	80
Boa constrictor	36
Cricket *(Gryllus domesticus)*	22
Lily *(Lilium longiflorum)*	24
Indian fern *(Ophioglossum reticulatum)*	1260

as broken into these four stages. Replication (duplication) of the genetic material occurs during the S-phase of the **cell cycle** (Figure 2–7). The timing of the four stages varies from species to species, from organ to organ within a species, and even from cell to cell within a given cell type.

Prophase

This stage of mitosis is characterized by a shortening and thickening of the chromosomes so that individual chromosomes become distinct. At this time also, the nuclear envelope (membrane) disintegrates; the **nucleolus** disap-

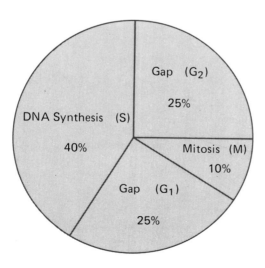

Figure 2–7.
Cell Cycle in the Broad Bean, *Vicia faba*
Total time is under 20 hours.

pears; the **centrioles,** when present, duplicate and migrate to opposite poles of the cell; and the **spindle** or **mitotic apparatus** forms (Figure 2–8). The nucleolus is a darkly staining body in the nucleus (Figure 2–3). It is involved in ribosome construction and forms around a point on one of the chromosomes called the **nucleolar organizer.** The number of nucleoli varies in different species, but in the simplest case there will be two nucleolar organizers per nucleus, one each on the two members of a homologous pair of chromosomes. Nucleoli are reformed after mitosis.

Centrioles are cylindrical organelles found in virtually every eukaryotic group except the higher plants. They organize the mitotic apparatus (Figure 2–9). The mitotic apparatus is a spindle-shaped conglomerate of **microtubules**—hollow cylinders made up of small subunits—which attach to the chromosomes at the centromeres and pull the chromosomes toward a pole of the spindle. The exact mechanism of movement is not known, but a model whereby the filaments of the spindle slide past each other is widely accepted.

As prophase progresses, each chromosome can be seen to be composed of identical subunits (Figure 2–4). These are referred to as **chromatids,** or **sister chromatids.** They are held together at the centromere and each will become a chromosome when the centromere divides. These chromatids are the visible manifestation of the previous replication process that has taken place in the S-phase of the cell cycle. Contraction of the chromosomes as the process of mitosis begins presumably is a mechanism to avoid physical damage to the chromosomes.

Metaphase

As prophase continues, spindle fibers can be seen to attach to the individual chromosomes at their centromeres (Figure 2–10). The actual attachment is

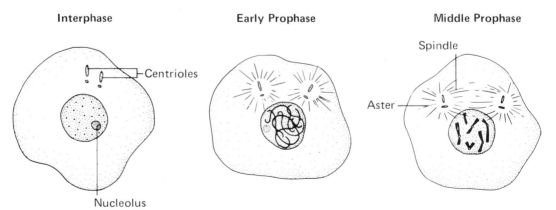

Figure 2-8.   Prophase, Mitosis

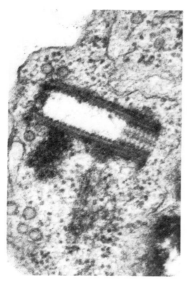

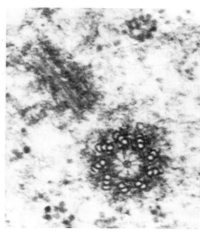

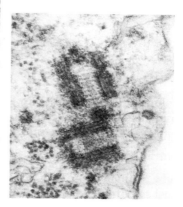

a. **Longitudinal Section**   b. **Cross Section Showing**   c. **Mother-Daughter Centrioles**
                                    Cartwheel Structure

Figure 2-9.   Centriole
magnification (a) 79,600X (b) 111,800X (c) 59,200X

Source: F. R. Turner, "An Ultrastructural Study of Plant Spermatogenesis: Spermatogenesis in *Nitella*," *Journal of Cell Biology* 37 (1968):381. Reproduced by permission.

at the kinetochores, two disk-shaped objects at each centromere. The kinetochores are on opposite sides of the centromere, and there is one associated with each sister chromatid. This geometry assures that the chromatids are separated from each other during the next stage of mitosis. The number of microtubules attaching to each kinetochore differs in different species. Four to seven microtubules attach per kinetochore in the cells of the rat fetus, whereas 70 to 150 attach in the plant *Haemanthus*. With the attachment of the spindle fibers and the completion of the spindle itself, the chromosomes are jockeyed into position in the plane of the equator of the spindle. This is called the **metaphase plate,** and alignment of the chromosomes on this plate marks the end of metaphase (Figure 2–10).

Anaphase

This stage is marked by the separation of sister chromatids (Figure 2–11). Almost simultaneously, all of the chromatids in the cell separate, with sisters being pulled to opposite poles of the cell. The dragging occurs at a uniform rate of speed, and the chromosomes appear to be pulled by the centromere.

Metaphase

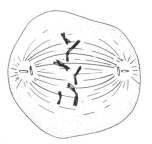

Figure 2-10.
Metaphase, Mitosis

This is in fact the case, as they are separated by the action of the spindle fibers attached at the kinetochore. Thus, metacentric chromosomes appear V shaped; subtelocentrics, J shaped; and telocentrics, rod shaped.

Telophase

At the end of anaphase, the separated sister chromatids (now chromosomes) have been pulled to opposite poles of the cell. The cell now, in essence, reverses the steps of prophase to return to the active interphase state (Figure 2–12). The chromosomes uncoil and begin to carry out their physiological functions (directing protein synthesis, replication, and so on). A nuclear envelope reforms about each set of chromosomes, nucleoli form, and cytokinesis takes place. The cell has now entered the G_1 phase of the cell cycle (Figure 2–7).

Significance of Mitosis

Mitosis succeeds in forming two daughter cells, each with genetic material identical to that of the one parent cell. There is an exact distribution of the genetic material, in the form of chromosomes, into the two daughter cells.

MEIOSIS

Gamete formation is an entirely new engineering problem to be solved. To form gametes in animals, and spores in plants, a diploid organism with its two copies of each chromosome must form daughter cells that have only one copy of each chromosome. That is, the genetic material must be reduced to half so that when gametes recombine to form zygotes, the original number of chromosomes is restored, not doubled.

   If we were to try to engineer this task, we would have to be able to rec-

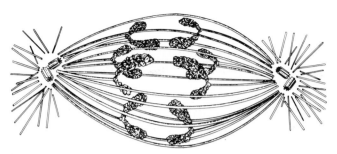

**Figure 2-11.
Mitotic Spindle
at Anaphase**

Source: E. J. DuPraw, *DNA and Chromosomes* (New York: Holt, Rinehart and Winston, 1970), p. 72. Reproduced with permission.

Ernest J. DuPraw
(1931–   )
Courtesy of Dr.
Ernest J. DuPraw

ognize homologous chromosomes. Being able to recognize the members of a homologous pair, we could then push one member into one daughter cell and the other member into the other daughter cell. Without being able to recognize homologues, we would not be able to ensure that each daughter cell received one and only one member of each chromosome pair. In accomplishing this task, the cell is faced with exactly the same problem. The problem is solved by having homologous chromosomes "find" each other and pair up during an extended prophase. This pairing of homologous chromosomes then allows the spindle apparatus to separate members of the pair. There is one complication. As in mitosis, cells entering meiosis have already replicated their chromosomes. Thus, two nuclear divisions are necessary in meiosis without an intervening chromosome replication. That is, meiosis is a two-division process that produces four cells for each parent cell. These two divisions are referred to as meiosis I and meiosis II. As you will see, meiosis II is similar to a mitosis.

Unlike mitosis, meiosis occurs in only certain kinds of cells. In animals meiosis takes place only in primary gametocytes; in plants the process takes place only in spore-mother cells in the sporophyte generation of higher plants, which show an alternation of generations. At the end of the chapter, we review the processes of gamete and spore formation and look at some typical life cycles of particular types of organisms to show when meiosis takes place.

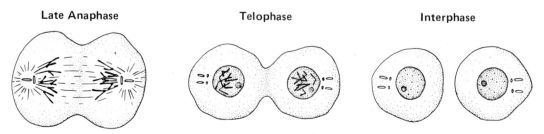

Late Anaphase   **Telophase**   **Interphase**

Figure 2-12. Telophase, Mitosis

Prophase of Meiosis I (Prophase I)

Cytogeneticists have divided this stage into five substages: **leptotene, zygo-tene, pachytene, diplotene,** and **diakinesis** (lepto-, thin; zygo- join; pachy-, thick; diplo-, double; dia-, apart). Since we are primarily interested in the genetic consequences of this process, we will not concern ourselves with the cytological details of these substages. A cell entering prophase I (leptotene) behaves similarly to one entering prophase of mitosis, with the centriole, spindle, nuclear envelope, and nucleolus behaving the same way. As chromosomes coil down in size, prior to being visible as chromatids, homologous chromosomes pair point-for-point along their lengths. This (zygotene) process is referred to as **synapsis.** Synapsis is mediated by a pro-teinaceous complex appearing between the homologous chromosomes; it is referred to as a **synaptinemal complex** (Figure 2–13). At this point the chromosome figures are referred to as **bivalents,** one bivalent per homolo-gous pair (Figure 2–14). As the chromosomes continue to shorten and thicken (pachytene), each chromosome can be seen to be made of two sister chromatids. At this point the chromosome figures are referred to both as bivalents and as **tetrads** because they are made up of four chromatids (Fig-ure 2–14). At about this time the synaptinemal complex disintegrates in most species. Further on in prophase I (diplotene), the chromosomes, while still shortening and thickening, appear to repel each other along most of their length. At this point one can see X-shaped configurations along the tetrads (Figure 2–14).

These X-shaped configurations are called **chiasmata** (singular: *chiasma*) and are of enormous significance because they represent physical indications of **crossing over,** a process whereby homologous chromosomes exchange parts. When two chromatids come to lie one over the other or directly next to each other, enzymes can break both chromatid strands at the same point and reattach them the alternate way (Figure 2–15). Thus, although genes have fixed positions on a chromosome, alleles that started out attached to a paternal centromere will be found attached to a maternal centromere. (The molecular mechanism of this process is examined in a later chapter.) This process can greatly increase the genetic variability present in gametes by associating alleles that were originally either just maternal or just paternal. As prophase I continues, spindle-fiber attachment takes place, and chiasmata terminalize, a process (at **diakinesis**) where they slip down the length of the chromosome until they reach the ends, freeing the chromo-somes to separate later.

Anaphase I

Metaphase takes place as in mitosis, with the tetrads being pulled to a metaphase plate by the spindle fibers. In anaphase I, homologous centro-

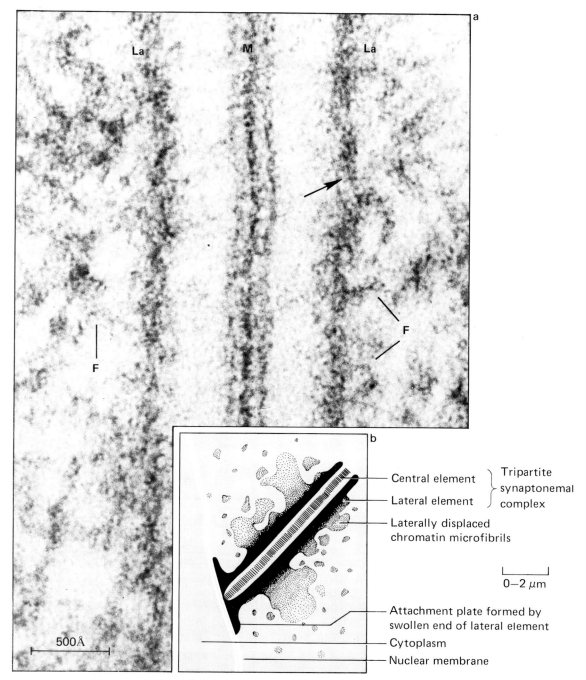

Figure 2-13. Synaptinemal Complex
In the electron micrograph, M is the central element, La are lateral elements, and F are chromosome fibers.
magnification 400,000X

Source: (a) R. Wettstein and J. R. Sotelo, "The Molecular Architecture of Synaptinemal Complexes," in E. J. DuPraw, ed., *Advances in Cell and Molecular Biology*, Vol. 1 (New York: Academic Press, 1971), p. 118. Reproduced by permission. (b) B. John and K. R. Lewis, *Chromosome Hierarchy* (London: Oxford University Press, 1975), p. 56. Reproduced by permission.

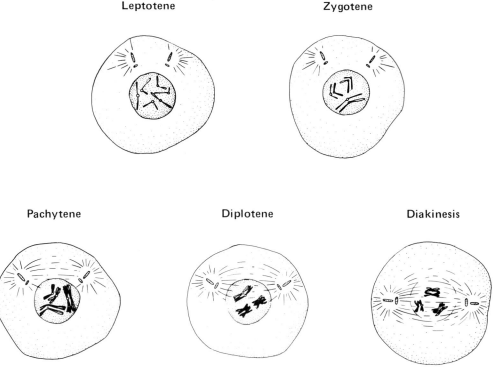

Figure 2-14. Prophase, Meiosis I

meres, each with its two chromatids attached (Figure 2–16), are pulled apart. This division is called **reductional division** because it reduces the number of chromosomes and centromeres to half the diploid number in each daughter cell. For every tetrad there is now one chromosome, in the form of a chromatid pair known as a **dyad,** at each pole of the cell. The initial requirement of meiosis, that of separating homologues into different daughter cells, is accomplished. However, since each dyad consists of two chromatids, a second division is required to reduce each chromosome to a single chromatid.

Depending on the organism, telophase I may or may not be greatly shortened in time. In some organisms the chromosomes spread into an interphase configuration and cytokinesis takes place—but no chromosome duplication takes place. Then, prophase II begins and meiosis II proceeds. In other organisms the late anaphase I chromosomes go almost directly into metaphase II and virtually skip prophase II. In any case, what follows, meiosis II, is basically a mitotic division in which the chromatids of each chromosome are pulled to opposite poles and form four cells at telophase II. This second meiotic division is called meiosis II and is referred to as an

Figure 2–15.   Crossing Over during Prophase, Meiosis I
Circles represent centromeres: one paternal, one maternal. One pair of
sister chromatids is represented at (a), the homologous pair at (b).

equational division; it does not further reduce the chromosome number
per cell (Figure 2–17). (Sometimes it is simpler to concentrate on the behavior of centromeres during meiosis rather than on chromosomes and chromatids. Meiosis I separates maternal from paternal centromeres and meiosis
II separates sister centromeres.)

Significance of Meiosis

Meiosis is significant for several reasons. First, the diploid number of chromosomes is reduced in such a way that each of the four daughter cells has
one complete haploid chromosome set. The diploid number is then reestablished at fertilization. Second, because of crossing over, there is an opportunity for increasing the allelic combinations of a population. Rather than
all maternal alleles staying together and all paternal alleles doing the same,
in each generation new combinations of maternal and paternal alleles can
form. New combinations are, of course, also introduced by the process itself,
which randomly combines maternal and paternal chromosomes in each
gamete.

The behavior of any tetrad follows the pattern of Mendel's rule of segregation. At gamete formation (meiosis), the diploid number of chromosomes is halved; each gamete receives only one chromosome from a homologous pair. The diploid number is then reestablished after each fertilization.
This process, of course, explains Mendel's rule of segregation. Independent

Metaphase I   Anaphase I

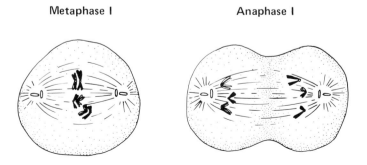

Figure 2–16.
Metaphase and
Anaphase of the
First Meiotic
Division

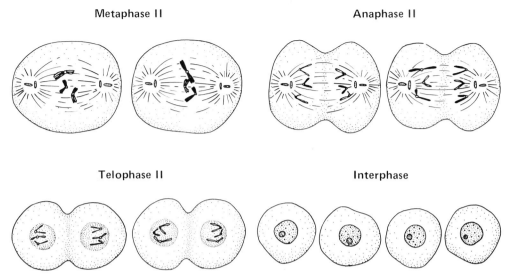

Figure 2-17.   Selected Views of Meiosis II

assortment is also explained by chromosome behavior at meiosis. When anaphase I takes place, the direction of separation is independent in different tetrads. While one pole may get the maternal centromere from chromosome pair number 1, it could get either the maternal or paternal centromere from chromosome pair number 2, and so on (Figure 2–18). This process explains Mendel's rule of independent assortment: Alleles at one locus segregate independently of alleles at other loci. Very shortly after rediscovery of Mendel's principles at the turn of the century, geneticists were quick to point this out.

MEIOSIS IN ANIMALS

In males each meiosis produces four equally sized **sperm** (Figure 2–19) in a process called **spermatogenesis.** In vertebrates a cell type in the testes known as a **spermatogonium** produces **primary spermatocytes** by mitosis. The primary spermatocytes undergo meiosis. After the first meiotic division, these cells are known as **secondary spermatocytes;** after the second meiotic division, they are known as **spermatids.** The spermatids mature into sperm—four sperm evolving from each primary spermatocyte. In humans and other vertebrates without a specific breeding season, the process of spermatogenesis is continuous through adult life. A normal human male may produce several hundred million sperm per day.

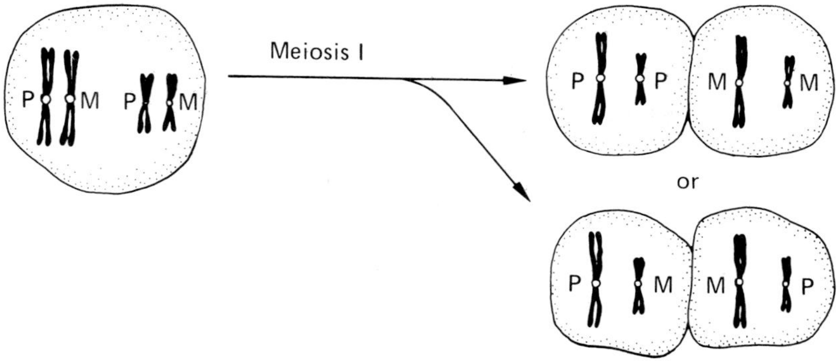

Figure 2–18.
Significance of
Meiosis to the Rule
of Independent
Assortment
P and M refer to
paternal and
maternal
centromeres.

In human females during embryological development, cells known as **oogonia** undergo numerous mitotic divisions to form cells known as **primary oocytes.** These begin the first meiotic division and then stop prior to birth of the female. A primary oocyte does not resume meiosis until past puberty, when ovulation occurs, a process usually occurring for only one oocyte per month during the female's reproductive life span (from about 12 to 45 years of age). Meiosis then proceeds. The two cells formed by meiosis I are termed **secondary oocytes** and are unequal in size. One contains almost all the nutrient-rich cytoplasm, and the other, a **polar body,** receives very little cytoplasm. The second meiotic division in the larger cell yields another polar body and an **ovum.** The first polar body divides to form two other polar bodies. Thus, **oogenesis** (Figure 2–20) produces cells of unequal size—an ovum and three polar bodies. The polar bodies disintegrate. Unequally sized cells are produced because the oocyte nucleus resides very close to the cell surface.

LIFE CYCLES

For eukaryotes the basic pattern of the life cycle is an alternation between a diploid and a haploid state (Figure 2–21). All life cycles are modifications of this general pattern. Most animals are diploids that form gametes by meiosis. The diploid number is restored by fertilization. Exceptions, however, are numerous. For example, in bees all males are haploid, and some fishes exist solely by parthenogenesis (the offspring come from unfertilized eggs). More complexity arises where copepods alternate sexual with parthenogenetic stages of their lives, and so on.

The general pattern of the life cycle in plants is one of an alternation of two distinct generations, each of which, depending on the species, may exist independently. In lower plants the haploid generation predominates, while in higher plants the diploid generation predominates. In flowering plants

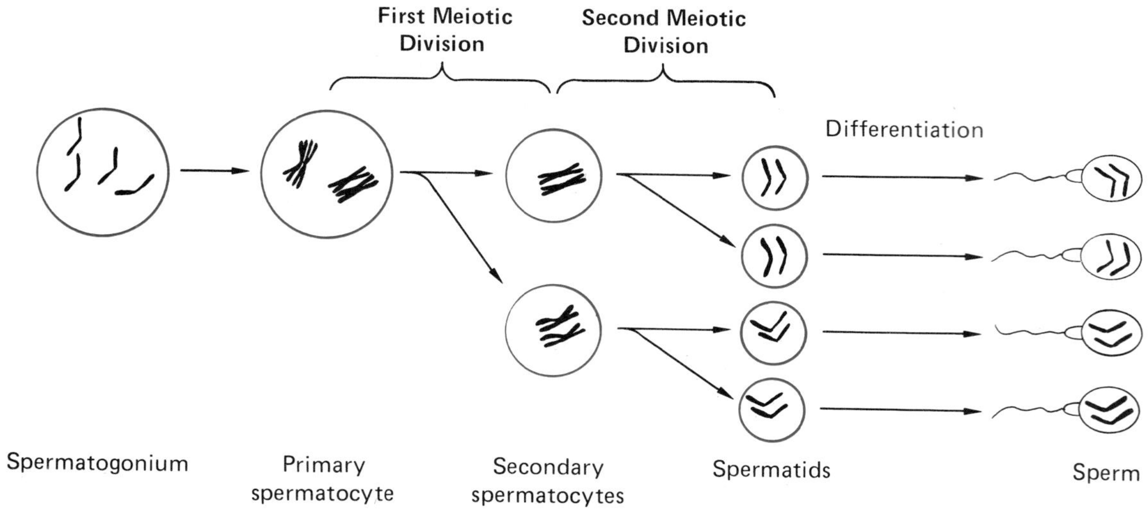

First Meiotic Division   **Second Meiotic Division**

Differentiation

Spermatogonium   Primary spermatocyte   Secondary spermatocytes   Spermatids   Sperm

Figure 2-19.   Spermatogenesis

(angiosperms), the plant that you see is the diploid sporophyte. It is referred to as a *sporophyte* because, through meiosis, it will give rise to spores. The spores germinate into the alternate generation, the haploid gametophyte, which produces gametes by mitosis. Fertilization then produces the diploid sporophyte. In lower plants the *gametophyte* has an independent existence; in angiosperms this generation is radically reduced. For example, in corn (Figure 2–22), an angiosperm, the corn plant is the sporophyte. In the male flowers microspores are produced by meiosis. After mitosis, three nuclei exist within each spore, a structure that we call a **pollen grain,** which is a gametophyte. In female flowers meiosis produces megaspores. Mitosis within a

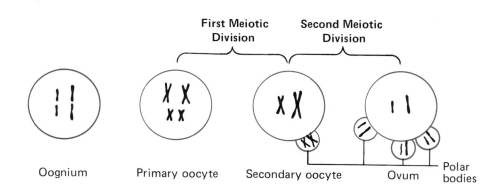

Figure 2-20. Oogenesis

Oognium   Primary oocyte   Secondary oocyte   Ovum   Polar bodies

First Meiotic Division   **Second Meiotic Division**

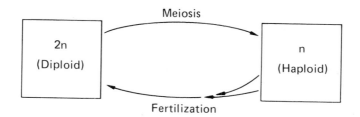

Figure 2-21.
Generalized Life
Cycle

megaspore produces an embryo sac of seven cells with eight nuclei. This is the female gametophyte. The egg cell is fertilized by a sperm nucleus. The two polar nuclei of the embryo sac are fertilized by a second sperm nucleus, producing a nutritive triploid $(3n)$ endosperm tissue. The sporophyte embryo grows from the diploid fertilized egg.

Many fungi and protista are haploid. Fertilization produces a diploid stage, which almost immediately undergoes meiosis to form haploid cells, these cells, in turn, increase in number by mitosis. More detail will be given when organisms such as *Neurospora,* the pink bread mold, are analyzed genetically.

Much of the knowledge of genetics derives from studies of certain specific organisms with unique properties. Pea plants were discussed earlier. Mendel found them useful because he could carefully control matings, their generation time was only a year, they were easily grown in his garden, and they had the discrete traits that he was looking for. Our interest in humans is obvious: We are humans. It is for this reason that so much effort has gone into human genetics. However, humans are a very difficult species to experiment with. They have a long generation time and a small number of offspring from matings that cannot be tailored for research purposes. The fruit fly, *Drosophila melanogaster,* is one of the organisms most extensively studied by geneticists. Fruit flies have a short generation time (12–14 days), which means that many matings can be carried out in a reasonable amount of time, they do exceptionally well in the laboratory, they have many easily observable discrete phenotypes, and in several organs they have giant banded chromosomes of great interest to cytologists. Other organisms will be examined extensively later on. As we make our way through this book and through other readings on genetics and as we come across studies involving new organisms, we should ask ourselves the question, "What are the properties of this organism that make it ideal for this type of research?"

CHROMOSOME THEORY OF HEREDITY

In a paper in 1903, Walter Sutton, a cytologist, firmly stated the concepts we have developed here: The behavior of chromosomes during meiosis explains Mendel's principles. Genes, then, must be located on chromosomes. This

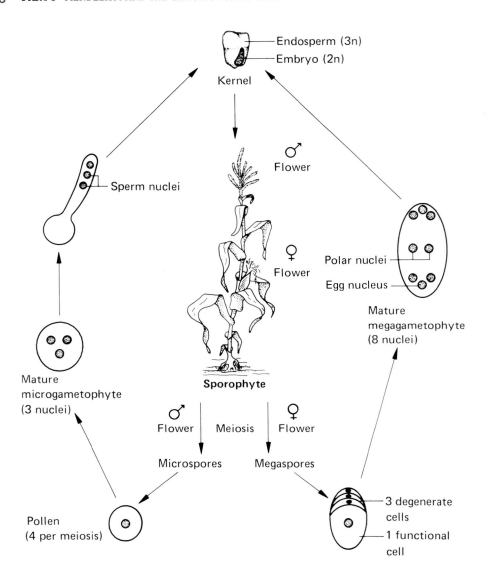

Figure 2-22.
Life Cycle of the
Corn Plant

idea was immediately accepted by biologists and ushered in the era of the **chromosomal theory of inheritance** wherein intensive effort was devoted to studying the relationship among genes on the same and on different chromosomes. The major portion of the first section of this book is devoted to the study of **linkage**—that is, to learning which gene is "linked" to which chromosome—and to the study of **mapping,** whereby it is possible to discover the sequence in which genes appear on a chromosome. This is basic information for a study of the structure and function of genes.

Here, we introduce a new term for gene. The term **locus** (plural: *loci*) is derived from the linear arrangement of the genes along the chromosome. Geneticists find this a useful and descriptive alternate term for the word *gene*.

CHAPTER SUMMARY

Mitosis and meiosis are the two processes in eukaryotes whereby the chromosomes are apportioned to daughter cells. The stages of cell division consist of a prophase, a metaphase, an anaphase, and a telophase. Mitosis is a one-step process in which the two sister chromatids making up each chromosome are separated into two daughter cells. Sex cells—gametes in animals and spores in plants—are produced by the two-step process of meiosis in which homologous chromosomes are separated into two daughter cells and a second (mitosis-like) division results in four cells, each with the full haploid chromosomal complement.

The spindle is the basic apparatus used by both processes, which are engineering marvels. Meiosis explains both of Mendel's principles: segregation and independent assortment.

This chapter ends by defining the chromosome theory of inheritance, the concept that shapes the first section of this book. Genes are located on chromosomes; their position and order on the chromosome can be discovered by mapping techniques described in later chapters.

EXERCISES AND PROBLEMS

1. You are working with a species with $2n = 6$, where one pair of chromosomes is telocentric; one pair, subtelocentric; and one pair, metacentric. The A, B, and C loci, each segregating a dominant and recessive allele (A and a, B and b, C and c), are each located on a different chromosome. Draw mitosis.

2. Given the same information as in Problem 1, diagram one of the possible meioses. How many different gametes can arise, excluding crossing over? What variation in gamete genotype is introduced by a crossover between the A locus and its centromere?

3. Given the following stages of nuclear division, identify the process, stage, and diploid number (for example, meiosis I, prophase, $2n = 10$). Keep in mind that one picture could possibly represent more than one process and stage.

a.

b.

c.
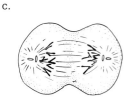

d.

e.
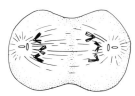

f.
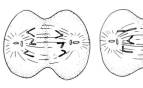

4. In humans, $2n = 46$. How many chromosomes would you find in:
   a. a brain cell?
   b. a red blood cell?
   c. a polar body?
   d. a sperm?
   e. a secondary oocyte?

5. Can you devise a method of chromosome partitioning during gamete formation that would not involve synapsis—that is, can you re-engineer meiosis?

6. If a dihybrid corn plant is self-fertilized, what genotypes of the triploid endosperm can result? If you know the endosperm genotype, can you determine the embryo sporophyte's genotype?

7. How many sperm come from ten primary spermatocytes? How many ova come from ten primary oocytes?

8. If inheritance were controlled primarily by the cytoplasm, what would be the relationship between an organism's phenotype and genotype and its parents' phenotypes and genotypes in:
   a. *Drosophila?*
   b. corn?
   c. *Neurospora?*

9. Given that a drone (male) bee is haploid and a queen (female) is diploid,

draw a testcross between a dihybrid queen and a drone. How many different kinds of sons and daughters result from this cross?

SUGGESTIONS FOR FURTHER READING

DuPraw, E. 1970. *DNA and Chromosomes.* New York: Holt, Rinehart and Winston.

John, B., and K. Lewis. 1975. *Chromosome Hierarchy.* New York: Oxford University Press.

Karp, G. 1979. *Cell Biology.* New York: McGraw-Hill.

Rhoades, M. M. 1961. Meiosis. In J. Brachet and A. E. Mirsky, eds., *The Cell,* vol. 3. New York: Academic Press.

Stubbe, H. 1972. *History of Genetics.* Cambridge, Mass.: MIT Press.

Sutton, W. S. 1903. The chromosomes in heredity. *Biol. Bull.* 4:231–251.

Swanson, C. P. 1957. *Cytology and Cytogenetics.* Englewood Cliffs, N.J.: Prentice-Hall.

Whittaker, R. H. 1969. New concepts of kingdoms of organisms. *Science* 163:150–160.

3 Probability and Statistics

In the experimental sciences, of which genetics is one, decisions about hypotheses must be made on the basis of data gathered during experiments. Geneticists must therefore develop an understanding of probability theory and statistical tests of hypotheses. Probability theory allows for accurate predictions of what to expect from a hypothesis. Statistical testing of hypotheses, particularly with the chi-square test, allows geneticists to have confidence in interpretations of the data gathered from experiments.

PROBABILITY

Part of the coming of age of genetics was Gregor Mendel's ability to work with simple mathematics. He was capable of turning counts into ratios and from them deducing the mechanism of inheritance. Taking numbers that did not *exactly* fit a ratio and rounding them off to fit is considered the crux of Mendel's deductive powers. The underlying rules that make the act of "rounding to a ratio" reasonable are the rules of probability.

Most of science deals with experimentation where manipulations are performed and results are predicted. The problem is that we live in a **stochastic** world, one dominated by random events. A bright new penny when flipped in the air twice in a row will not always give one "head" and one "tail." In fact, that penny if flipped 100 times could conceivably give 100 heads. In a stochastic world we can guess how often a coin should come down a head, but we cannot know for certain what the next toss will bring. We can guess how often a pea seed should be yellow in a given cross, but we cannot know for sure what the next pod will contain. Thus the need for **probability theory;** it tells us how to guess properly. This chapter closes with some thoughts on statistics, a branch of mathematics that helps us assess whether or not we did guess correctly.

The **probability** (p) that an event will occur is the number of favorable cases (a) divided by the total number of possible cases (n):

$$p = \frac{a}{n}$$

This probability can be determined in two ways: empirically or a priori. **Empirical probabilities** are calculated or discovered by observing a large number of cases and recording the number of times an event does (or does not) occur—the geometry of the event itself is not relevant. **A priori probabilities,** however, are determined by the nature of the event. An example of empirical probability is found in the life insurance business. It becomes more expensive to buy the same amount of life insurance as we get older. The reason is obvious: As we get older, we become more likely to die, and the insurance company becomes more likely to have to pay off on the policy within a relatively short amount of time. Insurance rates are calculated from actuarial, or life, tables. These tables are simple age-specific death rates for a given group of people. An insurance company may not know when a particular individual will die, but they can calculate the *probability* of death of that individual at a certain age. They can then equitably charge that individual a certain amount of money for life insurance, based on his or her age, so as to guarantee that they will make a profit yet be competitive with other companies. These probabilities are empirically determined and are based on how long people who have similar backgrounds and who come from similar communities actually do live.

A priori probabilities, the ones we are more familiar with, are the "odds" based on the geometry of an event. That is, a die (singular of dice) has six faces. When that die is tossed, there is no reason why one face should land "up" more often than any other. Thus, the probability of any one of the faces being up (for example, a four) is one-sixth:

$$p = \frac{a}{n} = \frac{1}{6}$$

Likewise, the probability of drawing the seven of clubs from a deck of cards is

$$p = \frac{1}{52}$$

The probability of drawing a spade from a deck of cards is

$$p = \frac{13}{52} = \frac{1}{4}$$

The probability (assuming a one-to-one sex ratio which actually is 1.06 males per female at birth among white Americans) of having a daughter is

$$p = \frac{1}{2}$$

And the probability that an offspring from a self-fertilized dihybrid will show the dominant phenotype is

$$p = \frac{9}{16}$$

Following from the preceding formula, we can say that an event that is a certainty has a probability of one and an event that is an impossibility has a probability of zero. If an event has a probability of p, all alternatives will have a probability of $q = 1 - p$; thus, $p + q = 1$. That is, the probability of the completely dominant phenotype in the F_2 of a selfed dihybrid is 9/16. The probability of any other phenotype is 7/16, which added to 9/16 equals 1.

Combining Probabilities

The basic principle of probability can be stated as follows: If one event has c possible outcomes and a second event has d possible outcomes, then there are cd possible outcomes of the two events. From this principle we obtain three rules that concern us as geneticists.

1. Sum Rule.   When the occurrence of one event precludes the occurrence of the other events, that is, when the *events are mutually exclusive,* the **sum rule** is used: *The probability of the occurrence of one of several of a group of mutually exclusive events is the sum of the probabilities of the individual events.* This is known as the *either/or* rule. For example, what is the probability, when throwing a die, of its showing *either* a four *or* a six? According to the sum rule:

$$p = \frac{1}{6} + \frac{1}{6} = \frac{2}{6} = \frac{1}{3}$$

2. Product Rule.   When the occurrence of one event is independent of the occurrence of other events, that is, when dealing with *independent events,* the **product rule** is used: *The probability of the occurrence of independent events is the product of their separate probabilities.* This is known as the *and* rule. For example, the probability of throwing a die two times and getting a four *and* then a six in that order is

$$p = \frac{1}{6} \times \frac{1}{6} = \frac{1}{36}$$

(*Note:* The probability of throwing two dice simultaneously and getting a four and a six is *not* 1/36; see rule #3.)

3. Binomial Theorem.   The **binomial theorem** is used for *unordered events*: *The probability of the occurrence of some arrangement of two mutually exclusive trials, where the final order is not specified, is defined by the binomial theorem.* For example, what is the probability when tossing two pennies simultaneously of getting a head and a tail?

USE OF RULES

There are several ways to calculate the probability just asked for. To put the problem in the form for rule #3 is the quickest method and we will dwell on it, but this problem can also be solved by using a combination of rules #1 and #2 in the following manner: For each penny the probability of getting a head (H) *or* a tail (T) is

$$\text{for H: } p = \frac{1}{2}$$
$$\text{for T: } q = 1 - \frac{1}{2} = \frac{1}{2}$$

Tossing the pennies one at a time, it is possible to get a head *and* a tail in two ways:

first head, then tail (HT)

or

first tail, then head (TH)

The results of each of the two tosses within a sequence are independent events. Thus, the probability for any one of the two ways involves the product rule (#2):

$$\frac{1}{2} \times \frac{1}{2} = \frac{1}{4} \quad \text{for HT or TH}$$

The two sequences are mutually exclusive. Thus, the probability of getting either of the two sequences is one of a set of mutually exclusive events and involves the sum rule (#1):

$$\frac{1}{4} + \frac{1}{4} = 2 \times \frac{1}{4} = \frac{1}{2}$$

Thus, for unordered events we can obtain the probability by a combination of rules #1 and #2. Rule #3 is the shorthand method.

To use rule #3, we must state it as follows: If the probability of an event (X) is p and an alternate (Y) is q, then the probability in n trials that event X will occur s times and Y will occur t times (where $s + t = n$ and $p + q = 1$) is

$$p = \frac{n!}{s!t!} p^s q^t$$

The symbol (!), as in $n!$, is called **factorial,** as in "n factorial," and is the product of all integers from n down to 1 (for example, $7! = 7 \times 6 \times 5 \times 4 \times 3 \times 2 \times 1$). Zero factorial equals 1, as does anything to the zero power ($0! = n^0 = 1$).

Now, what is the probability of tossing two pennies and getting one head and one tail? In this case, $n = 2$, s and $t = 1$, and p and $q = 1/2$. Thus,

$$\text{probability} = \frac{2!}{1!1!} \left(\frac{1}{2}\right)^1 \left(\frac{1}{2}\right)^1 = 2\left(\frac{1}{2}\right)^2 = 2\left(\frac{1}{4}\right) = \frac{1}{2}$$

This is, of course, our original answer. Now on to a few more genetically relevant problems. What is the probability that a family with 6 children will have precisely 5 girls and 1 boy? Since the order is not specified, we use rule #3 (we assume that the probability of a son equals the probability of a daughter and both equal ½),

$$\text{probability} = \frac{6!}{5!1!} \left(\frac{1}{2}\right)^5 \left(\frac{1}{2}\right)^1 = 6\left(\frac{1}{2}\right)^6 = \frac{6}{64}$$

What would happen if we asked for a specific family order, where first 4 girls were born, then a boy, and then a girl? This would entail rule #2; for a sequence of 6 independent events,

$$\text{probability} = \frac{1}{2} \times \frac{1}{2} \times \frac{1}{2} \times \frac{1}{2} \times \frac{1}{2} \times \frac{1}{2} = \frac{1}{64}$$

When no order is specified, the probability is six times larger than when it is; the reason is simply that there are six ways of getting 5 girls and a boy, and the sequence of 4 – 1 – 1 is only one of them. It is rule #3 that tells us that there are six ways. These are (letting B stand for boy and G for girl)

		Birth Order			
1	*2*	*3*	*4*	*5*	*6*
B	G	G	G	G	G
G	B	G	G	G	G
G	G	B	G	G	G
G	G	G	B	G	G
G	G	G	G	B	G
G	G	G	G	G	B

If two persons, heterozygous for albinism, are married, what is the probability that if they have 4 children, all 4 will be normal? The answer is simply $(\frac{3}{4})^4$ by rule #2. What is the probability that 3 will be normal and 1 albino? If we specify which of the 4 children will be albino (the second, for instance), then the probability is $(\frac{3}{4})^3(\frac{1}{4})^1$. If, however, we do not specify order,

$$\text{probability} = \frac{4!}{3!1!}\left(\frac{3}{4}\right)^3\left(\frac{1}{4}\right)^1 = 4\left(\frac{3}{4}\right)^3\left(\frac{1}{4}\right)^1$$

This is precisely four times the ordered probability because the albino child could have been born first, second, third, or last.

The formula for rule #3 is the formula of the **binomial expansion.** That is, if $(p + q)^n$ is expanded, the formula $(n!/s!t!)p^s q^t$ gives the probability for one of the terms, given that $p + q = 1$ and $s + t = n$. Since there are $(n + 1)$ terms in the binomial, the formula gives the probability for the term numbered $(t + 1)$. Two bits of useful information come from recalling that rule #3 is in reality the rule for the terms of the binomial expansion. First, if we have difficulty calculating the term, we can use Pascal's triangle to get the coefficients

```
            1
          1   1
        1   2   1
      1 3     3 1
    1 4   6   4 1
  1 5  10  10  5 1
```

Pascal's triangle is a trianglular array made up of coefficients in the binomial expansion and is calculated by starting any row with a 1, proceeding by adding the two adjacent terms in the row above, and then ending with a 1. For example, the next row would be

$$1, (1 + 5), (5 + 10), (10 + 10), (10 + 5), (5 + 1), 1 \quad \text{or} \quad 1, 6, 15, 20, 15, 6, 1$$

These numbers give us the combinations for any $p^s q^t$ term. That is, in our previous example, $n = 4$; so we use the $(n + 1)$, or fifth, row of Pascal's triangle. (The second number in any row of the triangle gives us the power of the expansion or n. Here, 4 is the second number of the fifth row.) We were interested in the case of 1 albino child in a family of 4 children, or $p^3 q^1$, where p is the probability of a normal child (3/4) and q is the probability of an albino child (1/4). Hence, we are interested in the $(t + 1)$—that is, the $(1 + 1)$—or second term of the fifth row of Pascal's triangle, which will tell us the number of ways of getting a 4-child family with 1 albino child. That number is 4. Thus, using Pascal's triangle, we see that the solution to the problem is

$$4 \left(\frac{3}{4} \right)^3 \left(\frac{1}{4} \right)^1$$

This is the same as the answer obtained the more conventional way.

The second gain from knowing that rule #3 is the binomial expansion formula is that we can now generalize to more than two events. The general form for **multinomial expansion** is $(p + q + r + \ldots)^n$ and the general formula for the probability is

$$\text{probability} = \frac{n!}{s! t! u! \ldots} p^s q^t r^u \ldots$$

where $s + t + u + \ldots = n$ and $p + q + r + \ldots = 1$. For example, our albino-carrying heterozygous parents may have wanted answers to the following question. If we have a family of 5, what is the probability that we will have 2 normal sons, 2 normal daughters, and 1 albino son? (This family will have no albino daughter.) By rule #2,

probability of a normal son = (3/4)(1/2) = 3/8

probability of a normal daughter = (3/4)(1/2) = 3/8

probability of an albino son = (1/4)(1/2) = 1/8

probability of an albino daughter = (1/4)(1/2) = 1/8

Thus,

$$\text{probability} = \frac{5!}{2! 2! 1! 0!} \left(\frac{3}{8} \right)^2 \left(\frac{3}{8} \right)^2 \left(\frac{1}{8} \right)^1 \left(\frac{1}{8} \right)^0$$

$$= 30(3/8)^4 (1/8)^1 = 30(3)^4 / (8)^5$$

$$= 2430/32{,}768$$

$$= 0.074$$

STATISTICS

In one of Mendel's experiments, he selfed F_1 long-stem (tall) heterozygous pea plants. In the next generation (F_2) he recorded 787 tall offspring and 277 dwarf offspring for a ratio of 2.84:1. Mendel rounded this to a 3:1 ratio, which supported his proposed rule of inheritance. In fact, is 787:277 "roundable" to a 3:1 ratio? From the brief discussion of probability, we expect some deviation from an exact 3:1 ratio (798:266), but how much of a deviation is acceptable? Would 786:278 still support Mendel's rule? Would 785:279 support it? Would 709:355 (a 2:1 ratio) or 532:532 (a 1:1 ratio)? Where do we draw the line? It is at this point that the discipline of statistics provides help. However, it does not provide certainty.

We can never speak with certainty about stochastic events. For example, in the case of Mendel's cross, although on the basis of Mendel's hypothesis a 3:1 ratio is expected, chance could give a 1:1 ratio in the data (532:532), and yet the mechanism could be the one that Mendel suggested. We could flip an honest coin 10 times and get 10 heads. Conversely, Mendel could have gotten exactly a 3:1 (798:266) ratio in his F_2 generation, and yet his hypothesis of segregation could have been wrong. The point is that any time we deal with probability there is some chance that the data will lead us to accept a bad hypothesis or reject a good one. Statistics quantifies these chances. We cannot say with certainty that a 2.84:1 ratio represents a 3:1 ratio; we can say, however, that we have a certain degree of confidence in the ratio. It is statistics that helps us ascertain these **confidence limits.**

Statistics is a branch of probability theory that helps the experimental geneticist in three ways. First, part of statistics is called **experimental design.** A bit of thought prior to the performance of an experiment may help design the experiment in the most efficient way. Although he did not know statistics, Mendel's experimental design was *very* good. The second way statistics is helpful is in the summarization of data. Such familiar terms as *mean, standard deviation,* and *coefficients of variation* are part of the body of descriptive statistics that takes large masses of data and reduces them to one or two meaningful values. We will examine further some of these terms and concepts in the chapter on quantitative inheritance.

Hypothesis Testing

The third way that statistics is valuable to geneticists is in the **testing of hypotheses:** determining whether to accept or reject a proposed hypothesis based on the likelihood (probability) that the hypothesis is correct. This area is most germane to our current discussion. For example, was a ratio of 787:277 really indicative of a 3:1 ratio? Since we now know that we cannot answer with an absolute "yes," how can we attach a degree of certainty to our answer? Statisticians would have us proceed as follows. To begin with, we need to determine what kind of variation to expect. This can be determined by calculating a **sampling distribution:**—the frequencies with which various possible events could occur in a particular experiment.

For example, if we selfed a heterozygous tall plant, we would expect a 3:1 ratio of tall and dwarf plants among the progeny. (The 3:1 ratio is our hypothesis based on the assumption of genetic control of height by one locus with two alleles.) If we looked only at the first 4 offspring, what is the probability of getting 3 tall and 1 dwarf plant: The answer is calculated using the formula for the terms of the binomial expansion:

$$\text{probability} = \frac{4!}{3!1!} \left(\frac{3}{4}\right)^3 \left(\frac{1}{4}\right)^1 = \frac{108}{256} = 0.42$$

Similarly, we can calculate the probability of getting all tall (81/256 = 0.32), 2 tall and 2 dwarf (54/256 = 0.21), 1 tall and 3 dwarf (12/256 = 0.05), and all dwarf (1/256 = 0.004). This distribution, as well as the distribution for samples of 8 and 40 progeny, is shown in Table 3–1. These distributions are graphed in Figure 3–1.

As sample sizes increase (from 4 to 8 to 40 in Figure 3–1), the sampling distribution takes on the shape of a smooth curve with a peak at the true ratio of 3:1 (75% tall progeny). That is, there is a high probability of getting very close to the true ratio. However, there is some chance the ratio will be fairly far off, and a very small part of the time our ratio will be very far off. An important result of this approach is to see that *any* ratio could arise in a given sample when the true ratio is 3:1. So where do we draw the line?

Statisticians have agreed on a convention. When all the frequencies are plotted, as in Figure 3–1, the area under the curve is taken as 1 unit, and we draw lines to include 95% of this area (Figure 3–2). Any ratios included within the 95% limits are considered acceptable for the hypothesis of a 3:1 ratio. Any ratio in the remaining 5% area is considered unacceptable. (Other conventions also exist, such as rejection within the outer 10% or 1% limits; we consider these at the end of the chapter.) Thus, experimental data in hand, it is possible to see whether it is acceptable as supporting our hypothesis (in this case the hypothesis of 3:1). One in 20 times we will make a **type I error:** We will reject a true hypothesis. A **type II error** is that of accepting a false hypothesis.

Is it necessary to calculate a sampling distribution every time we do an experiment? To determine whether to reject (or accept) a hypothesis, a frequency distribution must be derived for each *type* of experiment. Mendel could also have used the distribution shown in Figure 3–1 for seed coat or seed color, as long as he was expecting (hypothesized) a 3:1 ratio and had a similar sample size. What about independent assortment, where a 9:3:3:1 ratio is expected? A geneticist would have to calculate a new sampling distribution based on a 9:3:3:1 ratio and a particular sample size. Statisticians have devised a shortcut method by using *standardized distributions* from

TABLE 3-1. Sampling Distribution for Sample Sizes of 4, 8, and 40, Given a 3:1 Ratio of Tall and Dwarf Plants in This Experiment

No. Tall Plants	$n = 4$ Probability*	No. Tall Plants	$n = 8$ Probability*	No. Tall Plants	$n = 40$ Probability*
4	$\frac{81}{256} = (0.32)$	8	$\frac{6,561}{65,536} = (0.10)$	40	$\frac{1.2 \times 10^{19}}{1.2 \times 10^{24}} = (0.00001)$
3	$\frac{108}{256} = (0.42)$	7	$\frac{17,496}{65,536} = (0.27)$	39	$\frac{1.6 \times 10^{20}}{1.2 \times 10^{24}} = (0.0001)$
2	$\frac{54}{256} = (0.21)$	6	$\frac{20,412}{65,536} = (0.31)$	38	$\frac{1.1 \times 10^{21}}{1.2 \times 10^{24}} = (0.0009)$
1	$\frac{12}{256} = (0.05)$	5	$\frac{13,608}{65,536} = (0.21)$		$\cdots$
0	$\frac{1}{256} = (0.004)$	4	$\frac{5,670}{65,536} = (0.09)$	30	$\frac{1.7 \times 10^{23}}{1.2 \times 10^{24}} = (0.14)$
		3	$\frac{1,512}{65,536} = (0.02)$		$\cdots$
		2	$\frac{252}{65,536} = (0.004)$	2	$\frac{7020}{1.2 \times 10^{24}} = (0.59 \times 10^{-20})$
		1	$\frac{24}{65,536} = (0.0004)$	1	$\frac{120}{1.2 \times 10^{24}} = (0.10 \times 10^{-21})$
		0	$\frac{1}{65,536} = (0.00002)$	0	$\frac{1}{1.2 \times 10^{24}} = (0.83 \times 10^{-24})$

* Probabilities are calculated from the binomial theorem.

$$\text{probability} = (n!/s!t!)p^s q^t$$

where n = number of progeny observed
     s = number of progeny that are tall
     t = number of progeny that are dwarf
     p = probability of a progeny plant being tall (3/4)
     q = probability of a progeny plant being dwarf (1/4)

which to calculate probabilities. Many are in use, such as the t-distribution, binomial distribution, and chi-square distribution. Each is useful for particular kinds of data; geneticists tend to use the chi-square distribution to test hypotheses regarding breeding data.

Chi-Square

Only one statistical test will be discussed here. When sample subjects are distributed among discrete categories such as tall- and short-stem (dwarf)

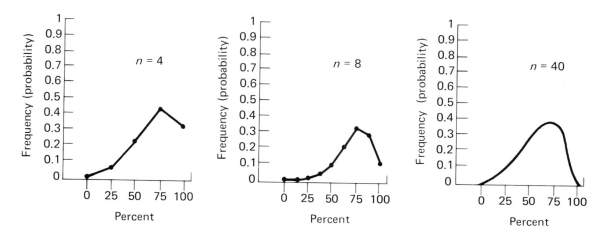

Figure 3-1. Sampling Distributions from an **Experiment** with an Expected Ratio of 3 Tall to 1 Dwarf Plant
As the sample size, n, gets larger, the distribution becomes smoother. These distributions are the plotted individual terms of the expanded binomial distribution (Table 3-1).

plants, the **chi-square distribution** is frequently used (Figure 3–3). We will examine degrees of freedom shortly. The formula for converting categorical experimental data to a chi-square value is

$$\chi^2 = \sum \frac{(O - E)^2}{E}$$

where χ is the Greek letter chi, O is the observed number for a category, E is the expected number for that category, and Σ means to sum the calculations for all the categories.

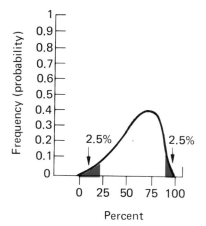

Figure 3-2. Sampling Distribution of Figure 3–1 ($n = \infty$) with 5% of the Area Marked off (2.5% at Each End)

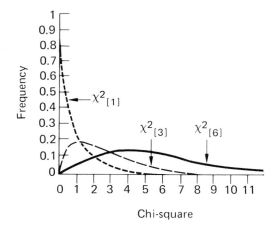

Figure 3–3.
Chi-Square
Distribution for
Several Different
Degrees of Freedom
It approaches a bell-
shaped curve when
the sample size gets
large.

A chi-square (χ^2) value of 0.60 is calculated in Table 3–2 for Mendel's data on the basis of a 3:1 ratio. If Mendel had originally expected a 1:1 ratio, he would have calculated a chi-square of 244.45 (Table 3–3). However, these values have little meaning of themselves: They are not probabilities. They must be converted to the probabilities under the chi-square distribution to determine how well the observed ratio compares to the expected. To do this, we usually use a chi-square table where probabilities have been calculated from the distribution curve (Table 3–4). Before we can use this table, we must define the concept of **degrees of freedom.**

Reexamination of the chi-square formula and Tables 3–2 and 3–3 will reveal that there is a contribution to the total chi-square value from each category: Chi-square is a summed value. We expect the chi-square value to increase as the total number of categories increases. That is, the more categories involved, the larger the chi-square even if the sample were a relatively good fit to the hypothesized ratio. Hence, we need some way of keeping track of categories. This is done with degrees of freedom, which is basically a count of *independent* categories. For our purposes here, degrees of freedom equal

TABLE 3–2. Chi-Square Analysis of One of Mendel's Experiments, Assuming a 3:1 Ratio

	Tall Plants	Dwarf Plants	Total
Observed numbers (O)	787	277	1064
Expected ratio	3/4	1/4	
Expected numbers (E)	798	266	
$O - E$	−11	11	
$(O - E)^2$	121	121	
$(O - E)^2/E$	0.15	0.45	$0.60 = \chi^2$

TABLE 3-3. Chi-Square Analysis of One of Mendel's Experiments, Assuming a 1:1 Ratio

	Tall Plants	Dwarf Plants	Total
Observed numbers (O)	787	277	1064
Expected ratio	1/2	1/2	
Expected numbers (E)	532	532	
$O - E$	255	-255	
$(O - E)^2$	65,025	65,025	
$(O - E)^2/E$	122.23	122.23	$244.45 = \chi^2$

the number of categories minus one. With Mendel's data, since the total was 1064 and 787 had tall stems, the short-stem category had to consist of 277 plants. Thus, in Mendel's plant experiment with two phenotypic categories, there is only one degree of freedom.

Table 3–4, the table of chi-square probabilities, is read as follows. Degrees of freedom are read in the left column, and in this case we would be interested in the first row where there is one degree of freedom. The numbers across the top are the probabilities. We are interested in the next-to-the-last column, headed by 0.05. The probabilities across the top are read as: *The*

TABLE 3-4. Chi-Square Values

Degrees of Freedom	Probabilities						
	0.99	0.95	0.80	0.50	0.20	0.05	0.01
1	0.000	0.004	0.064	0.455	1.642	3.841	6.635
2	0.020	0.103	0.446	1.386	3.219	5.991	9.210
3	0.115	0.352	1.005	2.366	4.642	7.815	11.345
4	0.297	0.711	1.649	3.357	5.989	9.488	13.277
5	0.554	1.145	2.343	4.351	7.289	11.070	15.086
6	0.872	1.635	3.070	5.348	8.558	12.592	16.812
7	1.239	2.167	3.822	6.346	9.803	14.067	18.475
8	1.646	2.733	4.594	7.344	11.030	15.507	20.090
9	2.088	3.325	5.380	8.343	12.242	16.919	21.666
10	2.558	3.940	6.179	9.342	13.442	18.307	23.209
15	5.229	7.261	10.307	14.339	19.311	24.996	30.578
20	8.260	10.851	14.578	19.337	25.038	31.410	37.566
25	11.524	14.611	18.940	24.337	30.675	37.652	44.314
30	14.953	18.493	23.364	29.336	36.250	43.773	50.892

Source: C. M. Thompson, **Biometrika** 32 (1941):188–189. Reprinted by permission of the Biometrika Trustees.

probability is 0.05 (for example) of getting a chi-square value this large or larger by chance alone given that the hypothesis is correct. If we examine this statement, we realize it is a formalization of what we have been talking about all along in our discussion of frequency distributions. Hence, we are interested in how large a chi-square will put us beyond that 95% acceptable area of the curve into the 5% unacceptable area. The *critical chi-square* (at $p = 0.05$, 1 degree of freedom) is 3.841. This is the value to which we compare our χ^2's (0.60 and 244.45). Since our chi-square for the 3:1 ratio (Table 3–2) is 0.60, which is less than the critical value, we accept our hypothesis. But since our χ^2 for the 1:1 ratio (Table 3–3) is 244.45, which is greater than the critical value, we reject the hypothesis of a 1:1 ratio. Notice, however, that once we did the chi-square test for the 3:1 ratio and accepted it, no other statistical tests were needed.

A word of warning for using chi-square: If the expected number for any category is less than 5, the conclusions are not reliable. In that case the experiment can be repeated to obtain a larger sample size, or categories can be combined. Note also that chi-square tests are always done on whole numbers, not on ratios or percentages.

Accepting Hypotheses

The hypothesis against which the data are tested is referred to as a **null hypothesis.** Hypothesis testing involves testing the assumption that there is no difference between the observed and expected samples. If the null hypothesis is accepted, then we say that the data are consistent with it, not that the hypothesis has been proved. (As previously discussed, there are built-in possibilities of accepting false hypotheses or rejecting true ones.) If, however, the hypothesis is rejected, as we rejected a 1:1 ratio for Mendel's data (Table 3–2), the only other choice is to accept the alternative that there is a difference between the observed and the expected values. The data then may be retested against some other hypothesis.

The use of the 0.05 level as a cutoff for rejecting a hypothesis is a convention. It is called the **level of significance.** When a hypothesis is rejected at that level, statisticians say that the result is *significant.* Other levels of significance are also used, such as 0.01. If a calculated chi-square is greater than the critical value in the table at the 0.01 level, the hypothesis is rejected with a *highly significant result.* Since the chi-square value at the 0.01 level is larger than the value at the 0.05 level, it is more difficult to reject a hypothesis at this level and hence more convincing when it is rejected.

Other levels of rejection are also set. For example, in clinical trials of medication, an attempt is made to make it very easy to reject the null hypothesis: A level of significance of 0.10 or higher is set. The rationale is that it is not desirable to throw away a drug or treatment that might be beneficial. Since the null hypothesis states that the drug has no effect—that is, the control and drug groups show the same response—clinicians would much rather be overly conservative. Accepting the null hypothesis means concluding that the drug has no effect. Rejecting the hypothesis means that the drug has some effect and it should be tested further. It is much better to have to retest some drugs that are actually worthless than to discard drugs that have potential value.

CHAPTER SUMMARY

We have examined the rules of probability theory relevant to genetic experiments. Probability theory allows us to properly predict the outcome of experiments. The probability of independent events is calculated by multiplying their separate probabilities. The probability of mutually exclusive events is calculated by adding their individual probabilities. And the probability of unordered events is defined by polynomial theory $(p + q + r + \dots)^n$:

$$\text{probability} = \frac{n!}{s!t!u! \dots} p^s q^t r^u \dots$$

In order to determine whether data gathered during an experiment actually fit a particular hypothesis, it is necessary to determine what the probability is of getting a particular data set when the null hypothesis is correct. We have considered the chi-square test

$$\chi^2 = \sum \frac{(O - E)^2}{E}$$

which is a method of quantifying the confidence we may have in the results obtained from typical genetic experiments.

The rules of probability and statistics allow us to devise hypotheses about inheritance and to test these hypotheses with experimental data.

EXERCISES AND PROBLEMS

1. The following are data from Mendel's original experiments. Suggest a hypothesis for each set and test this hypothesis with the chi-square test. Do you reach different conclusions with different levels of significance?

   a. Self-fertilization of round-seeded hybrids produced 5474 round seeds and 1850 wrinkled ones.

   b. One particular plant from part (a) yielded 45 round seeds and 12 wrinkled ones.

c. Of the 565 plants raised from F_2 round-seeded plants, 193 yielded only round seeds, while 372 gave both round and wrinkled seeds in a 3:1 proportion.

d. A violet-red flowered, long-stem plant was crossed with a white-flowered, short-stem plant with the following result:

47 violet-red, long-stem

40 white, long-stem

38 violet-red, short-stem

41 white, short-stem

2. Assuming a 1:1 sex ratio, what is the probability that a family of 5 children will consist of

a. 3 daughters and 2 sons?

b. alternating sexes, starting with a son?

c. alternating sexes?

d. all daughters?

e. all the same sex?

f. at least 4 daughters?

g. a daughter as the eldest child and a son as the youngest?

3. Phenylthiocarbamide (PTC) tasting is dominant (T) to nontasting (t). If a taster woman with a nontaster father married a taster man who, in a previous marriage, had a nontaster daughter, what would be the probability that

a. their first child would be a nontaster?

b. their first child would be a nontaster girl?

c. if they had 6 children, they would have 2 nontaster sons, 2 nontaster daughters, and 2 taster sons?

d. their fourth child would be a taster daughter?

4. Albinism is recessive, as are blue eyes. What is the probability that 2 brown-eyed persons, heterozygous for both traits, produce (remembering epistasis)

a. 5 albino children?

b. 5 albino sons?

c. 4 blue-eyed daughters and a brown-eyed son?

d. 2 sons like their father and 2 daughters like their mother?

5. On the average, about one child in every 10,000 live births in the United States has phenylketonuria (PKU). What is the probability that

a. the next child born in a Boston hospital will have PKU?

    b. after a PKU child is born, the next child will have PKU?

    c. two children born in a row will have PKU?

6. In fruit flies the diploid chromosome number is eight.

    a. What is the probability that a male gamete will contain only paternal centromeres or only maternal centromeres?

    b. What is the probability that an offspring *Drosophila* zygote will contain only centromeres from male grandparents? (Disregard the problems that the sex chromosomes may introduce.)

7. How many seeds should Mendel have tested to determine with complete certainty whether or not a plant with a dominant phenotype was heterozygous? With 99% certainty? With 95% certainty? With "pretty reliable" certainty?

8. Assuming that an "ideal" couple wants one son and one daughter, what chance do they have of achieving this goal? If all couples wanted at least one child of each sex, approximately what would the average family size be?

SUGGESTIONS FOR FURTHER READING

Ross, S. M. 1976. *A First Course in Probability*. New York: Macmillan.

Siegel, S. 1956. *Nonparametric Statistics for the Behavioral Sciences*. New York: McGraw-Hill.

Smith, J. M. 1968. *Mathematical Ideas in Biology*. Cambridge: Cambridge University Press.

Sokal, R. R., and F. J. Rohlf. 1969. *Biometry*. San Francisco: Freeman.

4 Sex Determination, Sex Linkage, and Pedigree Analysis

Since all organisms have many fewer chromosomes than genes, it follows that each chromosome has many loci. We must, then, study the pattern of inheritance of loci *linked* to each other on the same chromosome. Some chromosomes play a role in sex determination so that the inheritance of genes on these **sex chromosomes** shows patterns related to the sex of the offspring produced. Human traits are most often studied using pedigrees or family trees. This chapter starts a four-chapter sequence of the analysis of the relationship of genes to chromosomes. We begin with the study of sex determination.

SEX DETERMINATION

While sex chromosomes provide the most common method of sex determination in diploids, they are not the only mechanism. Sex can be determined by the ploidy of an individual, as in many hymenoptera (bees, ants, wasps) where males are haploid and females are diploid; by allelic mechanisms where sex is determined by a single allele or multiple alleles, as in some wasps; or by environmental factors, as in some marine worms and gastropods where the sex of the individual depends on the substrate on which an individual lands. The latter, in essence, have indeterminate genetic sex-determining mechanisms—each individual can develop into either sex.

An interesting example is the slipper shell, *Crepidula*, where individuals tend to live stacked up on one another (Figure 4–1). Young *Crepidula* are always male. However, as an individual ages, the male reproductive system degenerates. The reproductive system can then re-form as male or become female, depending on the sexes of the other organisms in the cluster. If the organism is attached to a female, the reproductive system will rede-

81

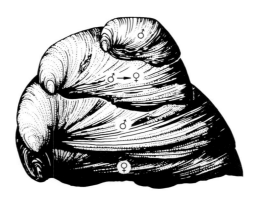

Figure 4-1.
A Cluster of
Crepidula **(Slipper Shells)**
The second organism from the top is in the process of changing from male to female. When it has changed to a female, it will be fertilized by the top organism, a male.

velop as male. Isolation or the presence of a large number of males will induce a male to become a female. Once a female, the individual will no longer change.

Sex Chromosomes

Basically, four types of chromosomal mechanisms exist. These are the XY, YX, XO, and compound chromosome mechanisms. In the XY case the females have a homomorphic pair of chromosomes, as in humans or fruit flies; males are heteromorphic. In the YX case males are homomorphic, while females are heteromorphic. In the remaining instances more or less than two sex chromosomes are involved. In the XO case there is only one sex chromosome, as in some grasshoppers and beetles (where females are usually XX and males XO). And in the compound-chromosome case, several X and Y chromosomes are involved in sex determination, as in bedbugs and some beetles.

The XY System.   The XY situation occurs in humans where females have 46 chromosomes arranged in 23 homologous, homomorphic pairs. Males, with the same number of chromosomes, have 22 homomorphic pairs and one heteromorphic pair. The heteromorphic pair is referred to as the XY pair: Females are homozygous for the X chromosome (Figure 4–2; notice that the Y chromosome is smaller than the X). During meiosis females will produce only gametes that contain the X chromosome, whereas males will produce two kinds of gametes: X- and Y-bearing (Figure 4–3). For this reason females are referred to as **homogametic** and males as **heterogametic.** As we

Figure 4-2.   Human Karyotype
Note the X and Y chromosomes. A female would have a second X chromosome in place of the Y.

Source: Reproduced courtesy of Dr. Thomas G. Brewster, Foundation for Blood Research, Scarborough, Maine.

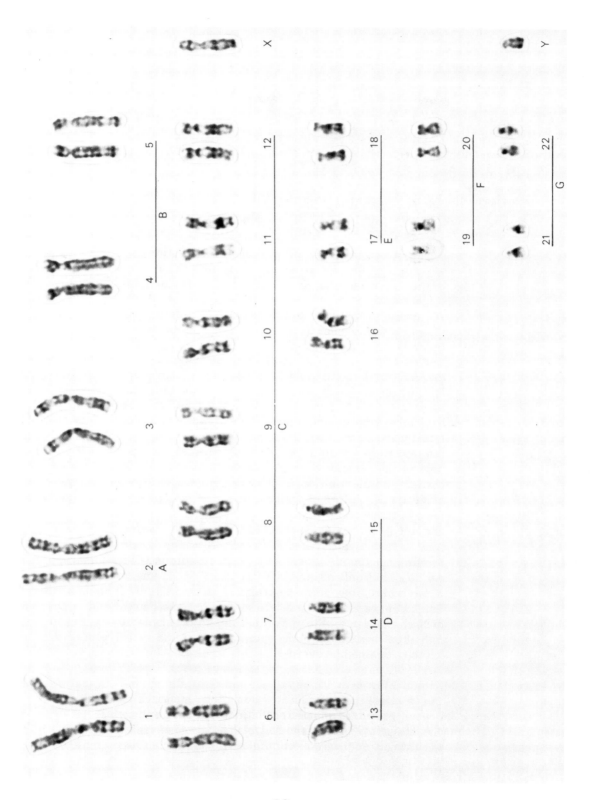

83

Sperm

One autosomal
set plus

Ovum	X	Y
One autosomal set plus X	two autosomal sets plus XX Daughter	two autosomal sets plus XY Son

Figure 4-3. Segregation of Human Sex Chromosomes during Meiosis, with Subsequent Zygote Formation

can see from Figure 4–3, in the human, fertilization will cause equal numbers of male and female offspring to be formed. In *Drosophila* the system is the same, but the X chromosome is smaller than the Y chromosome (Figure 4–4).

Since both human and *Drosophila* females normally have two X chromosomes while males have an X and a Y chromosome, it is not possible to know whether their maleness is determined by the presence of a Y chromosome or the absence of a second X chromosome. One way to resolve this problem would be to isolate individuals with odd numbers of chromosomes. In Chapter 7 we will examine the causes and outcomes of anomalous chromosome numbers. Here we will consider two facts from that chapter. First, sometimes (infrequently) individuals are formed (although they are not necessarily viable) with whole extra sets of chromosomes. These individuals are referred to as **polyploids** (*triploids* with 3n, *tetraploids* with 4n, and so on). Second, sometimes (also infrequently) individuals have more or less than the normal number of any one chromosome. These **aneuploids** come about by the failure of a pair of chromosomes to separate properly during meiosis (an occurrence called **nondisjunction**). The existence of polyploid and aneuploid individuals makes it possible to test whether the Y chromosome is male determining. For example, a human or a fruit fly that has all the proper nonsex chromosomes, or **autosomes** (44 in humans, 6 in *Drosophila*), but only a single X without a Y would answer our question. If the Y were absolutely male determining, then this XO individual should be female. However, if the sex-determining mechanism is a result of the number of X chromosomes, this individual should be a male. As it turns out, an XO individual is a human female and a *Drosophila* male. Each of these cases will be examined presently.

The XO System. The XO system naturally appears in many species of insects. Here the situation is as described for the XY mechanism except that instead of a Y chromosome the heterogametic sex (male) has only one X chromosome. This system is also sometimes referred to as an XO-XX system. Males produce gametes that contain either an X chromosome or no sex

Figure 4-4. Chromosomes of *Drosophila melanogaster*

Calvin B. Bridges
(1889–1938)
Genetics, 25 (1940):
frontispiece.

chromosome, while all the gametes from a female contain the X chromosome. The result of this arrangement is that females have an even number of chromosomes (all in homomorphic pairs), whereas males have an odd number of chromosomes (Figure 4–5).

The YX System. The YX system (sometimes referred to as the WZ system to distinguish it from the XY system) is identical to the XY system except that males are homogametic and females are heterogametic. This situation occurs in birds, some fishes, and moths. The compound chromosome systems tend to be complex. For example, in *Ascaris incurva,* a nematode, there are 8 X chromosomes and one Y. The species has 26 autosomes. Males have 35 chromosomes (26A + 8X + Y), while females have 42 chromosomes (26A + 16X). During meiosis the X chromosomes unite end-to-end and so behave as one unit.

Genic Balance in *Drosophila*

When the geneticist Calvin Bridges, working with *Drosophila,* crossed a triploid (3*n*) female with a normal male, he observed many combinations of autosomes and sex chromosomes in the offspring. Bridges suggested in 1922 that sex is determined in *Drosophila* by the balance (ratio) of autosomal alleles that favor maleness and alleles on the X chromosomes that favor femaleness. He then calculated a ratio of X chromosomes to autosomal sets in order to see if this ratio would predict the sex of a fly. An **autosomal set** (A) in *Drosophila* consists of three chromosomes. Table 4–1 which presents his results, shows that Bridges's **genic balance theory** of sex determination is correct. When the X:A ratio is 1.00 or greater, the organism is a female. When this ratio is 0.50 or less, the organism is a male. At 0.67 the organism is an **intersex. (Metamales** and **metafemales** are usually very weak and sterile. The metafemales usually do not even emerge from their pupal cases.)

Figure 4–5.
Mechanism
Determining the XO
Chromosome of
Some Insects

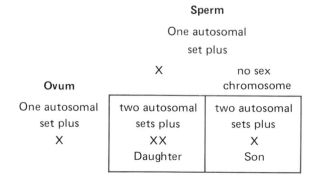

	Sperm	
	One autosomal	
	set plus	
	X	no sex
Ovum		chromosome
One autosomal	two autosomal	two autosomal
set plus	sets plus	sets plus
X	XX	X
	Daughter	Son

TABLE 4-1. Data Supporting Bridges's Theory of Sex Determination by Genic Balance in *Drosophila*

Number of X Chromosomes	Number of Autosomal Sets (A)	Total Number of Chromosomes	$\frac{X}{A}$ Ratio	Sex
3	2	9	1.50	Metafemale
4	3	13	1.33	Female
4	4	16	1.00	Female
3	3	12	1.00	Female
2	2	8	1.00	Female
1	1	4	1.00	Female
2	3	11	0.67	Intersex
1	2	7	0.50	Male
1	3	10	0.33	Metamale

The analysis so far should not give us a misleading view of sex determination, which is a very complex, multistage developmental process. In addition to the determinants already discussed, environment and alleles at several other loci can also influence the final sex of the fly. For example, the intersex fly with an X:A ratio of 0.67 can have its development altered by the temperature at which it is raised. At higher temperatures the flies tend toward the female end of the intersex spectrum; at lower temperatures the flies tend toward the male end of this spectrum. Also, there are autosomal loci with alleles that can override the chromosomal constituency. The recessive **doublesex** (*dsx*) allele converts males and females into developmental intersexes. **Transformer** (*tra*) is a recessive allele that converts chromosomal females into sterile males. Thus, the framework suggested by Bridges's genic balance theory can be influenced by the environment as well as by specific other loci.

Determinants of Sex in Humans

Since the XO genotype in humans is a female (Turner's syndrome), it seems reasonable to conclude that the Y chromosome is male determining in humans. This is verified by the fact that persons with Klinefelter's syndrome (XXY, XXXY, XXXXY) are all male while XXX, XXXX, and other multiple X karyotypes are all female. (Chapter 7 will present more details on these anomalies.) Recently, some exciting work by Stephen Wachtel and his colleagues at the New York Memorial Sloan-Kettering Cancer Center has provided new insights into sex determination in humans. It has been found that males of all mammal species so far tested have a surface protein on their cells that is not found in females. This protein has been called the *Histocom-*

patibility Y-antigen (**H-Y antigen**). The gene for this antigen is located on the short arm of the Y chromosome. Precisely what this antigen does is unknown, but its presence causes the undifferentiated gonad to become a functional testis. Further development of maleness, such as male secondary sexual characteristics, comes about through the influence of testosterone produced by the functional testis.

SEX LINKAGE

In an XY chromosomal sex-determining system, the pattern of inheritance for loci on the heteromorphic chromosomes must certainly differ from the pattern for loci on the homomorphic autosomal chromosomes because sex chromosome alleles of the heterogametic sex do not randomly assort into the offspring. Alleles on the male's X chromosome go to his daughters but not his sons because the presence or absence of the male's X chromosome determines the sex of his offspring. (His sons can of course get these alleles from their mother.) For example, the inheritance pattern of hemophilia (failure of blood to clot), where the common form is caused by an allele located on the X chromosome, has been known since the end of the eighteenth century. It was known that men got the disease while women could pass on the disease.

The first sex-linked trait studied experimentally was the white-eye phenotype in fruit flies. Before continuing, there is a simple distinction to be made. Since both X and Y chromosomes are involved, three different patterns of inheritance are possible, all called sex linked. In fact, the term **sex linked** usually refers to loci on the X chromosome, while the term **Y linked** is usually used to refer to traits on the Y chromosome. No particular term exists for loci found on both the X and Y chromosomes. In humans there are over 100 loci known to be on the X chromosome; there are only a few known loci on the Y chromosome.

X-Linkage in *Drosophila*

Thomas Hunt
Morgan
(1866–1945)
Genetics, 32 (1947):
frontispiece.

T. H. Morgan demonstrated the X-linked pattern of inheritance in *Drosophila* in 1910 when a white-eyed male appeared in a culture of wild-type (red-eyed) flies. This male was crossed with a wild-type female. All the offspring were wild type. When these F_1 individuals were crossed with each other, their offspring fell into two categories (Figure 4–6). All the females and half the males were wild type while the remaining half of the males were white eyed. Morgan interpreted this to mean that the white-eye locus was on the X chromosome. We can thus redraw Figure 4–6 to include the sex chromosomes of Morgan's flies (Figure 4–7). We denote the white-eye allele as X^w because it appears on the X chromosome. Similarly, X^+ is the wild-type allele, and Y is the Y chromosome, which has no allele at this locus.

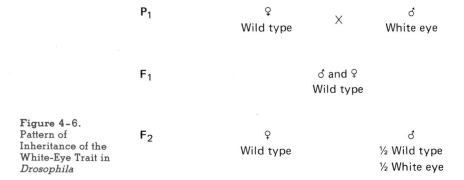

P_1 ♀   ×   ♂
        Wild type   White eye

F_1 ♂ and ♀
                        Wild type

Figure 4-6.
Pattern of
Inheritance of the
White-Eye Trait in
Drosophila

F_2 ♀   ♂
        Wild type   ½ Wild type
                                    ½ White eye

Figure 4–7 indicates an interesting property of sex linkage. Since females have two X chromosomes, they can have normal homozygous and heterozygous gene combinations. But males, with only one copy of the X

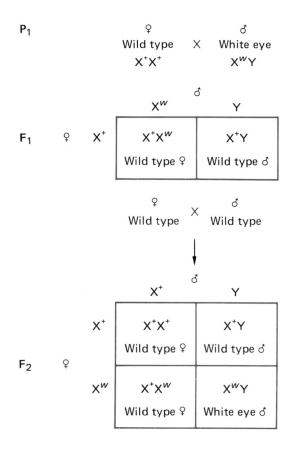

Figure 4–7.
Crosses of Figure
4–6 Redrawn to
Include the Sex
Chromosomes

chromosome, can be neither homozygous nor heterozygous. Instead, the term **hemizygous** is used for X-linked traits in males. Also, since only one copy of an allele is present, a recessive will determine the male phenotype. This phenomenon is called **pseudodominance.** Because a male with the w allele is white eyed, the allele is acting like a dominant; the allele determines the phenotype when only one copy is present. This is the same way that a dominant would determine the phenotype in a normal diploid.

Nonreciprocity

The X-linked pattern has long been referred to as a **criss-cross pattern of inheritance** because a father passes a trait to his daughters, who pass it on to their sons. That this analysis is correct and that the inheritance pattern is not reciprocal are shown in Figure 4–8 where the white-eyed female is

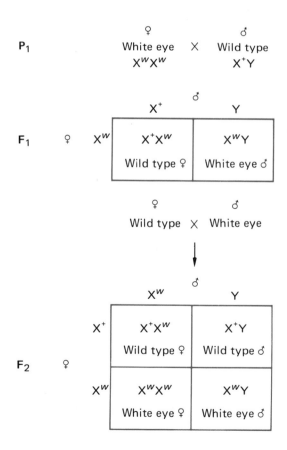

Figure 4–8.
Reciprocal Cross to
That in Figure 4–7

crossed with a wild-type male. If this were an autosomal trait (one where the locus is on an autosomal chromosome), there should be reciprocity with the cross shown in Figure 4–7. That is, half the F_2 males should be white eyed, whereas the other half and all the females should be wild. This is not the case. Here the F_1 males are white eyed, and 50% of each sex in the F_2 are white eyed. Such nonreciprocity suggests sex linkage, and the criss-cross pattern confirms it.

Figure 4–9 shows the inheritance pattern of a sex-linked trait in chick-

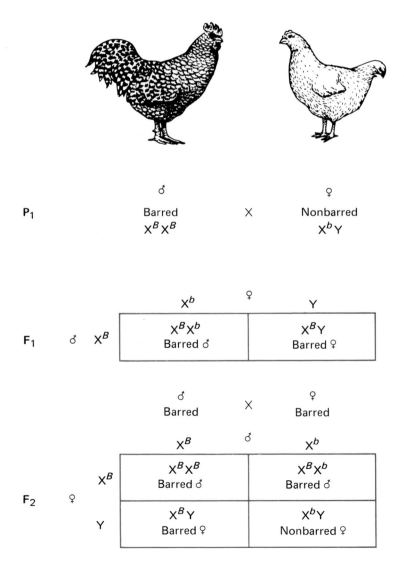

Figure 4-9.
Inheritance Pattern
of Barred Plumage
in Chickens

ens where the male is the homogametic sex. Barred is X-linked and dominant to nonbarred. If we substitute white eyed for nonbarred and male for female, we get the same pattern as in fruit flies (Figure 4–7) where, of course, females are homogametic.

The bobbed locus (*bb*) is an interesting one in *Drosophila*. As a homozygous recessive, it causes bristles to be shortened. The locus occurs on *both* the X and Y chromosomes. (The Y is also homologous to the X for the nucleolar organizer. In addition, the Y is known to carry at least seven loci required for male fertility.) Figures 4–10 and 4–11 show the results of reciprocal crosses involving bobbed. In both cases 1/4 of the F_2 are bobbed. In one cross it is males, and in the other it is females.

Sex Linkage in Humans

In humans the H-Y antigen locus, as well as the locus for the testis-determining factor (TDF), is known to occur on the Y chromosome. There is

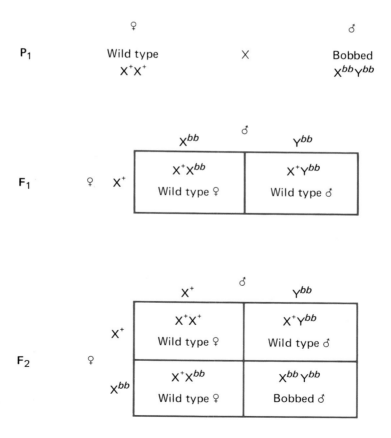

Figure 4–10.
Inheritance Pattern
of the Bobbed Locus
in *Drosophila*

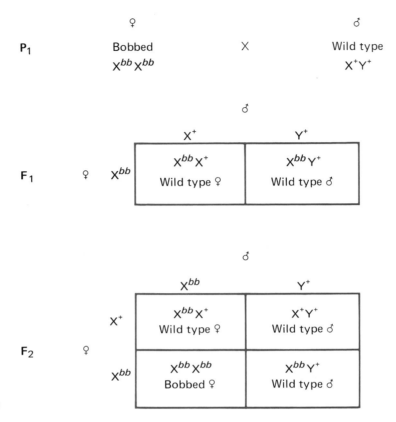

Figure 4–11.
Reciprocal Cross to
That in Figure 4–10

evidence that there are Y-linked genes for height—XYY men are taller than XY men who are taller than XO females. But the customary assignment of the locus for "hairy ears" to the Y chromosome is most likely incorrect. This trait (Figure 4–12), the presence of bristly hairs in the rim of ears of certain men, has long been suggested to be a Y-chromosome gene because of the male-to-male pattern of inheritance. It is, however, probably an autosomal dominant that expresses itself only in men. (Often the dominance and linkage relations of human traits have been misclassified. Another "classic" example, shown in Figure 4–13, is tongue rolling. Originally classified as an autosomal dominant, it is probably not genetically controlled.)

Hairy ears may be an example of one of three other inheritance patterns that show nonreciprocity without actually being controlled by loci on the sex chromosomes: **Sex-limited genes** are autosomal genes whose phenotypes are expressed in only one sex, although the genes are present in both sexes. Besides hairy ears, other examples would be plumage in birds (where in many species the male is brightly colored); horns found only in males of cer-

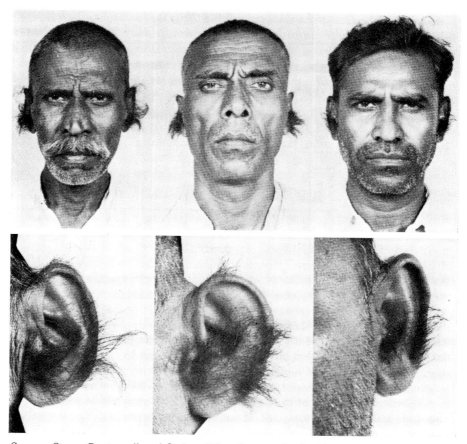

Figure 4–12.
"Hairy Ears," an
Inherited Trait
Common in Parts of
India

Source: Stern, Centerwall and Sarkar, "New Data on the Problem of Y-linkage of Hairy Pinnae," *American Journal of Human Genetics* 16 (1964):467. Copyright © 1964 by the University of Chicago Press. Reprinted by permission.

tain sheep species; and milk yield in cattle, which shows phenotypically only in females. **Sex-influenced traits** appear in both sexes, but their dominance is influenced by the sex of the individual and they therefore occur more frequently in one sex or the other. Pattern baldness in humans is an example of a sex-influenced trait. It appears to be controlled by an allele that acts as a dominant in men but a recessive in women. If the baldness allele is B and the normal allele is b, then a Bb heterozygote will be a bald male but a nonbald female. **Sex-controlled traits** also appear in both sexes and also occur more frequently in one sex than in the other, but they are not controlled by a simple dominant recessive relation as is baldness. For example, cleft lip, which is more frequent in men than in women, is a sex-controlled trait. The mechanism is not known.

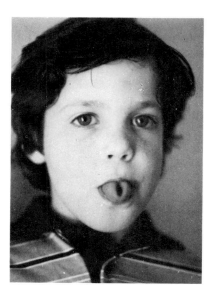

Figure 4-13.
Tongue Rolling

PEDIGREE ANALYSIS

Inheritance patterns in many organisms are relatively easy to determine because crucial crosses can be made to test hypotheses about the genetic control of a particular trait. Many of these same organisms produce an abundance of offspring so that numbers large enough to compute ratios can be gathered. Recall Mendel's work with garden peas, where his 3:1 F_2 ratio led him to suggest segregation of discrete genes. If Mendel's sample sizes had been smaller, he might not have seen the ratio. He could have further tested his idea by self-fertilizing the F_2. He predicted that all pea plants with the recessive phenotype would breed true (the 1/4 category) and that among the 3/4 category 1/3 should breed true whereas 2/3 should produce a phenotypic 3:1 ratio in their offspring. That this did occur confirmed his hypothesis about the mechanisms of inheritance. Think of what Mendel's difficulty would have been had he decided to work with humans. In fact, these same problems are faced by human geneticists today. The occurrence of a trait in one of four people or even in two of eight people is not a valid basis for predicting an overall 3:1 ratio.

We cannot do crucial crosses to confirm the nature of many traits because we cannot manipulate humans the way we do experimental organisms. Family size is also a problem. Humans do not produce offspring in numbers large enough for adequate genetic analysis. Thus, to determine the inheritance pattern of many human traits, human geneticists often have little more to go on than family trees (pedigrees) that many times do not include the critical mating combinations. Another difficulty encountered by the human geneticist is lack of **penetrance** as well as different degrees of **expressivity.** Both are aspects of the expression of a phenotype.

Penetrance

Penetrance refers to the expected appearance in the phenotype of traits determined by the genotype. Unfortunately for geneticists, not all genotypes "penetrate" into the phenotype. For example, a person could have the genotype that specifies vitamin-D-resistant rickets and yet not have rickets. This disease is caused by a sex-linked dominant gene and is distinguished from normal vitamin D deficiency by its failure to respond to low levels of vitamin D. It does, however, respond to very high levels of vitamin D and is thus curable. In any case, family trees are known where affected children are born from normal parents. This would violate the rules of dominant inheritance because one of the parents must have had the allele yet did not show it. That the parent actually had the allele is demonstrated by the occurrence of low blood phosphorous levels, another aspect of the same allele. The low-phosphorous aspect of the phenotype is always fully penetrant.

We thus see that certain genotypes, especially those for many developmental pathways, are not always fully penetrant. Most genes, however, penetrate most of the time—that is, they are fully penetrant. For example, there are no known cases of homozygous albinos who were not albinistic. Vitamin-D-resistant rickets illustrates another phenomenon, in which a phenotype that is not genetically determined mimics a phenotype that is a genetically controlled trait. This **phenocopy** is the result of dietary deficiency or environmental trauma. A dietary deficiency of vitamin D, for example, will produce rickets that is virtually indistinguishable from the genetically caused rickets (the vitamin-D-resistant kind).

Many developmental traits not only sometimes fail to penetrate but also show a variable pattern of expression, from very mild to very extreme, when they do. For example, harelip is a trait that shows both variable penetrance and variable expressivity. Once the genotype penetrates, the severity of the impairment varies considerably from a very mild external cleft to a very severe clefting of the hard and soft palates.

Family Tree

One way to examine a pattern of inheritance is to draw a family tree. The symbols used in constructing a family tree, or pedigree, are defined in Figure

4–14; a pedigree is shown in Figure 4–15. The circles represent females and the squares represent males. Symbols that are filled in represent individuals who have the trait under study; they are said to be **affected.** The hollow circles represent those who do not have the trait. The direct horizontal lines between two individuals (one male, one female) are called marriage lines. Children are attached to a marriage line by vertical lines. All the brothers and sisters (**siblings** or *sibs*) from the same parents are connected by a horizontal line above their symbols. Siblings are numbered below their symbols according to birth order, and generations are numbered on the right in Roman numerals. When the sex of a child is unknown, the symbol used is diamond shape (the children of III-1 and III-2, for example). A number within a symbol represents the number of siblings not separately listed. Twins are denoted by their "vertical" lines coming from the same point on the horizontal sib line. IV-7 and IV-8 are nonidentical twins, whereas III-3 and III-4 are identical, as shown by the short vertical line connecting their individual lines.

When other symbols occur in a pedigree, they are usually defined in the legend. The individual V-5 is called a **propositus** (or *proposita*) or **proband.** The arrow pointing to V-5 in Figure 4–15 indicates that it was through this individual that the pedigree was ascertained. Usually a pedigree is ascertained by a physician or clinical investigator.

On the basis of the information in a pedigree, it is possible to determine the mode of inheritance of a trait. There are two types of questions the pedigree may answer. First, are there consistencies within the pedigree that support a particular mode of inheritance. Second, are there inconsistencies within the pedigree that would suggest that a particular mode of inheritance is not possible. As will often happen, however, one will be left with possibil-

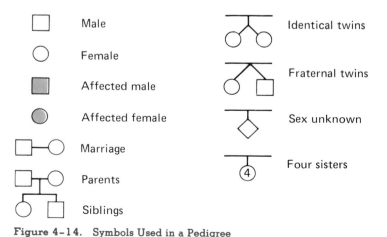

Figure 4-14. Symbols Used in a Pedigree

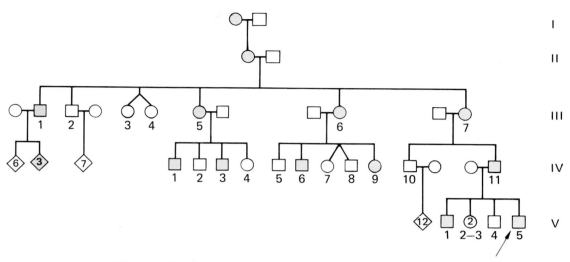

Figure 4-15. Part of a Pedigree for Polydactyly
After Lucas, 1881.

ities rather than certainties with respect to any particular trait. As of 1977, McKusick and Ruddle report that the mode of inheritance of about 1200 loci in humans is known with confidence. This group includes about 900 loci that cause disorders usually rare in the population. The group of 1200 is made up of autosomal dominants, autosomal recessives, and sex-linked traits.

Dominant Inheritance

In looking again at the pedigree of Figure 4–15, several points emerge. First, polydactyly (Figure 4–16) occurs in every generation: There are no generations skipped—that is, every affected child has an affected parent. This indicates dominant inheritance. Second, the trait occurs about equally among the sexes. There are six affected males and six affected females. This indicates autosomal rather than sex-linked inheritance. Thus, so far we would categorize the trait as an autosomal dominant. Note also that individual IV-11, a male, passed on the trait to two of his three sons. This would rule out sex linkage. Remember that a male gives his X chromosome to all of his daughters but none of his sons. His sons receive his Y chromosome. Since IV-11 had affected sons, as well as normal sons and normal daughters, we could rule out sex linkage. In fact, polydactyly is an autosomal dominant.

Polydactyly is interesting because it also shows variable penetrance and expressivity in the heterozygote. In the actual pedigree individual III-1 was normal. Similarly, many of the affected individuals had varying degress of expressivity of the trait. The most extreme manifestation of the trait is the

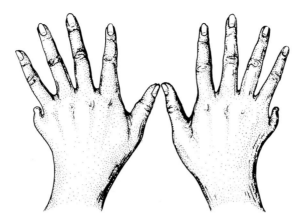

Figure 4-16.
Hands of a Person
with Polydactyly

occurrence of an extra digit on each hand and one or two extra toes on each foot (Figure 4–16). However, some individuals had only extra toes, some had extra fingers, and some had an asymetrical distribution (such as six toes on one foot and seven on the other).

Recessive Inheritance

Figure 4–17 is a pedigree with a different pattern of inheritance. Here affected individuals are not found in each generation. The three affected daughters (identical triplets) come from normal parents; in fact, they are the first appearance of the trait in the pedigree. A telling point here is that the triplets' parents are first cousins. A marriage between relatives is referred to as **consanguineous.** If the degree of relatedness is closer than the law permits, the union is called **incestuous.** In all states brother-sister and mother-son marriages are forbidden; and in all states except Georgia, father-daughter marriages are forbidden. Thirty states prohibit the marriage of first cousins. Consanguineous marriages often produce offspring that have rare recessive and often deleterious traits. The reason is that through common ancestry (for example, first cousins have a pair of grandparents in common) an allele found in an ancestor can be inherited through both sides of the pedigree and become homozygous in the child. The occurrence of a trait in a pedigree with common ancestry is often good evidence for an autosomal recessive mode of inheritance. Consanguinity by itself does not guarantee that the trait being examined is an autosomal recessive; all modes of inheritance are to be found in consanguineous pedigrees. However, it is especially suggestive of autosomal recessiveness when a trait first shows up in the offspring of related individuals.

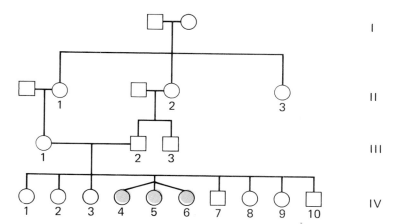

Figure 4-17.
Part of a Pedigree of
Hypotrichosis

Sex-Linked Inheritance

Figure 4–18 is a pedigree of the children of Queen Victoria of England. Through these children hemophilia, a defect in the blood-clotting system, was passed on to most of the royal houses of Europe. Several interesting aspects of this pedigree help to confirm the method of inheritance. First, generations are skipped. While Alexis (1904–1918) was a hemophiliac, neither his parents nor grandparents were. This pattern occurs in several other places in the pedigree and indicates a recessive mode of inheritance. From other pedigrees and from the biochemical nature of the defect, it has been determined that hemophilia is a recessive trait. (There are several different inherited forms of hemophilia known, each deficient in one of the steps in the pathway of the formation of fibrinogen, the blood clot protein. Two of these forms, "classic" hemophilia A and hemophilia B, which is called Christmas disease, are both sex linked. Other hemophilias are autosomal.)

Further inspection of the pedigree in Figure 4–18 will reveal that all the affected individuals are sons. This is strongly suggestive of sex linkage—since males are hemizygous for the X chromosome, more males than females should have the phenotype of a sex-linked recessive trait. If this mode is correct, we can make several predictions. First, since all males get their X chromosomes from their mothers, affected males should be the offspring of carrier (heterozygous) females. A female is a carrier if her father had the disease. She has a 50% chance of being a carrier if her brother has the disease but not her father (her mother was a carrier) and a 25% chance of being a carrier if her mother's brother has the disease. There is no place in the pedigree where the trait is passed on from father to son. This would defy the route of an affected X chromosome. We can conclude from the pedigree that hemophilia is a sex-linked recessive. One other interesting point about this

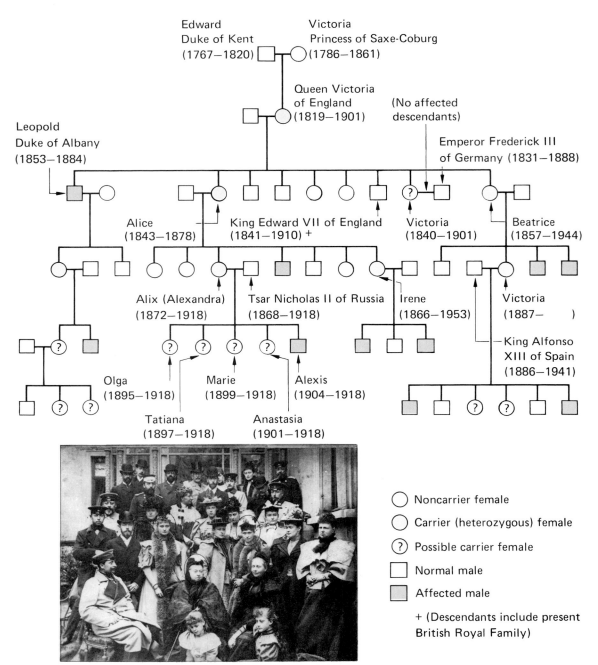

Figure 4-18. Hemophilia in the Pedigree of Queen Victoria of England
In this photograph of the Queen and some of her descendants, three carriers—Queen Victoria, Princess Irene of Prussia (right), and Princess Alix (Alexandra) of Hesse (left)—are indicated.

Source: Reprinted courtesy of the Photography Collection, HRC, University of Texas at Austin.

pedigree is that the disease seems to have arisen *de novo* by mutation in one of the gametes of Queen Victoria's parents. Prior to this pedigree there had been no evidence of the disease in Queen Victoria's family; yet she was obviously a heterozygote (one affected son and two known carrier daughters). Thus, from a homozygous normal mother and a hemizygous normal father, one of Queen Victoria's X chromosomes had the hemophilia allele. This could only have happened if there had been a change (mutation) in one of the gametes that formed Queen Victoria. We will explore this in the chapter on mutation.

Another pedigree is shown in Figure 4–19, where dominant inheritance is suspected because no generations are skipped. This pedigree shows an interesting distribution of the trait (defective tooth enamel) among the sexes. The male in generation I is affected as are all his daughters; yet none of his sons are affected. This pattern is the pattern of the X chromosome. A male passes it on to all of his daughters but to none of his sons. Affected daughter II-5 would be expected to pass the X chromosome from her father to half her sons and half her daughters. In fact, one of two sons is affected and one of three daughters is also. Although this pedigree is in good agreement with a sex-linked dominant mode of inheritance, the sample is too small to rule out autosomal inheritance.

There is also the slight possibility that the trait is a recessive. This could be true if the wife in generation I and the husband of II-5 were both heterozygotes. If it is a rare trait, then such a possibility is very slight. Other pedigrees must be examined before a definitive ruling on the inheritance of this trait can be made. While there are not many sex-linked dominant traits, defective enamel leading to brown teeth, is known to be sex-linked dominant and may be the trait followed by this pedigree. The expected patterns for various types of inheritance in human pedigrees can be grouped into four categories.

Autosomal dominant:

1. Should not skip generations (unless reduced penetrance).
2. An affected married to a normal should produce a 1:1 ratio of affected to normal offspring.
3. Distribution of the trait among sexes should be almost equal.

Autosomal recessive:

1. Generations are skipped.
2. There should be an equal distribution among sexes.
3. Often found in consanguineous marriages.
4. If both parents are affected, all children should be affected.
5. Most cases of normals married to affecteds produce all normal children. When a child is affected (indicating that the normal parent is

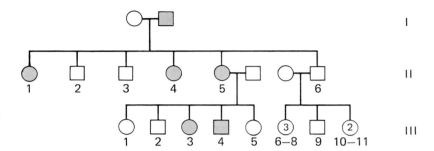

Figure 4-19.
Possible Pedigree of
Defective Tooth
Enamel

heterozygous), then half the children should be affected.

6. Most affected individuals have normal parents.

Sex-linked recessives:

1. Most affected are males.
2. Affected males result from carrier (heterozygote) mothers who are known by having affected brothers, fathers, or maternal uncles.
3. Affected males do not come from affected fathers.
4. Affected females come from affected fathers and carrier mothers.
5. Half the sons of carrier females should be affected.

Sex-linked dominants:

1. No generations are skipped.
2. Affected males come from affected mothers.
3. Half an affected female's sons and daughters are affected.
4. Affected females come from affected mothers or fathers.
5. All the daughters but none of the sons of an affected father are affected.

DOSAGE COMPENSATION

Recall that in the XY chromosome system females have two X chromosomes while males have only one. Thus females have twice the dose of X-linked genes as males. The question arises as to how we compensate for this dosage difference between the sexes, given that there is a potential for serious abnormality: An incorrect number of autosomes usually presents problems (Chapter 7).

In humans and other mammals, the necessary **dosage compensation** is accomplished by inactivation of one of the X chromosomes in females so that both males and females have only one functional X chromosome per

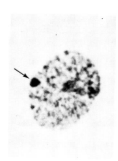

Figure 4 – 20.
Barr Body in the
Nucleus of a Cell of
a Normal Female

cell. The inactive X in females becomes tightly coiled into **heterochroma-tin,** a condensed form of chromatin visible as a dark spot, or **Barr body,** in the nucleus of female cells (Figure 4–20*). This condensed (and therefore visible) chromatin was first observed by Barr, who noted that normal females show a single Barr body whereas males show none and who, therefore, referred to the body as sex chromatin. Mary Lyon then suggested that this Barr body represented an inactive X chromosome. Several lines of evidence support the **Lyon hypothesis.** First, XXY males have a Barr body whereas XO females have none. Then, persons with varying numbers of X chromosomes have one less Barr body than they have X chromosomes per cell: XXX females have two Barr bodies and XXXX females have three (Figure 4–21*).

Proof of the Lyon Hypothesis

Direct proof of the Lyon hypothesis came when cytological analysis identified the Barr body in normal females as an X chromosome. Genetic evidence also supports the Lyon hypothesis: Females heterozygous for a locus on the X chromosome show a unique pattern of phenotypic expression. It is now known that in humans the X chromosome is inactivated in each cell at about the twelfth day of embryonic life and further that it is randomly determined which X is inactivated in a given cell. Thus, these heterozygous females show a **mosaicism** at the cellular level for X-linked traits—that is, instead of being typically heterozygous, they are hemizygous for one or the other of the X chromosome alleles.

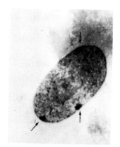

Figure 4 – 21.
Three Barr Bodies
(See Arrows) in the
Nucleus of an XXXX
Female's Cell

Glucose-6-phosphate dehydrogenase (G-6-PD) is an enzyme controlled by a locus that occurs on the X chromosome. The enzyme occurs in several different allelic forms, which differ by single amino acids. Thus, two forms (A and B) will both dehydrogenate glucose-6-phosphate—both are fully functional enzymes—but they are different by an amino acid and can be detected by their rate of migration in an electrical field (one form moves faster than another). This electrical separation, termed **electrophoresis,** is carried out by placing samples of the enzymes being tested in a supporting gel, usually starch, polyacrylamide, or cellulose acetate (Figure 4–22). After electric current is applied for several hours, the gel is stained in such a way as to show the distance traveled by each enzyme through the gel. The first three samples in the electrophoretic gel of Figure 4–22 are of blood from (1) a female homozygous for the A form of G-6-PD, (2) a female homozygous for the B form, and (3) a female heterozygous for the A and B forms. Slots (4)–(10) received clones of cells from the heterozygous female. (A clone is a group

Source: Thomas G. Brewster and Park S. Gerald, "Chromosome Disorders Associated with Mental Retardation," *Pediatric Annals* 7, No. 2 (1978). Reproduced courtesy of Dr. Thomas G. Brewster, Foundation for Blood Research, Scarborough, Maine.

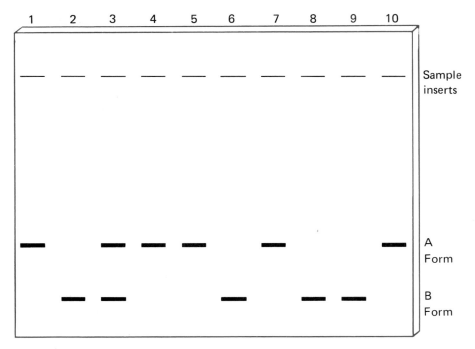

Figure 4-22.   Electrophoretic Gel Stained for Glucose-6-Phosphate Dehydrogenase

of daughter cells derived from a single cell by mitosis.) Observe that the blood, which is a conglomerate of cells, of the heterozygote female produces both the A and B bands, while any single cell has only one or the other band. Since the gene for G-6-PD is carried on the X chromosome, this electrophoretic display indicates that only one X is active in any particular cell.

The Lyon hypothesis has been demonstrated with many X-linked loci, but the most striking examples are those for the color phenotypes in some mammals. For example, the calico pattern of cats is due to the inactivation of X chromosomes (Figure 4–23). Calico cats are normally females heterozygous for the yellow and black alleles of the X-linked color locus. They exhibit patches of these two colors, indicating that at a certain stage in development one of the X chromosomes was inactivated and all of the ensuing daughter cells in that line kept the same X chromosome inactive. The result is broad patches of coat color rather than the microscopic color pattern to be expected if every new cell produced randomly had one of its X chromosomes inactivated. In human females also, when one X is inactivated, all the daughter cells of that cell have the same inactive X chromosome. (Recent research has challenged the view that the whole X chromosome is inactivated, or "Lyonized." At least a short segment of the X, containing the loci for the enzyme steroid sulfatase and the red cell antigen Xg^a, seems to be active in both X chromosomes in women tested.)

Figure 4-23.
Calico Cat

Source: Reproduced courtesy of Neil Todd.

Dosage Compensation for *Drosophila*

The mechanism of dosage compensation is not known for fruit flies. However, it is apparently different than in mammals because no Barr bodies are found in fruit flies. In addition, genetic evidence indicates that both X chromosomes are active in the cells of female *Drosophila*.

CHAPTER SUMMARY

Sex determination in animals is often based on chromosomal difference. In humans and fruit flies, females are homogametic (XX) and males are heterogametic (XY). In humans the Y chromosome determines maleness; in *Drosophila* sex is determined by the balance between the X chromosome and the autosomes. Since different chromosomes are normally associated with each sex, inheritance of loci located on these chromosomes shows specific, nonreciprocal patterns. The white-eye locus in *Drosophila* was the first case of a closely analyzed locus assigned to the X chromosome. Over 100 sex-linked loci are now known in humans.

Pedigree analysis is used in human genetic studies because it is impossible to carry out large-scale, controlled crosses. From pedigrees we can often infer the mode of inheritance of a particular trait by its pattern. However not all traits determined by the genotype are apparent in the phenotype, and this lack of penetrance can create problems in pedigree analysis.

Problems of dosage compensation for loci on the X chromosome are solved in different ways in different organisms. In humans one of the female X chromosomes is Lyonized, or inactivated. Lyonization in female humans leads to cellular mosaicism for most loci on the X chromosome.

ISOZYMES

The technique of *electrophoresis* has opened up new and exciting areas of research in both population and biochemical genetics. The reason for this is that electrophoresis allows us to see enzymes, the gene products, directly. It has thus allowed us to see variations in large parts of the genotype previously difficult or impossible to sample. In population genetics (see Chapters 17–19), electrophoresis has made it possible to estimate the amount of variability occurring in natural populations. The resulting discovery that a great deal of heterozygosity occurs in nature was a revelation and has opened up whole new areas of theoretical as well as empirical study. In biochemical genetics electrophoretic techniques can be used to study enzyme pathways, to sequence nucleotides (see discussion in Chapter 9), and to assign various loci to particular chromosomes.

Electrophoresis entails placing a sample—often blood, a homogenized organ, or a cell homogenate—at the top of a gel prepared from hydrolyzed starch, polyacrylamide, cellulose acetate, etc., and a suitable buffer system. An electrical current is passed through the gel (Figure 1), and the gel is then treated with a dye that will stain the protein. In the simplest case, if a protein is homogeneous (the product of a homozygote), it will form a single band on the gel. If it is heterogeneous (the product of a heterozygote), it will form two bands when the two allelic proteins differ by an amino acid so that they have different electrical charges (and therefore travel through the gel at different rates; see Figure 4–22).

A sample of mouse blood serum that has been stained for general protein is shown in Figure 2. With the general protein stain used, most of the staining is accounted for by albumins and β-globulins (transferrin). Because they are present in very small concentrations, the many enzymes present are invisible, but a stain that is specific for a particular enzyme will make that enzyme visible on the gel. For example, lactate dehydrogenase (LDH) can be stained for in the following way: LDH catalyzes the reaction

$$\text{lactic acid} + \text{NAD} \overset{\text{LDH}}{\rightleftharpoons} \text{pyruvic acid} + \text{NADH}$$

Thus, we can specifically stain for the LDH enzyme by adding the substrates of the enzyme (lactic acid and NAD) and a suitable stain system specific for a product of the enzyme reaction (pyruvic acid or NADH). For example, if lactic acid and NAD are added to the system, only LDH will convert them to pyruvic acid and NADH. We can then test for the presence of NADH by having it reduce the dye, nitro blue tetrazolium, to the blue precipitate, formazan, an electron carrier (phenazine methosulfate is an intermediary in this reaction). We then add all the preceding reagents and look for blue bands on the gel (Figure 3).

In addition to its uses in population genetics and chromosome mapping, electrophoresis has been extremely useful in determining the structure of many proteins and for studying developmental pathways. As we can see from the LDH gel in Figure 3, five bands can occur. In a homozygote these bands occur roughly in a ratio of 1:4:6:4:1. This can come about if the enzyme is a tetramer whose four subunits are random mixtures of two gene products (A and B). Thus we would get:

- AAAA
- AAAB
- AABB
- ABBB
- BBBB

This mode has been verified by protein chemists. In this way electrophoresis has helped us determine the structure of several enzymes. It has also been shown that the five forms differ in their various concentrations in different tissues of the body (Figure 4). This has led to various hypotheses as to how the production of the enzyme is developmentally controlled.

A bonus from this study has been of clinical diagnostic value. In various diseases there is a cell destruction that causes cell contents to be dumped into the blood stream. Thus the LDH pattern of certain cell types is found in the blood in certain disease states (Figure 5). Hence examination of blood LDH is often a diagnostic test used to pick up early signs of heart and liver diseases (among others).

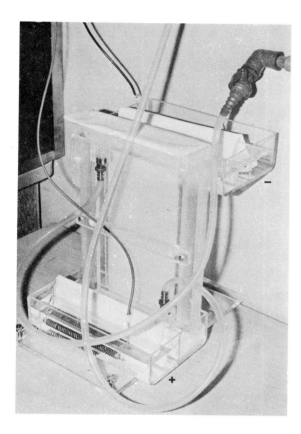

Figure 1.   Vertical Starch Gel Apparatus
Current flows from the upper buffer chamber to the lower one by way of the paper wicks and the starch gel. Cooling water flows around the system.

Source: R. P. Canham, "Serum Protein Variations and Selection in Fluctuating Populations of Cricetid Rodents," Ph.D. thesis, University of Alberta, 1969. Reproduced by permission.

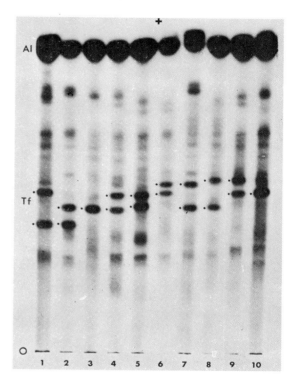

Figure 2.   Ten Samples of Deer Mouse (*Peromyscus*) Blood Studied for General Protein
Al is albumin and Tf is transferrin, the two most abundant proteins in mammalian blood. Note the transferrin variations in the different samples.

Source: R. P. Canham, "Serum Protein Variations and Selection in Fluctuating Populations of Cricetid Rodents," Ph.D. thesis, University of Alberta, 1969. Reproduced by permission.

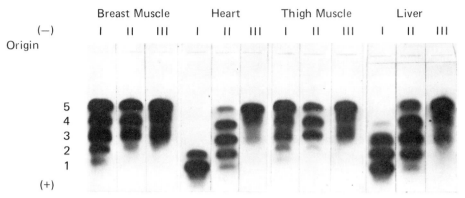

Figure 3. LDH Isozyme Patterns in Pigeons
Note the five bands for some individual samples.
Source: W. H. Zinkham et al., "A Variant of Lactate Dehydrogenase in Somatic Tissues of Pigeons: Physicochemical Properties and Genetic Control," *Journal of Experimental Zoology* 162, no. 1 (June 1966):45–46. Reproduced by permission.

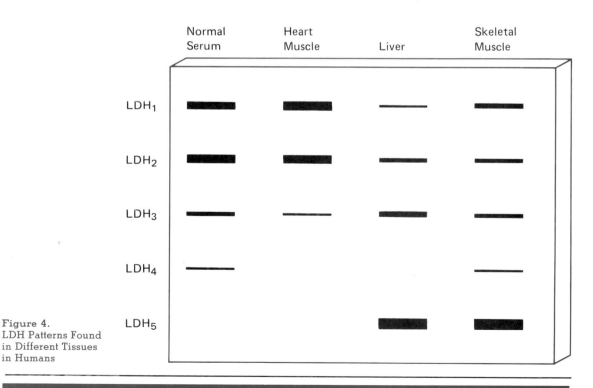

Figure 4.
LDH Patterns Found in Different Tissues in Humans

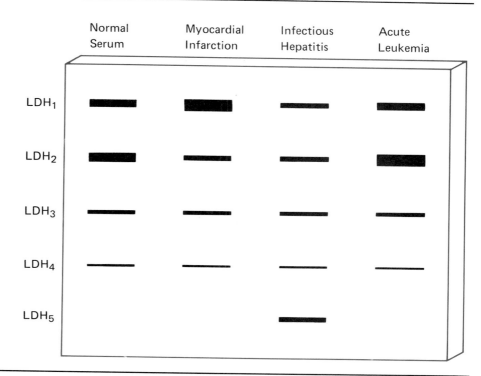

Figure 5.
LDH Pattern of
Normal Human
Serum and Various
Disease States

1. In *Drosophila* the lozenge phenotype, caused by a sex-linked recessive *(lz)*, is of narrow eyes. Diagram to the F_2 a cross of a lozenge male and a homozygous normal female. Diagram the reciprocal cross.

2. Sex linkage was originally detected in 1906 in moths with YX sex-determining mechanism. In the currant moth a pale color *(p)* is recessive to the wild type and located on the X chromosome. Diagram reciprocal crosses to the F_2 generation in these moths.

3. The following electrophoretic gel shows activity for a particular enzyme. Slot 1 is a "fast" homozygote. Slot 2 is a "slow" homozygote. In slot 3 the blood from the first two was mixed. Slot 4 comes from one of the children of the two homozygotes. Can you guess the structure of the enzyme? If this were an X-chromosome trait, what pattern would you expect from a heterozygous female:

a. whole blood?

b. individually cloned cells?

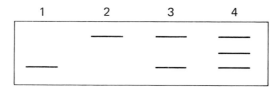

4. How many Barr bodies would you see in the nuclei of persons with the following sex chromosomes:

   a. X?   e. XXX?

   b. XX?   f. XXXXX?

   c. XY?   g. XX/XY mosaic?

   d. XXY?

   What would the sexes of these persons be? If these were the sex chromosomes of individual *Drosophila* that were diploid for all other chromosomes, what would their sexes be?

5. In the following pedigrees of rare human traits, including twin production, determine which modes of inheritance are most probable, possible, or impossible.

a.

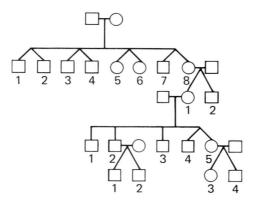

b.

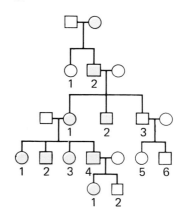

c.

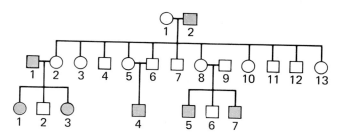

d.

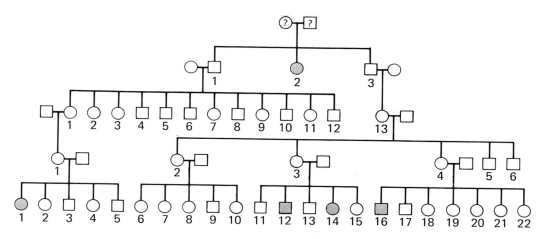

6. Hairy ears shows reduced penetrance (less than 100%). Mechanisms proposed include Y linkage, autosomal dominance, and autosomal recessiveness. Construct a pedigree for each of these mechanisms.

7. What does the fact that calico cats have large patches of colored fur indicate about the age of onset of Lyonization (early or late as compared to humans)?

8. Construct pedigrees for traits that *could not be*
   a. autosomal recessive   c. sex-linked recessive
   b. autosomal dominant   d. sex-linked dominant.

9. In *Drosophila* cut wings are controlled by a recessive sex-linked allele (*ct*) and a fuzzy body is controlled by a recessive, autosomal allele (*fy*.) A fuzzy female is mated with a cut male and all the F_1 are wild type. What are the proportions of F_2 phenotypes, by sex?

SUGGESTIONS FOR FURTHER READING

Bridges, C. 1932. The genetics of sex in *Drosophila*. In *Sex and Internal Secretions, A Survey of Recent Research,* edited by E. Allen, pp. 53–93. Baltimore, Md.: Williams & Wilkins.

Farrow, M., and R. Juberg. 1969. Genetics and laws prohibiting marriage in the United States. *J. Amer. Med. Assoc.* 209:534–538.

Lindsley, D., and E. Grell. 1968. *Genetic Variations of* Drosophila melanogaster. Washington, D.C.: Carnegie Institute.

Lyon, M. 1962. Sex chromatin and gene action in the mammalian X-chromosome. *Amer. J. Hum. Genet.* 14:135–148.

McKusick, V. 1975. *Mendelian Inheritance in Man,* 4th ed. Baltimore, Md.: Johns Hopkins University Press.

Morgan, T. 1910. Sex limited inheritance in *Drosophila. Science* 32:120–122.

Rao, D. 1972. Hypertrichosis of the ear rims. *Acta Genet. Med. Gemellol.* 21:216–220.

Shapiro, L., et al. 1979. Noninactivation of an X-chromosome locus in man. *Science* 204:1224–1226.

Swanson, C. 1957. *Cytology and Cytogenetics.* Englewood Cliffs, N.J.: Prentice-Hall.

Wachtel, S. 1977. H-Y antigen and the genetics of sex determination. *Science* 198:797–799.

White, M. 1973. *Animal Cytology and Evolution,* 3rd ed. Cambridge: Cambridge University Press.

5 Linkage and Mapping in Eukaryotes

Each chromosome of an organism contains many genes. If alleles were locked together permanently on a chromosome, all loci on that chromosome would segregate together; they would not assort independently. However, at meiosis crossing over between loci allows the alleles of these loci to show some measure of independent assortment. Crossing over between loci can be used by the geneticist as a tool to determine how close one locus actually is to another on a chromosome, and thus it is possible to map an entire chromosome and the entire genome of an organism. How do we know that each chromosome of an organism contains more than one gene? Simple. Organisms contain more genes than chromosomes.

For example, a conservative estimate places between 5,000 and 10,000 genes in fruit flies, yet the flies have only 8 chromosomes. So, at least some of the chromosomes must contain more than one gene. What then about Mendel's rule of independent assortment? It is predicated on the fact that alleles of different loci segregate independently. This will only be the case if the loci can behave independently of each other during meiosis. Presumably, if they are on the same chromosome, they will not behave independently. Indeed, there are numerous cases of loci that do not follow the rules of independent assortment. It is said that loci carried on the same chromosome are *linked* to each other. There are as many **linkage groups** as there are homologous pairs of chromosomes. *Drosophila* has 4 linkage groups ($2n = 8$, $n = 4$), while humans have 23 linkage groups ($2n = 46$, $n = 23$).

THE NOBEL PRIZE

On December 10th each year, the Nobel prizes are awarded by the King of Sweden at the Stockholm Concert Hall. The date is the anniversary of Alfred Nobel's death. Awards are given annually in physics, chemistry, medicine, physiology, literature, economics, and peace. Each award is currently worth about $180,000, although they are sometimes split among two or three recipients. The prestige is priceless.

Winners of the Nobel prize are chosen according to the will of Alfred Nobel, a wealthy Swedish inventor and industrialist, who held over 300 patents when he died in 1896 at the age of 63. Nobel developed a detonator and processes for detonation for nitroglycerine, a substance invented by the Italian chemist Ascanio Sobrero in 1847. In the form developed by Nobel, the explosive was patented as dynamite. Nobel invented several other forms of explosive. He was a benefactor of Sobrero, hiring him as a consultant and paying his wife a pension after he died.

Nobel believed that dynamite would be so destructive that it would serve as a deterrent to war. Later, realizing that this would not come to pass, he instructed that his fortune be invested and the interest used to fund the awards. The first ones were given in 1901. Each award consists of a diploma, medal, and check.

More prizes have been awarded to Americans than to members of any other nation (Table 1).

Highlights of Nobel laureate achievements in genetics are shown in Table 2.

Figure 1. The Nobel Medal
The medal is half a pound of 23-karat gold, measures about two and a half inches across, and has Nobel's face and the dates of his birth and death on the front. The diplomas that accompany the awards are individually designed.

Source: Reproduced by permission of the Nobel Foundation.

TABLE 1. Distribution of Nobel Awards According to Country for the Top Five Nations with Recipients (as of 1980)

	Physics	Chemistry	Medicine & Physiology	Peace	Literature	Economics	Total
United States	50	25	61	17	8	9	170
Britain	20	22	20	8	6	3	79
Germany	13	23	10	4	6	0	56
France	9	6	7	8	11	0	41
Sweden	3	4	4	4	7	2	24

TABLE 2. Some Nobel Laureates in Genetics (Physiology or Medicine)

Name	Year	Nationality	Cited for
Thomas Hunt Morgan	1933	USA	Discovery of the way that chromosomes govern heredity
Hermann J. Muller	1946	USA	X-ray inducement of mutations
George W. Beadle	1958	USA	Genetic regulation of biosynthetic pathways
Edward L. Tatum	1958	USA	
Joshua Lederberg	1958	USA	Bacterial genetics
Severo Ochoa	1959	USA	Discovery of enzymes that synthesize nucleic acids
Arthur Kornberg	1959	USA	
Francis H. C. Crick	1962	British	Discovery of the structure of DNA
James D. Watson	1962	USA	
Maurice Wilkins	1962	British	
François Jacob	1965	French	Regulation of enzyme biosynthesis
Andre Lwoff	1965	French	
Jacques Monod	1965	French	
Robert W. Holley	1968	USA	Unraveling of the genetic code
H. Gobind Khorana	1968	USA	
Marshall W. Nirenberg	1968	USA	
Max Delbrück	1969	USA	Viral genetics
Alfred Hershey	1969	USA	
Salvador Luria	1969	USA	
Renato Dulbecco	1975	USA	Tumor viruses
Howard Temin	1975	USA	Discovery of reverse transcriptase
David Baltimore	1975	USA	
Werner Arber	1978	Swiss	Discovery of, sequencing of, and mapping with restriction
Hamilton Smith	1978	USA	endonucleases
Daniel Nathans	1978	USA	
Walter Gilbert	1980	USA	Techniques of sequencing DNA
Frederick Sanger	1980	British	
Paul Berg	1980	USA	Pioneer work in recombinant DNA

DIPLOID MAPPING

Two-Point Cross

In *Drosophila* the recessive band gene (*bn*) causes a dark transverse band to be on the thorax, and the detached gene (*det*) causes the crossveins of the wings to be either detached or absent (Figure 5–1). A banded fly was crossed with a detached fly to produce a wild-type dihybrid. Females were then test-

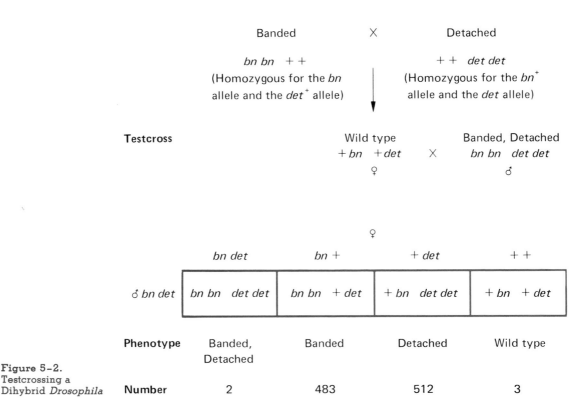

Figure 5–1.
Wild-Type (+) and
Detached (*det*)
Crossveins in
Drosophila

Wild type   Detached

crossed to banded, detached males (Figure 5–2). (In fruit flies there is no crossing-over in males; experiments designed to detect linkage must cross heterozygous females—where crossing over will occur—with recessive males.) If the loci were assorting independently, a 1:1:1:1 ratio of the four possible phenotypes would be expected. However of the first 1000 offspring scored, a ratio of 2:483:512:3 was recorded.

Several points emerge from the data in Figure 5–2. First, no simple ratio is apparent. If we divide through by 2, we get a ratio of 1:241.5:256:1.5. While

Banded   ×   Detached

bn bn + +   + + *det det*
(Homozygous for the *bn*   (Homozygous for the bn^+
allele and the det^+ allele)   allele and the *det* allele)

Testcross

Wild type   Banded, Detached
+ *bn* + *det*   ×   *bn bn* *det det*
♀   ♂

♀

	bn det	*bn* +	+ *det*	+ +
♂ *bn det*	*bn bn* *det det*	*bn bn* + *det*	+ *bn* *det det*	+ *bn* + *det*

Phenotype	Banded, Detached	Banded	Detached	Wild type
Number	2	483	512	3

Figure 5–2.
Testcrossing a
Dihybrid *Drosophila*

the first and last categories seem about equal, as do the middle two categories, there appears to be no simple numerical relation between the middle and end categories. Second, the two categories in very high frequency have the same phenotypes as the original parents in the cross (Figure 5–2). That is, banded and detached were the original parents as well as the testcross offspring in very high frequency. We call these **parentals,** or **nonrecombinants.** The testcross offspring in low frequency combine the phenotypes of the two original parents. These two categories are referred to as **nonparentals,** or **recombinants.** The simplest explanation for these results is that the banded and detached loci are located near each other on the same chromosome or linkage group and therefore move together during meiosis.

The original cross can be revisualized by drawing the loci as points on a chromosome. This is done in Figure 5–3, which shows that 99.5% of the

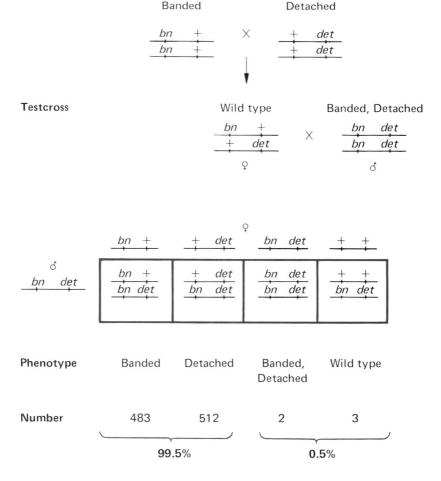

Figure 5–3.
Chromosomal Arrangements of the Two Loci in the Crosses of Figure 5–2
A line arbitrarily represents the chromosomes on which these loci are actually situated.

testcross offspring (the nonrecombinants) come about through the simple linkage of the two loci. The remaining 0.5% (the recombinants) must have arisen through a crossover of homologues (from a chiasma at meiosis) between the two loci (Figure 5–4). Note that since it is not possible to tell

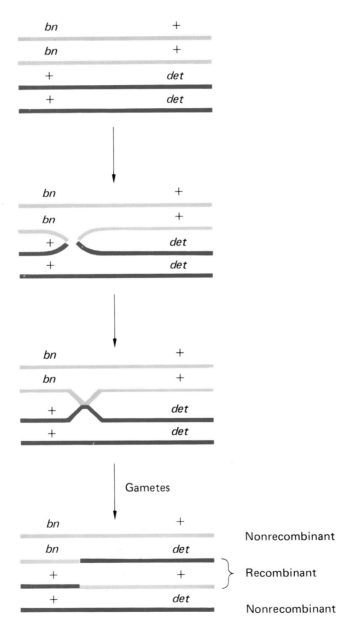

Figure 5–4.
Crossover of
Homologues during
Meiosis between the
bn and *det* Loci in
the Tetrad of the
Dihybrid Female

from these crosses which chromosome the loci are on or where the centromere is in relation to the loci, the centromeres are not included in the figures. The crossover event is viewed as a breakage and reunion of two chromatids lying adjacent to each other during prophase I of meiosis. Later in this chapter we will find cytological proof that this is correct; in Chapter 14 we will explore the molecular mechanisms of this breakage and reunion phenomenon.

From the testcross of Figures 5–2, 5–3, and 5–4, we can see that 99.5% of the gametes produced by the dihybrid are nonrecombinant, whereas only 0.5% are recombinant. This very small frequency of recombinants indicates that the two loci lie very close to each other on their particular chromosome. In fact, we can use the percentage recombination as an estimate of distance between loci on a chromosome: 1% recombination is referred to as one **map unit** (or one **centimorgan** in honor of the geneticist T. H. Morgan, the first geneticist to win the Nobel Prize). While this unit does not provide a physical distance along a chromosome, it does provide a relative distance and thereby makes it possible to know the order of loci on a chromosome.

The arrangement of the *bn* and *det* alleles in the dihybrid of Figure 5–3 is termed **repulsion** because each chromosome carries one mutant and one wild-type allele. The alternative arrangement, where one chromosome carries both mutants and the other chromosome carries both wild-type alleles (Figure 5–5), is referred to as **coupling.**

A cross involving two loci is usually referred to as a **two-point cross;** it is a powerful tool for dissecting the makeup of a chromosome. The next step in our analysis is to look at three loci simultaneously so that we can determine the relative order of the loci on the chromosome.

Three-Point Cross

Analysis of three loci, each segregating two alleles, is referred to as a **three-point cross.** We will examine wing morphology, body color, and eye color in *Drosophila*. Black body (*b*), purple eyes (*pr*), and curved wings (*c*) are all recessive traits. Since the most efficient way of studying linkage is with the testcrossing of a multihybrid, we will study these three loci by means of the crosses shown in Figure 5–6. A point should be clarified in this figure: There are alternative ways of denoting the alleles of a multihybrid. Since the organisms are diploid, they have two alleles at each locus. Various ways are used

Figure 5–5.
Coupling and
Repulsion
Arrangements of
Dihybrid
Chromosomes

Repulsion		Coupling	
bn	+	*bn*	*det*
+	*det*	+	+

by geneticists of presenting this situation. For example, the homozygous recessive can be pictured as

(a) $b\,b\ pr\,pr\ c\,c$

(b) $b/b\ pr/pr\ c/c\ \left(\dfrac{b\ pr\ c}{b\ pr\ c}\right)$

(c) $b\ pr\ c/b\ pr\ c\ \left(\dfrac{b\ pr\ c}{b\ pr\ c}\right)$

A slash or a rule is used to separate alleles on the same chromosome. Thus (a) is used tentatively when we do not know the linkage arrangement of the loci, (b) would be used to indicate that the three loci are on different chromosomes, and (c) would indicate that all three loci are on the same chromosome. We will start by using (a) in this example and switch to (c) when we indicate that all three loci are linked.

In Figure 5–6 the trihybrid is testcrossed. If there were independent assortment, the eight types of gametes produced by the trihybrid should appear with equal frequencies, and thus the eight phenotypic classes would each be 1/8 of the offspring. However, if there were *complete linkage* (where the loci are so close together on the same chromosome that virtually no crossing over takes place), we would expect the trihybrid to produce only two gamete types in equal frequency and yield two phenotypic classes identical to the original parentals. This would occur because under complete linkage

Black, Purple, Curved   X   Wild type
$b\,b\ \ pr\,pr\ \ c\,c$ $+\,+\ \ +\,+\ \ +\,+$

Testcross the Trihybrid   Wild type (Trihybrid)   X   Black, Purple, Curved
$+\,b\ \ +\,pr\ \ +\,c$ $b\,b\ \ pr\,pr\ \ c\,c$

If Unlinked   **If Completely Linked**

1/8 $b\,b\ \ pr\,pr\ \ c\,c$ 1/2 $b\,b\ \ pr\,pr\ \ c\,c$
1/8 $b\,b\ \ pr\,pr\ \ +\,c$ 1/2 $+\,b\ \ +\,pr\ \ +\,c$
1/8 $b\,b\ \ +\,pr\ \ c\,c$
1/8 $b\,b\ \ +\,pr\ \ +\,c$
1/8 $+\,b\ \ pr\,pr\ \ c\,c$
1/8 $+\,b\ \ pr\,pr\ \ +\,c$
1/8 $+\,b\ \ +\,pr\ \ c\,c$
1/8 $+\,b\ \ +\,pr\ \ +\,c$

Figure 5–6.
Alternative Results
in the Testcross
Progeny of the
b–pr–c Trihybrid

the trihybrid would be able to produce only two chromosome types: the *b pr c* type from one parent and the + + + type from the other. Crossing over between linked loci would produce the eight phenotypic classes in various proportions depending on the map distance between loci. The data are presented in Table 5–1.

The data in the table are arranged in reciprocal classes. That is, the wild-type class of 5701 is grouped with the black, purple, curved class of 5617. These represent the two nonrecombinant classes. (Two classes are reciprocal if between them they contain each mutant phenotype just once. The purple, curved and the black classes are also reciprocal, but the purple, curved and the black, purple are not. As we shall see, reciprocal classes arise from the same meiotic recombinant event.) The purple, curved class of 378 is grouped with the black class of 357. These two would be the products of a crossover between the *b* and *pr* loci if we assume that the three loci are linked and that the gene order is *b pr c* (Figure 5–7). The next two classes, 1402 and 1373, would result from a crossover between *pr* and *c*, and the last set, 90 and 82, would result from two crossovers, one between *b* and *pr* and the other between *pr* and *c* (Figure 5–8). Groupings according to these crossover positions are shown at the right of Table 5–1. We should also note that the classes are reciprocal by the fact that they occur in roughly equal frequencies within a pair (5701–5617, 378–357, 1402–1373, 90–82). How would we determine if they actually did occur in equal frequencies within the range of random sampling?

TABLE 5–1. Results of Testcrossing Female *Drosophila* Trihybrid for Black Body Color, Purple Eye Color, and Curved Wings (+ *b* + *pr* + *c*)

Phenotype	Genotype	Number	Alleles from Trihybrid Female	Alleles from Homozygous Recessive Male	Recombinant between b–pr	pr–c	b–c
Wild type	+ *b* + *pr* + *c*	5701	+ + +	*b pr c*			
Black, purple, curved	*bb prpr cc*	5617	*b pr c*	*b pr c*			
Purple, curved	+ *b prpr cc*	378	+ *pr c*	*b pr c*	378		378
Black	*bb* + *pr* + *c*	357	*b* + +	*b pr c*	357		357
Curved	+ *b* + *pr cc*	1402	+ + *c*	*b pr c*		1402	1402
Black, purple	*bb prpr* + *c*	1373	*b pr* +	*b pr c*		1373	1373
Purple	+ *b prpr* + *c*	90	+ *pr* +	*b pr c*	90	90	
Black, curved	*bb* + *pr cc*	82	*b* + *c*	*b pr c*	82	82	
Total		15,000			907	2947	3510
Percent					6.0	19.6	23.4

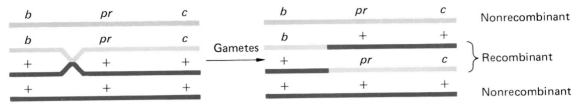

Figure 5-7.   Results of a Crossover between Black and Purple Loci in *Drosophila*

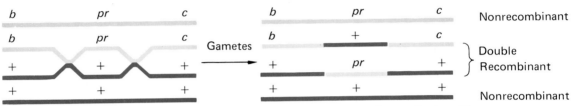

Figure 5-8.   Results of a Double Crossover in the *b–pr–c* Region of the *Drosophila* Chromosome

Map Distances.   The totals in Table 5–1 reveal that 6.0% (907/15,000) of the offspring resulted from recombination between *b* and *pr,* 19.6% between *pr* and *c,* and 23.4% between *b* and *c.* This allows us to form a tentative map of the three loci (Figure 5–9). There is, however, a discrepancy. The distance between *b* and *c* can be calculated in two ways. By adding the two distances, *b–pr* and *pr–c,* we get 6.0 + 19.6 = 25.6 map units; yet by directly counting the recombinants, we get a distance of 23.4 map units. What causes this difference of 2.2 map units?

Returning to Table 5–1, we observe that the double crossovers (90 and 82) are not counted in the *b–c* category; yet each actually represents two crossovers in this region. The reason they are not counted is simply that if only the end loci of this chromosome segment could be observed, the double crossovers would not be detected; the first one causes a recombination between the two end loci, whereas the second one returns these outer loci to the original configuration of the nonrecombinant (Figure 5–8). If we took the 3510 recombinants between *b* and *c* and added in twice the total of double recombinants, or 344, we would get a total of 3854. This is 25.7 map units— within rounding error of 25.6, which is the more precise figure calculated before. As loci get farther and farther apart on a chromosome, more and more double crossovers can occur. Double crossovers tend to mask recombinants, as in our example, so that distantly linked loci sometimes appear

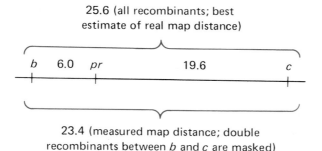

Figure 5–9.
Tentative Map of the Black, Purple, and Curved Chromosome in *Drosophila* Numbers are map units.

closer than they really are. The most accurate map distances are those established on very closely linked loci.

Gene Order.   The data of Table 5–1 confirm our assumption that the gene order is *b pr c*. Of the four reciprocal classes in Table 5–1, one class has the highest frequency and one the lowest. The class with the highest frequency is always the nonrecombinant class. The one with the lowest frequency is always the double recombinant class. A comparison of the double recombinant class with the parental, nonrecombinant class will show which gene is in the middle and by extension the gene order. Since + + + was one of the nonrecombinant gametic combinations and + *pr* + was one of the double recombinant gametic combinations, the *pr* stands out as the "odd" locus, or the one in the middle, since both end loci behave the same. In a similar manner, comparing *b* + *c* with + + + would also point to *pr* as the inside locus (or **inside marker**).

Table 5–1 also supplies information regarding the association of alleles in the original parents—that is, since the table resulted from a trihybrid testcross, the original parental genotypes are reflected by the high-frequency nonrecombinant classes. In this case, one is the result of a + + + gamete; the other, of a *b pr c* gamete. Thus the trihybrid had the genotype of *b pr c*/+ + +: The alleles were in coupling.

Coefficient of Coincidence.   The next question in our analysis of this three-point cross is: Are all crossovers strictly a function of distance between loci? Since at this point it is not possible to actually map the loci on the basis of physical distance, we are dependent on the recombination data to determine distances between genes. Strictly speaking, we cannot answer the question without becoming involved in circular reasoning. Since we are defining the map distances on the basis of empirical data, we cannot then turn around and use these same data to verify the distances—that is, on the basis of mapping, a locus may seem equidistant between two others, yet on the basis of physical distance, the middle marker may be much closer to one locus than the other. We are assuming that map units are a function of actual distance when we perform the analysis.

We can, however, answer a related question: Are crossovers occurring independently of each other? That is, are the observed number of double recombinants equal to the expected number? In the example there were 172/15,000 double crossovers, or 1.15%. The expected number is based on the independent occurrence of crossing over in the two regions measured. That is, if 6.0% of the time there is a crossover in the *b–pr* region and 19.6% of the time there is a crossover in the *pr–c* region, then a double crossover should occur as the product of the two percentages (when expressed as proportions or probabilities): 0.060 × 0.196 = 0.0117, or 1.17% of the gametes should be double recombinants. There is a slight discrepancy: The observed number of double crossovers is less than the expected (1.15% observed, 1.17% expected). This implies a **positive interference** where the occurrence of the first crossover reduces the chance of the second. We can express this as a **coefficient of coincidence** defined as

$$\text{coefficient of coincidence} = \frac{\% \text{ observed double crossovers}}{\% \text{ expected double crossovers}}$$

In the example the coefficient of coincidence is 1.15/1.17 = 0.983. In other words, 98.3% of the expected double crossovers occurred. Often this value is quite low. Sometimes this reduced quantity of double crossovers is measured as the degree of interference defined as

$$\text{interference} = 1 - \text{coefficient of coincidence}$$

In the example the interference is 1.7%.

Let us work out one more example, where neither the middle gene nor the coupling–repulsion state of the alleles is known. On the third chromosome of *Drosophila,* hairy (*h*) causes extra hairs on the body, thread (*th*) causes a thread-shaped arista (antenna tip), and rosy (*ry*) causes the eyes to be reddish brown. All three traits are recessive. Trihybrid females are testcrossed and results from 1000 offspring are given in Table 5–2. At this point it should be possible to determine from the table what the parental genotypes were, which gene is in the middle, the map distances, and the coefficient of coincidence. The table presents the data as it would have been recorded by an experimenter. The phenotypes are tabulated, and from these the genotypes can be reconstructed. Notice that the data can be put into the form found in Table 5–1; there are a large reciprocal set (359 and 351), a small reciprocal set (4 and 6), and large and small intermediate sets (98 and 92, 47 and 43).

First, from the data presented, is it obvious that the three loci are

TABLE 5-2. Trihybrid ($h+$ $ry+$ $th+$) Testcross in
Drosophila

Phenotype	Genotype (order unknown)	Number
Thread	+ + th/h ry th	359
Rosy, thread	+ ry th/h ry th	47
Hairy, rosy, thread	h ry th/h ry th	4
Hairy, thread	h + th/h ry th	98
Rosy	+ ry +/h ry th	92
Hairy, rosy	h ry +/h ry th	351
Wild type	+ + +/h ry th	6
Hairy	h + +/h ry th	43

linked? The pattern, as just mentioned, is identical to that of the previous
example where the three loci were linked. (What pattern would appear if two
of the loci were linked while one assorted independently?) Second, what are
the coupling and repulsion states of the alleles in the trihybrid? The parentals, or nonrecombinants, can be determined from the data because they are
the reciprocal pair of highest frequency. Table 5–2 shows that thread and
hairy, rosy offspring are the nonrecombinants. Thus, the nonrecombinant
gametes of the trihybrid are $+ + th$ and h ry $+$. This is the chromosomal
complement of the trihybrid (with the actual order still unspecified): $+ +$
th/h ry $+$. (What were the genotypes of the parents, assuming they were
homozygotes?) Third, which gene is in the middle? From Table 5–2 we know
that h ry th and $+ + +$ are the double recombinant gametes of the trihybrid because they occur in such low numbers. Comparison of either of these
chromosomes with either of the nonrecombinant chromosomes ($+ + th$ or
h ry $+$) will show that the thread *(th)* locus is in the middle. We now know
that the original trihybrid had the following chromosomal composition: $+$
th $+/h$ $+$ ry. The h and ry alleles are coupled, with the th in repulsion.

We can now compare the trihybrid gamete in each of the eight offspring
categories to the parentals and tabulate them for the region that had a crossover. This is done in Table 5–3. We can see that the h–th distance is 20 map
units, the th–ry distance is 10 map units, and the apparent h–ry distance is
28 map units (Figure 5–10). As in the earlier example, the h–ry discrepancy
is from not counting the double crossovers twice each: $280 + 20 = 300$,
which is 30 map units and the more accurate figure. Lastly, we wish to know
what the coefficient of coincidence is. The observed frequency of double
crossovers is 10/1000 or 1%. The expected percentage is $0.200 \times 0.100 =$

TABLE 5-3. Data from Table 5–2 Arranged to Show Recombinant Regions

Trihybrid Gamete	Number	h–th	th–ry	h–ry
+ th +	359			
h + ry	351			
h th +	98	98		98
+ + ry	92	92		92
+ th ry	47		47	47
h + +	43		43	43
h th ry	4	4	4	
+ + +	6	6	6	
Total	1000	200	100	280

0.020, or 2%. Thus,

$$\text{coefficient of coincidence} = 1.0/2.0 = 0.50$$

Only 50% of the expected double crossovers occur.

From three-point crosses of this type, the chromosomes of many eukaryotic organisms have been mapped out—*Drosophila* is probably the most extensively studied. *Drosophila* and other species of flies have giant **polytene** salivary gland chromosomes, which arise as a result of **endomitosis.** In this process chromosomes replicate but the cell does not divide. In the salivary gland of the fruit fly, homologous chromosomes synapse and then replicate to about 1000 copies, forming very thick structures with a distinctive banding pattern (Figure 5–11). In methods to be discussed in Chapter 7, many loci have been mapped to particular bands. The *Drosophila* map is presented in Figure 5–12. Locate the loci we have mapped so far to verify the map distances.

Figure 5-10. Map of the *h–th–ry* Region of the *Drosophila* Chromosome, with Numerical Discrepancy in Distances Numbers are map units.

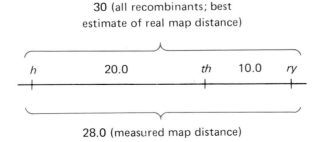

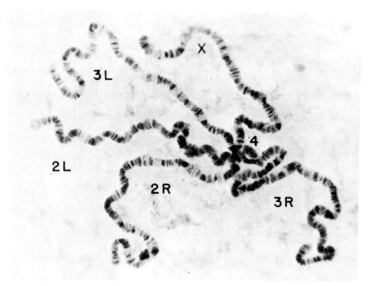

Figure 5-11. Giant Salivary Gland Chromosomes of *Drosophila*
X, 2, 3, and 4 are the four nonhomologous chromosomes. L and R
indicate the left and right arms (in relation to the centromere).

Source: B. P. Kaufman, "Induced Chromosome Rearrangements in *Drosophila melanogaster*," *Journal of Heredity* 30 (1939):178–190. Reproduced by permission.

Cytological Demonstration of Crossing Over

If we are correct that a chiasma during meiosis results from the crossover process, then we should be able to demonstrate that genetic crossing over is accompanied by a cytological crossing over. That is, recombination should entail the exchange of parts of nonsister chromatids. This can be demonstrated if we can distinguish between two homologous chromosomes, a technique first used by Creighton and McClintock in maize (corn) and by Stern with *Drosophila*. We will look at Creighton and McClintock's experiments.

Creighton and McClintock worked with chromosome number 9 in maize ($2n = 20$). In one strain of maize, they found a chromosome with abnormal ends. One end had a knob and the other an added piece of chromatin from another chromosome (Figure 5–13). This knobbed, interchange chromosome was thus clearly different from the normal homologue. It also carried a colored allele (C) and a waxy (wx) texture allele. Following mapping studies in which it was established that C was very close to the knob and wx was close to the added piece of chromatin, Creighton and McClintock made the cross shown in Figure 5–13. The dihybrid heteromorph was crossed to the homo-

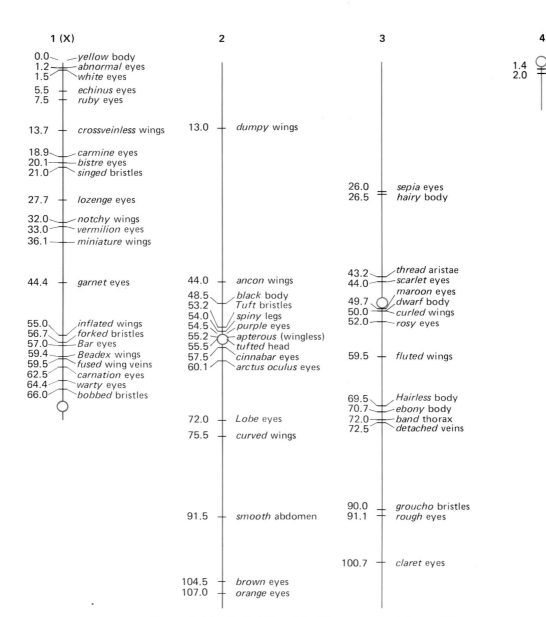

Figure 5-12. Partial Map of the Chromosomes of *Drosophila*
Data from C. B. Bridges.

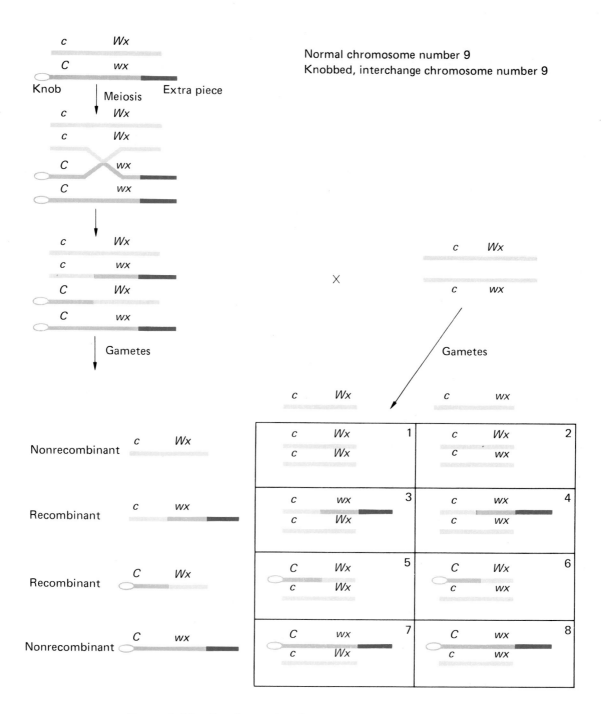

Figure 5-13. Creighton and McClintock Experiment in Maize

morph that had the genotype of c Wx/c wx (colorless and nonwaxy). If a crossover occurred during meiosis in the heteromorph in the region between C and wx, there should also be a cytological crossover where the knob would become associated with an otherwise normal chromosome and the extra piece of 9 should be associated with a knobless chromosome; the result would be four types of gametes (Figure 5–13).

Of 28 offspring examined, all were consistent with the predictions. Those of class 8 with the colored, waxy phenotype all had a knobbed interchange chromosome as well as a normal homologue. Those with the colorless, waxy phenotype (class 4) had a knobless interchange chromosome. All of the colored, nonwaxy phenotypes (classes 5, 6, and 7) had a knobbed, normal chromosome, which indicated that only classes 5 and 6 were in the sample. Of two that were tested, both were Wx Wx, indicating that they were of class 5. The remaining classes (1, 2, and 3) were of the colorless, nonwaxy phenotype. All were knobless. Of those that contained only normal chromosomes, some were Wx Wx (class 1) and some were heterozygotes (Wx wx, class 2). Of those containing interchange chromosomes, two were heterozygous, representing class 3. Two were homozygous, Wx Wx, yet interchange-normal heteromorphs. These represent a crossover in the region between the waxy locus and the extra piece of chromatin. This would give a knobless-c-Wx-extra-piece chromosome that when combined with a c-Wx-normal chromosome, would give these anomalous genotypes. The sample size was not large enough to pick up the reciprocal event. Creighton and McClintock concluded, "Pairing chromosomes, heteromorphic in two regions, have been shown to exchange parts at the same time they exchange genes assigned to these regions."

Number of Chromatids Involved in Crossing Over

When we view tetrads in meiosis, it appears that only two of the four chromatids are involved in any given chiasma and thus probably only two are involved in a recombination event. Can we prove this? To do so, we can make use of organisms that have unique properties in their life cycles. Fungi of the class Ascomycetes retain all the products of their meiosis in a sac called an **ascus.** By dissecting out the haploid **ascospores** and growing each one separately into a haploid colony, we can determine how many of the four meiotic products are recombinant. In yeasts and pink bread molds *(Neurospora)*, it is found that when a single recombination has taken place, it involves only two of the four products—that is, it involves only two of the four chromatids of a tetrad. For *Drosophila* and other diploid eukaryotes, genetic analysis such as that considered earlier in this chapter is referred to as **random strand analysis** and cannot be used to answer questions whose answers require the four products of a single meiosis. In diploid eukaryotes, sperm, each of which carry only one chromatid of a tetrad, unite with eggs that also carry only one chromatid from a tetrad. Thus, zygotes are a result of the random uniting of chromatids. We cannot, in general, use these types of organisms to answer questions requiring a comparison of the total products of meiosis. Luckily, however, nature provides us with organisms that do allow us to answer these types of questions.

Double Crossovers

As an introduction to the next topic, mapping functions, we need to examine double crossovers in a bit more detail. So far, we have only looked at double crossovers involving the same two chromatids. Actually, double crossovers can occur in several different ways involving nonsister chromatids and the events occur more or less randomly. The crossovers considered so far are called **two-strand double crossovers.** Double crossovers can also be three-strand and four-strand (Figure 5–14). This figure may be examined with the following points in mind. First, we are primarily interested in recombination between the outside loci (**outside markers,** a and c; the b locus gives an indication of the crossover events that have taken place between a and c. With the help of the b locus, we see that the two-strand double has in fact taken place; without it we would see no recombination between a and c.

Second, we start with a single crossover between a and b and then add a crossover between b and c. We assume that the loci are close enough so that these are the major recombination events. Thus, the two-strand and four-strand doubles can occur only one way, while the three-strand double can occur two equally likely ways. What conclusions regarding the effect of multiple crossovers on map distance can we draw from this diagram? With only a single crossover, half the chromatids would be recombinant and thus contribute to the map distance between a and c. With double crossovers we should see an increase in recombination and a concomitant increase in measured map distance *because we measure map distance as the percentage recombination between loci.* The four tetrads shown in Figure 5–14 should occur in equal frequencies. It can be seen that 8/16 or 50% of the chromatids are recombinant. This is the same percentage of recombinant chromatids as is seen when only a single crossover occurred. Hence, when there is a double crossover between two loci, there is twice the actual amount of recombination as in a single crossover; yet the same apparent recombination is tabulated.

In addition, two other phenomena occur. First, interference occurs and there are fewer double crossovers than expected. Second, more multiple crossovers can occur as the loci get farther and farther apart. All of this reinforces what was said before about the necessity of using closely linked loci to obtain accurate map distances. It also gives us a feeling for the relation of actual recombination versus measured map distance. This relation is referred to as a **mapping function.**

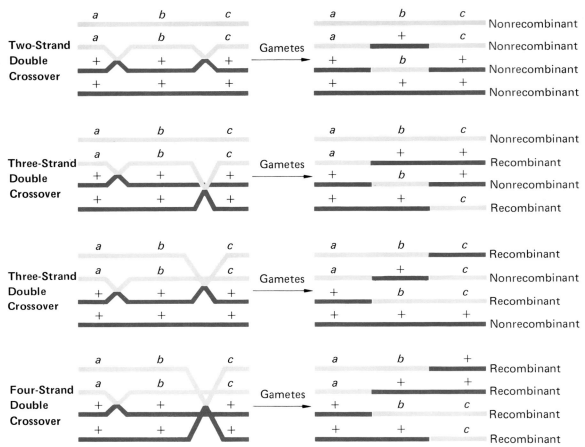

Figure 5-14. Genetic Consequences to Outside Markers of Double Crossovers

Recombinant and nonrecombinant refer to outside loci only.

Mapping Function

The relation of actual recombination (real map distance) and measured map distance when only two loci are considered is shown in Figure 5–15. As the figure shows, the undetected multiple recombinants cause an apparent decrease in the measured map distance. What is the maximum distance detectable between two loci? The answer is 50. This is the same value as that for loci on separate chromosomes (50% of the F_2 will be parental and 50% will be recombinant). As we analyze loci that are farther and farther apart on a chromosome, they begin to appear as if they were assorting independently. In fact, they are assorting independently.

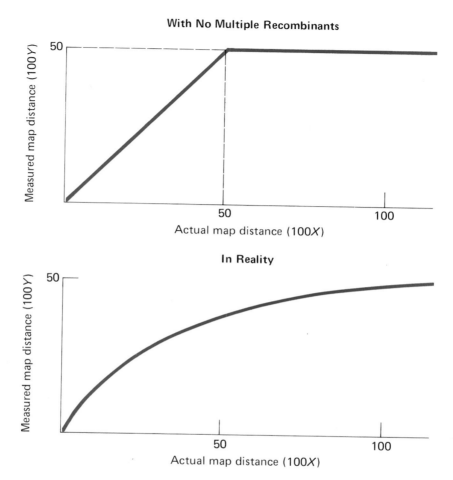

Figure 5–15.
Mapping Function

The graph in Figure 5–15 is a result of multiple crossovers and varying amounts of interference. Geneticists have made several attempts to put a mathematical formula to this curve. Kosambi, who assumed that coincidence was proportional to the measured map distance derived the following equation:

$$Y = 1/2 \tanh 2X$$

where tanh is the hyperbolic tangent function. Haldane, assuming the statistical Poisson distribution, derived

$$Y = 1/2 \, (1 - e^{-2X})$$

For example, if we measured a recombination frequency of 0.25 (25 map units), what is the actual map distance? According to

$$\text{Kosambi's function:} \quad 0.25 = 1/2 \tanh 2X \quad \text{or} \quad \tanh 2X = 0.50$$

By looking at a tanh table, we find that $X = 0.28$. According to

$$\text{Haldane's function:} \quad 0.25 = 1/2\,(1 - e^{-2X}) \quad \text{or} \quad e^{-2X} = 0.50$$

By looking at an e^X table, or by solving with logarithms, we get $X = 0.35$. Thus an apparent map distance of 25 centimorgans is actually in the vicinity of 28 to 35 centimorgans, depending on the various assumptions that went into each mapping function.

HAPLOID MAPPING (TETRAD ANALYSIS)

Several groups of lower eukaryotes retain the four (or eight) haploid products of meiosis. These organisms, which usually exist as haploids, provide a unique opportunity to look at the total product of meiosis in a tetrad. Unique techniques are used for these analyses. We will look at two fungi, the pink bread mold, *Neurospora crassa,* and the common baker's yeast, *Saccharomyces cerevisiae,* both of which retain the products of meiosis as ascospores. *Neurospora* has ordered spores whereas yeast does not. *Neurospora*'s life cycle is shown in Figure 5–16. Fertilization takes place within an immature fruiting body after a spore or filament of a mating type contacts a special filament extending from the fruiting body of the opposite mating type. The zygote nucleus undergoes meiosis without any intervening mitosis.

Ordered Spores

Meiosis proceeds in a unique fashion. Since the cell is narrow, the spindle is forced to lie with the cell's long axis. The two nuclei then undergo a second meiosis, which is also oriented along the long axis of the ascus. The result is that the spores are ordered according to centromeres. That is, if we label one centromere *A* for the *A* mating type and the other *a* for the *a* mating type, a tetrad at meiosis I will consist of one *A* and one *a* centromere, each with two chromatids attached. During the first division of meiosis, the two centromeres are separated (anaphase I); they then divide during the second meiotic division (anaphase II). Thus, the four ascospores at the end of meiosis in *Neurospora* are in the order *A A a a* or *a a A A* in regard to centromeres. (We talk more simply of centromeres rather than chromatids

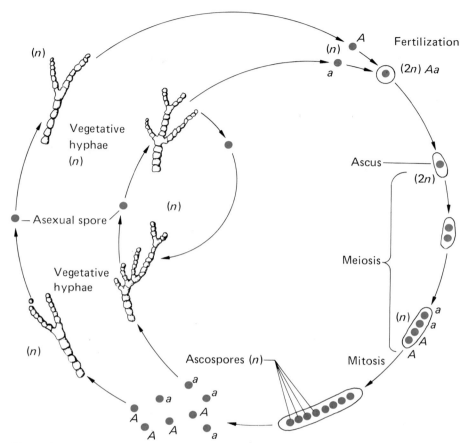

Figure 5-16. Life Cycle of *Neurospora*
A and *a* are mating types; *n* is a haploid stage; 2*n* is diploid.

because of the complications that crossing over adds. A type *A* centromere is always a type *A* centromere whereas a chromatid attached to that centromere may be part from the type *A* parent and part from the type *a* parent.) *Neurospora* adds one complication; before maturation of the ascospores, a mitosis takes place in each nucleus so that we have four pairs (eight) of spores rather than just four spores. With the exception of phenomena such as mutation or gene conversion, to be discussed later in the book, pairs will always be identical (Figure 5-17). As mentioned earlier, *Neurospora* is a good organism with which to show that crossing over occurs in only two of the four chromatids in a tetrad at meiosis. As we will see in a moment, by the nature of ordered spores, *Neurospora* provides us with an extra locus every time we do a cross.

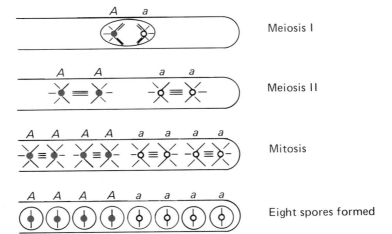

Figure 5-17.
Meiosis in
Neurospora
Although
Neurospora has
seven pairs of
chromosomes at
meiosis, only one
pair is shown. *A* and
a, the two mating
types, represent the
two centromeres of
the tetrad.

Meiosis I

Meiosis II

Mitosis

Eight spores formed

Phenotypes of *Neurospora*

At this point we might wonder what the phenotypes of *Neurospora* are that
allow us to do genetic analysis. In general, microorganisms have phenotypes
that fall into three broad categories: colony morphology, drug resistance, and
nutritional requirements. Microorganisms are cultured in petri plates or test
tubes that contain a supporting medium such as agar, to which various components can be added (Figure 5–18). Wild-type *Neurospora* is the familiar
pink bread mold. Various mutants exist that change colony morphology. For
example, fluffy (*fl*), tuft (*tu*), dirty (*dir*), and colonial (*col 4*) are all mutants
of the basic growth form. Also, wild-type *Neurospora* is sensitive to the sulfa
drug sulfonamide, whereas one of the mutants (*Sfo*) actually requires sulfonamide in order to survive and grow.

Nutritional-requirement phenotypes have provided great insight not
only into genetic analysis but also into biochemical pathways of metabolism.
Wild-type *Neurospora* can grow on a medium containing only sugar, a nitrogen source, some inorganic acids and salts, and the vitamin biotin. This is
referred to as *minimal medium*. However, there are several mutant types
(mutant strains) of *Neurospora* that cannot grow on this minimal medium
until some essential nutrient is added. For example, strain *x* will not grow
on minimal medium but will grow if one of the amino acids, arginine, is
added (Figure 5–19). From this we can infer that the wild type, +, has the
normal enzyme in the synthesis pathway of arginine. The arginine-requiring
mutant has an allele that specifies an enzyme that is incapable of converting
one of the intermediates in the pathway directly into arginine or, alternatively, into one of the precursors to arginine. We can see that if a synthesis
pathway is long, there may be many different loci with alleles that cause the

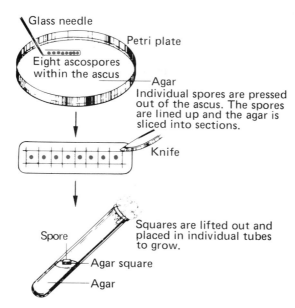

Figure 5-18.
Technique of Spore
Isolation in
Neurospora

strain to be arginine requiring (Figure 5–20). This, in fact, happens and the different loci are usually named arg_1, arg_2, and so on. There are numerous biosynthetic pathways in *Neurospora*, and mutants exhibit many different nutritional requirements. Experimental mutants can be induced by irradiation. These, then, are the tools with which we analyze and map the chromosomes of *Neurospora*. (The techniques will be expanded in the next chapter.)

First and Second Division Segregation.   Recall that there is a 4:4 segregation of the centromere in the ascus. Two kinds of patterns will be found among the loci on these chromosomes. These patterns depend on whether or not there is a crossover between the locus and its centromere (Figure 5–21). If there is no crossover between the locus and its centromere, the chromosome patterns will be essentially the same as the centromeric patterns, which are referred to as **first division segregation (FDS)** because alleles are separated from each other at meiosis I. If, however, there has been a crossover between the locus and its centromere, patterns of a different type will emerge (2:4:2 or 2:2:2:2), which are referred to as **second division segregation (SDS)**. The centromere is acting as a locus because of the ordered nature of the spores. If so, then we should be able to map the distance of a locus to its centromere. The method is straightforward. Every SDS configuration has two recombinant and two nonrecombinant chromatids (actually 4 and 4 because of the mitosis). Thus, half of the chromatids in an SDS ascus

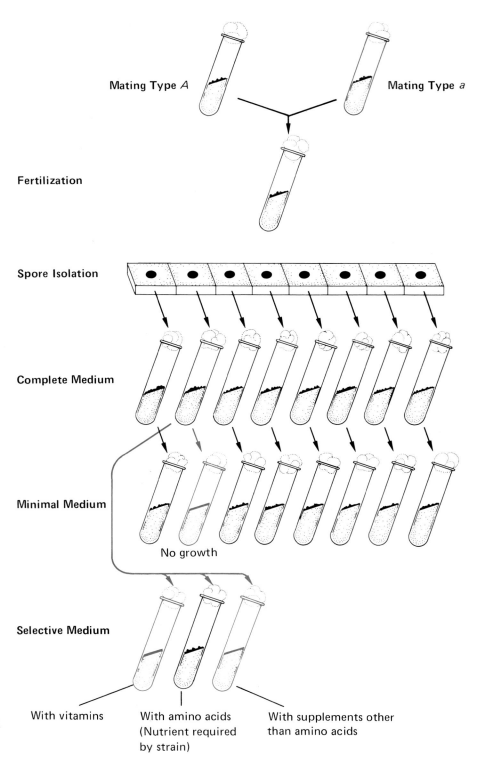

Mating Type *A*

Mating Type *a*

Fertilization

Spore Isolation

Complete Medium

Minimal Medium

No growth

Selective Medium

With vitamins

With amino acids
(Nutrient required
by strain)

With supplements other
than amino acids

Figure 5-19.
Isolation of
Nutritional-
Requirement
Mutants in
Neurospora
Mutants are induced
in mating type *a* by
irradiation.

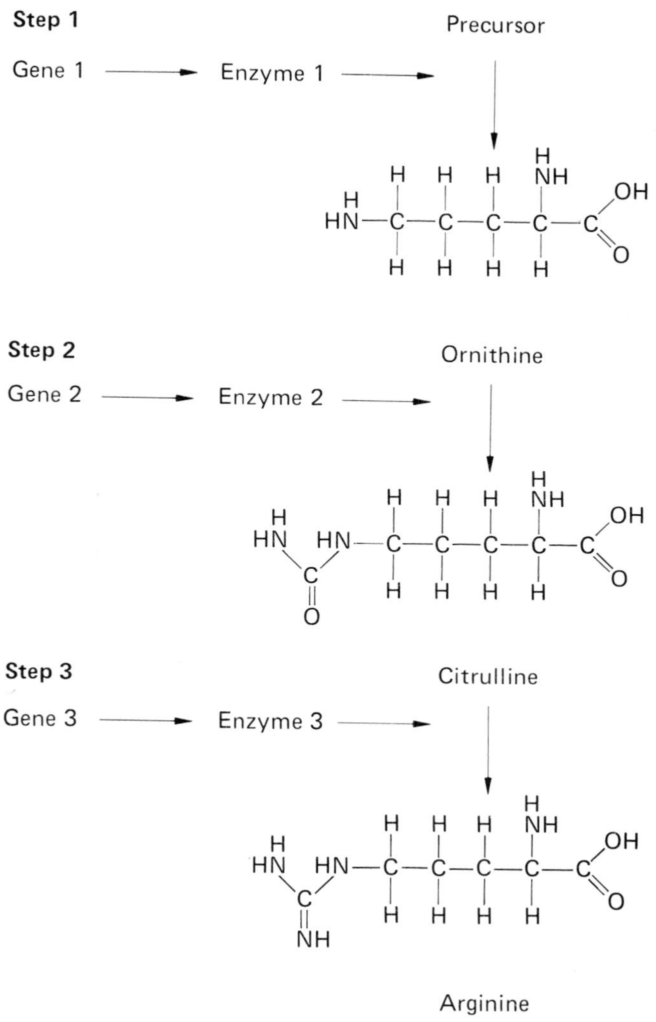

Figure 5-20.
Arginine
Biosynthetic
Pathway of
Neurospora with
Various Mutants

are recombinant. Therefore, map distance = 1/2(% SDS asci). An example is given in Table 5–4.

Three-point crosses in *Neurospora* can also be done. Let us map two loci and their centromere. For simplicity, we will use the *a* and *b* loci. A dihybrid is formed from fused mycelia (*a b* × + +), which then undergo meiosis. One thousand asci are scored for the two loci, keeping the spore order intact. Before presenting the data, we should consider the way in which it will be grouped. Given that each locus can show 6 different patterns (Figure 5–21), we may expect the data to be in the form of 36 possible arrangements (Table 5–5). Some thought, however, will tell us that many of these

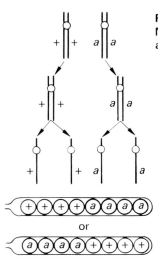

First Division Segregation with No Crossover between *a* Locus and Centromere

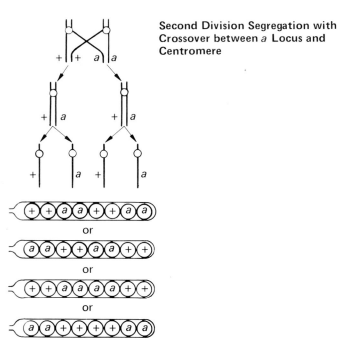

Second Division Segregation with Crossover between *a* Locus and Centromere

Figure 5-21.
The Six Possible
Ascospore Patterns
in *Neurospora*

patterns are really random variants of each other. The tetrad in meiosis is a three-dimensional entity rather than a flat, four-rod object as it is usually drawn. At the first meiotic division, either centromere can go to the left or the right. At the second meiotic division, either sister chromatid can go to

TABLE 5-4. Genetic Patterns Following Meiosis in an $a/+$ Heterozygous *Neurospora*. Ten asci are examined.

1	2	3	4	5	6	7	8	9	10
a	a	+	a	a	+	a	+	+	+
a	a	+	a	a	+	a	+	+	+
a	a	+	+	+	+	a	a	a	+
a	a	+	+	+	+	a	a	a	+
+	+	a	+	a	a	+	a	+	a
+	+	a	+	a	a	+	a	+	a
+	+	a	a	+	a	+	+	a	a
+	+	a	a	+	a	+	+	a	a
FDS	FDS	FDS	SDS	SDS	FDS	FDS	SDS	SDS	FDS

Map distance (a locus to centromere) = 1/2 (% SDS)

= 1/2 (40%)

= 20 map units

or:

total ascospores = $8 \times 10 = 80$

SDS ascospores = $8 \times 4 = 32$

recombinants = 1/2 SDS ascospores

= (1/2)(32) = 16

Map distance (a locus to centromere) = 16/80 $\times$ 100 = 20%

= 20 map units

the left or right in either of the two cells (look back at Figure 5–21). Thus, arrangements 3, 4, 5, and 6 are actually representatives of exactly the same genetic event. (We should be able to prove this to ourselves by drawing a tetrad and following it through meiosis to show that those four arrangements are equally likely, given that both loci are on the same chromosome, a is closer to the centromere, and a single crossover has occurred between a and b.) The 36 possible patterns then reduce to only 7 unique patterns as shown in Table 5–6.

Gene Order. We must now determine the distance from each locus to its centromere and the linkage arrangement of the loci if they are both linked

TABLE 5-5. Six of the Thirty-Six Possible Patterns of Asci Scored for Two Loci

1	2	3	4	5	6
a b	a +	a b	a b	a +	a +
a b	a +	a b	a b	a +	a +
a b	a +	a +	a +	a b	a b
a b	a +	a +	a +	a b	a b
+ +	+ b	+ b	+ +	+ +	+ b
+ +	+ b	+ b	+ +	+ +	+ b
+ +	+ b	+ +	+ b	+ b	+ +
+ +	+ b	+ +	+ b	+ b	+ +

to the same centromere. We proceed as follows. We can establish by inspection that the two loci are linked to each other and therefore to the same centromere. This is done by examining classes 1 and 2. If the two loci are unlinked, these two categories would represent two equally likely alternative events when no crossover takes place. Since category 1 is almost 75% of all the asci, we can be sure the two loci are linked. (A convincing way to demonstrate this is to draw a meiosis with two tetrads to show that 1 and 2 should be equally likely.)

To determine the distance of each locus to the centromere, we calculate 1/2 the percentage of SDS for each locus. For the *a* locus, classes 4, 5, 6, and 7 are SDS. For the *b* locus, classes 3, 5, 6, and 7 are SDS. Therefore, the distances to the centromere, in map units, are

$$\text{for locus } a: \left(\frac{1}{2}\right) \frac{9 + 150 + 1 + 8}{1000} \times 100 = 8.4 \text{ centimorgans}$$

$$\text{for locus } b: \left(\frac{1}{2}\right) \frac{101 + 150 + 1 + 8}{1000} \times 100 = 13.0 \text{ centimorgans}$$

(It should now be possible to describe exactly what type of crossover event produced each of the 7 classes of Table 5–6.)

Unfortunately these two distances do not provide a unique solution to the gene order. Figure 5–22 shows that two equally likely alternatives are possible. How do we determine between these? The simplest way is to find out what happens to the *b* locus when a recombination occurs between the *a* locus and its centromere. If order (i) is correct, the SDS for *a* should have no effect on *b*, whereas if (ii) is correct, most of the SDS for *a* should include an SDS for *b*. That is, for order (ii) most of the crossovers that move the *a* locus should also move the *b* locus because the latter is only a short distance farther down the chromosome.

The answer lies with classes 4, 5, 6, and 7, which include all the SDS for

TABLE 5-6. The Seven Unique Classes of Asci Resulting from Meiosis in a Dihybrid *Neurospora*, a+ b+

	1	2	3	4	5	6	7
	a b	a +	a b	a b	a b	a +	a b
	a b	a +	a b	a b	a b	a +	a b
	a b	a +	a +	+ b	+ +	+ b	+ +
	a b	a +	a +	+ b	+ +	+ b	+ +
	+ +	+ b	+ +	+ +	+ +	+ b	+ b
	+ +	+ b	+ +	+ +	+ +	+ b	+ b
	+ +	+ b	+ b	a +	a b	a +	a +
	+ +	+ b	+ b	a +	a b	a +	a +
	729	2	101	9	150	1	8
SDS for a locus				9	150	1	8
SDS for b locus			101		150	1	8

the *a* locus. Of 168 asci, 150 of them (class 5) have a similar SDS for the *b* locus. Thus 89% of the time that there is a crossover that affects the *a* locus it also affects the *b* locus, compelling evidence in favor of alternative (ii). (What form would the data take if the (i) alternative were correct?)

Given that alternative (ii) is correct, we can improve the accuracy of our measurement by including multiple crossovers that were originally missed when we did not know the order of the loci. Figure 5–23 categorizes the seven classes of asci on the basis of the type of crossover event that caused them. We then generate more precise map distances by summing up all of the recombinations to calculate the percentage of chromatids that have a cross-over, as shown at the bottom of Figure 5–23. To get the centromere-to-locus-*b* distance, we have to take into account that most probably one crossover occurred in the centromere-to-*b* region to give classes 3 and 5. Thus half the chromatids are recombinant and we count only half the percent of the occur-

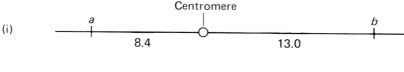

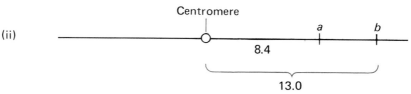

Figure 5-22. Two Alternative Arrangements of the *a* and *b* Loci and Their Centromere Distances are in map units.

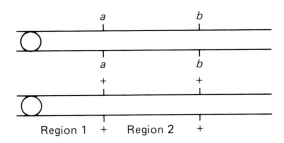

Class   **Type of Event**

1   Nonrecombinant
2   Four-strand double in region 2
3   Single in region 2
4   Two-strand double in regions 1, 2; four-strand double in regions 1, 2
5   Single in region 1
6   Four-strand triple in regions 1, 2, 2
7   Three-strand double in regions 1, 2

More Precise Map Distances

Region 1 (centromere to locus a) =
½ (%) (class 4 + 5 + 6 + 7) = ½(16.8) = 8.4 map units

Regions 1 and 2 (centromere to locus b) =
(%) class 2 + ½(%) class 3 + (%) class 4 + ½(%) class 5 + 1½(%)
class 6 + (%) class 7 = 14.6 map units

∴ Region 2 (locus a to locus b) =
14.6 − 8.4 = 6.2 map units

Figure 5-23.
Assignment of
Crossover Events to
the Asci of Table
5–6

rence of these classes. Classes 2, 4, and 7 probably had double crossovers in this region; so we count these classes fully for map distance because all chromatids are recombinant. Since class 6 requires three crossovers in the centromere-to-b region, we count the percentage occurrence of this class 1½ times because on the average each chromatid has one and a half crossovers. This is as far as we will go with this analysis. The values are still not perfect for several reasons. First, there are known "hidden" events. If a four-strand double can occur in region 2 to generate class 2, then a four-strand double can occur in region 1, which will generate nonrecombinants. The occurrence of this hidden double would be in proportion to the centromere–a distance. Second, some of the classes could have been generated by more complex events of which we were unaware. Since a triple crossover occurred to produce class 6, presumably other triple crossovers could have occurred to generate some of the other classes. (Try to give an example.) Lastly, we have not made use of the mapping function because in these relatively short map distances we are better off determining the occurrence of multiple crossovers, as we did in Figure 5–23, rather than guessing at them as a mapping function would do. To use a mapping function for the centromere–b distance, we would calculate the SDS for b and plug that directly into one of the functions. Actually, as more work were done on this chromosome, accuracy would be improved by the discovery of intervening loci. With shorter distance between loci, accuracy improves.

Unordered Spores

In fungi such as the baker's yeast, *Saccharomyces cerevisiae,* the ascus shape places no restriction on the arrangement of spores in meiosis. All the products of meiosis are contained, but not in order. Here unless, as in some cases, loci are known that are very close to the centromere, the centromere cannot be used as a marker. Figure 5–24 shows the life cycle of yeast. Let us look at a mapping problem, using the a and b loci for convenience. Yeast have phenotypes similar to *Neurospora.*

When an a b spore (or gamete) fuses with a $+$ $+$ spore (or gamete) and the diploid then undergoes meiosis, the spores can be isolated and grown as haploid colonies, which are then noted for the two loci. The only three types that can occur are shown in Table 5–7. Class 1 has two types of spores, which are identical to the parent haploid spores. This ascus type is referred to as a **parental ditype (PD).** The second class also has only two spore types, but they are of the recombinant type. This ascus type is referred to as a **nonparental ditype (NPD).** The third class has all four spore types and is referred to as a **tetratype (TT).**

All three ascus types can be generated whether or not the two loci are linked. Figure 5–25 demonstrates this. Since all three ascus types can be generated through either linkage or independent assortment, how do we distinguish between the two processes? The answer lies in the relative frequencies of the various types. As Figure 5–25 shows, if the loci are linked, PDs come from the lack of a crossover whereas NPDs come about from four-strand double crossovers. We would thus expect PDs to be more numerous. However, if the loci are not linked, both PDs and NPDs come about through independent assortment—and they should occur in equal frequency. We can therefore determine whether or not the loci are linked by comparing PDs and NPDs. In Table 5–7 the PDs greatly outnumber the NPDs; the two loci are, therefore, linked. What, then, is the map distance between the loci?

A return to Figure 5–25 shows that in a NPD all four chromatids are

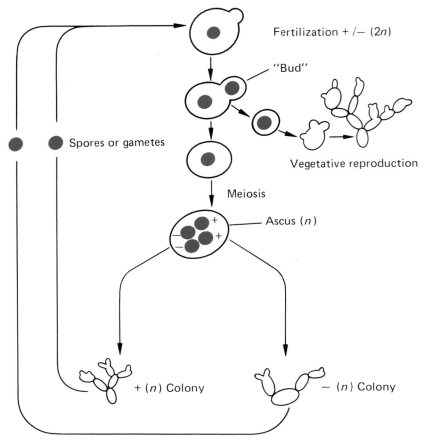

Figure 5-24. Life Cycle of Yeast
Mature cells are mating types + or −; n is the haploid stage; $2n$ is diploid.

TABLE 5-7. The Three Ascus Types in Yeast Resulting from Meiosis in a Dihybrid, $a+\ b+$

1 (PD)		2 (NPD)		3 (TT)	
a	b	a	+	a	b
a	b	a	+	a	+
+	+	+	b	+	b
+	+	+	b	+	+
75		5		20	

146

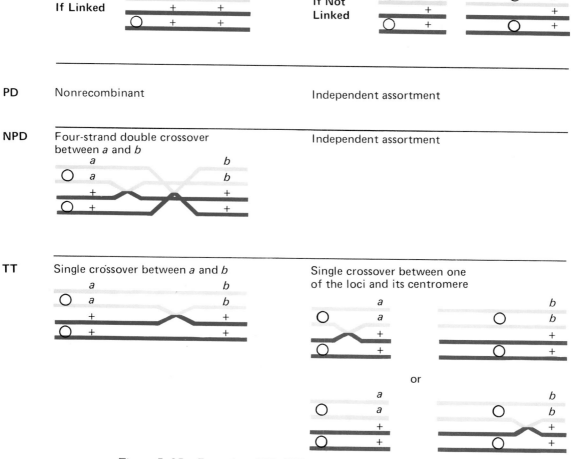

PD	Nonrecombinant	Independent assortment
NPD	Four-strand double crossover between *a* and *b*	Independent assortment
TT	Single crossover between *a* and *b*	Single crossover between one of the loci and its centromere

Figure 5–25. Formation of PD, NPD, and TT at Meiosis in a Dihybrid Yeast through Linkage or Independent Assortment

recombinant, whereas in a TT only half the chromatids are recombinant. We can use the following formula:

$$\text{map units} = \frac{(1/2)\text{TT} + \text{NPD}}{\text{total asci}} \times 100$$

Thus, for the data of Table 5–7,

$$\text{map units} = \frac{10 + 5}{100} \times 100 = 15$$

147

Yeast asci can be analyzed like ordered spores by finding loci that are very close to the centromere. This can be done by first finding loci that are not linked and then making matings until no tetratypes result. Since TT occur because of a crossover between one of the loci and its centromere, the lack of TT would indicate that the two loci are so close to their respective centromeres that crossover is a rare event. The cross would produce only PD and NPD in equal numbers. Once these loci close to the centromere (**centromere markers**) are found, it is possible to order the spores as if they had been inside a *Neurospora* ascus. For example, if the *a* locus of our previous example were a centromere marker, we could reexamine the data of Table 5–7 and look only for SDS (second division segregation) in the *b* locus. The table has already been set up that way: All the *a*-locus information has been ordered into an FDS pattern. Then, when we remember the *Neurospora* analysis, it would become apparent that classes 1, 2, and 3, of Table 5–7 are the same as classes 1, 2, and 3 of Table 5–6. Since class 2 comes about from a four-strand double and class 3 from a single crossover, we would use half the percentage of class 3 and all of class 2 to obtain the same result as before: 15 map units.

In a similar manner, the methodology used with yeast could be applied to *Neurospora*. The classes of Table 5–6 could be tabulated as PD, NPD, and TT and map distances calculated using the yeast formula (Table 5–8).

TABLE 5-8. Reclassification of the Data of Table 5–6 on the Basis of PD, NPD, and TT

1 PD	2 NPD	3 TT	4 TT	5 PD	6 NPD	7 TT
a b	a +	a b	a b	a b	a +	a b
a b	a +	a b	a b	a b	a +	a b
a b	a +	a +	+ b	+ +	+ b	+ +
a b	a +	a +	+ b	+ +	+ b	+ +
+ +	+ b	+ +	+ +	+ +	+ b	+ b
+ +	+ b	+ +	+ +	+ +	+ b	+ b
+ +	+ b	+ b	a +	a b	a +	a +
+ +	+ b	+ b	a +	a b	a +	a +
729	2	101	9	150	1	8

$$\text{Map units} = \frac{(1/2)\text{TT} + \text{NPD}}{\text{total asci}} \times 100$$
$$= \frac{(1/2)118 + 3}{1000} \times 100 = 6.2$$

This gives us the a–b distances directly and agrees with the corrected value calculated earlier (6.2 map units in each case).

SOMATIC (MITOTIC) CROSSING OVER

Crossing over is also known to occur in somatic cells as well as during meiosis. It apparently occurs by a similar reciprocal mechanism whereby two nonsister chromatids come to lie next to each other and a breakage and reunion follows. Unlike meiosis there is no synaptinemal complex formed. Since mitotic chromosomes do not normally lie side-by-side, the occurrence of mitotic crossing over is relatively rare. In the fungus *Aspergillus nidulans,* mitotic crossing over occurs about once in every 100 cell divisions.

Mitotic recombination was discovered in 1936 by Stern, who noticed the occurrence of *twin spots* in fruit flies that were dihybrids for the yellow allele of body color (y) and the singed allele (sn) for bristle morphology (Figure 5–26). This could be explained by mitotic crossing over between sn and the centromere (Figure 5–27). A crossover in the sn–y region would produce only a yellow spot, whereas a double crossover, one between y and sn and the other in the sn–centromere region, would produce only a singed spot. (We should verify this for ourselves.) These three phenotypes were found in the relative frequencies expected. That is, given that the gene locations are drawn to scale in Figure 5–27, we would expect double spots to be most common, followed by yellow spots, with singed spots rarest of all because they require a double crossover. This in fact occurred and no other explanation was consistent with these facts.

Mitotic crossing over has been used extensively in fungal genetics as a supplemental, or even a primary, method for determining linkage relations. While gene orders are consistent between mitotic and meiotic mapping, relative distances usually are not. This is not totally unexpected. We know that neither meiotic nor mitotic crossing over is uniform along a chromosome. Apparently, the factors that cause deviation from uniformity differ in the two processes.

Curt Stern (1902–   )
Courtesy of the
Science Council of
Japan

HUMAN CHROMOSOME MAPS

Conceptually, human chromosomes can be mapped just as for any other organism. Realistically, the problems mentioned earlier (the inability to make specific crosses coupled with the relatively small family size in humans) make human chromosome mapping very difficult. However, there has been some progress based on pedigrees, especially in assigning genes to the X chromosome. As the pedigree analysis in the previous chapter has shown, X-chromosome traits have a unique inheritance pattern and those

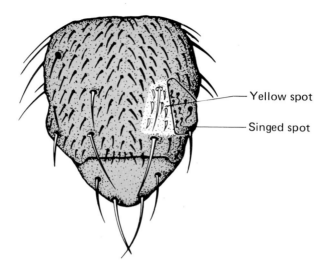

Figure 5-26.
Twin Spots on the
Thorax of a Female
Drosophila

loci on the X chromosome are easily identified. Currently, there are more than 100 loci known to be on the X chromosome. It has been estimated, by several different methods, that there are about 50,000 loci on human chromosomes.

X-Linkage

After determining that a gene is X-linked, the next problem is to determine the position of the locus on the X chromosome and to determine map units between loci. This can be done with the proper pedigrees, where crossing over can be ascertained. An example—what is referred to as the "grandfather method"—is shown in Figure 5-28. In this example a grandfather is found who has one of the traits in question (here, color blindness). We then find that he has a grandson who is G6PD deficient. From this we can infer that the mother was dihybrid in the repulsion configuration. That is, she received the color-blindness allele on one of her X chromosomes from her father, and she must have received the G6PD-deficiency allele on the other X chromosome from her mother (why?). Thus, the two sons on the left of Figure 5-28 are nonrecombinant and the two on the right are recombinant. Theoretically, we can determine map distance by simply totalling the recombinant grandsons and dividing by the total number of grandsons. Of course, the methodology is the same if the grandfather were both color blind and G6PD deficient. The mother would then be dihybrid in the coupling configuration and the sons would be tabulated in the reverse manner. The point is that the grandfather's phenotype gives us information from which to infer

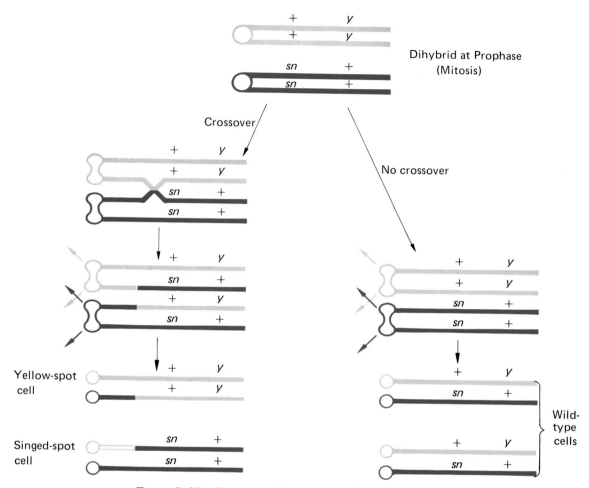

Figure 5-27. Formation of Twin Spots and Single Mutant Spots through Somatic Crossing Over

that the mother was dihybrid, as well as telling us the coupling–repulsion arrangement of her alleles. We can then score her sons as either recombinant or nonrecombinant.

Autosomal Linkage

From this we can see that it is relatively easy to deal with the X chromosome. The autosomes are another story. Since we have 22 autosomal linkage groups (22 pairs of nonsex chromosomes), it is virtually impossible to determine

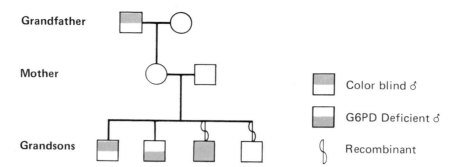

Figure 5-28. "Grandfather Method" of Determining Crossing Over between Loci on the Human X Chromosome

from simple pedigrees which chromosome two loci are on. Pedigrees can tell us if two loci are linked to each other, but not on which chromosome (Figure 5-29). The nail-patella syndrome includes, among other things, abnormal nail growth coupled with the absence or underdevelopment of knee caps. It is a dominant trait. The male in generation II of Figure 5-29 is dihybrid with the I^A allele of the ABO system associated with nail-patella, and the I^B allele with the normal. Thus, only one child (III-5) is recombinant. Actually, the map distance is about 10%. It should be noted that, in general, map distances appear greater in females than males because there is more crossing over in females.

The first locus that was definitely established to be on a particular autosome was the Duffy blood group on chromosome 1. This was ascertained in 1968 from a family that had a morphologically odd, or "uncoiled," chromosome 1. Inheritance in the Duffy blood group system followed the pattern of inheritance of the "uncoiled" chromosome. Real strides have been made since then; over 110 loci have been placed on specific chromosomes. This placement has come about only in the last several years with the advent of two techniques. One is chromosome banding and the other is somatic-cell hybridization.

Chromosome Banding. Chromosome banding techniques were developed around 1970 and involve certain histochemical stains that produce repeatable banding patterns on the chromosomes. (Giemsa staining is one technique—the resulting bands are called **G-bands.**) Prior to these techniques, human and other mammalian chromosomes were grouped into general size categories because of the difficulty of distinguishing many of the chromosomes (Figure 4-2). With banding techniques came the ability to positively identify each human chromosome in a karyotype.

Somatic-Cell Hybridization. This ability to distinguish each human chromosome is required for the latest technique in human chromosome mapping: somatic-cell hybridization. In general, human and mouse (or hamster)

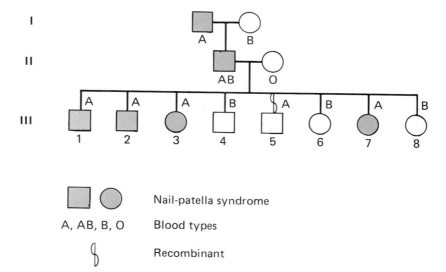

Figure 5-29.
Linkage of the Nail–
Patella Syndrome
Locus and the ABO
Locus

cells are fused to form a hybrid. The fusion is usually mediated by a chemical or by an inactive virus such as the Sendai virus, which has the property of being able to bind to more than one cell at the same time. Because of this property, the mediating cell draws several cells close together and facilitates their fusion. When this hybrid is formed, it tends to preferentially lose human chromosomes through succeeding generations. The end product is a line of cells with one or more human chromosomes in addition to the original mouse or hamster chromosome complement. Since banding techniques allow the differentiation of human chromosomes from those of other species, these lines of cells can be characterized by the specific human chromosomes that they carry. A geneticist then looks for specific human phenotypes, such as enzyme products, and can then assign the phenotype to a small group of chromosomes. Two particular tests are used.

A **synteny test** (same linkage group) determines whether two loci are in the same linkage group if the phenotypes of the two loci are either always together or always absent in various hybrid cell lines. An **assignment test** determines which chromosome a particular locus is on by the concordant appearance of the phenotype whenever that particular chromosome is in a cell line or by the lack of the particular phenotype when a particular chromosome is absent from a cell line. The first autosomal synteny test was performed in 1970 and demonstrated that the B locus of lactate dehydrogenase (LDH_B) was linked to the B locus of peptidase (PEP_B). (Both enzymes are formed from subunits controlled by two loci each. In addition to the B loci, both proteins have subunits controlled by A loci.) Later, by assignment, these loci were shown to reside on chromosome 12.

TABLE 5-9. Three Hybrid Cell Lines (A, B, and C), Each Containing Four Human Chromosomes. A plus (+) indicates that the chromosome is present.

Clone Designation	Human Chromosome Present							
	1	*2*	*3*	*4*	*5*	*6*	*7*	*8*
A	+	+	+	+	−	−	−	−
B	+	+	−	−	+	+	−	−
C	+	−	+	−	+	−	+	−

An optimal set of hybrid lines is shown in Table 5–9. (Optimal, in that every possible chromosome combination is present.) In these three clones it is found, for example, that loci for human hydroxyacyl-Co A dehydrogenase (HADH) and human mitochondrial malate dehydrogenase (MDH-2) are found in clone C but not in clones A and B. From this it can be inferred that the loci for HADH and MDH-2 are linked (synteny) and located on chromosome 7 (assignment). The human map as we know it now (about 210 assigned loci of about 1200 known) is shown in Figure 5–30 and Table 5–10. At the present time geneticists studying human chromosomes feel hampered not by a lack of techniques but by a lack of marker loci. When a new locus is discovered, it is now relatively easy to assign it to its proper chromosome.

The problem still exists of determining exactly where on a chromosome a particular locus belongs. This is being facilitated by particular cell lines with chromosomes that have been broken such that parts are either missing or have been translocated to other chromosomes. These processes reveal new linkage arrangements and make it possible to determine on which region of a particular chromosome a particular locus is situated.

Figure 5-30.   Human G-Banded Chromosomes with Their Accompanying Assigned Loci
See Table 5-10 for key.

Source: Courtesy of Dr. Victor McKusick.

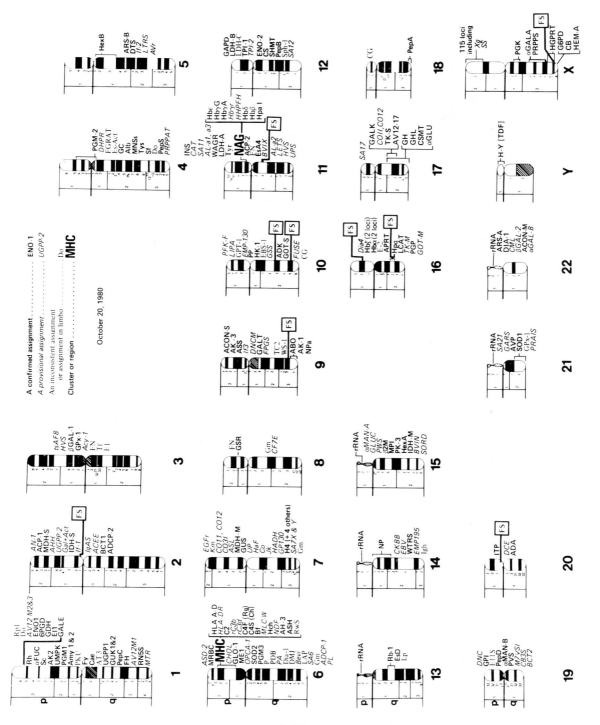

The human gene map, October 20, 1980.

A confirmed assignment ENO-1
A provisional assignment *UGPP-2*
An inconsistent assignment
or assignment in limbo .... *Do*
Cluster or region .... **MHC**

October 20, 1980

TABLE 5-10. Key to the Loci Shown in the Chromosome Map of Figure 5–30

ABO	ABO blood group (chr. 9)
ACEE	Acetylcholinesterase expression (chr. 2)
ACON-M	Aconitase, mitochondrial (chr. 22)
ACON-S	Aconitase, soluble (chr. 9)
ACP1	Acid phosphatase-1 (chr. 2)
ACP2	Acid phosphatase-2 (chr. 11)
ACY1	Aminoacylase-1 (chr. 3)
ADA	Adenosine deaminase (chr. 20)
ADCP1	Adenosine deaminase complexing protein-1 (chr. 6)
ADCP2	Adenosine deaminase complexing protein-2 (chr. 2)
ADK	Adenosine kinase (chr. 10)
AH3	Adrenal hyperplasia III (21-hydroxylase deficiency) (chr. 6)
AHH	Aryl hydrocarbon hydroxylase (chr. 2)
AK1	Adenylate kinase-1 (soluble) (chr. 9)
AK2	Adenylate kinase-2 (mitochondrial) (chr. 1)
AK3	Adenylate kinase-3 (mitochondrial) (chr. 9)
AL	Lethal antigen: 3 loci (chr. 11)
Alb	Albumin (chr. 4)
AMY1	Amylase, salivary (chr. 1)
AMY2	Amylase, pancreatic (chr. 1)
An1	Aniridia, type 1 (chr. 2)
APRT	Adenine phosphoribosyltransferase (chr. 16)
ARS-A	Arylsulfatase A (chr. 22)
ARS-B	Arylsulfatase B (chr. 5)
ASD2	Atrial septal defect, secundum type (chr. 6)
ASH	Asymmetric septal hypertrophy (chr. 6)
ASL	Argininosuccinate lyase (chr. 7)
ASS	Argininosuccinate synthetase (chr. 9)
AT3	Antithrombin III (chr. 1)
AV12M1	Adenovirus-12 chromosome modification site-1 (chr. 1)
AV12M2	Adenovirus-12 chromosome modification site-2 (chr. 1)
AV12M3	Adenovirus-12 chromosome modification site-3 (chr. 1)
AV12-17	Adenovirus-12 chromosome modification site-17 (chr. 17)
AVP	Antiviral protein (chr. 21)
AVr	Antiviral state regulator (chr. 5)
β2M (B2M)	Beta-2-microglobulin (chr. 15)
BCT-1	Branched chain amino acid transferase-1 (chr. 2)
BCT-2	Branched chain amino acid transferase-2 (chr. 19)
BEVI	Baboon M7 virus infection (chr. 6)
BF	Properdin factor B (chr. 6)
BVIN	BALB virus induction, N-tropic (chr. 15)

156

BVIX	BALB virus induction, xenotropic (chr. 11)
C2	Complement component-2 (chr. 6)
C4F	Complement component-4 fast (chr. 6)
C4S	Complement component-4 slow (chr. 6)
Cae	Cataract, zonular pulverulent (chr. 1)
CAT	Catalase (chr. 11)
CB	Colorblindness (deutan and protan) (X chr.)
CB3S	Coxsackie B3 virus susceptibility (chr. 19)
CF7E	Clotting factor VII expression (chr. 8)
CG	Chorionic gonadotropin (chr. 10 and 18; chr. 5 or 6)
Ch	Chido blood group (same as C4S)
CHOL	Hereditary hypercholesterolemia (chr. 6)
CKBB	Creatine kinase, brain type (chr. 14)
CML	Chronic myeloid leukemia (chr. 22)
Co	Colton blood group (chr. 7)
CO11	Collagen I alpha-1 chain (chr. 7 and 17)
CO12	Collagen I alpha-2 chain (chr. 7 and 17)
CO31	Collagen III alpha-1 chain (chr. 7)
CS	Citrate synthase, mitochondrial (chr. 12)
CSMT (or CSH)	Chorionic somatomammotropin (chr. 17)
DCE	Desmosterol-to-cholesterol enzyme (chr. 20)
DHPR	Quinoid dihydropteridine reductase (chr. 4)
Dia-1	NADH-diaphorase (chr. 22)
DIA-4	Diaphorase-4 (chr. 16)
DMJ	Juvenile diabetes mellitus (chr. 6)
DNC	Lysosomal DNA-ase (chr. 19)
DNCM	Cytoplasmic membrane DNA (chr. 9)
Do	Dombrock blood group (chr. 1 or 4)
DTS	Diphtheria toxin sensitivity (chr. 5)
E1	Pseudocholinesterase-1 (chr. 3)
E2	Pseudocholinesterase-2 (chr. 16)
E11S	Echo 11 sensitivity (chr. 19)
EBS1	Epidermolysis bullosa, Ogna type (chr. 10)
EBV	Epstein-Barr virus integration site (chr. 14)
EGFR	Epidermal growth factor, receptor for (chr. 7)
E11	Elliptocytosis-1 (chr. 1)
EMP130	External membrane protein-130 (chr. 10)
EMP195	External membrane protein-195 (chr. 14)
ENO1	Enolase-1 (chr. 1)
ENO2	Enolase-2 (chr. 12)
Es-Act	Esterase activator (chr. 4 or 5)
EsA4	Esterase-A4 (chr. 11)
EsD	Esterase D (chr. 13)

TABLE 5-10. (continued)

FGPS	Folylpolyglutamate synthetase (chr. 9)
FGRAT	Formylglycinamide ribotide amidotransferase (chr. 4 or 5)
FH	Fumarate hydratase (chr. 1)
FN	Fibronectin (chr. 8, 11)
FS	Fragile site, observed in cultured cells, with or without folate deficient medium, or BrdU (chr. 2, 9, 10, 11, 16, 20, X)
αFUC (FUCA)	Alpha-L-fucosidase (chr. 1)
FUSE	Polykaryocytosis inducer (chr. 10)
Fy	Duffy blood group (chr. 1)
Gal+-Act	Galactose + activator (chr. 2)
αGALA	Alpha-galactosidase A (Fabry disease) (X chr.)
αGALB	Alpha-galactosidase B (chr. 22)
βGAL-1	Beta-galactosidase-1 (chr. 3)
βGAL-2	Beta-galactosidase-2 (chr. 22)
GALE	Galactose-4-epimerase (chr. 1)
GALK	Galactokinase (chr. 17)
GALT	Galactose-1-phosphate uridyltransferase (chr. 9)
GAPD	Glyceraldehyde-3-phosphate dehydrogenase (chr. 12)
GARS	Glycinamide ribonucleotide synthetase (chr. 21)(s)
GC	Group-specific component (chr. 4)
GDH	Glucose dehydrogenase (chr. 1)
GH	Growth hormone (chr. 17)
GHL	Growth hormone like (chr. 17)
αGLU (GLUA)	Alpha-glucosidase (chr. 17)
GLUC	Neutral alpha-glucosidase C (chr. 15)
GLO1	Glyoxylase I (chr. 6)
GOT-M	Glutamate oxaloacetate transaminase, mitochondrial (chr. 16)
GOT-S	Glutamate oxaloacetate transaminase, soluble (chr. 10)
G6PD	Glucose-6-phosphate dehydrogenase (X chr.)
GP130	Granulocyte glycoprotein (chr. 7)
GPI	Glucosephosphate isomerase (chr. 19)
GPT1	Glutamate pyruvate transaminase, soluble (chr. 10)
GPx1	Glutathione peroxidase-1 (chr. 3)
GSR	Glutathione reductase (chr. 8)
GSS	Glutamate-gamma-semialdehyde synthetase (chr. 10)
Gm	Immunoglobulin heavy chain (chr. 6, 7, 8)
GUK1 & 2	Guanylate kinase-1 & 2 (chr. 1)
GUS	Beta-glucuronidase (chr. 7)
H4	Histone H4 and 4 other histone genes (chr. 7)
HADH	Hydroxyacyl-CoA dehydrogenase (chr. 7)
HaF	Hageman factor (chr. 7)

Hbα	Hemoglobin alpha chain (chr. 16)
Hbβ	Hemoglobin beta chain (chr. 11)
Hbδ	Hemoglobin delta chain (chr. 11)
HbγA,G	Hemoglobin gamma chains, ala or gly as AA 136 (chr. 11)
Hbγr	Hemoglobin gamma regulator (chr. 11)
Hbε	Hemoglobin epsilon chain (chr. 11)
Hbζ	Hemoglobin zeta chain (chr. 16)
Hch	Hemochromatosis (chr. 6)
HEM-A	Classic hemophilia (X chr.)
HexA	Hexosaminidase A (chr. 15)
HexB	Hexosaminidase B (chr. 5)
HGPRT	Hypoxanthine-guanine phosphoriboryltransferase (X chr.)
HHPFH	Heterocellular hereditary persistence of fetal hemoglobin (chr. 11)
HK1	Hexokinase-1 (chr. 10)
HLA(A-D)	Human leukocyte antigens (chr. 6)
HLA-DR	Human leukocyte antigen, D-related (chr. 6)
Hpα	Haptoglobin, alpha (chr. 16)
HpaI	Hpa I restriction endonuclease polymorphism (chr. 11)
HVS	Herpes virus sensitivity (chr. 3 and 11)
H-Y	Y histocompatibility antigen (Y chr.)
IDH-M	Isocitrate dehydrogenase, mitochondrial (chr. 15)
IDH-S	Isocitrate dehydrogenase, soluble (chr. 2)
If1	Interferon-1 (chr. 2)
If2	Interferon-2 (chr. 5)
If3	Interferon-3 (chr. 9)
IgAS	Immunoglobulin heavy chains attachment site (chr. 2)
Igh	Immunoglobulin heavy chains (mu, gamma, alpha) (chr. 14)
Ins	Insulin (chr. 11)
ITP	Inosine triphosphatase (chr. 20)
Jk	Kidd blood group (chr. 7)
Km	Kappa immunoglobulin light chains, Inv (chr. 7)
LAP	Laryngeal adductor paralysis (chr. 6)
LCAT	Lecithin-cholesterol acyltransferase (chr. 16)
LDH-A	Lactate dehydrogenase A (chr. 11)
LDH-B	Lactate dehydrogenase B (chr. 12)
LDH-C	Lactate dehydrogenase C (chr. 12)
LIPA	Lysosomal acid lipase-A (chr. 10)
Lp	Lipoprotein—Lp (chr. 13)
LTRS	Leucyl-tRNA synthetase (chr. 5)
β2M (B2M)	Beta-2-microglobulin (chr. 15)
M7VS1	Baboon M7 virus sensitivity-1 (chr. 19)
αMAN-A	Cytoplasmic alpha-D-mannosidase (chr. 15)

TABLE 5-10. (continued)

αMAN-B	Lysosomal alpha-D-mannosidase (chr. 19)
MDH-M	Malate dehydrogenase, mitochondrial (chr. 7)
MDH-S	Malate dehydrogenase, soluble (chr. 2)
ME1	Malic enzyme, soluble (chr. 6)
MHC	Major histocompatibility complex (chr. 6)
MLC-W	Mixed lymphocyte culture, weak (chr. 6)
MNSs	MNSs blood group (chr. 4)
MPI	Mannosephosphate isomerase (chr. 15)
MRBC	Monkey red blood cell receptor (chr. 6)
MTR	5-Methyltetrahydrofolate: L-homocysteine S-methyltransferase, or tetrahydropteroyl-glutamate methyltransferase (chr. 1)
NAG	Non-alpha globin region (chr. 11)
NDF	Neutrophil differentiation factor (chr. 6)
NP	Nucleoside phosphorylase (chr. 14)
NPa	Nail-patella syndrome (chr. 9)
OPCA1	Olivopontocerebellar atrophy I (chr. 6)
P	P blood group (chr. 6)
PA	Plasminogen activator (chr. 6)
PDB	Paget disease of bone (chr. 6)
PepA	Peptidase A (chr. 18)
PepB	Peptidase B (chr. 12)
PepC	Peptidase C (chr. 1)
PepD	Peptidase D (chr. 19)
PepS	Peptidase S (chr. 4)
PFK-F	Phosphofructokinase, fibroblast (chr. 10)
6PGD	6-Phosphogluconate dehydrogenase (chr. 1)
PGK	Phosphoglycerate kinase (X chr.)
PGM1	Phosphoglucomutase-1 (chr. 1)
PGM2	Phosphoglucomutase-2 (chr. 4)
PGM3	Phosphoglucomutase-3 (chr. 6)
PGP	Phosphoglycolate phosphatase (chr. 16)
PK3	Pyruvate kinase-3 (chr. 15)
PKU	Phenylketonuria (chr. 1)
PL	Prolactin (chr. 6)
PP	Inorganic pyrophosphatase (chr. 10)
PRPPAT	Phosphoribosylpyrophosphate amidotransferase (chr. 4)
PRPPS	Phosphoribosylpyrophosphate synthetase (X chr.)
PRAIS	Phosphoribosylaminoimidazole synthetase (chr. 21)
PVS	Polio virus sensitivity (chr. 19)
PWS	Prader-Willi syndrome (chr. 15)
RB1	Retinoblastoma-1 (chr. 13)

rC3b	Receptor for C3b (chr. 6)
rC3d	Receptor for C3d (chr. 6)
Rg	Rodgers blood group—same as C4F
Rh	Rhesus blood group (chr. 1)
RN5S	5S RNA gene(s) (chr. 1)
RP1	Retinitis pigmentosa-1 (chr. 1)
rRNA	Ribosomal RNA (chr. 13, 14, 15, 21, 22)
RwS	Ragweed sensitivity (chr. 6)
SA6	Surface antigen 6 (chr. 6)
SA7	Surface antigen 7 (chr. 7)
SA11	Surface antigen 11 (chr. 11)
SA12	Surface antigen 12 (chr. 12)
SA17	Surface antigen 17 (chr. 17)
SA21	Surface antigen 21 (chr. 21)
Sc	Scianna blood group (chr. 1)
Sf	Stoltzfus blood group (chr. 4)
SHMT	Serine hydroxymethyltransferase (chr. 12)
SOD1	Superoxide dismutase, soluble (chr. 21)
SOD2	Superoxide dismutase, mitochondrial (chr. 6)
SORD	Sorbitol dehydrogenase (chr. 15)
Sph 1	Spherocytosis, Denver type (chr. 8 or 12)
SS	Steroid sulfatase (X chr.)
TC2	Transcobalamin II (chr. 9)
TDF	Testis determining factor—prob. same as H-Y (F)
Tf	Transferrin (chr. 3)
TK-M	Thymidine kinase, mitochondrial (chr. 16)
TK-S	Thymidine kinase, soluble (chr. 17)
TPI-1 & 2	Triosephosphate isomerase-1 & 2 (chr. 12)
tsAF8	Temperature-sensitive (AF8) complement (chr. 3)
Tyr	Tyrosinase (chr. 11)
Tys	Sclerotylosis (chr. 4)
UGPP1	Uridyl diphosphate glucose pyrophosphorylase-1 (chr. 1)
UGPP2	Uridyl diphosphate glucose pyrophosphorylase-2 (chr. 2)
UMPK	Uridine monophosphate kinase (chr. 1)
UP	Uridine phosphorylase (chr. 7)
UPS	Uroporphyrinogen I synthase (chr. 11)
WAGR	Wilms tumor—aniridia/ambiguous genitalia/mental retardation (chr. 11)
WTRS	Tryptophanyl-tRNA synthetase (chr. 14)
WS1	Waardenburg syndrome-1 (chr. 9)
Xg	Xg blood group (X chr.)

Source: Courtesy of Dr. Victor McKusick

CHAPTER SUMMARY

The principle of independent assortment is violated by loci lying near each other on the same chromosome. Recombination between loci on the same chromosome results from the crossing over of chromosomes at meiosis. The amount of recombination provides a measure of the distance between these loci. One map unit (centimorgan) equals one percent recombination. This value can be determined by testcrossing a dihybrid and recording the percentage of recombinant offspring in the F_2. If three loci are used (a three-point cross), then more complex events such as double crossovers, can be discovered. A coefficient of coincidence, the ratio of observed to expected double crossovers, can be calculated to determine if one crossover changes the probability of a second one nearby.

That the chiasma seen during prophase I of meiosis represents both a physical and a genetic crossing over can be demonstrated when homologous chromosomes with morphological distinctions are used. It has also been demonstrated genetically that crossover events occur at the four-strand stage of meiosis.

Because of multiple crossovers, the measured percentage recombination underestimates the true map distance, especially for loci relatively far apart. Mapping functions can be used to estimate true map distance, but the best map estimates come from using very closely linked loci.

Organisms that retain all the products of meiosis are mapped by techniques known as tetrad analysis (haploid mapping). Map units between a locus and its centromere in organisms with ordered spores, such as *Neurospora,* are calculated as

$$\text{map units} = \frac{1/2(\text{SDS})}{\text{total asci}} \times 100$$

With unordered spores, such as in yeast, we use

$$\text{map units} = \frac{\text{NPD} + (1/2)\text{TT}}{\text{total asci}} \times 100$$

When there are centromere markers in organisms with unordered spores, these mapping techniques are interchangeable. In *Neurospora* several techniques are used to determine map distances between loci other than the centromere.

Crossing over also occurs during mitosis but at a much reduced rate. Somatic (mitotic) crossing over can be used to map loci.

Human chromosomes can be mapped. Recombination distances can be established by pedigrees, and loci can be attributed to specific chromosomes by synteny and assignment tests in hybrid cell lines.

EXERCISES AND PROBLEMS

1. A homozygous groucho fly (gro = bristles clumped above the eyes) is crossed with a homozygous rough fly (ro = eyes rough). The F_1 is test-crossed, with the following offspring produced:

Groucho	518
Rough	471
Groucho, rough	6
Wild type	5
	1000

   a. What is the linkage arrangement of these loci?

   b. What offspring would result if the F_1 were selfed instead of being testcrossed?

2. A female fruit fly with abnormal eyes (abe = rough eyes) of a brown color (bis = bistre) is crossed with a wild-type male. Her sons have abnormal, brown eyes; her daughters are of the wild type. When these F_1 are crossed among themselves, the following offspring are produced:

	Sons	Daughters
Abnormal, brown	219	197
Abnormal	43	45
Brown	37	35
Wild type	201	223

   What is the linkage arrangement of these loci?

3. In *Drosophila* the loci inflated (if = small, inflated wings) and warty (wa = rough eyes) are about 10 map units apart on the X chromosome. Construct a data set that would allow you to determine this linkage arrangement. What differences would be involved if the loci were located on an autosome?

4. A homozygous claret (ca = ruby eye color), curled (cu = upcurved wings), fluted (fl = creased wings) fruit fly is crossed to a pure-breeding wild type. The F_1 are testcrossed with the following results:

Fluted	4
Claret	173
Curled	26
Fluted, claret	24
Fluted, curled	167

Claret, curled   6
Fluted, claret, curled 298
Wild type   302

a. Are the loci linked?

b. If so, give gene order, map distances, and coefficient of coincidence.

5. The following three recessive markers are known in rats: a = albinism; b = bent tail; c = crossed eyes. A trihybrid of unknown origin is test-crossed, producing the following offspring:

Albino, bent, crossed   357
Albino, bent   74
Crossed   66
Bent   79
All wild type   343
Albino, crossed   61
Bent, crossed   11
Albino   9
                       ────
                       1000

a. If the genes are linked, determine the relative order and the map distance between genes.

b. What was the allele arrangement in the trihybrid parent?

c. Is there any crossover interference? If yes, how much?

6. The following three recessives are found in corn: b = brittle endosperm; g = glossy leaf; r = red aleurone. A trihybrid of unknown origin is test-crossed with the following offspring:

Brittle, glossy, red   236
Brittle, glossy   241
Red   219
Glossy   23
Wild type   224
Brittle, red   17
Glossy, red   21
Brittle   19
                      ────
                      1000

a. If the genes are linked, determine the relative order and map distances.

b. Reconstruct chromosomally the trihybrid.

c. Is there any crossover interference? If yes, how much?

7. In *Drosophila* ancon (*an* = legs and wings short), spiny legs (*sple* = irregular leg hairs), and arctus oculus (*at* = small narrow eyes) have the following linkage arrangement on chromosome 3:

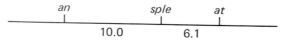

a. Devise a data set, with no interference, that would yield these map units.

b. What changes would yield the same map units but with a coefficient of coincidence of 0.60?

8. Given that ancon (*an*) and spiny legs (*sple*), of problem 7, are 10 map units apart. Notchy (*ny* = wing tips nicked) is on the X chromosome (chromosome 1). Create a data set that would result if you were making crosses to determine the linkage arrangement of these three loci. How would you know that notchy is on the X chromosome?

9. The Duffy blood group with alleles Fy^a and Fy^b was localized to chromosome 1 in humans when an "uncoiled" chromosome was associated with it. Construct a pedigree that would verify this.

10. Given the following cross in *Neurospora: ab* × +. Give the results showing that crossing over is at the four-strand stage rather than the two-strand stage.

11. Apply the two mapping functions in this chapter to the map distances calculated in problems 1 and 2. Try the mapping functions on two loci that assort independently.

12. A strain of yeast requiring both tyrosine (*t*) and arginine (*a*) is crossed to the wild type. After meiosis, the following 10 asci are dissected. Classify each ascus as to segregational type (PD, NPD, TT). What is the linkage relationship of these two loci?

(1)	*a*	*t*	+	+	+	+	*a*	*t*
(2)	+	+	+	+	*a*	*t*	*a*	*t*
(3)	*a*	+	*a*	+	+	*t*	+	*t*
(4)	*a*	*t*	*a*	*t*	+	+	+	+
(5)	*a*	*t*	*a*	+	+	*t*	+	+
(6)	+	+	+	+	*a*	*t*	*a*	*t*
(7)	*a*	*t*	+	+	*a*	+	+	*t*
(8)	+	+	+	+	*a*	*t*	*a*	*t*
(9)	+	+	*a*	*t*	*a*	*t*	+	+
(10)	*a*	*t*	+	+	+	+	*a*	*t*

13. A certain haploid strain of yeast was deficient for the synthesis of tryptophan (t) and methionine (m). It was crossed to the wild type and meiosis occurred. The asci were analyzed for their tryptophan and methionine requirements. The following results with the inevitable lost spores were obtained:

(1)	t m	? ?	? ?	t m		(7)	+ +	+ m	? ?	t m
(2)	? ?	t m	+ +	+ +		(8)	+ +	t m	? ?	+ +
(3)	t +	t m	+ m	+ +		(9)	t +	+ m	t +	+ m
(4)	t m	+ +	? ?	t m		(10)	t m	+ +	t m	+ +
(5)	t +	? ?	? ?	+ m		(11)	+ +	+ +	? ?	? ?
(6)	+ +	+ +	t m	t m		(12)	? ?	+ m	? ?	t +

a. Classify each ascus as to segregational type (note that some asci may not be classifiable).

b. Are the genes linked?

c. If so, how far apart are they?

14. Assume in the previous example that tryptophan marks the yeast centromere. Calculate the distance of methionine from its centromere.

15. In *Neurospora* a haploid strain requiring arginine (a) is crossed to the wild type $(+)$. Meiosis occurs and 10 asci are dissected with the following spore orders. Give a chromosome map of *Neurospora*.

(1)	+ + a a + + a a				(6)	+ + a a a a + +
(2)	a a + + a a + +				(7)	a a + + + + a a
(3)	+ + + + a a a a				(8)	+ + + + a a a a
(4)	+ + + + a a a a				(9)	a a + + + + a a
(5)	a a a a + + + +				(10)	a a a a + + + +

16. A certain haploid strain of *Neurospora* had a fuzzy colony morphology (f) and was crossed to the wild type $(+)$. The asci were analyzed in the order in which spores were isolated after meiosis. The following results with the inevitable lost spores were obtained:

(1)	? f f ? ? + + +				(7)	+ + f f f f + +
(2)	f f + + + + f f				(8)	f f f ? ? + + +
(3)	f ? ? ? + ? ? ?				(9)	+ ? ? ? ? ? f f ?
(4)	+ ? ? ? f f f f				(10)	f f + + f f + +
(5)	f f ? ? ? + ? +				(11)	f f f f + + + +
(6)	? f f ? ? ? ? ?				(12)	f f ? ? ? ? + +

a. Classify each ascus as to segregational type and note which asci cannot be classified.

b. Draw a map of the chromosome containing the *f* locus and give all relevant measurements.

17. In yeast the *a* and *b* loci are 12 map units apart. Construct a data set to demonstrate this.

18. In *Neurospora* the *a* locus is 12 map units from its centromere. Construct a data set to show this.

19. An *a b Neurospora* was crossed with a + + form. Meiosis occurred and 1000 asci were dissected. Using the classes of Table 5–6, the following data resulted:

Class 1 700
     2   0
     3 190
     4   90
     5   5
     6   5
     7   10

What is the linkage arrangement of these loci?

20. Given the following linkage arrangement in *Neurospora*,

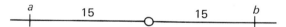

construct a data set similar to Table 5–6 that is consistent with this linkage arrangement.

21. The following diagram shows three man/mouse hybrid clones (A, B, and C) with the human chromosomes they contain.

	Human Chromosomes Present							
	1	2	3	4	5	6	7	8
A	+	+	+	+	−	−	−	−
B	+	+	−	−	+	+	−	−
C	+	−	+	−	+	−	+	−

Five enzymes, a, b, c, d, and e, were tested in each of the clones and the results were as follows:

Enzyme a   Activity only in clone B
      b   Activity in all three clones
      c   Activity only in clones A and B

d   Activity only in clone C

e   No activity in any clone

Where are the loci responsible for these enzymes located? Are these synteny or assignment tests?

22. A man with X-linked color blindness and X-linked glucose-6-phosphate-dehydrogenase deficiency (G6PD⁻) marries a normal woman and has a normal daughter who marries a normal man and produces 10 sons (as well as 8 normal daughters). Of the sons, 5 were normal, 3 were like their grandfather, one was only color blind and one was only G6PD⁻. From these data what can you say about the two X-linked loci?

SUGGESTIONS FOR FURTHER READING

Creighton, H. S., and B. McClintock. 1931. A correlation of cytological and genetical crossing over in *Zea mays. Proc. Nat. Acad. Sci.* 17:492–497.

Crow, J. F. 1976. *Genetics Notes,* 7th ed. Minneapolis, Minn.: Burgess Publishing.

Fincham. J. R. S., and P. R. Day. 1971. *Fungal Genetics,* 3rd ed. Oxford: Blackwell.

Haldane, J. B. S. 1919. The combination of linkage values and the calculation of distance between loci of linked factors. *J. Genet.* 8:299–309.

Kosambi, D. 1943. The estimation of map distances from recombination values. *Ann. Eugen.* 12:172–175.

Lindsley, D., and E. Grell. 1968. *Genetic Variations of* Drosophila melanogaster. Washington, D.C.: Carnegie Institution.

McKusick, V. 1975. *Mendelian Inheritance in Man,* 4th ed. Baltimore, Md.: Johns Hopkins University Press.

———, and F. Ruddle. 1977. The status of the gene map of the human chromosomes. *Science* 196:390–405.

Morgan, T. H. 1919. *The Physical Basis of Heredity.* Philadelphia: Lippincott.

Plough, H. 1917. The effect of temperature on crossing over in *Drosophila. J. Exp. Zool.* 24:147–209.

Stern, C. 1931. Zytologisch-genetische Untersuchungen als Beweise für die Morgansche Theorie des Faktorenaustauchs. *Biol. Zentralbl.* 51:547–587.

———, 1936. Somatic crossing over and segregation in *Drosophila melanogaster. Genetics* 21:625–730.

6 Linkage and Mapping in Prokaryotes and Viruses

All organisms, including the viruses, have genes located sequentially in the genetic material; and all, with the possible exception of a small group of viruses, can have recombination between homologous pieces of genetic material. Because such recombination does occur, it is possible to map the location and sequence of genes along the chromosomes of all organisms and viruses. The unique properties of the life cycles of bacteria and viruses require special mapping techniques. We will dwell on these techniques in this chapter because of the enormous importance bacteria and viruses have assumed in genetics in the past three decades. It is through work with bacteria and viruses that we have entered the modern era of molecular genetics, the subject of the second section of this book.

Bacteria, along with blue-green algae, make up the prokaryotes. The true bacteria can be classified according to shape: A spherical bacterium is called a **coccus;** a rod-shaped bacterium is called a **bacillus;** and a spiral bacterium is called a **spirillum** (Figure 6–1). Prokaryotes do not undergo mitosis or meiosis but simply divide in half after their chromosome, a circle of DNA, has replicated. Viruses do not even divide; they are mass-produced within the host cell. Several properties of bacteria and viruses have made them especially suitable for molecular genetic research.

BACTERIA AND VIRUSES IN GENETIC RESEARCH

First, bacteria and viruses have a very short generation time. One virus can become 100 in about an hour; an *E. coli* cell (*Escherichia coli,* the common intestinal bacterium, discovered by Theodor Escherich in 1885) doubles every 20 minutes. In contrast, there is a generation time of 14 days in fruit flies, one year in corn, and 25 years in humans.

Second, bacteria and viruses have much less genetic material than do

169

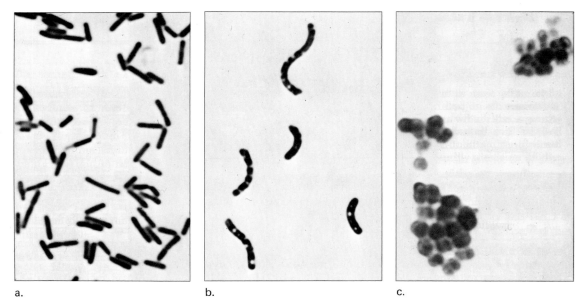

a.   b.   c.

Figure 6-1. The Three Typical Bacterial Forms
(a) Rods; (b) Spirals; (c) Spheres. The average bacterium is 1 to 2 μm long.

Source: © Carolina Biological Supply Company. Reproduced by permission.

eukaryotes, and the organization of this material is much simpler. The term *prokaryote* arises from the fact that these organisms do not have true nuclei (*pro* means before and *karyon* means kernel or nucleus); they have no nuclear membranes (Figure 2–2) and only a single "naked" chromosome. Viruses are even simpler. They consist almost entirely of genetic material surrounded by a protein coat. Or, more precisely, the bacterial viruses in which we are interested, the **bacteriophages**—or just **phages** (phage = one who eats)—are exclusively genetic material surrounded by a protein coat (Figure 6–2). Some animal viruses are more complex. In this chapter our viral interests will focus exclusively on phage.

Viruses can be classified by their host preference (animal, plant, or bacteria) and by the nature of their genetic material (RNA *or* DNA; the details of genetic material will be discussed in Chapter 9). Figure 6–3 shows some examples of animal and plant viruses. Viruses are obligate parasites. Outside of a host, they are inert molecules. Once their genetic material penetrates a host, they take over the metabolism of that host and construct multiple copies of themselves. We will examine this in detail later.

A *third* reason for the use of bacteria and viruses in genetic study is their ease of handling. Millions and millions of bacteria can be handled in a single culture with a minimal amount of work compared with the effort required to grow the same number of eukaryotic organisms such as fruit flies

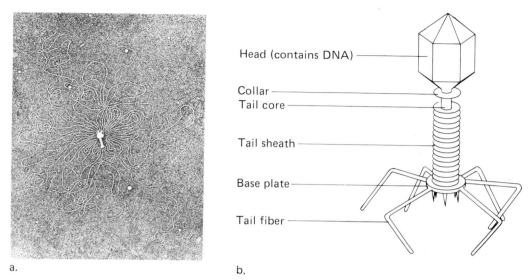

a.   b.

Figure 6-2. Phage T2 and Its Chromosome
(a) The chromosome, which is about 50 μm long, has burst from the head.
(b) The intact phage

Source: A. K. Kleinschmidt et al., "Darstellung und Längenmessungen des gesamten Desoxyribosenucleinsäure-Inhaltes von T2-Bacteriophagen," *Biochimica et Biophysica Acta* 61 (1962):857–864. Reproduced by permission.

or corn. (Some eukaryotes, such as yeast or *Neurospora,* can, of course, be handled using prokaryotic techniques. See Chapter 5.) The following discussion is an expansion of the techniques that were introduced in Chapter 5 and that are used in bacterial and viral studies.

TECHNIQUES OF CULTIVATION

Since different groups of bacteria have different nutritional requirements, different media have been developed on which they are grown in the laboratory. All organisms need an energy source, a carbon source, nitrogen, sulphur, phosphorus, several metallic ions, and water. Those that require an organic form of carbon are termed **heterotrophs.** Those that can utilize carbon as carbon dioxide are termed **autotrophs.** All bacteria obtain their energy either by photosynthesis or chemical oxidation. Bacteria are usually grown in or on a chemically defined or **synthetic medium** either in liquid or in test tubes or petri plates using an agar base to supply rigidity. Petri plates are circular dishes of glass or plastic about 2 cm deep with matching circular covers. When one cell is placed on the medium in the plate, it will begin to divide. After incubation, often overnight, a colony, or clone, will

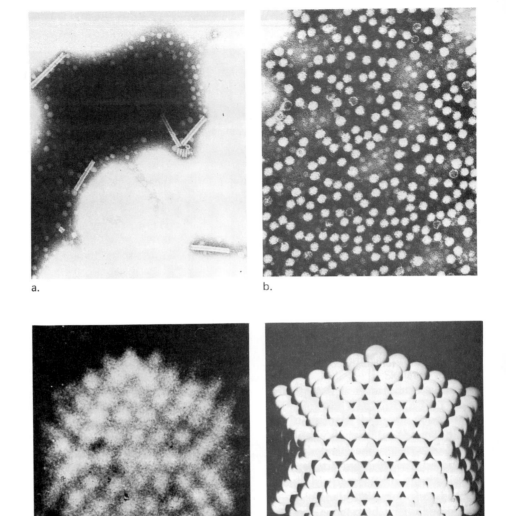

a.

b.

c.

d.

Figure 6-3. Plant and Animal Viruses
(a) Tobacco rattle virus. magnification 178,600X. (b) Turnip yellow mosaic virus particles. magnification 427,500X. (c) Electron micrograph of an adenovirus particle, a DNA animal virus. (d) Ping-pong ball model of an adenovirus particle. Each capsomere (spherical subunit) is about 50 Å in diameter.

Source: (a) and (b) are reproduced courtesy of Dr. T. C. Allen, Jr.; (c) and (d) are reproduced with permission from R. W. Horne et al., "The Icosahedral Form of an Adenovirus," *Journal of Molecular Biology* 1 (1959):84–86. Copyright by Academic Press Inc. (London) Ltd.

exist where there was previously only one cell. Overlapping colonies form a solid lawn of growth (Figure 6–4). A culture medium that has only the bare minimal necessities required by the bacterial species being grown is referred to as a **minimal medium.** Table 6–1 shows a minimal medium for growing *E. coli.*

Alternatively, we could grow bacteria on a medium that supplies not only the minimal requirements but also all the requirements of a bacterium, including amino acids, vitamins, and so on. A medium of this kind will allow the growth of strains of bacteria that have specific nutritional requirements. (These strains are **auxotrophs** as opposed to the parent or wild types, which are **prototrophs.**) For example, a strain that has an enzyme defect in the pathway of the production of the amino acid histidine will not grow on a minimal medium because it has no way of obtaining histidine. If, however, histidine were provided in the medium, the organisms could grow. This his-tidine-requiring auxotrophic mutant could thus grow on an **enriched,** or **complete, medium,** whereas the parent prototroph could grow on a mini-mal medium. Media are often enriched by adding complex mixtures of organic substances such as blood; beef extract; yeast extract; or peptone, a digestion product. Many media are made up of a minimal medium with the addition of only one other substance, such as an amino acid or a vitamin. These are called **selective media;** their uses will be discussed later in the chapter. In addition to minimal, complete, and selective media, other media exist for purposes such as aiding in counting colonies, helping maintain cells in a nongrowth phase, and so on.

The experimental cultivation of viruses is somewhat different. Since viruses are obligate parasites, they can only grow in living cells. Thus, for the cultivation of phages, petri plates of appropriate media are inoculated with enough bacteria to form a continuous cover, or **bacterial lawn.** This bac-terial culture serves as a medium for the growth of viruses added to the plate.

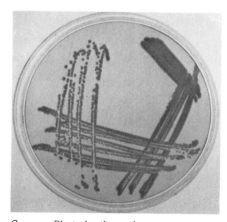

Figure 6–4.
Bacterial Colonies
on a Petri Plate

Source: Photo by the author.

TABLE 6–1. Minimal Synthetic Medium for Growing *E. coli,* a Heterotroph

Component	Quantity
$NH_4H_2PO_4$	1 g
Glucose	5 g
NaCl	5 g
$MgSO_4 \cdot 7H_2O$	0.2 g
K_2HPO_4	1 g
H_2O	1000 ml

Source: Data from M. Rogosa et al., *J. Bac-teriol.* 54 (1947): 13.

THE INFLUENZA VIRION

The cause of animal influenza is a *virion* (virus particle) of extreme complexity compared to phage. The influenza virion is about 100 nanometers in diameter and covered with spikes of two types. The H spikes, so called because of their hemagglutinin ability (cause red cells to clump), allow the virions to attach to the host cells. The other type of spike is an N spike, so called because it is the enzyme neuraminidase. Presumably this enzyme allows the virions to get out of the host cells. Immunity, primarily to the H spikes, protects a person from being reinfected by the same strain of influenza.

The H and N spikes are embedded in a lipid bilayer which surrounds a protein matrix. Within the virion are eight segments of single-stranded RNA. Each segment is capable of directing the synthesis of one of the virion's proteins.

The exact sequence of events during an infection is not precisely known; however, much is understood, especially regarding the complex structure of the virion.

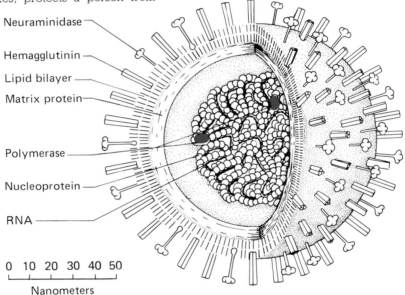

Neuraminidase

Hemagglutinin

Lipid bilayer

Matrix protein

Polymerase

Nucleoprotein

RNA

0 10 20 30 40 50
Nanometers

Figure 1.
Structure of the
Influenza Virion

Source: From Martin M. Kaplan and Robert G. Webster, "The Epidemiology of Influenza." Copyright © 1977 by Scientific American, Inc. All rights reserved. Reprinted by permission.

Since the virus attack eventually results in rupture, or **lysis,** of the bacterial cell, addition of the virus produces clear spots, known as **plaques,** on the petri plates (Figure 6–5). Different types of bacteria can be used to determine growth potentials of the various viral strains under study.

174

Figure 6-5.
Viral Plaques (λ) on
a Bacterial Lawn (*E. coli*)

Source: Photo by author.

BACTERIAL PHENOTYPES

Bacterial phenotypes fall into three general classes: colony morphology, nutritional requirements, and drug and infection resistance.

Colony Morphology

The first of these classes, colony morphology, relates simply to the form, color, and size of the colony that grows from a single cell. A bacterial cell growing on an agar slant or petri plate divides as frequently as once every 20 minutes. Thus, the number of cells doubles every 20 minutes. Each original cell will give rise to a colony, or **clone,** at the site of its original position. In a relatively short amount of time (overnight, for example), the colonies will consist of enough cells to be seen with the unaided eye. The different morphologies observed among the colonies are under genetic control (Figure 6–6).

Nutritional Requirements

The second basis for classifying bacteria—nutritional requirements—reflects the failure of one or more enzymes in the biochemical pathways of

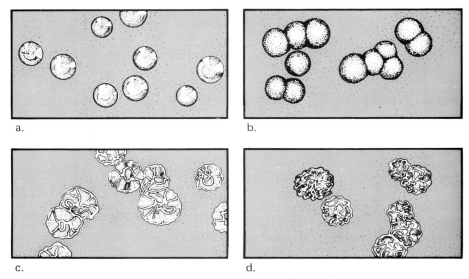

a.   b.

c.   d.

Figure 6-6. Various Bacterial Colony Forms on Petri Plates
(a) Smooth, circular, raised surface (b) Granular, circular raised surface
(c) Elevated folds on a flat colony with irregular edges (d) Irregular
elevations on a raised colony with an undulating edge

the bacteria. If an auxotroph has a requirement for the amino acid cysteine
that the parent strain (prototroph) does not have, then that auxotroph most
likely has a nonfunctional enzyme in the pathway for the synthesis of cysteine. Figure 6–7 shows five steps in cysteine synthesis; it also shows that
each step is controlled by a different enzyme. All enzymes are proteins, and
the sequence in the strings of amino acids that make up those proteins are
determined by information in one or more genes (Chapter 11). A normal or
wild-type allele controls a normal, functional enzyme. The alternate allele
can control a nonfunctional enzyme. The *one-gene-one-enzyme* rule,
although not strictly correct (see Chapter 15), is a useful rule of thumb at
this point.

Screening Techniques. Replica plating, a technique devised by Lederberg and Lederberg, is a rapid **screening technique** that makes it possible to quickly determine if a given strain of bacteria is auxotrophic for a
particular metabolite. In this technique a petri plate of complete medium is
inoculated with bacteria. The resulting growth will have a certain configuration of colonies. This plate of colonies is pressed onto a piece of sterilized
velvet. Then, any number of petri plates, each containing a medium that
lacks some specific metabolite, can be pressed onto this velvet to pick up
inocula in the same pattern as the growth on the original plate (Figure 6–8).
If a colony grows on the complete medium but does not grow on a plate with

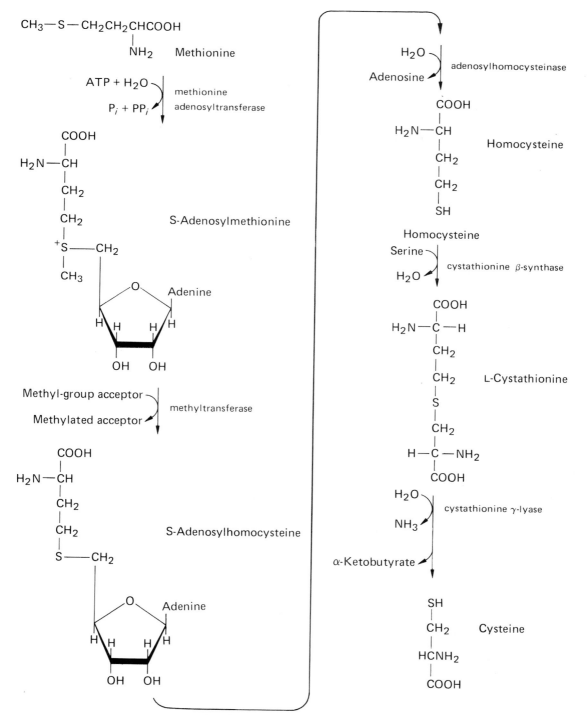

Figure 6-7. Five-Step Conversion of Methionine to Cysteine
Each of the five steps is controlled by a different enzyme.

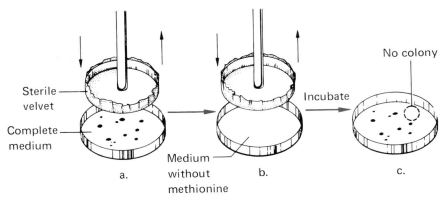

Figure 6-8.   Technique of Replica Plating
(a) A pattern of colonies from a plate of complete medium is transferred
(b) to a second plate of medium that lacks methionine. (c) Where colonies
fail to grow on the second plate, we can infer that the original colony in
that location was a methionine-requiring auxotroph.

a medium in which a metabolite is missing, the inference is that the colony
growing in that location on the complete medium is an auxotroph that
requires the metabolite absent from the second plate. This bacterial strain
can be isolated from the complete medium for further study. Its nutritional
requirement is its phenotype. The methionine-requiring auxotroph of Figure
6–8 would be designated as Met⁻ (methionine-minus or Met-minus).

Resistance and Sensitivity

The third class of phenotypes in bacteria involves resistance and sensitivity
to drugs, phages, and other environmental insults. For example, penicillin,
an antibiotic that prevents the final stage of cell-wall construction in bacteria, will kill growing bacterial cells. Nevertheless, we frequently find a small
number of cells that do grow in the presence of penicillin. These colonies are
resistant to the drug. This resistance is under simple genetic control. The
phenotype is penicillin resistant or Penʳ as compared to penicillin sensitive
(Penˢ), the normal condition. Many phenotypes of bacteria are resistances
and sensitivities to various antibacterial agents (Table 6–2).

Screening Techniques.   Drug sensitivity provides another rapid screening technique for isolating nutritional mutations. For example, if we were
looking for mutants that lacked the ability to synthesize a particular amino
acid (e.g., methionine), we could grow large quantities of bacteria (prototrophs) and then place them on a medium that lacked methionine but had

TABLE 6-2. Some Antibiotics and Their Antibacterial Mechanisms

Antibiotic	Microbial Origin	Mode of Action
Penicillin G	*Penicillium chrysogenum*	Blocks cell-wall synthesis
Tetracycline	*Streptomyces aureofaciens*	Blocks protein synthesis
Streptomycin	*Streptomyces griseus*	Interferes with protein synthesis
Terramycin	*Streptomyces rimosus*	Blocks protein synthesis
Erythromycin	*Streptomyces erythraeus*	Blocks protein synthesis
Bacitracin	*Bacillus subtilis*	Blocks cell-wall synthesis

penicillin. Here, any *growing* cells would be killed. But methionine auxotrophs would not grow and, therefore, they would not be killed. The penicillin could then be washed out and the cells reinoculated onto a complete medium. The only colonies that would result should be methionine auxotrophs (Met⁻).

Screening for resistance to phage is similar to screening for drug resistance. When bacteria are placed in a medium containing phages, only those bacteria that are resistant to the phage will grow and produce colonies. They can thus be easily isolated and studied.

Bacteriophages

In regard to viral phenotypes, we will consider only bacteriophage phenotypes, which fall roughly into two categories: plaque morphology and growth characteristics on different bacterial strains. For example, T2, an *E. coli* phage, produces a characteristic plaque (genotype r^+). Rapid-lysis mutants (genotype r) produce large, smooth-edged plaques (Figure 6–9). Similarly, T4 rapid-lysis mutants produce large, smooth-edged plaques on *E. coli* B but will not grow at all on *E. coli* K12, a different strain. Rapid-lysis mutants illustrate both colony morphology and strain growth restrictions as phenotypes of phages.

SEXUAL PROCESSES IN BACTERIA AND VIRUSES

While bacteria and viruses are ideal subjects for biochemical analysis, they would be useless for genetic study if they did not undergo sexual processes.

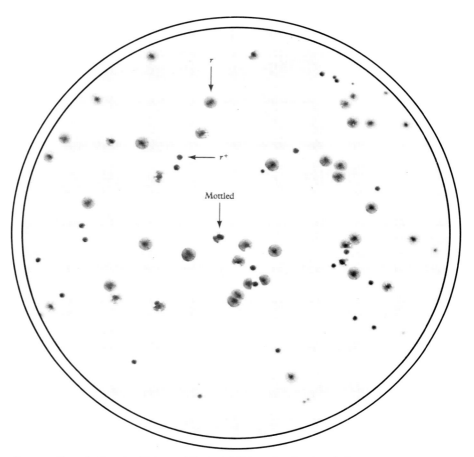

Figure 6–9.
Normal (r^+) and
Rapid-Lysis (r)
Mutants of Phage T2
Mottled plaques
occur when r and r^+
phages grow
together.

Source: From *Molecular Biology of Bacterial Viruses* by Gunther S. Stent.
W. H. Freeman and Company. Copyright © 1963. Reproduced by
permission.

If we consider the purpose of the sexual processes of the eukaryotes to be
the combining of genetic material between individuals, then the life cycles
of bacteria and viruses can be said to include sexual processes. Although they
do not undergo sexual reproduction by means of the fusion of haploid
gametes, bacteria and viruses do undergo processes in which genetic material
from one cell can be incorporated into another cell and recombinants are
formed. Actually, bacteria use four different methods to gain access to foreign genetic material: **transformation, conjugation, transduction,** and
sexduction. Phages can exchange genetic material when a bacterium is
infected by more than one virus particle **(virion).** During the process of viral
infection, the genetic material of the two different phage strains can

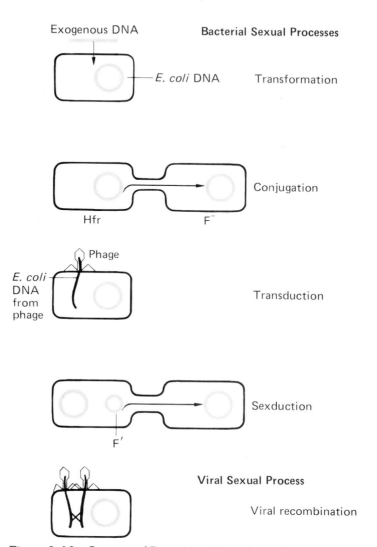

Figure 6-10. Summary of Bacterial and Viral Sexual Processes

exchange parts (recombine) (Figure 6–10). We will examine the exchange processes in bacteria and then in bacteriophages and proceed to the use of these methods for mapping bacterial and viral chromosomes. (Chromosome refers to the structural entity in a cell or virus made up of the genetic material. In viruses and prokaryotes it is pure genetic material—circular double-stranded DNA in prokaryotes, and any combination of linear or circular, single- or double-stranded, RNA or DNA, in viruses.)

Oswald T. Avery
(1877–1955)
Courtesy of the
National Academy of
Sciences

Transformation

Transformation was first observed in 1928 by Griffith and later (1944) examined at the molecular level by Avery and his colleagues, who demonstrated that DNA was the genetic material. The details of these experiments are presented in Chapter 9. In transformation a cell takes up extraneous DNA found in the environment and incorporates it into its **genome** (genetic complement) through recombination. Not all cells are competent to be transformed, and not all extracellular DNA is competent to transform. To be competent to transform, the extracellular DNA must be double stranded and relatively large. To be competent to be transformed, a cell must have the surface protein **competence factor,** which binds the extracellular DNA in an energy-requiring reaction.

As the extracellular DNA is brought into the cell, one of the strands is hydrolyzed by an intracellular DNAase (DNA-degrading enzyme) that apparently uses the energy of hydrolysis to pull the remaining strand into the cell. This single strand brought into the cell can then be incorporated into the host genome by crossing over (Figure 6–11). The molecular mechanisms of crossing over will be discussed in Chapter 14.

Transformation is a very efficient method of mapping in some bacteria, especially those that are inefficient in other mechanisms of DNA intake (such as transduction). For example, a good deal of the mapping in *Bacillus subtilis* has been done through the process of transformation. *Escherichia coli,* however, is inefficient in transformation while possessing very efficient transducing phages (see discussion of transduction mapping later in this chapter). For this reason, very little mapping in *E. coli* has been done through transformation.

Transformation Mapping. The general procedure for transformation mapping is as follows. Two strains of bacteria are selected such that one strain is mutant at two loci. For example, strain B might have wild-type alleles for histidine (*his*$^+$) and methionine synthesis (*met*$^+$). Strain A should then be auxotrophic for both of these amino acids (*his*$^-$ and *met*$^-$). The DNA is isolated from strain A and put in the culture medium with strain B.

Exogenous genetic material

Figure 6–11.
Exogenous Linear
Genetic Material
Can Be Incorporated
into a Circular
Bacterial
Chromosome by Two
Crossovers

Bacterial chromosome

After a time interval for transformation to take place, the B strain is tested for its methionine and histidine properties (Figure 6–12).

Since we are interested in transformants (auxotrophs), the nontransformed cells can be eliminated by culturing on minimal medium with penicillin. The transformant auxotrophs will not grow and hence will not be killed. The nontransformed prototrophs (his^+ met^+), however, will be killed. After the penicillin is washed away, the remaining cells can be plated on complete medium; and, by replica plating onto media lacking histidine and methionine, it is possible to determine the phenotype of each transformed cell. If several of these experiments are run in parallel with various concentrations of DNA, we would expect the ratios of single and double transformants to be different depending on whether or not the two genes are close to each other.

If the genes are close to each other (closely linked), we would expect double transformations and single transformations to decline at the same rate as the concentration of transforming DNA decreases. However, if the two genes are far apart on the bacterial chromosome, we would expect a different picture. A double transformant will come about only by the simultaneous occurrence of two separate transformation events because the loci are too far apart to be incorporated in one piece of transforming DNA. Thus, since the probability of two separate transformation events in the same cell will drop drastically as the concentration of the transforming DNA decreases, we expect a very rapid drop in the occurrence of the double transformant. These dynamics are shown in Figure 6–13.

Relative map distances between loci can be obtained by comparing the

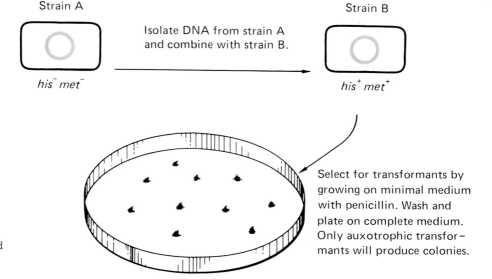

Figure 6–12.
Procedure for
Isolating Bacterial
Transformants
The two strains of
bacteria used differ
in their histidine and
methionine
requirements.

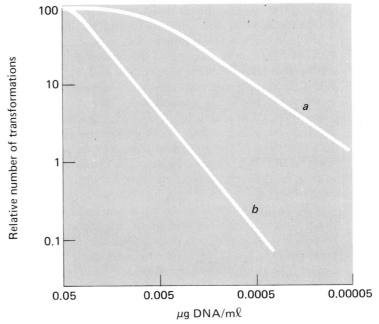

Figure 6-13.
Diagram of
Transforming
Dynamics
Single transformants
(a) and double
transformants far
apart (b).

Source: S. H. Goodgal, "Studies on Transformation of *Hemophilus influenzae*. IV Linked and Unlinked Transformations," *Journal of General Physiology* 45 (1961):211, fig. 3. Reprinted by permission.

number of single transformants (those with a crossover between the two loci) with the total number of transformants. (Double transformants presumably do not have a crossover between the loci.) For example, given the following numbers obtained in the experiment of Figure 6–12:

34 $his^- met^+$ transformants

28 $his^+ met^-$ transformants

194 $his^- met^-$ transformants

Then, relative recombination frequency would be

$$\frac{\text{number of single transformants}}{\text{total number of transformants}} = \frac{34 + 28}{34 + 28 + 194} = 0.24$$

By systematically examining many loci, relative order can be obtained. For example, if A is closely linked to B and B to C, we can establish the order $A–B–C$. As in most bacterial analysis, it is not possible by this method to determine exact order for very closely linked genes. For this information we need to rely upon transduction, which we will consider shortly. However, transformation has allowed us to determine that the map of *Bacillus subtilis* is circular, a phenomenon found in all prokaryotes and many phages.

Conjugation

In 1946 Joshua Lederberg and E. L. Tatum (later to be Nobel laureates) discovered that *Escherichia coli* cells can exchange genetic material through the process of conjugation. They mixed two auxotrophic strains of *E. coli*. One strain was *met⁻ bio⁻* (methionine and biotin requiring), and the other was *thr⁻ leu⁻* (threonine and leucine requiring). This cross is shown in Figure 6–14. Remember that if a strain is *met⁻ bio⁻*, it is, without saying, wild type for all other loci. Thus, *met⁻ bio⁻* is actually *met⁻ bio⁻ thr⁺ leu⁺*. Likewise, the *thr⁻ leu⁻* strain is actually *met⁺ bio⁺ thr⁻ leu⁻*.

Lederberg and Tatum used multiple auxotrophs in order to rule out spontaneous reversion (mutation). About one in 10^6 *met⁻* cells will spontaneously become prototrophic (*met⁺*) every generation. However, with multiple auxotrophs the probability of a spontaneous reversion (e.g., *met⁻* → *met⁺*) of several loci simultaneously becomes vanishingly small. (In fact, the control plates of Figure 6–14 showed no growth for the double mutants used

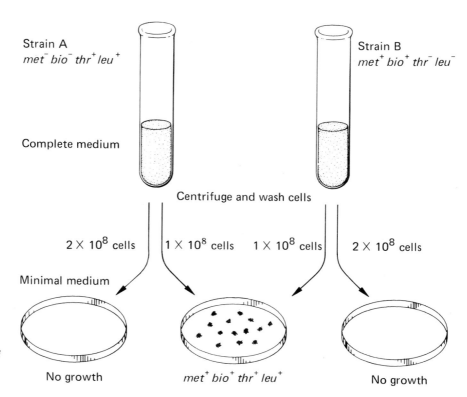

Figure 6-14. Lederberg and Tatum's Cross Showing That *E. coli* Undergoes Genetic Recombination

Strain A
met⁻ bio⁻ thr⁺ leu⁺

Strain B
met⁺ bio⁺ thr⁻ leu⁻

Complete medium

Centrifuge and wash cells

2×10^8 cells 1×10^8 cells 1×10^8 cells 2×10^8 cells

Minimal medium

No growth

met⁺ bio⁺ thr⁺ leu⁺

No growth

as parents.) After mixing the strains, Lederberg and Tatum found that about one cell in 10^7 was prototrophic (met^+ bio^+ thr^+ leu^+). Transformation was ruled out as an explanation by conducting several types of experiments showing that in this case direct cell-to-cell contact was required for genetic exchange recombination. The best known of these was Davis's U-tube experiment, in which one strain was put in each arm of a U-tube at the bottom of which was a sintered glass filter (Figure 6–15). The liquid and large molecules, including DNA, were mixed by alternate application of pressure and suction to one arm of the tube; whole cells did not pass through the filter. The result of this mixture was that the fluids surrounding the cells, as well as any large molecules (e.g., DNA), could be freely mixed while the cells were kept separate. After cell growth stopped in the two arms (in complete medium), the contents were plated out on minimal medium. There were no prototrophs in either arm.

At first, Lederberg and Tatum interpreted their results in light of conventional sexual process where two cells fuse forming a diploid zygote that then undergoes meiosis. Then this conventional view of bacterial sexuality was shown to be incorrect. In bacteria, conjugation is a one-way transfer with one strain acting as a donor and another as a recipient. This was demonstrated by experiments where one or the other strain was killed just prior to mixing. When one strain was killed, the experiment failed (there was no recombination). When the other strain was killed, the experiment still worked (recombination occurred). It was later shown that the strain that must be alive contributes most of the other alleles to the recombinant offspring. Thus, since one cell acts as a donor of some of its genetic material

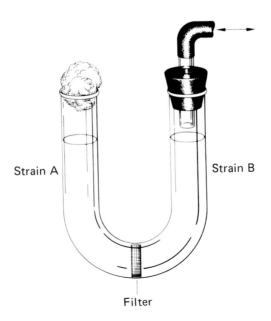

Strain A   Strain B

Figure 6–15.
The U-Tube
Experiment of Davis

Filter

while the other is the recipient, the donor can be killed prior to the process and will still transfer some of its genetic material. The recipient cannot be killed if recombination is to occur.

F Factor. It was shown that sometimes, if stored for a long time, donor cells can lose the ability to be donor cells but can regain the donor ability if they are mated with other donor strains. This led to the hypothesis that a **fertility factor,** F, made any strain that carried it a male (donor) strain, termed F^+. The strain that did not have the F factor, referred to as a female (or F^-) strain, served as recipient for genetic material during conjugation.

The F factor is a plasmid or episome. Jacob and Wollman's term, **episome,** refers to a genetic particle that not only can exist independently in the cytoplasm of the cell but also can become integrated into the host's chromosome. The term, **plasmid,** coined by Lederberg, is a more general term for independent particles that may or may not have the ability to become integrated into the chromosome. Plasmids may exist in multiple copies of 2 to 50 per cell. Other examples of plasmids include the **resistance transfer factors** (R factors), which confer on the host cell a simultaneous resistance to several different antibiotics, such as streptomycin, tetracycline, and ampicillin. **Colicinogenic factors** are plasmids that are responsible for producing antibiotic substances used by one strain of bacteria to kill other strains. Plasmids are at the heart of recombinant DNA work, which will be discussed in detail in later chapters, especially in Chapter 12.

The F factor has been conclusively shown to be composed of DNA (just like the *E. coli* chromosome and about 2% of its size) by introducing it into a different bacterium, *Serratia marcescens*. The DNA of *Serratia* was then isolated and the presence of a new piece in the DNA of the F^+ cells showed that the F factor was in fact DNA. The F factor plasmid contains at least 30 loci of which a minimum of 13 are responsible for the formation of F-pili. About two thirds of the F factor is unmapped.

It was found that transfer of the F factor occurred far more frequently than did transfer of other genetic material. That is, during conjugation, there was about one recombinant in 10^7 cells, while transfer of the F factor occurred at a rate of about one conversion of F^- to F^+ in every five conjugations. An *E. coli* strain was then discovered that transferred its genetic material at a rate about 1000 times that of the normal F^+ strain. This strain was called **Hfr,** for high frequency of recombination. Several other phenomena occurred simultaneously with this high rate of transfer. First, the ability to transfer the F factor itself dropped to almost zero in this strain. Second, not all loci were transferred at the same rate. Some loci were transferred much more frequently than others.

Escherichia coli cells are normally coated with hair-like pili (**fimbriae**). The F^+ and Hfr cells have several additional pili called **F-pili,** or sex pili. These can be seen under the electron microscope (Figure 6–16). During conjugation these pili form a connecting bridge between the F^+ (or Hfr) and F^- cells (Figure 6–17). It has not been absolutely established that the genetic

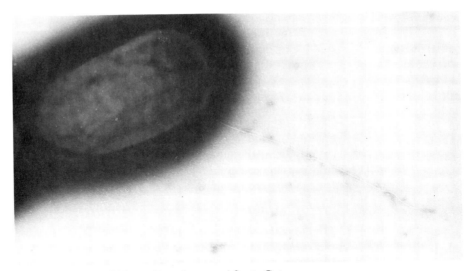

Figure 6–16.
Electron Micrograph
of an *E. coli* Cell
with a Sex Pilus
magnification
20,000X

Source: Courtesy of Wayne Rosenkrans and Sonia Guterman.

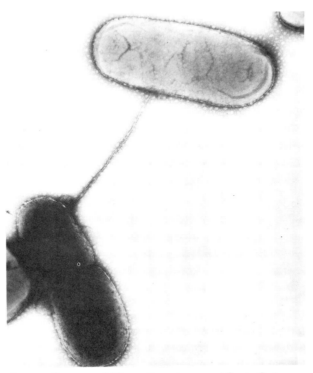

Figure 6–17.
Electron Micrograph
of Conjugation
between an F⁺
(upper right) and F⁻
(lower left) Cell with
the F-Pilus between
Them
magnification
9,000X

Source: Courtesy of Wayne Rosenkrans and Sonia Guterman.

material flows through the F-pili; they may merely act as anchors, securing the two conjugating cells to each other, with another bridge forming for recombination.

In the transfer process of conjugation, the donor cell does not lose its F factor or its chromosome because the material transferred is a copy of the donor's genetic material. Presumably a locus on the F factor specifies the initiation of DNA replication. A replicate then enters the F⁻ cell. (The process of DNA replication is described in Chapter 9.) For a short while the F⁻ cell has two copies of whatever loci were transferred. Having these two copies, the cell is a partial diploid, or a **merozygote.** The new chromosomal DNA (**exogenote**) can be incorporated into the host chromosome (**endogenote**) by an even number of breakages and reunions between itself and the host chromosome. The unincorporated linear DNA is soon degraded by enzymes. This process is diagrammed in Figure 6–18.

Elie Wollman
(1917–)
Photo: J. Mainbourg

Interrupted Mating.   To demonstrate that the transfer of genetic material from the donor to the recipient cell during conjugation was a linear event, Jacob and Wollman devised the technique of **interrupted mating.** In this technique an F⁻ and an Hfr strain are mixed together in a food blender. After a specific amount of time, the blender is turned on. The spinning separates cells that were conjugating and thereby interrupts mating. Then, the F⁻ cells are tested for various alleles originally in the Hfr cell. In an experiment like this, the Hfr strain is usually sensitive to an antibiotic such as streptomycin. After interrupted mating the cells are plated on complete medium with streptomycin. This will kill all the Hfr cells. Then the F⁻ cells can be tested by replica plating for their genotypes.

Figure 6–19 shows the results of the following mating with this procedure. In the food blender an Hfr strain sensitive to streptomycin *(str^s)* but resistant to azide *(azi^r)*, resistant to T1 *(tonA^r)*, and prototrophic for leucine

Exogenous genetic material   Degraded

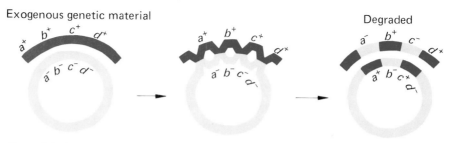

Bacterial chromosome

Figure 6–18.   Incorporation of External DNA into the Host Chromosome
This incorporation is accomplished only after an even number of breakage and reunion events occur after the exogenate lines up (synapses) with the identical (homologous) region of the host chromosome.

(leu^+), galactose (gal_b^+), and lactose (lac^+) was added to an F$^-$ strain that was resistant to streptomycin (str^r), sensitive to azide (azi^s), sensitive to T1 ($tonA^s$), and auxotrophic for leucine, galactose, and lactose (leu^-, gal_b^-, and lac^-). After a specific number of minutes (ranging from zero to 60), the food blender was turned on. The cell suspension was plated on streptomycin to kill all the Hfr cells. The cells remaining were then plated on a medium without leucine. The only colonies that resulted were F$^-$ recombinants. They must have received the leu^+ allele from the Hfr in order to grow on a medium lacking leucine. Hence all colonies had been selected to be F$^-$ recombinants. Then, by replica plating onto specific media, the azi, $tonA$, lac, and gal_b alleles were determined and the percentage of recombinant colonies that had the original Hfr allele was noted.

The graph of Figure 6–19 shows that as time of mating increases, two things happen. First, new alleles enter the F$^-$ cells from the Hfr cells. The $tonA^r$ allele first appears among recombinants after 10 minutes of mating, while gal_b^+ first enters the F$^-$ cells after about 25 minutes. This suggests a sequential entry of loci into the F$^-$ cells from the Hfr (see Figure 6–20). Secondly, as time proceeds, the percentage of recombinants with a given allele from the Hfr increases. At 10 minutes, $tonA^r$ is first found among recombinants. After 15 minutes, about 40% of recombinants have the $tonA^r$ allele

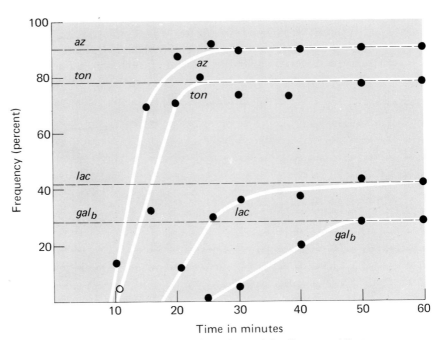

Figure 6-19. Frequency of Hfr Genetic Characters among Recombinants after Interrupted Mating

Source: F. Jacob and E. L. Wollman, *Sexuality and the Genetics of Bacteria* (New York: Academic Press, 1961). Reprinted by permission.

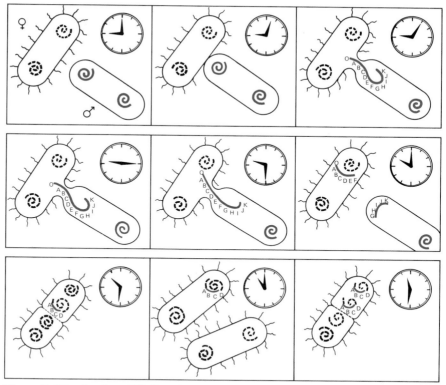

Figure 6-20.   Jacob and Wollman's Interpretation of Conjugation
Capital letters represent the position of various genes on the donor chromosome, with 0 representing the origin of transfer. The female is F⁻; the male is Hfr.

Source: F. Jacob and E. L. Wollman, *Sexuality and the Genetics of Bacteria* (New York: Academic Press, 1961). Reprinted by permission.

from the Hfr; and after about 25 minutes, about 80% of the recombinants have the *tonA*^{r} allele. This limiting percentage does not increase with additional time. It does, however, decrease with later entering loci, a fact explained by assuming that even without the food blender, mating is eventually interrupted before completion by normal agitation alone.

The scheme presented in Figure 6–20 was suggested by Jacob and Wollman shortly after the first tentative map of the *E. coli* chromosome was worked out. There are thus several aspects of the diagram that we now know are incorrect. The *E. coli* chromosome is circular, not linear as shown. Although actively growing cells are continually replicating their genetic material, there need only be one copy per cell, not the two shown. And, when genetic material is transferred across the conjugation bridge, it is one strand of the Hfr chromosome, not the whole chromosome as shown; one strand, replicated, is left behind in the Hfr cell.

Mapping and Conjugation.   Lederberg, using conjugation to map genes, had found that some genes did not fit a linear pattern. For example, three loci appeared to be linked (very closely associated) to *met* without being linked to each other. This observation prompted Lederberg to suggest a chromosome with three branches at the *met* locus (Figure 6–21). But the work of Jacob and Wollman indicated that the bacterial chromosome was linear. The breakthrough here occurred when they did interrupted matings with several different Hfr strains. These strains were all of independent origin. The results were quite striking (Table 6–3).

If we ponder this table for a short while, one fact will become obvious. The relative order of the loci is always the same. What differs is the point of origin and the direction of transfer. These findings led Jacob and Wollman to suggest that the *E. coli* chromosome was circular. This not only fit perfectly with their data, it also solved Lederberg's problem of a nonlinear map.

Jacob and Wollman then proposed that normally the F factor is an independent circular DNA entity in the F$^+$ cell and that during conjugation only the F factor is passed to the F$^-$ cell. Since it is a small fragment of DNA, it can be entirely passed in a high proportion of the conjugants prior to spontaneous separation of the cells. Every once in a while, however, the F factor becomes integrated into the chromosome of the host, which then becomes an Hfr cell. The point of integration can be different in different strains. However, once the F factor is integrated, it determines the initiation point of transfer of the *E. coli* chromosome, as well as the direction of transfer.

The F factor is the last part of the *E. coli* chromosome to be passed from the Hfr cell. This explains why an Hfr, in contrast to an F$^+$, rarely passes the F factor itself. Basically, the cells remain in contact for too short a period of time. In the original work of Lederberg and Tatum, the one recombinant in 10^7 cells was most likely from a conjugation between an F$^-$ cell and an F$^+$ cell, which then spontaneously became an Hfr. Integration of the F factor is diagrammed in Figure 6–22. The F factor can also reverse this

Figure 6-21. *E. coli* Chromosome Model Suggested by Lederberg et al. in 1951 to Explain Odd Linkage Arrangements

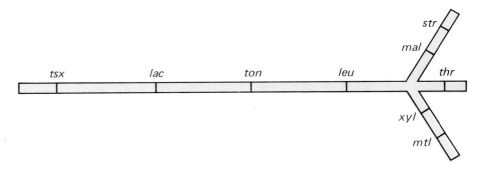

TABLE 6–3. Gene Order of Various Hfr Strains Determined by Means of Interrupted Mating

Types of Hfr	Hfr H	1	2	3	4	5	6	7	AB 311	AB 312	AB 313
0	0	0	0	0	0	0	0	0	0	0
	T	L	Pro	Ad	B_1	M	Isol	T_1	H	Sm	Mtl
	L	T	T_1	Lac	M	B_1	M	Az	Try	Mal	Xyl
	Az	B_1	Az	Pro	Isol	T	B_1	L	Gal	Xyl	Mal
	T_1	M	L	T_1	Mtl	L	T	T	Ad	Mtl	Sm
	Pro	Isol	T	Az	Xyl	Az	L	B_1	Lac	Isol	S-G
	Lac	Mtl	B_1	L	Mal	T_1	Az	M	Pro	M	H
	Ad	Xyl	M	T	Sm	Pro	T_1	Isol	T_1	B_1	Try
	Gal	Mal	Isol	B_1	S-G	Lac	Pro	Mtl	Az	T	Gal
	Try	Sm	Mtl	M	H	Ad	Lac	Xyl	L	L	Ad
	H	S-G	Xyl	Isol	Try	Gal	Ad	Mal	T	Az	Lac
	S-G	H	Mal	Mtl	Gal	Try	Gal	Sm	B_1	T_1	Pro
	Sm	Try	Sm	Xyl	Ad	H	Try	S-G	M	Pro	T_1
	Mal	Gal	S-G	Mal	Lac	S-G	H	H	Isol	Lac	Az
	Xyl	Ad	H	Sm	Pro	Sm	S-G	Try	Mtl	Ad	L
	Mtl	Lac	Try	S-G	T_1	Mal	Sm	Gal	Xyl	Gal	T
	Isol	Pro	Gal	H	Az	Xyl	Mal	Ad	Mal	Try	B_1
	M	T_1	Ad	Try	L	Mtl	Xyl	Lac	Sm	H	M
	B_1	Az	Lac	Gal	T	Isol	Mtl	Pro	S-G	S-G	Isol

*Order of Transfer of Genetic Characters**

*The O refers to origin of transfer.
Source: F. Jacob and E. L. Wollman, *Sexuality and the Genetics of Bacteria* (New York: Academic Press, 1961). Reprinted by permission.

process and loop out of the *E. coli* chromosome, a process we will examine in detail shortly (under sexduction).

We could now diagram the *E. coli* chromosome and show the map location of all the known loci. The map units would be in minutes, having been obtained by interrupted mating. However, at this point the map would not be complete. Interrupted mating is most accurate in giving the relative position of loci that are not very close to each other. With this method alone there would be a good deal of ambiguity as to the specific order of very close genes on the chromosome. The two remaining sexual processes in bacteria, sexduction and transduction, will provide the details unattainable by interrupted mating or transformation.

Sexduction (F-duction)

In Figure 6–22 we saw how the F factor can become integrated into the host genome. It leaves the host genome in the reverse process, one of excision, or

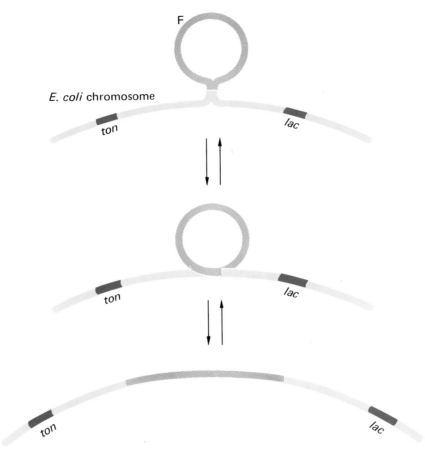

Figure 6–22. Integration of the F Factor

A simultaneous breakage in both the F factor and the *E. coli* chromosome is followed by a reunion of the two broken circles to make one large circle, the Hfr chromosome. In this case, integration is between *ton* and *lac*.

looping out. Occasionally, however, the process of looping out is not precise: The F factor takes with it some of the cell's genome (Figure 6–23). This new F factor is referred to as F′ and it endows the bacterial cells with certain interesting characteristics. First, F′ cells transfer their genes at a very high rate, even though they are not from an Hfr strain: This makes sense because we know that F⁺ cells transfer their F factor at a high rate during conjugation.

Second, the F′ factor has a much higher rate of spontaneous integration; and, unlike a normal F factor, it usually integrates at the same point that it

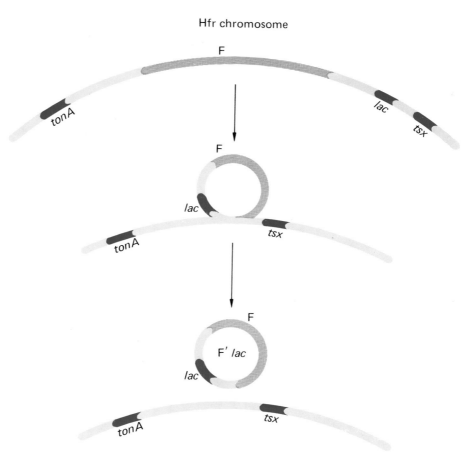

Figure 6-23. Occasional Imprecise Looping Out of the F Factor with Part of the Cell's Genome Included in the Loop

The circular F factor is freed by a single recombination (crossover) at the loop point.

originally occupied when in an Hfr cell. This too makes sense since a transferred F′ will have a region of homology with the new *E. coli* chromosome and will, therefore, tend to pair at that point. A single crossover will then reintegrate it at its original point. The first F′ discovered carried the *lac* locus (Figure 6–23). Since then, many F′ factors have been isolated, each with a different *E. coli* region incorporated (Figure 6–24). Thus, "co-sexduction" is an additional tool available for mapping. That is, two loci must be close to each other if they both are on the same F′ factor.

More will be said about the details of mapping in the section on transduction, the most common way of determining gene order of very closely linked genes. The same methods used for transduction mapping are used in

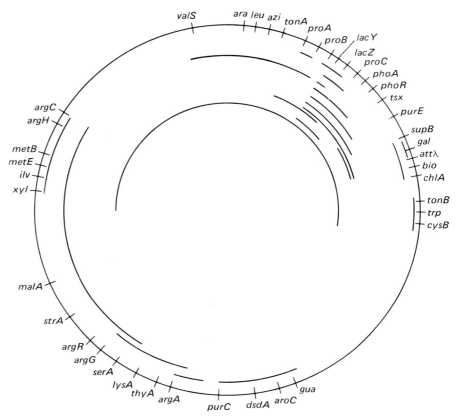

Figure 6-24.
The Circular
Chromosome of *E.
coli* with Arcs
Showing the Genes
Carried by Various
F′ Particles
The loci are
identified in Table
6-7.

Source: A. Campbell, *Episomes* (New York: Harper & Row, 1969), p. 65.
Reprinted by permission.

sexduction mapping and will not be repeated here. Sexduction has other uses besides mapping because partial diploids, or merozygotes, are formed. Their existence allows the study of the interaction of alleles in a normally haploid organism. Transduction, the final method of getting foreign DNA into a bacterial cell, is a phage-mediated process. Discussion of transduction must, therefore, be deferred until after some of the phage characteristics have been described.

LIFE CYCLES OF BACTERIOPHAGE

Phages are obligate intracellular parasites that adsorb to cell surfaces. Phage genetic material enters the bacterial cell by a variety of mechanisms. Once inside, the viral genetic material takes over the metabolism of the host cell.

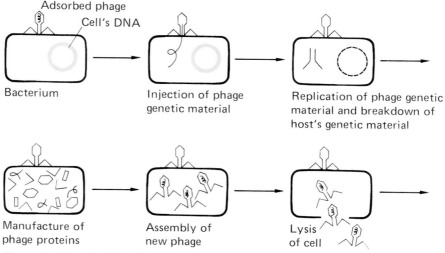

Figure 6-25. Viral Life Cycle Using T4 Infection of *E. coli* as an Example

During the infection process, the cell's own genetic material is destroyed while the genetic material of the virus is replicated many times. Viral genetic material controls the mass production of various protein components of the virus. New virus particles are then assembled, within the host cell, which bursts open—is lysed—releasing a **lysate** of upwards of several hundred viral particles to infect other bacteria. This life cycle is shown in Figure 6-25.

Recombination

The primary genetic work on phages has been done with a group of seven *E. coli* phages called the T series (T-odd: T1, T3, T5, and T7; T-even: T2, T4, and T6) and several others including phage λ (lambda) (Figure 6-26). The complex structure of T2 was shown in Figure 6-2. The details of its assembly process are given in Chapter 12. The phage can undergo recombination processes when a cell is infected with two virions that are genetically distinct. Hence, the phage genome can be mapped by recombination. As an example the host-range and rapid-lysis loci will be mapped here. Rapid-lysis mutants (r) of the T-even phage produce large, sharp-edged plaques. The wild type produce a smaller, more fuzzy-edged plaque (Figure 6-9).

Alternate alleles are known also for host-range loci, which determine which strains of bacteria a phage can infect. For example, T2 can infect *E. coli* cells. These phages can be designated as $T2h^+$ for the normal host range.

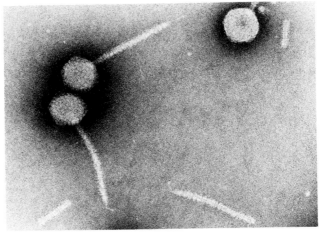

Source: Gunther S. Stent, *Molecular Genetics* (San Francisco: W. H. Freeman, 1963), p. 421. Reproduced courtesy of Dr. Robley C. Williams.

Figure 6-26.
λ Phage
magnification
239,000X

The *E. coli* would then be Tto^s, referring to their sensitivity to the T2 phage. In the course of evolution, an *E. coli* mutant has arisen that is resistant to the normal phage. This mutant is Tto^r for T2 resistance. In the further course of evolution, the phages have produced mutant forms that can grow on the Tto^r strain of *E. coli*. These phage mutants are designated as T2h for host-range mutant.

In 1945 Max Delbrück (a 1969 Nobel laureate) developed the technique of mixed indicators, which can be used to demonstrate four phage phenotypes on the same petri plate (Figure 6–27). A bacterial lawn of mixed Tto^r and Tto^s is grown. On this lawn, the rapid-lysis phage mutants (r) produce large plaques, whereas the wild type (r$^+$) produce smaller plaques. Plaques of the host-range mutants (h) are clear because the phage can infect both Tto^r and Tto^s bacteria. Since phages with the wild-type host-range allele (h$^+$) can only infect the Tto^s bacteria, they produce turbid, darker plaques. The Tto^r bacteria can grow in these plaques and produce the turbidity.

From the wild stock of phage, we can isolate host-range mutants by looking for plaques on a Tto^r bacterial lawn. Only h mutants will grow. These phages can then be tested for the r phenotype. Hence the double mutants can be isolated. Once the two strains (double mutant and wild type) are available, they can be added in large numbers to sensitive bacteria. Large numbers of phage are used in order to ensure that each bacterium is infected by at least one of each phage type *(multiplicity of infection)*, which then provides the possibility for recombination within the host bacterium. After

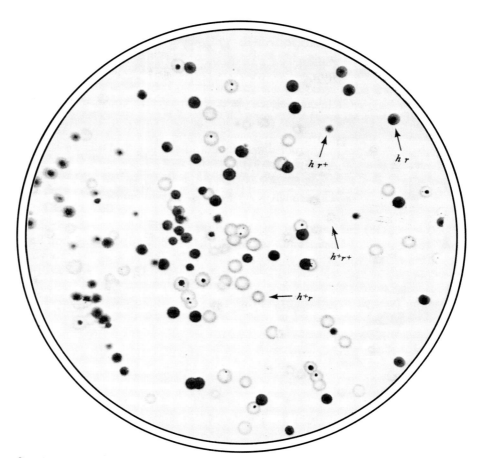

Figure 6-27.
Four Types of
Plaques Produced
on a Mixed Lawn of
E. coli by Mixed
Phage T2

Source: From *Molecular Biology of Bacterial Viruses* by Gunther S. Stent.
W. H. Freeman and Company. Copyright © 1963. Reproduced by
permission.

a round of phage multiplication, the phages are isolated and plated out on
Delbrück's mixed-indicator stock. From this growth the phenotype (and
hence genotype) of each phage can be recorded. The percentage of recombinants can be read directly from the plate. For example, on a given petri
plate (e.g., Figure 6-27) there might be

$h\ r$	40	$h^+\ r^+$	52
$h^+\ r$	54	$h\ r^+$	18

The first two, $h\ r$ and $h^+\ r^+$, are the original, or parental, phage genotypes.
The second two categories result from recombination between the h and r

loci on the phage chromosome. A single crossover in this region will produce the recombinants. The proportion of recombinants is

$$\frac{54 + 18}{40 + 52 + 54 + 18} = \frac{72}{164} = 0.44 \qquad \text{or} \qquad 44\%$$

This percentage recombination is the map distance, which, as in eukaryotes, is a relative index of distance between loci: The greater the physical distance, the greater the recombination and thus the larger the map distance. One map unit (one centimorgan) is equal to 1% of recombinant offspring; with transduction a different index is used.

Lysogeny

Certain phages are capable of acting as episomes. Some of the time these phages will replicate in the host cytoplasm and cause destruction of the host cell. At other times these phages are capable of integrating into the host chromosome. The host is then referred to as **lysogenic** and the phage as **temperate.** Lysogeny is the phenomenon of integration of viral and bacterial host genomes.

The majority of research on this process has been done on phage λ, or lambda (Figure 6–26). Phage λ, unlike the F factor, attaches at a specific point, termed *att*λ. This locus can be mapped on the *E. coli* chromosome and lies between *gal* (for galactose) and *bio* (for biotin). When the phage is integrated, it protects the host from superinfection by other λ phage. The integrated phage is termed a **prophage.** Presumably it becomes integrated by a single crossover between itself and the host after apposition of the two at the *att*λ site. (This process resembles the F-factor integration shown in Figure 6–22.)

A prophage can become virulent by a process of **induction,** which involves the excision of the prophage followed by the virulent or lytic stage of the viral life cycle. We will consider the control mechanisms in detail in Chapter 12. Induction can take place through a variety of mechanisms including UV irradiation and passage of the prophage during conjugation **(zygotic induction).** The complete life cycle of a temperate phage is shown in Figure 6–28.

TRANSDUCTION

Prior to cell lysis, when phage DNA is being packaged into phage heads, an occasional error occurs, where bacterial DNA is incorporated into the phage head. When this happens, bacterial genes can be transferred from one bacterium to another via the phage coat. This process is called **transduction**

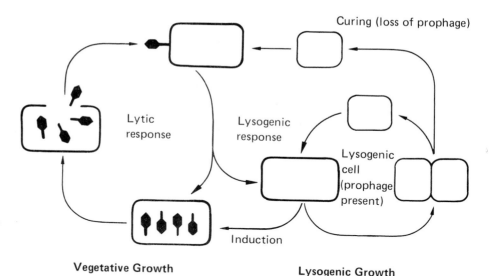

Figure 6-28.
Alternate Life-Cycle
Stages of a
Temperate Phage
(Lysogenic and
Vegetative Growth)

and has been of great use in mapping the bacterial chromosome. Transduction occurs in two general patterns: specialized (restricted) and generalized.

Specialized Transduction

The process of specialized, or restricted, transduction was first discovered in phage λ by Lederberg and his students. It is less useful than the process of generalized transduction for mapping of the bacterial chromosome, but it is nevertheless an interesting phenomenon. Specialized transduction is completely analogous to sexduction—it depends upon a mistake made during the looping out of an episome. In sexduction the episome is the F factor. In specialized transduction the episome is the λ prophage. Figure 6–29 shows the λ prophage looping out incorrectly to create a defective phage carrying the adjacent *gal* locus.

The defective λ phages, upon infection of new *E. coli* cells, will carry the *gal* locus with them. Lederberg and his students induced strain K12 (λ), which is prototrophic, to produce a lysate of λ phages. These phages were then used to infect K12 *gal⁻* strains, which were nonlysogenic. About one in 10^6 of the newly infected *gal⁻* cells became *gal⁺*. (Obviously, restricted transduction is a very low-efficiency event.) In the process the *gal⁻* cells were not simply changed into *gal⁺* cells but became partial heterozygotes of the *gal* region (*gal⁺/gal⁻*), as well as being lysogenic, or at least protected from superinfection by λ phage. That is, the transducing λ may contain all of the original λ genes, in which case the transduced bacterium would become lysogenic. If the transducing λ phage were defective, the transduced bacterium

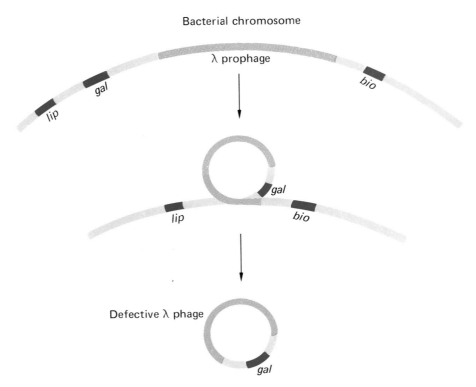

Figure 6-29.
Excision, or Looping
Out, of the λ
Prophage Resulting
in a Defective Phage
Carrying *gal*

would not be lysogenic. However, it would be protected from superinfection if it had the part of the λ responsible for protection (Chapter 12). The mechanism of partial diploidy is shown in Figure 6–30.

In specialized transduction only loci adjacent to the phage attachment site can be transduced. In the case of phage λ, only the *gal* locus, adjacent to *att*λ, can be studied by transduction. For this reason, specialized transduction has not proven generally useful for mapping the host chromosome.

Generalized Transduction

Generalized transduction, discovered by Zinder and Lederberg, was the first mode of transduction discovered. The lysogenic bacterium was *Salmonella typhimurium* and the temperate phage was P22. Zinder and Lederberg discovered that in the process of transduction, virtually any locus can be transduced. The mechanism, therefore, does not depend on a faulty excision, but, rather on the random inclusion of a piece of the host chromosome within the phage protein coat.

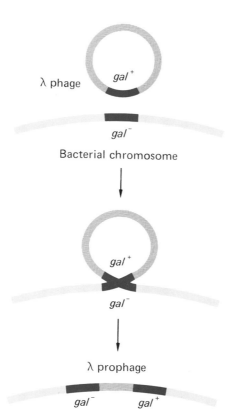

Figure 6–30. Transduction of a K12 *gal⁻* Strain by a λ Phage Containing the *gal⁺* Allele Synapsis of homologous regions (the *gal* region) occurs, followed by a single crossover. This produces a λ prophage bordered by the two *gal* regions.

Generalized transduction is studied by infecting a nonlysogenic donor strain with the phage. The lysate is then harvested and used to infect a recipient strain that is either lysogenic or not. If the recipient strain is lysogenic, then all cells infected with the transducing phage will survive. If the recipient strain is nonlysogenic, then descendant colonies of bacteria will only result from those cells that survived the attack. A cell can survive because it escaped infection by a phage or because it entered into lysogeny with an infecting phage. Alternatively, the cell could have been attacked by a defective phage, one that carried bacterial DNA rather than phage DNA. This type of phage package is called a *transducing particle*. Transduction will be complete if the incoming DNA enters the host chromosome by recombination.

For P22, the rate of transduction is about once for every 10^5 infecting phages. Since a transducing phage can only carry 2 to 2.5% of the host chromosome, only genes very close to each other can be transduced together (**cotransduced**). Cotransduction can thus be used to fill in the details of gene order over short distances after the general pattern has been ascertained by interrupted mating.

Mapping with Transduction

Transduction can be used to establish gene order and map distance. Gene order can be established by two-factor transduction. For example, if gene A is cotransduced with gene B and B with gene C, but A is *never* cotransduced with C, then we have established the order $A–B–C$ (Table 6–4). This would also apply to quantitative differences in cotransduction. For example, if E is often cotransduced with F and F often with G, but E is very rarely cotransduced with G, then we have established the order $E–F–G$.

However, more valuable is a three-factor transduction, where gene order and relative distance can be established simultaneously. Three-factor transduction is especially valuable when the three loci are so close as to make ordering decisions on the basis of two-factor transduction very difficult. For example, if genes A, B, and C are usually cotransduced, we can find the order and relative distance by taking advantage of the rarity of multiple crossovers. Let us use the prototroph ($A^+B^+C^+$) to make transducing phages that then infect the $A^-B^-C^-$ stock. Transduced cells are recovered by growth in complete medium minus the nutritional requirements of A, B, or C. These cultures are then checked for cotransduction by replica plating.

For example, colonies that grow on complete medium without A are replica plated onto complete medium without B and then onto complete medium without C. In this way, each transductant can be scored for all three loci (Table 6–5). Now let us take all those transductants for which the A allele was brought in (A^+). These can be of four categories: $A^+B^+C^+$, $A^+B^+C^-$, $A^+B^-C^+$, and $A^+B^-C^-$. We now compare the relative numbers of

TABLE 6 – 4.   Gene Order Established by Two-Factor Cotransduction. An $A^+B^+C^+$ strain of bacteria is infected with phage. The lysate is used to infect an $A^-B^-C^-$ strain. The transductants are scored for the wild-type alleles they contain. The data below only include those bacteria transduced for two or more of the loci. Since there are AB cotransductants and BC cotransductants, but no AC types, the order of $A–B–C$ is inferred.

Transductants	Number
A^+B^+	30
A^+C^+	0
B^+C^+	25
$A^+B^+C^+$	0

TABLE 6-5. Method of Scoring Three-Factor Transductants. The plus indicates growth and the minus indicates lack of growth. An $A^- B^- C^-$ strain was transduced by phage from an $A^+ B^+ C^+$ strain.

Colony Number	Minimal Medium			Genotype
	Without A Requirement	Without B Requirement	Without C Requirement	
1	+	+	−	$A^+ B^+ C^-$
2	+	−	−	$A^+ B^- C^-$
3	+	−	−	$A^+ B^- C^-$
4	−	+	+	$A^- B^+ C^+$
5	−	−	+	$A^- B^- C^+$
.	.	.	.	.
.	.	.	.	.
.	.	.	.	.

each of these four categories. The rarest category will be caused by the event that brings in the outer two markers but not the center one because this event requires four crossovers (Figure 6–31). Thus, by looking at the numbers of the various categories, we can determine the gene order to be A–B–C (Table 6–6) since the $A^+ B^- C^+$ category is the rarest.

The relative cotransductance frequencies are calculated in Table 6–6. These values are inversely related to actual distance. That is, the greater the

TABLE 6-6. Numbers of Transductants in the Experiment Used to Determine the A–B–C Gene Order (Table 6–5). Relative cotransduction frequencies are also given.

Class	Number
$A^+ B^+ C^+$	50
$A^+ B^+ C^-$	75
$A^+ B^- C^+$	1
$A^+ B^- C^-$	300
	426

Relative Cotransductance

A–B: $(50 + 75)/426 = 0.29$

A–C: $(50 + 1)/426 \ \ = 0.12$

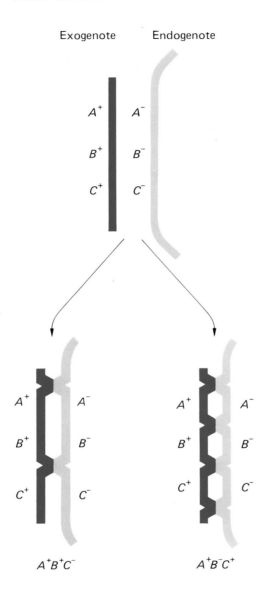

Figure 6–31.
The Rarest
Transductant
Requiring Four
Crossovers

cotransductance rate, the closer the two loci are. The data of Table 6–6 cannot be used to calculate B–C cotransductance because the data given are selected values, all of which are A^+. They are not the total data. From these sorts of transduction experiments, it is possible to round out the details of map relations for which the overall picture is obtained by interrupted mating. The map of *E. coli* is presented in Figure 6–32. Definitions of loci can be found in Table 6–7.

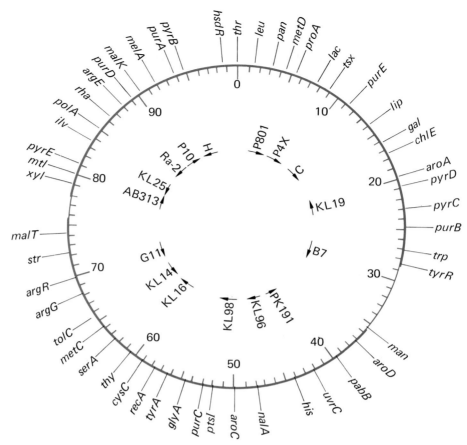

Figure 6-32. Selected Loci on the Circular Map of *E. coli*
Definitions of loci not found in the text can be found in Table 6-7. Units on the
map are in minutes. Arrows within the circle refer to Hfr-strain transfer starting
points, with directions. The two thin regions are the only areas not covered by
P1 cotransducing phages.

Source: B. J. Bachmann, K. B. Low, and A. L. Taylor "Recalibrated Linkage Map
of *Escherichia coli* K-12," *Bacteriological Reviews* 40 (1976):116–167.
Reprinted Courtesy of Dr. Barbara Bachmann and Dr. K. B. Low.

TABLE 6-7. Symbols Used in the Gene Map of the *E. coli* Chromosome

Genetic Symbols	Mutant Character	Enzyme or Reaction Affected
araD	Cannot use the sugar arabinose as a carbon source	L-Ribulose-5-phosphate-4-epimerase
araA		L-Arabinose isomerase
araB		L-Ribulokinase
araC		
argB		N-Acetylglutamate synthetase
argC		N-Acetyl-γ-glutamokinase

TABLE 6-7 (*continued*)

Genetic Symbols	Mutant Character	Enzyme or Reaction Affected
argH argG argA argD argE argF	Requires the amino acid arginine for growth	N-Acetylglutamic-γ-semialdehyde dehydrogenase Acetylornithine-d-transaminase Acetylornithinase Ornithine transcarbamylase Argininosuccinic acid synthetase Argininosuccinase
argR	Arginine operon regulator	
aroA, B, C	Requires several aromatic amino acids and vitamins for growth	Shikimic acid to 3-enolpyruvyl-shikimate-5-phosphate
aroD		Biosynthesis of shikimic acid
azi	Resistant to sodium azide	
bio	Requires the vitamin biotin for growth	
carA	Requires uracil and arginine	Carbamate kinase
carB		
chlA–E	Cannot reduce chlorate	Nitrate-chlorate reductase and hydrogen lysase
cysA cysB cysC	Requires the amino acid cysteine for growth	3-Phosphoadenosine-5-phosphosulfate to sulfide Sulfate to sulfide; 4 known enzymes
dapA dapB	Requires the cell-wall component diaminopimelic acid	Dihydrodipicolinic acid synthetase N-Succinyl-diaminopimelic acid deacylase
dap + hom	Requires the amino acid precursor homoserine and the cell-wall component diaminopimelic acid for growth	Aspartic semialdehyde dehydrogenase
dnaA–Z	Mutation, DNA replication	DNA biosynthesis
Dsd	Cannot use the amino acid D-serine as a nitrogen source	D-Serine deaminase
fla	Flagella are absent	
galA galB galD	Cannot use the sugar galactose as a carbon source	Galactokinase Galactose-1-phosphate uridyl transferase Uridine-diphosphogalactose-4-epimerase
glyA	Requires glycine	Serine hydroxymethyl transferase
gua	Requires the purine guanine for growth	
H	The H antigen is present	
his	Requires the amino acid histidine for growth	10 known enzymes[a]
hsdR	Host restriction	Endonuclease R
ile	Requires the amino acid isoleucine for growth	Threonine deaminase
ilvA ilvB ilvC	Requires the amino acids isoleucine and valine for growth	α-Hydroxy-β-keto acid rectoisomerase α,β-dihydroxyisovaleric dehydrase[a] Transaminase B

TABLE 6-7 (*continued*)

Genetic Symbols	Mutant Character	Enzyme or Reaction Affected
ind (indole)	Cannot grow on tryptophan as a carbon source	Tryptophanase
λ *(att*λ*)*	Chromosomal location where prophage λ is normally inserted	
lacI	*Lac* operon regulator	
lacY	Unable to concentrate β-galactosides	Galactoside permease
lacZ	Cannot use the sugar lactose as a carbon source	β-Galactosidase
lacO	Constitutive synthesis of lactose operon proteins	Defective operator
leu	Requires the amino acid leucine for growth	3 known enzymes[a]
lip	Requires lipoate	
lon (long form)	Filament formation and radiation sensitivity are affected	
lys	Requires the amino acid lysine for growth	Diaminopimelic acid decarboxylase
lys + met	Requires the amino acids lysine and methionine for growth	
λ *rec, malT*	Resistant to phage λ and cannot use the sugar maltose	Regulator for 2 operons
malK	Cannot use the sugar maltose as a carbon source	Amylomaltase(?)
man	Cannot use mannose sugar	Phosphomannose isomerase
melA	Cannot use melibiose sugar	Alpha-galactosidase
metA–M	Requires the amino acid methionine for growth	10 or more genes
mtl	Cannot use the sugar mannitol as a carbon source	Mannitol dehydrogenase (?)
muc	Forms mucoid colonies	Regulation of capsular polysaccharide synthesis
nalA	Resistance to nalidixic acid	
O	The O antigen is present	
pan	Requires the vitamin pantothenic acid for growth	
pabB	Requires *p*-aminobenzoate	
phe A, B	Requires the amino acid phenylalanine for growth	
pho	Cannot use phosphate esters	Alkaline phosphatase
pil	Has filaments (pili) attached to the cell wall	
plsB	Deficient phospholipid synthesis	Glycerol 3-phosphate acyltransferase
polA	Repairs deficiencies	DNA polymerase I
proA *proB* *proC*	Requires the amino acid proline for growth	

TABLE 6-7 (*continued*)

Genetic Symbols	Mutant Character	Enzyme or Reaction Affected
ptsI	Defective phosphotransferase system	Pts-system enzyme I
purA	Requires certain purines for growth	Adenylosuccinate synthetase
purB		Adenylosuccinase
purC, E		5-Aminoimidazole ribotide (AIR) to 5-aminoimidazole-4-(N-succino carboximide) ribotide
purD		Biosynthesis of AIR
pyrB	Requires the pyrimidine uracil for growth	Aspartate transcarbamylase
pyrC		Dihydroorotase
pyrD		Dihydroorotic acid dehydrogenase
pyrE		Orotidylic acid pyrophosphorylase
pyrF		Orotidylic acid decarboxylase
R gal	Constitutive production of galactose	Repressor for enzymes involved in galactose production
R1 pho, R2 pho	Constitutive synthesis of phosphatase	Alkaline phosphatase repressor
R try	Constitutive synthesis of tryptophan	Repressor for enzymes involved in tryptophan synthesis
RC (RNA control)	Uncontrolled synthesis of RNA	
recA	Cannot repair DNA radiation damage or recombine	
rhaA–D	Cannot use the sugar rhamnose as a carbon source	Isomerase, kinase, aldolase, and regulator
rpoA–D	Problems of transcription	Subunits of RNA polymerase
serA	Requires the amino acid serine for growth	3-Phosphoglycerate dehydrogenase
serB		Phosphoserine phosphatase
str	Resistant to or dependent on streptomycin	
suc	Requires succinic acid	
supB	Suppresses ochre mutations	t-RNA
tonA	Resistant to phages T1 and T5 (mutants called B/1, 5)	T1, T5 receptor sites absent
tonB	Resistant to phage T1 (mutants called B/1)	T1 receptor site absent
T6, colK rec	Resistant to phage T6 and colicine K	T6 and colicine receptor sites absent
T4 rec	Resistant to phage T4 (mutants called B/4)	T4 receptor site absent
tsx	T6 resistance	
thi	Requires the vitamin thiamine for growth	
tolC	Tolerance to colicine E1	
thr	Requires the amino acid threonine for growth	
thy	Requires the pyrimidine thymine for growth	Thymidylate synthetase

TABLE 6-7. (*continued*)

Genetic Symbols	Mutant Character	Enzyme or Reaction Affected
trpA trpB trpC trpD trpE	Requires the amino acid tryptophan for growth	Tryptophan synthetase, A protein Tryptophan synthetase, B protein Indole-3-glycerolphosphate synthetase Phosphoribosyl anthranilate transferase Anthranilate synthetase
tyrA tyrR	Requires the amino acid tyrosine for growth	Chorismate mutase T-prephenate dehydrogenase Regulates 3 genes
uvrA–E	Resistant to ultraviolet radiation	Ultraviolet-induced lesions in DNA are reactivated
valS	Cannot charge Valyl-tRNA	Valyl-tRNA synthetase
xyl	Cannot use the sugar xylose as a carbon source	

^aDenotes enzymes controlled by the homologous gene loci of *Salmonella typhimurium*.

Source: B. J. Bachmann and K. B. Low, "Linkage Map of *Escherichia coli* K-12, Edition 6," *Microbiological Reviews* 44 (1980):1–56. Reprinted by permission.

CHAPTER SUMMARY

Prokaryotes (bacteria and blue-green algae) have a single circular chromosome of DNA. The bacterial sexual processes—transformation, conjugation, transduction, and sexduction—can be used to map the bacterial chromosome. A bacteriophage consists of a chromosome wrapped in a protein coat. The chromosome can be circular or linear, double- or single-stranded, DNA or RNA. The phage chromosome can be mapped by recombination techniques after a bacterium has been infected by two strains of the virus carrying different alleles.

Phenotypes of bacteria include colony morphology, nutritional requirements, and drug resistance. Phage phenotypes include plaque morphology and host range. Replica plating is a rapid screening technique for assessing the phenotype of a bacterial clone.

In transformation a competent bacterium can take up relatively large pieces of double-stranded DNA from the medium. This DNA can be incorporated into the bacterial chromosome if homologous regions come to lie adjacent to each other and an even number of crossovers then take place.

During the process of conjugation, the fertility factor, F, is passed from an F$^+$ to an F$^-$ cell. If the F factor, an episome, integrates into the host chromosome, the cell becomes an Hfr, which can pass its entire chromosome into an F⁻ cell. The F factor is the last region to cross into the F⁻ cell. The bacterial chromosome can be mapped by the process of interrupted mating, which times the entry of loci from the Hfr into the F⁻ cell.

The F factor can loop out of the host chromosome. In some cases it takes part of the host chromosome with it. When this defective F, or F′, is passed to an F⁻ cell, recombination of the original host chromosome segment with the F⁻ chromosome can take place. In transduction a phage protein coat containing some of its host chromosome is passed to a new host bacterium. Again recombination with this new chromosomal segment can take place (generalized transduction); or the new DNA, with part or all of the phage chromosome, can be incorporated into the new host chromosome (specialized transduction).

In *E. coli,* mapping is most efficiently done via interrupted mating and transduction. The former provides information on general gene arrangement and the latter provides fine details.

EXERCISES AND PROBLEMS

1. In conjugation experiments the Hfr strain should carry a locus for some sort of sensitivity (e.g., *azi*ˢ or *str*ˢ) so that the Hfr donors can be eliminated on selective media after conjugation has taken place. Should this locus be near to or far from the origin of transfer point of the Hfr chromosome? What are the consequences of its being near?

2. Diagram the step-by-step events required to integrate foreign DNA into a bacterial chromosome in the four processes outlined in the chapter. Do the same for viral recombination.

3. An *E. coli* cell is placed on a petri plate containing λ phages. It produces a colony overnight. By what mechanisms might it have survived?

4. Three Hfr strains of *E. coli* are individually mated with an auxotrophic F⁻ strain, using interrupted mating techniques. Using these data alone, construct a map of the *E. coli* chromosome, including distances in minutes.

	Approximate Time of Entry		
Donor Loci	Hfr P4X	Hfr KL98	Hfr Ra-2
gal^+	11	67	70
thr^+	94	50	87
xyl^+	73	29	8
lac^+	2	58	79
his^+	38	94	43
ilv^+	77	33	4
$argG^+$	62	18	19

How many different petri plates and selective media are needed?

5. A petri plate with complete medium has six colonies growing on it after one of the conjugation experiments described earlier. The colonies are numbered and the plate is used as a master to replicate onto plates of minimal medium with various combinations of additives. From the following data showing the presence or absence of growth, give your best assessment of the genotypes of the six colonies:

On Minimal Medium Plus	Colony					
	1	2	3	4	5	6
Nothing	−	−	+	−	−	−
Xylose + Arginine	+	−	+	+	−	−
Xylose + Histidine	−	−	+	−	−	−
Arginine + Histidine	−	+	+	+	-	−
Galactose + Histidine	−	−	+	−	−	+
Threonine + Isoleucine + Valine	−	−	+	−	+	−
Threonine + Valine + Lactose	−	−	+	−	−	−

6. Lederberg and his colleagues (Nester, Schafer, and Lederberg, 1963, *Genetics* 48:529) determined gene order and relative distance using 3 markers in *Bacillus subtilis*. DNA from a prototrophic strain ($trp_2^+ \, his_2^+ \, tyr^+$) was used to transform the auxotroph. The seven classes of transformants, with their count, are tabulated as follows:

trp^+	trp^-	trp^-	trp^+	trp^+	trp^-	trp^+
his^-	his^+	his^-	his^+	his^-	his^+	his^+
tyr^-	tyr^-	tyr^+	tyr^-	tyr^+	tyr^+	tyr^+
2600	418	685	1180	107	3660	11,940

Outline the techniques used to obtain these data. Taking the loci in pairs, calculate recombinant distances. Construct the most consistent linkage maps of these loci.

7. In *E. coli* the three loci *ara*, *leu*, and *ilvH* are within 1/2 minute map distance. To determine the exact order and relative distance, the prototroph (ara^+, leu^+, $ilvH^+$) was infected with transducing phage P1. The lysate was used to infect the auxotroph (ara^-, leu^-, $ilvH^-$). The seven classes of transductants were counted to produce the following tabulation:

ara^+	ara^-	ara^-	ara^+	ara^+	ara^-	ara^+
leu^-	leu^+	leu^-	leu^+	leu^-	leu^+	leu^+
$ilvH^-$	$ilvH^-$	$ilvH^+$	$ilvH^-$	$ilvH^+$	$ilvH^+$	$ilvH^+$
32	3	11	9	0	28	340

Outline the specific techniques used to isolate the various transduced classes. What is the gene order and relative distance between genes? Why do some classes occur so infrequently?

8. Outline an experiment to demonstrate that two phages do not undergo recombination until a bacterium is infected simultaneously with both.

9. In T4 phages the linkage map is circular. However, the chromosome itself is linear. How can this be?

10. Doermann (1953, *Cold Spr. Harb. Symp. Quant. Biol.* 18:3) mapped three loci of phage T4: minute, rapid lysis, and turbid. He infected *E. coli* cells with both the triple mutant ($m\ r\ tu$) and the wild type ($m^+\ r^+\ tu^+$) and obtained the following data:

m	m^+	m	m	m^+	m^+	m	m^+
r	r	r^+	r	r^+	r	r^+	r^+
tu	tu	tu	tu^+	tu	tu^+	tu^+	tu^+
3467	474	162	853	965	172	520	3729

What is the linkage relationship among these loci? Include gene order, relative distance, and coefficient of coincidence.

SUGGESTIONS FOR FURTHER READING

Bachmann, B., K. Low, and A. Taylor. 1976. Recalibrated linkage map of *Escherichia coli K-12. Bact. Rev.* 40:116–167

Cairns, J., G. Stent, and J. Watson, eds. 1966. Phage and the origins of molecular biology. *Cold Spr. Harb. Symp. Quant. Biol.* 31.

Dyson, R. 1978. *Cell Biology, A Molecular Approach,* 2nd ed. Boston: Allyn and Bacon.

Jacob, F., and E. Wollman. 1961. *Sexuality and the Genetics of Bacteria.* New York: Academic Press.

Kaplan, M., and R. Webster. 1977. The epidemiology of influenza. *Sci. Amer.* (December):88–106.

Lederberg, J., et al. 1951. Recombination analysis of bacterial heredity. *Cold Spr. Harb. Symp. Quant. Biol.* 16:413–443.

Pelczar, M., R. Reid, and W. Chan. 1977. *Microbiology,* 4th ed. New York: McGraw-Hill.

Stent, G., and R. Calendar. 1978. *Molecular Genetics, An Introductory Narrative,* 2nd ed. San Francisco: Freeman.

7 Variation in Chromosome Structure and Number

Our understanding of the chromosomal theory of genetics was derived primarily through the mapping of loci, a technique that requires alternate allelic forms, or mutants, of these loci. Changes in the genetic material also occur at a much coarser level—changes in large parts of chromosomes or changes in chromosome numbers. This chapter investigates how these alterations happen and what their consequences are to the organism.

VARIATION IN CHROMOSOME STRUCTURE

In general, chromosomes can break spontaneously or be broken by ionizing radiation, physical stress, or chemical compounds. Spontaneous events may in reality be caused by all these mechanisms.

Chromosomal breaks can occur at either the chromatid or the chromosome level. When a break in the chromosome occurs prior to DNA replication during the S-phase of the cell cycle, the break itself is replicated. In this case we have a chromosome-level break involving both sister chromatids at the same point. After the S-phase, each break occurs independently in a single chromatid—that is, it is a chromatid-level break.

For every break in a chromatid two ends are produced. These ends have been described as "sticky," meaning simply that enzymatic processes of the cell tend to reunite them. Broken ends do not attach to the terminal ends of other chromosomes. If ends are not brought into apposition, they can remain broken. But, if broken chromatid ends are brought into apposition, there are several alternate ways in which they can be rejoined. First, the two broken ends of a single chromatid can be reunited. Second, the broken end of one chromatid can be fused with the broken end of another chromatid, resulting in an exchange of chromosomal material and a new combination of alleles. Multiple breaks can lead to a variety of alternative recombinations. These

217

chromosomal aberrations have important genetic, evolutionary, and medical consequences. The types of breaks and reunions discussed in this chapter can be summarized as follows:

I. Noncentromeric break
1. Single break (chromatid) a. Restitution
   (chromosome) b. Dicentric bridge
2. Two breaks (same chromatid)
   a. Deletion
   b. Inversion
3. Two breaks (nonsister chromatids)
   a. Translocation, reciprocal
   b. Translocation, nonreciprocal
II. Centromeric breaks
1. Fission
2. Fusion

Single Breaks: Chromatid

If the break is of the chromatid type, the broken ends may be rejoined. When the broken ends of a single chromatid are rejoined (restitution), there is no consequence of the break. If they are not rejoined, the result is an **acentric fragment,** without a centromere, and a **centric fragment,** with a centromere (Figure 7–1). The centric fragment will migrate normally during the division process because it has a centromere. The acentric fragment, however, is soon lost. It is subsequently excluded from the nuclei formed and is eventually degraded. The $h–i–j$ fragment is missing from the centric segment in Figure 7–1; that is, the viable centric part of the chromosome has suffered a deletion of the $h–i–j$ region (the chromosome has an $h–i–j$ defi-

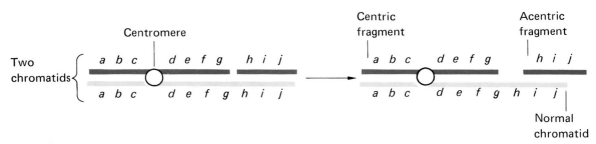

Figure 7–1.   Consequence of a Single Chromatid Break

ciency). After mitosis, the daughter cell that receives the **deletion chromosome** may show several effects.

Pseudodominance may be observed. (This term was used in Chapter 4 when we described alleles located on the X chromosome. A recessive allele in males, with only one copy of the locus present, shows itself in the phenotype as if it were a dominant—hence, the term pseudodominance.) The normal chromosome homologous to the deletion chromosome will have loci in the *h–i–j* region and recessive alleles will show pseudodominance. A second effect is that, depending on the length of the deleted segment and the specific loci lost, the imbalance created in a daughter cell by a deletion chromosome may be *lethal.* If the deletion occurs during meiosis, the same effects may be observed in the zygote or in succeeding cells (pseudodominance, some lethality). Because of the process of synapsis, the deletion can be observed under the microscope. This is discussed in the next section.

Single Breaks: Chromosomal

A single break can have another effect if it is a chromosomal break. Occasionally the two centric fragments of a single chromatid may join, forming a two-centromere or **dicentric chromosome** and leaving the two acentric fragments to join or, alternatively, remain as two fragments (Figure 7–2). The acentric fragments will be lost, as mentioned before. The dicentric fragment is pulled to opposite ends of a mitotic cell because the centromeres are on sister chromatids. This forms a bridge during mitosis or during the second meiotic division (Figure 7–3). The ultimate fate of this bridge is to be broken

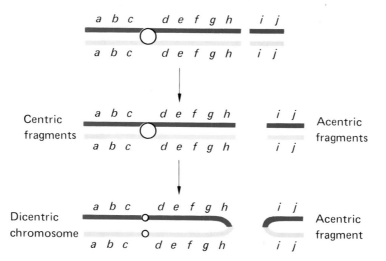

Figure 7–2. Chromosomal Break with Subsequent Reunion to Form a Dicentric Chromosome and an Acentric Fragment

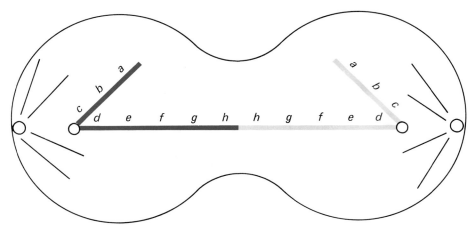

Figure 7-3.   At Mitosis or Meiosis II, the Dicentric Chromosome Forms a Bridge across the Dividing Cell

by the spindle fibers pulling the centromeres to opposite poles (or possibly to be excluded from new nuclei if the bridge is not broken).

The dicentric is not necessarily broken in the middle, and subsequent processes will exacerbate the imbalance created by an off-center break: Duplications will occur on one strand while additional deletions will occur on the other (Figure 7–4). In addition, the "sticky" ends produced on both

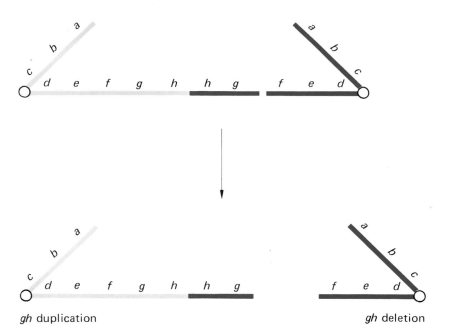

Figure 7-4.
Breakage of a
Dicentric Bridge
Causes Duplications
and Further
Deficiencies

gh duplication

gh deletion

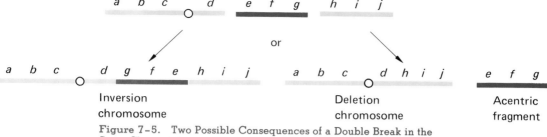

Inversion
chromosome

Deletion
chromosome

Acentric
fragment

Figure 7–5.   Two Possible Consequences of a Double Break in the
Same Chromatid

fragments by the break increase the likelihood of repeating this *breakage–fusion–bridge* cycle each generation. The great imbalances resulting from the duplications and deletions usually result in the death of the cell line within several cell generations.

Two Breaks in the Same Chromatid

Figure 7–5 shows two of the possible results when two breaks occur in the same chromatid. One alternative is a reunion that omits an acentric fragment, which will be lost. The centric piece, missing the acentric fragment (e–f–g in Figure 7–5), is a deletion chromosome. An organism having this chromosome and a normal homologue will have, during meiosis, a bulge in the tetrad if the deleted section is large enough (Figure 7–6). The bulge can usually be seen in the giant salivary gland chromosomes of *Drosophila* (Figure 7–7), unless, of course, its effects are lethal.

Inversion.   Two breaks in the same chromosome can also lead to **inversion,** where the middle section is reattached but in the inverted configuration (Figure 7–5). An inversion has several interesting properties. To begin with, a stock of fruit flies homozygous for an inverted chromosome will show new linkage relations when mapped. An outcome of this new linkage arrangement is the possibility of a **position effect,** a change in the expression of a gene due to a changed linkage arrangement. Position effects are

Figure 7–6.
Bulge in a Meiotic
Tetrad Indicates
That a Deletion Has
Occurred

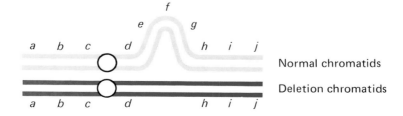

Normal chromatids

Deletion chromatids

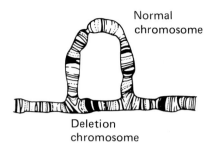

Normal
chromosome

Deletion
chromosome

Figure 7–7.
Deletion
Heterozygote in
Drosophila Salivary
Gland X
Chromosome.

either stable, as in *Bar* eye of *Drosophila* (to be discussed), or variegated as with *Drosophila* eye color. A normal female fly that is heterozygous (X^w/X^+) has red eyes. If, however, the *white* locus is moved, through an inversion, so that it comes to lie next to heterochromatin (Figure 7–8), the fly will show a **variegation**—patches of the eye will be white. Whether this is caused by a product of the heterochromatic region or whether the tight coiling in the heterochromatin "turns off" the allele is not known. It is known, however, that this type of position effect is limited to loci placed in proximity to heterochromatin.

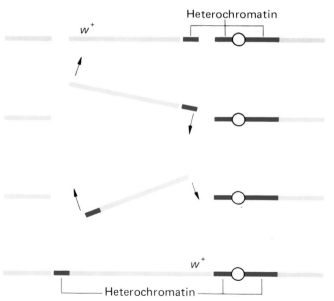

Figure 7–8.   Inversion in the X Chromosome of *Drosophila* That Will Produce a Variegation in Eye Color in a Female if Her Other Chromosome Is Normal and Carries the White-Eye Allele (X^w)

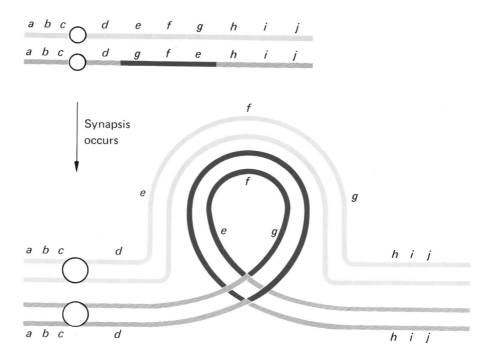

Figure 7-9.
Tetrad at Meiosis
Showing the Loop
Characteristic of an
Inversion
Heterozygote

When an inversion heterozygote synapses, either at meiosis or in the *Drosophila* salivary gland during endomitosis, a loop is often formed to accommodate the point-for-point pairing process (Figures 7–9 and 7–10). An interesting outcome of this looping tendency is "crossover suppression." That is, an inversion heterozygote shows very little recombination for alleles within the inverted region. The reason is not that crossing over is actually suppressed but rather that the products of recombination within a loop are usually lost. A crossover within a loop is shown in Figure 7–11. The two non-sister chromatids not involved in a crossover in the loop will result in normal gametes (carrying either the normal chromosome or the intact inverted chromosome). The products of the crossover, rather than being a simple recom-

Figure 7-10.
Drosophila Salivary
Gland Inversion
Heterozygote
Chromosome Loop

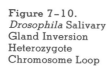

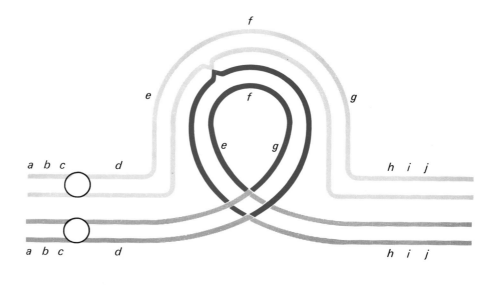

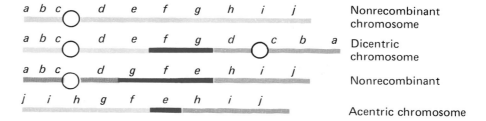

		Nonrecombinant chromosome
		Dicentric chromosome
		Nonrecombinant
		Acentric chromosome

Figure 7–11. Consequences of a Crossover in an Inversion Heterozygote

bination of alleles, are a dicentric and an acentric chromatid. The acentric is not incorporated into a gamete nucleus, whereas the dicentric begins a breakage–fusion–bridge cycle that creates a genetic imbalance in the gametes: These gametes carry duplications and deficiencies.

The inversion pictured in Figure 7–11 is a **paracentric inversion,** one in which the centromere is outside the inversion loop. A **pericentric inversion** is one where the inverted section contains the centromere. It too "suppresses" crossovers, but for a slightly different reason (Figure 7–12). All four chromatid products of a single crossover within the loop have centromeres and are thus incorporated into the nuclei of gametes. However, the two recombinant chromatids are unbalanced—they both have duplications and deficiencies. One is duplicate for $a–b–c–d$ and is deficient for $h–i–j,$ whereas the other is the reciprocal of this—deficient for $a–b–c–d$ and duplicate for $h–i–j.$ These duplication–deletion gametes tend to form inviable zygotes. The result is, as with the paracentric inversion, the apparent suppression of crossing over.

Results of Inversion.   Crossing over within inversion loops also causes **semisterility.** Almost all gametes that contain dicentric or imbalanced chromosomes form inviable zygotes. Thus a certain proportion of the progeny of inversion heterozygotes are not viable.

There are several evolutionary ramifications of inversions. To begin with, various alleles become associated with each other. Those alleles originally together in the noninversion chromosome and those found together in the inversion tend to stay together because of the low rate of recombination within the loop. Second, there tends to be an evolution toward the association of certain genes within loops. These are called **supergenes** because they are transmitted as if they were one gene. Supergenes can be beneficial when they involve favorable gene complexes. However, at the same time, their inversion structure prevents the formation of new complexes. Supergenes, therefore, have evolutionary advantages and disadvantages. More about these topics will be considered in the chapters on population and evolutionary genetics.

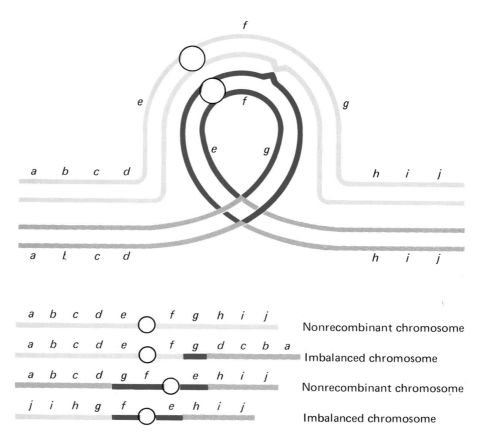

Figure 7-12.
Consequences of a
Crossover in a
Pericentric Inversion
Loop

Sometimes an outcome of the inversion process is a record of the evolutionary history of a group of species. As species evolve, inversions can occur on preexisting inversions. This leads to very complex arrangements of loci as compared to the original. These patterns are readily studied in Diptera by, for example, noting the changed patterns of bands in salivary gland chromosomes. Since certain arrangements can only come about by a specific sequence of inversions, it is possible to decide which species evolved from which other species.

In summary then, inversions result in suppressed crossing over, semi-sterility, variegated position effects, and new linkage arrangements. They are useful to evolutionary biologists.

Breaks in Nonhomologous Chromosomes

Breaks can occur simultaneously in two nonhomologous chromosomes. There are various ways in which reunion can take place. The most interesting one is the situation where the ends of two nonhomologous chromosomes are translocated to each other. This is called a **reciprocal translocation** (Figure 7–13). The organism in which this has happened (a reciprocal translocation heterozygote) has all the genetic material of the normal homozygote. Two outcomes of this event, as in an inversion, are new linkage arrangements in a homozygote (an organism with only translocation chromosomes) and variegation position effects.

During synapsis, either at meiosis or endomitosis, a point-for-point pairing in the translocation heterozygote can be accomplished by the formation of a cross-shaped figure (Figure 7–13). Such a figure is diagnostic for a reciprocal translocation. Unlike the case for an inversion heterozygote, in a reciprocal translocation heterozygote, a single crossover will not produce chromatids that are imbalanced. However, nonviable progeny are produced by reciprocal translocation heterozygotes. Problems can arise when centromeres separate at the first meiotic division.

Segregation after Translocation.   Since there are two homologous pairs of chromosomes (normally two tetrads at meiosis), we have to take into account the normal independent segregation of the centromeres of the two tetrads. There are two common possibilities (Figure 7–14). The first, called **alternate segregation,** occurs when the first centromere assorts with the fourth centromere, leaving the second and third centromeres to go to the opposite pole. The end result will be balanced gametes. One gamete will be completely normal while the other will be a reciprocal translocation gamete. Equally likely is the **adjacent-1** type of segregation, where the first and third centromeres segregate together in the opposite direction from the second and fourth centromeres. Here, both types of gametes are unbalanced,

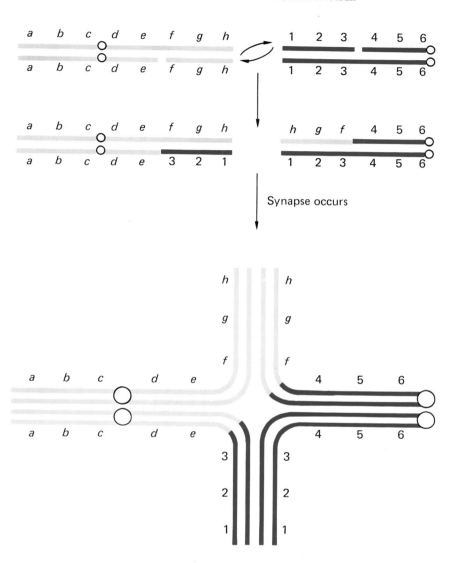

Figure 7-13.
Reciprocal
Translocation
Heterozygote

carrying duplications and deficiencies that are usually lethal. Since both of these types of segregation are about equally likely, a significant amount of sterility results from the translocation (about 50%). However, of the viable gametes so formed, half will reconstitute the translocation heterozygote in the next generation. Thus this type of chromosomal rearrangement propagates itself, albeit poorly.

There is also an **adjacent-2** type of segregation where homologous centromeres go to the same pole. This is rare and usually the result of the double tetrad opening out into a circle prior to anaphase.

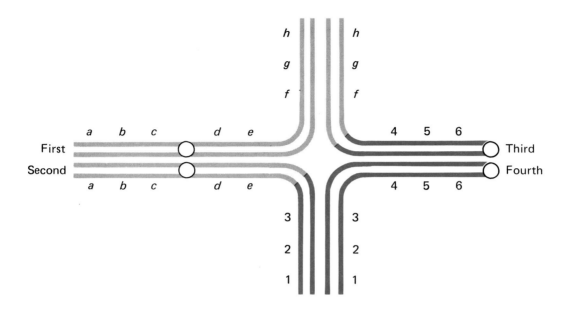

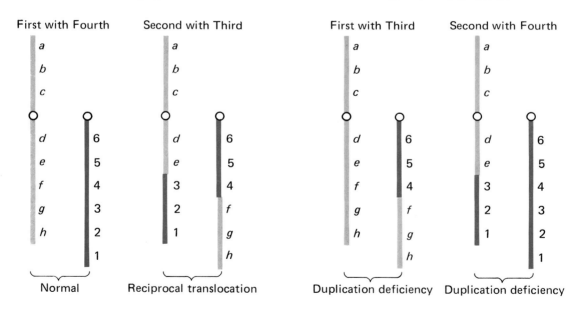

Figure 7-14. Two Common Possibilities of Chromatid Separation in the Meiosis of a Reciprocal Translocation Heterozygote

Mating Results after Translocation. In a population with a translocation in relatively high frequency, many of the crosses will be of translocation heterozygotes mating with each other. Here, more than half the offspring will be inviable because only certain combinations of gametes will produce genetically balanced young. Table 7–1 presents these possibilities. Six of the 16 possibilities are balanced (1 normal homozygote, 1 translocation homozygote, 4 translocation heterozygotes). The remaining 10 possibilities are unbalanced: 8 of the 10 (without asterisks) have 3 copies of some genes with only 1 copy of others. The two imbalances with asterisks are the cases where some genes are present four times while others are totally absent. Thus crosses with a translocation heterozygote as one parent cause 50% mortality and those involving both parents cause 62.5% mortality, assuming that all imbalances are lethal. A translocation homozygote mated to a normal homozygote will produce all viable young—all translocation heterozygotes.

In summary then, reciprocal translocations result in new linkage arrangements, variegated position effects, a cross-shaped figure during synapsis, and semisterility.

Ralph E. Cleland
(1892–1971)
Courtesy of National
Academy of
Sciences

Oenothera. An interesting situation involving multiple translocations occurs in *Oenothera*, the evening primrose, studied by DeVries and most recently reviewed by Cleland. Almost all species in this genus have 7 pairs of chromosomes ($2n = 14$). However, in some races there appears in meiosis one giant ring or many combinations of tetrads and smaller rings. It has been hypothesized that the rings result from translocations. The cross-shaped reciprocal translocation of Figure 7-13 becomes a small circle during the latter stages of prophase I of meiosis when the chromosomes tend to show some mutual repulsion. If the reciprocal translocation involves 3 nonhomologous

TABLE 7–1. Results of Matings between Reciprocal Translocation Heterozygotes

	Alternate ♂		Adjacent-1 ♂	
	First with Fourth	Second with Third	First with Third	Second with Fourth
Alternate ♀				
First with Fourth	Normal homozygote	Balanced heterozygote	Imbalanced	Imbalanced
Second with Third	Balanced heterozygote	Translocation homozygote	Imbalanced	Imbalanced
Adjacent-1 ♀				
First with Third	Imbalanced	Imbalanced	Imbalanced*	Balanced heterozygote
Second with Fourth	Imbalanced	Imbalanced	Balanced heterozygote	Imbalanced*

*Some genes in four copies, others absent.

A CASE HISTORY OF THE USE OF INVERSIONS TO DETERMINE EVOLUTIONARY SEQUENCE

In 1966 David Futch finished an interesting study of the chromosomes of the fruit fly, *Drosophila ananassae,* which is widely distributed throughout the tropical Pacific. The study was originally designed to determine something about the species status of various melanic forms of the fly. In the course of this work, Futch looked at the salivary gland chromosomes of flies from 12 different localities. He discovered 12 paracentric inversions, 3 pericentric inversions, and 1 translocation. Because of the precise banding patterns of these chromosomes, it was possible to determine the points of breakage for each inversion.

Figure 1. Photomicrographs of the Left Arm of Chromosome 2 (2L) from Larvae Heterozygous for Various Complex Gene Arrangements

(a) Pairing when heterozygous for standard gene sequence and overlapping inversions (2LC; 2LD) and inversion 2LB. (Standard × Tutuila light)
(b) Pairing when heterozygous for standard gene sequence and single inversion 2LC and overlapping inversions (2LE;2LB). (Standard × New Guinea) (c) Pairing when heterozygous for overlapping inversions (2LD;2LE;2LF). (Tutuila light × New Guinea)

Source: David G. Futch, "A Study of Speciation in South Pacific Populations of *Drosophila ananassae,*" in Marshall R. Wheeler, ed., *Studies in Genetics,* no. 6615 (Austin: University of Texas Press, 1966). Reproduced by permission.

a.

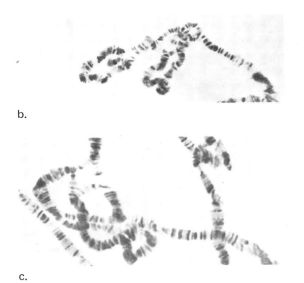

b.

Figure 2. Chromosome Maps of 2L

(a) Standard gene sequence. (b) Ponape: Break-points of 2LC and 2LB are indicated and the segments are shown inverted. (c) Tutuila light: Break-points of 2LD are indicated. 2LC and 2LB are inverted. 2LD, which overlaps 2LC, is also shown inverted. (d) New Guinea: Break-points of 2LE, 2LF, and 2LG are indicated. 2LC, 2LB, and 2LE are shown inverted. Note: Only the break-points of 2LF and 2LG are shown; neither of these inversions is inverted in the map.

Source: David G. Futch, "A Study of Speciation in South Pacific Populations of *Drosophila ananassae,*" in Marshall R. Wheeler, ed., *Studies in Genetics,* no. 6615 (Austin: University of Texas Press, 1966). Reprinted by permission.

c.

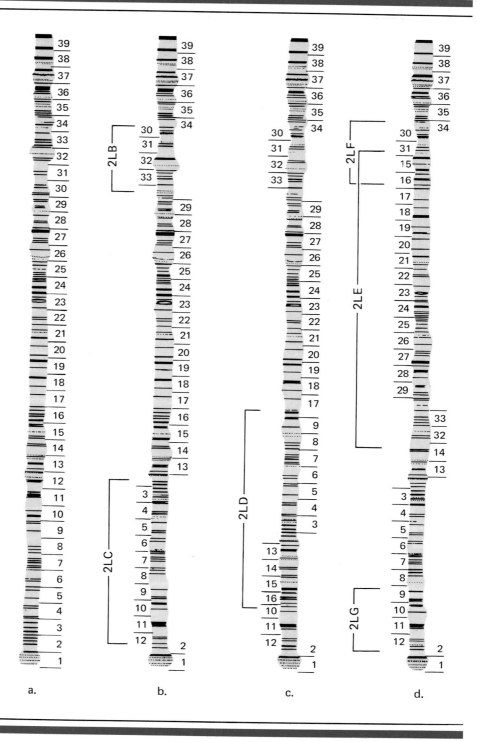

a.

b.

c.

d.

Observation of several populations, which have had sequential changes in their chromosomes, makes it possible to determine the sequence of events if each successive alteration occupied part of the previous change. By knowing the sequence of change in different populations of *Drosophila ananassae* and knowing the geographic location of these populations, it is possible to determine the history of colonization of these tropical islands by the flies. *D. ananassae* is particularly suited to this type of work because it is believed to be a recent, commensal invader with humankind to most of the Pacific islands that it occupies. It is of interest to know something about the spread of this species as an adjunct to studies of human migration in the Pacific islands because *D. ananassae* is commensal with humans.

Some of Futch's results are shown in the four accompanying figures, which diagram the left and right arms of the second chromosome, as well as the synaptic patterns. We can see vividly the sequence of change where one inversion occurs after a previous inversion has already taken place. In the figures the standard (a) gave rise to (b), which then gave rise independently to (c) and (d). The standard is from Majuro in the Marshall Islands and is believed to be in the ancestral group of the species. (b) is Ponape, (c) is Tutuila (Eastern Samoa), and (d) is New Guinea. Thus the sequence, given that the standard is at the beginning rather than the end, is: Majuro to Ponape and from there the same stock was transferred to Tutuila as well as New Guinea. This type of analysis has been useful in the *Drosophila* group throughout its range but especially in the Pacific Island populations and in the southwestern United States.

a.

b.

c.

Figure 3. Photomicrographs of Right Arm of Chromosome 2 (2R) from Larvae Heterozygous for Various Complex Gene Arrangements
(a) Pairing when heterozygous for standard gene sequence and overlapping inversions (2RA;2RB). (Standard × Tutuila light) (b) Pairing when heterozygous for standard gene sequence and overlapping inversions (2RA;2RC) and inversion 2RD. (Standard × New Guinea) (c) Pairing when heterozygous for overlapping inversions 2RB, 2RC, and 2RD. Inversion 2RA is homozygous. (Tutuila light × New Guinea)

Source: David G. Futch, "A Study of Speciation in South Pacific Populations of *Drosophila ananassae*," in Marshall R. Wheeler, ed., *Studies in Genetics*, no. 6615 (Austin: University of Texas Press, 1966). Reproduced by permission.

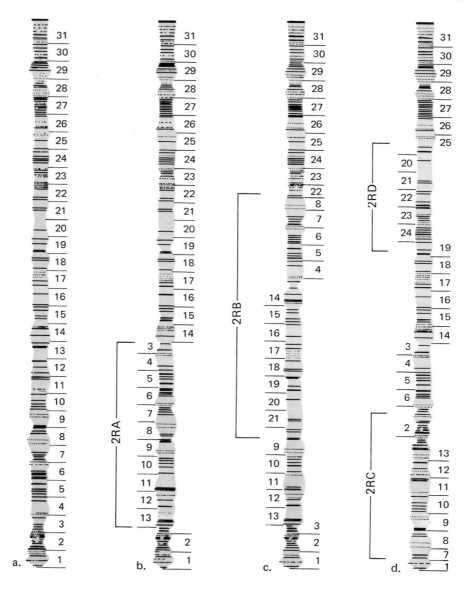

Figure 4. Chromosome Maps of 2R
(a) Standard gene sequence. (b) Ponape: Break-points of 2RA are indicated and the segment is shown inverted. (c) Tutuila light: Break-points of 2RB are indicated. 2RA is inverted and 2RB, which overlaps it, is also shown inverted. (d) New Guinea: Break-points of 2RC and 2RD are indicated. 2RA is inverted; 2RC, which overlaps 2RA, and 2RD are shown inverted.

Source: David G. Futch, ''A Study of Speciation in South Pacific Populations of *Drosophila ananassae*,'' in Marshall R. Wheeler, ed., *Studies in Genetics*, no. 6615 (Austin: University of Texas Press, 1966). Reprinted by permission.

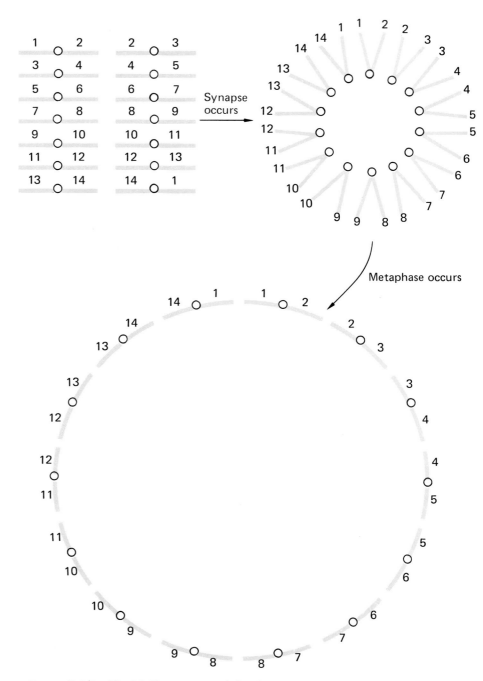

Figure 7-15. The 14 Chromosomes of *Oenothera biennis*, with Numbers to Denote Each Arm

chromosomes, then a larger ring will be formed, and so on up to a ring of 14 chromosomes (Figure 7–15).

An interesting feature of these *Oenothera* translocation rings is that the pattern of centromere segregation is almost exclusively of the alternate type (Figure 7–16). Thus, the chromosome complex of this species falls into two linkage groups. These gametic combinations have been given specific names, such as *johansen, velans, gaudans,* and so on, or generally termed **Renner complexes** after the man who devised the terminology. Further surprising findings are that: (1) races breed true for particular heterozygote combination—no homozygous complexes are formed; and (2) about half the seeds produced by self-fertilization of these heterozygous races are sterile. These facts make sense if we assume that each complex has different recessive lethal alleles. Thus a hybrid of *a* and *b* complexes produces gametes that are of either the *a* or *b* type. In self-pollination an *aa* or *bb* organism would be homozygous for the recessive lethals and would die. The heterozygous combinations (*ab* or *ba*) would reconstitute the parent. Thus the original combination would be maintained and the homozygous combinations would be sterile. This system is an example of a **balanced lethal system.**

Fusions.   Another interesting variant of the simple reciprocal translocation occurs when two acrocentric chromosomes become joined together at the centromere. The process is called a **Robertsonian fusion** and produces a decrease in the number of chromosomes while the same amount of genetic material is maintained. Often, closely related species will have markedly different chromosome numbers without any significant difference in the quantity of their genetic material. Such situations could be the result of Robert-

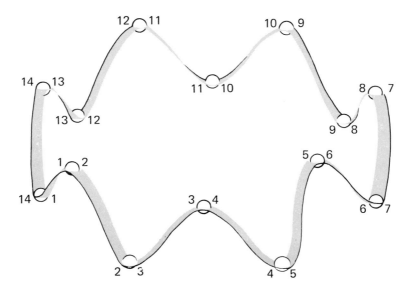

Figure 7–16
Alternate
Segregation of
Centromeres in
Oenothera biennis

sonian fusions. Therefore, cytotaxonomists frequently count the number of chromosome arms rather than the number of chromosomes to get a more accurate picture of species affinities. The number of arms is referred to as the **fundamental number** or **NF.** In a similar fashion, **centromeric fission** will increase the chromosome number without changing the NF.

Duplications

Duplications of chromosome segments can occur, as we have just seen, by the breakage–fusion–bridge cycle or by crossovers within the loop of an inversion. There is also another way that duplications arise, specifically, duplications of small adjacent regions of a chromosome. A particularly interesting example is the *Bar* eye phenotype in *Drosophila* (Figure 7–17). The wild-type fruit fly has about 800 facets in each eye. The *Bar* (*B*) homozygote has about 70 (20–120). Another allele, *Doublebar* (*BB:* sometimes referred to as *Ultrabar*, B^U), brings the facet number of the eye down to about 25 when homozygous. Around 1920 it was shown that about 1 progeny in 1600 from *Bar* homozygous females was *Doublebar*. This is much more frequent than we expect from mutation.

Sturtevant found that in every *Doublebar* fly there was a crossover between loci on either side of the *Bar* locus. He suggested that the change to *Doublebar* was due to unequal crossing over rather than to a simple mutation of one allele to another (Figure 7–18). If the homologous chromosomes do not line up exactly during synapsis, a crossover will result in an

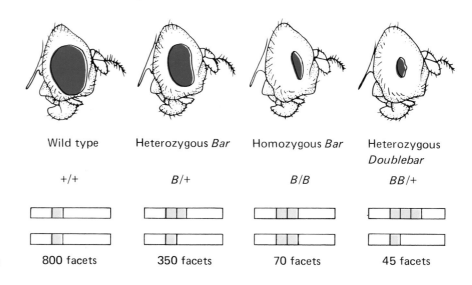

Wild type	Heterozygous *Bar*	Homozygous *Bar*	Heterozygous *Doublebar*
+/+	B/+	B/B	BB/+
800 facets	350 facets	70 facets	45 facets

Figure 7–17. Bar Eye in *Drosophila* Females

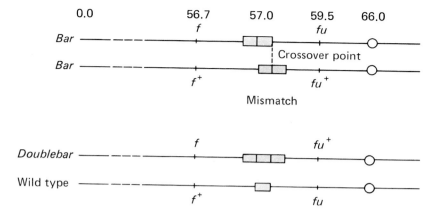

Figure 7-18. Unequal Crossing Over in a Female *Bar*-Eyed Homozygous *Drosophila* as a Result of Improper Pairing
The production of *Doublebar* (and concomitant reversion to wild type) is accompanied by crossover between forked (*f*) and fused (*fu*), two flanking loci.

unequal distribution of chromosomal material. Later, an analysis of the banding pattern of the salivary glands confirmed Sturtevant's position. It was found that *Bar* is a duplication of six bands in the 16A region of the X chromosome (Figure 7–19). *Doublebar* is a triplication of the segment.

There is an interesting position effect in the *Bar* system. A *Bar* homozygote (*B/B*) and a *Doublebar*/wild-type heterozygote (*BB/+*) both have four copies of the 16A region. It would therefore be reasonable to expect that both genotypes would produce the same phenotype. However, the *Bar* homozygote has about 70 facets in each eye while the other has about 45. Thus, not only the amount of genetic material but also its configuration determines the extent of the phenotype. *Bar* eye was the first position effect discovered.

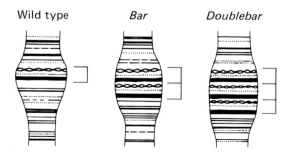

Wild type   Bar   Doublebar

Figure 7-19.
Bar Region of the X
Chromosome of
Drosophila

VARIATION IN CHROMOSOME NUMBER

Anomalies of chromosome number occur as two types—**euploidy** and **aneuploidy.** Euploidy involves changes in whole sets of chromosomes; aneuploidy involves changes in individual chromosomes (the change is less than an addition or deletion of a whole set). Aneuploidy is discussed first.

Aneuploidy

The terminology of aneuploid change is given in Table 7–2. A diploid cell missing a single chromosome is a **monosomic.** If it is missing both copies of that chromosome, it is a **nullisomic.** A cell missing two nonhomologous chromosomes is called a double monosomic. A similar terminology exists for extra chromosomes. Aneuploidy results from nondisjunction in meiosis or by chromosomal lagging whereby one chromosome moves more slowly than the others during anaphase and is then lost. Here, nondisjunction is illustrated using the sex chromosomes in an XY organism such as humans or fruit flies. We will look at four examples: nondisjunction in either the male or female at either the first or second meiotic divisions (Figure 7–20). The types of zygotes that can result when these nondisjunction gametes fertilize normal gametes are shown in Figure 7–21. All of the offspring produced are chromosomally abnormal. The names and kinds of these imbalances in humans are detailed later in this chapter.

Drosophila.   The occurrence of nondisjunction in *Drosophila* was first shown by Bridges in 1916 with crosses involving the white-eye locus. When a white-eyed female was crossed with a wild-type male, typically the daughters were wild type and the sons were white eyed. However, occasionally (1 or 2 per 1000), a white-eyed daughter or a wild-type son appeared (Figure 7–22). This could be explained most easily by a nondisjunction event in the

TABLE 7 – 2.   Partial List of Terms to Describe Aneuploidy, Using *Drosophila* as an Example (8 chromosomes: X, X, 2, 2, 3, 3, 4, 4)

Type	Formula	Number of Chromosomes	Example
Normal	$2n$	8	X/X; 2/2; 3/3; 4/4
Monosomic	$2n - 1$	7	X/X; 2/2; 3; 4/4
Nullisomic	$2n - 2$	6	X/X; 2/2; 4/4
Double monosomic	$2n - 1 - 1$	6	X/X; 2; 3; 4/4
Trisomic	$2n + 1$	9	X/X; 2/2; 3/3; 4/4/4
Tetrasomic	$2n + 2$	10	X/X; 2/2; 3/3/3/3; 4/4
Double trisomic	$2n + 1 + 1$	10	X/X; 2/2/2; 3/3/3; 4/4

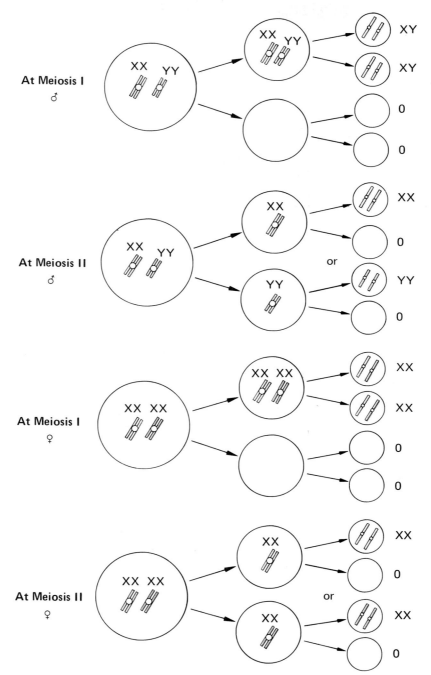

Figure 7-20. Nondisjunction of the Sex Chromosomes in *Drosophila* or Humans

	Nondisjunction ♂				Nondisjunction ♀	
Normal	XY	XX	YY	0	XX	0
♀X	XXY	XXX	XYY	X0		
♂X					XXX	X0
♂Y					XXY	Y0

Figure 7-21. Results of Fusion of a Nondisjunction Gamete by a Normal Gamete

white-eyed females, where X^wX^w and 0 eggs (without sex chromosomes) were formed. If an X^wX^w egg were fertilized by a Y-bearing sperm, the offspring would be an X^wX^wY white-eyed daughter. If the 0-bearing egg were fertilized by a normal X^+ bearing sperm, the result would be an X^+ wild-type son. After 1916 these exceptional individuals were found by cytological examination to have precisely the predicted chromosome types (XXY daughters and X sons). The other types produced by this nondisjunctional event are the XX egg fertilized by an X-bearing sperm and the 0 egg fertilized by the Y-bearing sperm. The XXX zygotes are genotypically $X^wX^wX^+$, or viable wild-type daughters, whereas the Y flies die.

Datura. Nondisjunction can also occur in the autosomes. In humans, trisomy 21 (Down's syndrome) is caused this way. Among plants, the jimsonweed *(Datura stramonium)*, which has 12 pairs of chromosomes ($2n = 24$), has been studied extensively. Blakeslee noted a mutant, globe, that had an

P — White eye ♀ X^wX^w × Wild type ♂ X^+Y

F_1 — Normally: Wild type ♀ X^+X^w; White eye ♂ X^wY

or

Occasionally (nondisjunction): White eye ♀ X^wX^wY; Wild type ♂ X^+

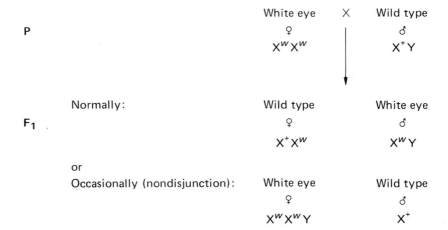

Figure 7-22. Crosses Showing Nondisjunction of the *Drosophila* Sex Chromosomes

unusual inheritance pattern. This mutation was inherited primarily through females, but with distorted segregation ratios. Eventually, 12 mutants with this type of distorted inheritance pattern were found. This number corresponds to the haploid chromosome number of *Datura*. Belling, in 1920, discovered that each of these 12 mutants was trisomic for a different one of the 12 chromosomes. The seed capsule for each of these trisomics is shown in Figure 7–23. The distorted pattern of inheritance was due to the fact that pollen with an extra chromosome did not grow as fast as normal pollen. Thus the trisomic tended to be inherited only through the females. The odd ratios (1:3 for a dominant) seemed to be caused by both a loss of some of these extra chromosomes during meiosis and a lowered viability of the trisomics.

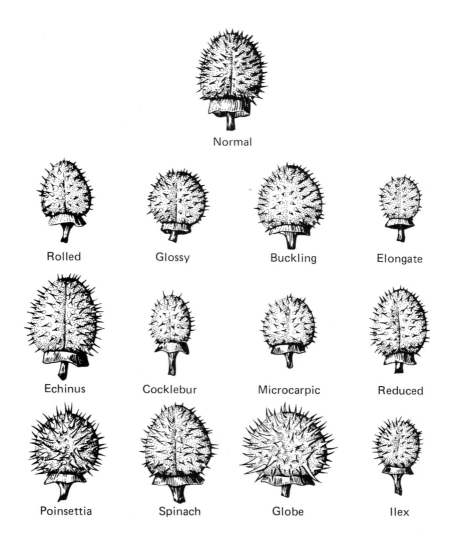

Figure 7–23.
Seed Capsules for
the Normal and the
12 Trisomic *Datura*
Types

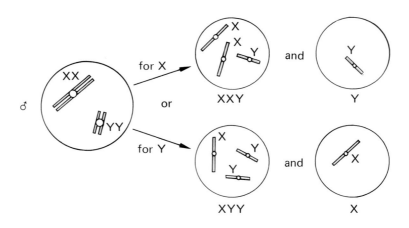

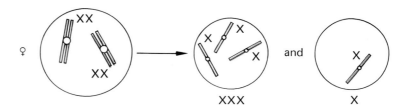

Figure 7-24.
Mitotic
Nondisjunction of
the Sex
Chromosomes

Mosaicism

In many species some individuals are made up of several cell lines, each with different chromosome numbers. These individuals are referred to as **mosaics** or **chimeras.** Such conditions are the result of nondisjunction or chromosome lagging during mitosis in the zygote or cleavage nuclei, in a manner

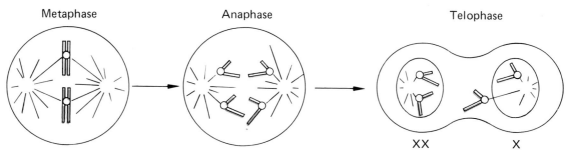

Figure 7-25. Chromosome Lagging at Mitosis in the X Chromosome of Female *Drosophila*

similar to that found in meiosis. This is demonstrated, again for sex chromosomes, in Figure 7–24. A lagging chromosome is shown in Figure 7–25, where the X chromosome is shown to be lost in one of the dividing somatic cells; this results in an XX cell line and an X0 cell line. In *Drosophila*, if this chromosome lagging occurs early in development, an organism that is part male (X) and part female (XX) develops. Figure 7–26 shows a fruit fly in which chromosome lagging has occurred at the one-cell stage, causing half the fly to be male and half female. Mosaics of this type, involving male and female phenotypes, have a special name—**gynandromorphs.** Many sex chromosome mosaics are known in humans, including XX/X, XY/X, XX/XY, and XXX/X.

Aneuploidy in Humans

In humans, 50% of spontaneous abortions (miscarriages) in the United States are found to have some chromosomal abnormality, of which about half are autosomal trisomics. About one in 160 human live births has some sort of chromosomal anomaly, of which most are balanced translocations, autosomal trisomics, and sex-chromosome aneuploids.

A normal human chromosome complement is 46,XX for a female and 46,XY for a male. In the standard system of nomenclature, the total chromosome number is given first, followed by the description of the sex chromosomes, and finally, a description of autosomes if some autosomal anomaly is evident. For example, a person with an extra X chromosome would be 47,XXY. A person with a single X chromosome would be 45,X. Since all the autosomes are numbered, we describe their changes by referring to their addition (+) or deletion (−). For example, a female with trisomy 21 would

Figure 7-26. *Drosophila gynandromorph*
Left side is wild-type XX female. Right side is XO male, hemizygous for white eye and miniature wing.

be 47,XX,+21. The short arm of a chromosome is designated p; the longer arm, q. When a change in part of the chromosome occurs, a + after the arm indicates an increase in the length of that arm, whereas a − indicates a decrease in its length. For example, a translocation where part of the short arm of 9 is transferred to the short arm of 18 would be: 46,XX,t(9p−;18p+). The semicolon indicates that both chromosomes kept their centromeres.

Trisomy 21 (Down's Syndrome), 47,XX,+21 or 47,XY,+21. Down's syndrome affects about 1 in 1000 live births (Figures 7–27, 7–28). Most affected individuals are mildly to moderately mentally retarded and have congenital heart defects and a very high (1/100) risk of acute leukemia. They are also usually short and have a broad, short skull; hyperflexibility of joints; and excess skin on the back of the neck. Langdon Down first described this syndrome in 1866. It was the first human syndrome found to be due to chromosome disorder. An interesting aspect of this syndrome is the increased incidence among the children of older mothers (Figure 7–29). Since the future ova are set aside prior to birth, all ova are the same age as the female. Presumably older ova have an increased likelihood of nondisjunction of chromosome 21. (Most recently, it has been determined that 20 to 25% of the

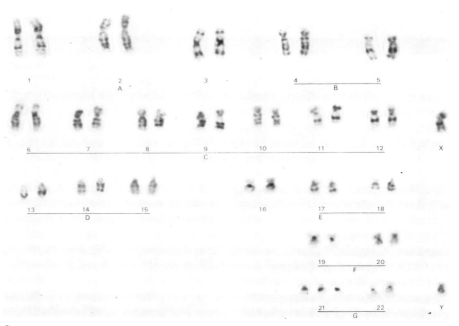

Figure 7–27. Karyotype of an Individual with Trisomy 21

Source: Reproduced courtesy of Dr. Thomas G. Brewster, Foundation for Blood Research, Scarborough, Maine.

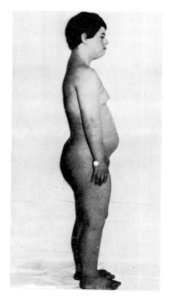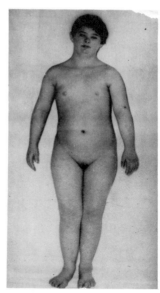

Figure 7-28.
Individual with
Trisomy 21

Source: Reproduced courtesy of Dr. Jérôme Lejeune, Institut de Progenese, Paris.

cases of Down's syndrome are due to the father, and the effect increases with the father's age.)

Trisomy 18 (Edward's Syndrome), 47,XX,+18 or 47,XY,+18. Edward's syndrome affects 1 in 7500 live births (Figure 7–30). Most affected individuals are female, with 80 to 90% mortality by two years of age. The affected usually have an elfin appearance with small nose and mouth, a receding lower jaw, abnormal ears, and a lack of distal flexion creases on the fingers. There is limited motion of the distal joints and a characteristic posturing of the fingers where the little and index fingers overlap the middle two. The syndrome is usually accompanied by severe mental retardation.

Trisomy 13 (Patau's Syndrome), 47,XX,+13 or 47,XY,+13, and Other Trisomic Disorders. Patau's syndrome affects 1 in 15,000 live births. Diagnostic features are cleft palate, cleft lip, congenital heart defects, polydactyly, and severe mental retardation. Mortality is very high in the first year of life.

Other autosomal trisomies are known but are extremely rare. These include trisomy 8, 47,XX or XY,+8, and cat's eye syndrome, a trisomy of an unknown, small acrocentric, 47,XX or XY(+acrocentric). Several sex-chromosome related aneuploids are known.

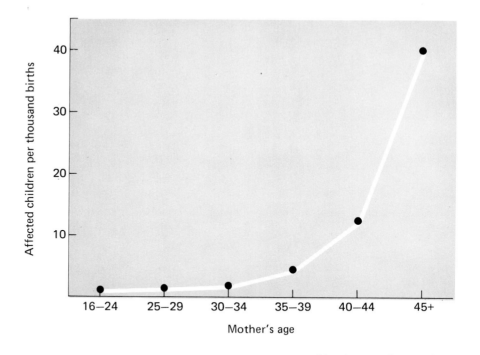

		Number per thousand
16−24	1/1700	0.58
25−29	1/1100	0.91
30−34	1/770	1.30
35−39	1/250	4
40−44	1/80	12.5
45+	1/25	40

Figure 7–29.
Increased Risk of
Trisomy 21
Attributed to the
Age of the Mother

Source: Data from E. Hook, "Estimates of Maternal Age-Specific Risks of a Down's-Syndrome Birth in Women Age 34–41," *Lancet* 2 (1976):33–34.

Turner's Syndrome, 45,X. About 1 in 2000 live female births is a Turner's syndrome infant. This and 45,XX,−22 and 45,XY,−22 are the only nonmosaic, viable monosomics recorded in humans (Figure 7–31). Turner's syndrome individuals have about normal intelligence but under-developed ovaries, abnormal jaws, webbed necks, and shield-like chests.

XYY Syndrome, 47,XYY. About 1 in 1000 live births is an XYY male. These individuals possess normal intelligence; they are taller than normal. There has been some controversy surrounding this syndrome because it was originally reported that there was an abundance of this karyotype in a group of mentally subnormal males in a hospital for prisoners requiring special treatment for violent crimes. Seven XYY males were found among 197 inmates whereas only 1 in about 2000 control men were XYY. This study has subsequently been expanded and corroborated. Although it is now fairly well established that there is about a twentyfold higher incidence of XYY males in prison than in society at large, the statistic is somewhat misleading: The overwhelming number of XYY men lead perfectly normal lives. At most, about 4% of XYY men end up in penal or mental institutions, where they make up about 2% of that population.

Research on this syndrome has produced its own problems. A research project at Harvard University on XYY males came under intense public pressure and was eventually terminated. The project, under the direction of Stanley Walzer (a psychiatrist) and Park Gerald (a geneticist) involved screening all newborn boys at the Boston Hospital for Women and following the development of those with chromosomal anomalies. The criticism of this work centered mainly on the necessity of informing parents that their sons had an XYY karyotype and that there could be behavioral problems. Opponents of this work claimed that telling the parents would constitute a self-

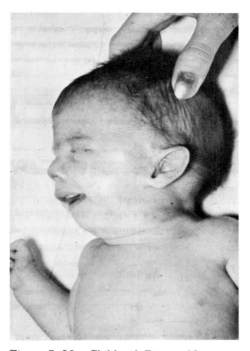 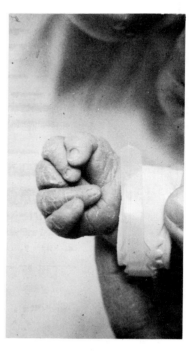

Figure 7-30. Child with Trisomy 18
Source: Reproduced courtesy of Dr. Jérôme Lejeune, Institut de Progenese, Paris.

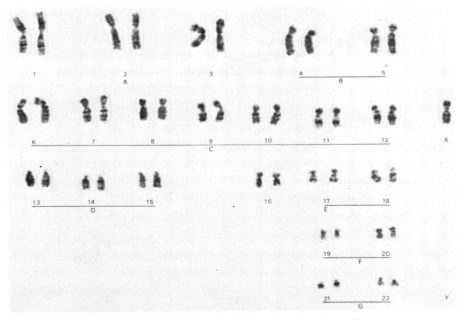

Figure 7-31.
Karyotype of a
Person with Turner's
Syndrome (XO)

Source: Reproduced courtesy of Dr. Thomas G. Brewster, Foundation for
Blood Research, Scarborough, Maine.

fulfilling prophecy. That is, parents who were told that their children were not normal and "might" cause trouble would have children that caused trouble. The opponents claimed that the risks of this research outweighed the benefits. The project was shut down in 1975 because of the harassment that Walzer received.

Klinefelter's Syndrome, 47,XXY. The incidence of Klinefelter's syndrome is about 1 in 2000 live births. Tall stature and infertility are common. Diagnosis is usually by buccal smear to ascertain the presence of a Barr body with a male phenotype (indicative of XXY). Some behavioral and speech development problems have been associated with this syndrome.

Triple-X Female, 47,XXX, and Other Sex-Chromosomal Aneuploid Disorders. A triple-X female appears in about 1 in 2000 live births. Fertility can be normal, but there is usually a mild mental retardation. Delayed growth, as well as nonspecific congenital malformations, has been noted. Other sex-chromosomal aneuploids, including XXXX, XXXXX, and XXXXY, are extremely rare. All seem to be characterized by mental retardation and growth deficiencies.

Chromosome Rearrangements in Humans

In addition to the aneuploids mentioned, several human syndromes and abnormalities are the result of chromosomal rearrangements, including deletions and translocations. The most common disorders are described here. Three points should be kept in mind. First, all are extremely rare. Second, the deletion syndromes are often found to be caused by a balanced translocation in one of the parents. And third, about 1 in 500 live births contains a *balanced* rearrangement of some kind—either a reciprocal translocation or inversion.

Cri-du-Chat Syndrome, 46,XX, (or XY,)5p−. The syndrome known as *cri du chat* is so called because of the cat-like cry that about half the affected show as infants. Microcephaly and congenital heart disease are also common (Figure 7–32). Mental retardation is severe. This disorder arises from a deletion in chromosome 5 (Figure 7–33). Most of the deletions studied (4p−, 13q−, 18p−, 18q−) result in microcephaly and severe mental retardation.

Familial Down's Syndrome. Trisomy 21 has been described earlier as the result of a nondisjunction event in oogenesis. This is a function of mater-

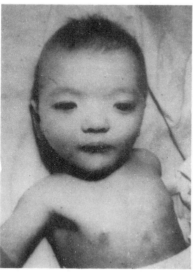

Figure 7-32.
Child with *Cri-du-Chat* Syndrome

Source: Reproduced courtesy of Dr. Jérôme Lejeune, Institut de Progenese, Paris.

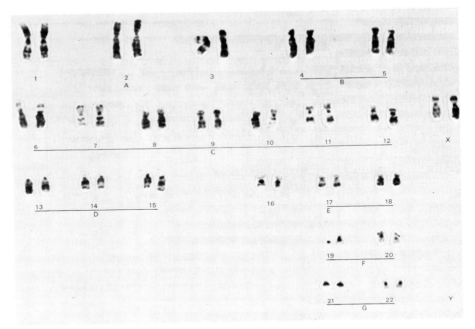

**Figure 7-33.
Karyotype of
Individual with** *Cri-
du-Chat* **Syndrome**
Partial deletion of
the short arm of
chromosome 5
(5p−).

Source: Reproduced courtesy of Dr. Thomas G. Brewster, Foundation for
Blood Research, Scarborough, Maine.

nal age and is not inherited. (Although about half the children of a trisomy
21 person will have trisomy 21 because of aneuploid gamete production, the
possibility of an unaffected relative of the trisomy 21 person having abnormal children is no greater than for a person of the same age chosen at random from the general population.) However, a very small fraction of those
with Down's syndrome have been found to have a translocation of chromosome 21, usually associated with chromosomes 14, 15, or 22. The translocation and nontranslocation types of Down's syndrome have identical symp-
toms; but, of course, a balanced translocation could be passed on to offspring
(see Figure 7–14). Alternate separation of centromeres in the translocation
heterozygote will produce either a normal gamete or one carrying the balanced translocation. Adjacent separation will cause partial trisomy for certain chromosome parts. When this occurs for most of the 21 chromosome,
Down's syndrome results.

Attached-X Chromosomes

In 1922 L. V. Morgan discovered a strain of *Drosophila* where nondisjunction occurred 100% of the time—that is, white-eyed females mated to red-

eyed males produced only white-eyed daughters and red-eyed sons. Thus, instead of the expected crisscross pattern of sex linkage, these matings produced daughters exactly like their mothers and sons like their fathers. It seemed that daughters received both of their X chromosomes from their mothers and sons received their X chromosomes directly from their fathers. This could be explained if the two X chromosomes of the female were attached and inherited as a unit (Figure 7–34). Females in this stock would be attached-X, (XX)Y, while males would be the normal XY. Females would produce (XX) gametes and Y-bearing gametes. The fertilization possibilities are shown in Figure 7–34. (XX)X females usually die, as do all YY males. This leaves (XX)Y females and XY males. The females are white eyed like their mothers; the males, red eyed like their fathers.

Later cytological evidence has shown that, in fact, the attached-X chromosomes are connected to each other at the centromere. (Since the X chromosome in *Drosophila* is telocentric, two chromosomes can be attached to each other at their centromeres by a Robertsonian fusion. The resulting chromosome, with two identical halves, is called an **isochromosome.**) Discovery of the attached-X chromosome came at an important time in the development of the chromosome theory of inheritance and gave added support to that theory, in that cytological evidence verified the earlier prediction. Strains of *Drosophila* with attached-X chromosomes have been very useful experimentally because both X chromosomes (homologues) are recovered together. This has made it possible to demonstrate that crossing over in *Drosophila* is at the four-strand stage, a fact previously demonstrated in *Neurospora.* This attached-X strain has also been useful in mutation work and studies of chromosomal anomalies and recombinational events.

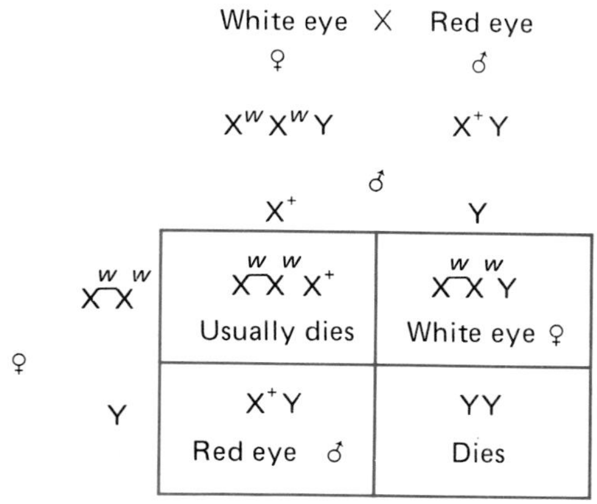
White eye × Red eye
♀
♂
X^w X^w Y
X^+ Y
♂
X^+
Y
X^w X^w
X^w X^w X^+
Usually dies
X^w X^w Y
White eye ♀
♀
Y
X^+ Y
Red eye ♂
YY
Dies

Figure 7–34. Inheritance Pattern for the Attached-X Chromosomes in *Drosophila*

Euploidy

Euploid organisms have varying numbers of complete haploid chromosome sets. We are already familiar with haploids (n) and diploids ($2n$). Organisms with higher numbers of sets are called **polyploids,** such as **triploids** ($3n$) and **tetraploids** ($4n$). Three kinds of problems plague polyploids. First, there is a potential for a general imbalance in the organism due to the extra genetic material in each cell. Second, if there is a chromosomal sex-determining mechanism, it will be distorted by polyploidy. And third, meiosis produces unbalanced gametes in many polyploids.

If the polyploidy is of odd sets of chromosomes, such as triploidy, two of the three homologues will tend to pair, producing a bivalent and a univalent at prophase I of meiosis. The bivalent will separate normally, but the third chromosome will go separately to one of the poles. This separation results in some aneuploid gametes and therefore tends to cause sterility. An alternative to the bivalent–univalent type of synapsis is the formation of trivalents, which have similar problems (Figure 7–35). Even-numbered poly-

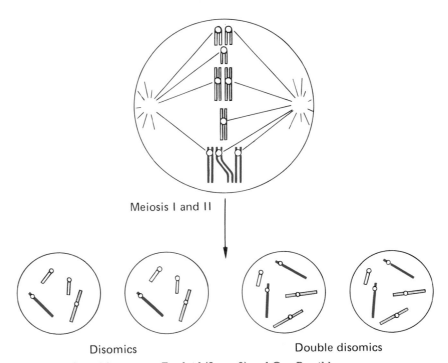

Meiosis I and II

Disomics   Double disomics

Figure 7 – 35.   Meiosis in a Triploid ($3n = 9$) and One Possible
Resulting Arrangement of Gametes
The probability of a "normal" gamete is $(1/2)^n$ where n equals the haploid chromosome number. Here, $n = 3$ and $(1/2)^3 = 1/8$.

ploids, such as tetraploids, can do better during meiosis. Often, however, the multiple copies of the chromosomes form complex figures during synapsis, including monovalents, bivalents, trivalents, and quadrivalents. This too tends to result in aneuploid gametes and sterility. If the segregation of centromeres is 2 and 2, balanced gametes can result. Some groups of organisms, primarily plants, have exploited polyploidy in their evolutionary development; they have many polyploid members. For example, the genus of wheat (*Triticum*) has members with 14, 28, and 42 chromosomes. Because the basic *Triticum* chromosome number is $n = 7$, these forms are $2n$, $4n$, and $6n$ species. Chrysanthemums have species of 18, 36, 54, 72, and 90 chromosomes. With a basic number of $n = 9$, these species represent a $2n$, $4n$, $6n$, $8n$, and $10n$ series. In both these examples, the even-numbered polyploids are viable species but the odd-numbered polyploids are not.

Autopolyploidy.   Polyploidy can come about in two different ways. In **autopolyploidy** all of the chromosomes come from within the same species. In **allopolyploidy** the chromosomes come from the hybridization of two different species. Autopolyploidy occurs in several different ways. The fusion of nonreduced gametes will create polyploidy. For example, if a diploid gamete fertilizes a normal haploid gamete, the result will be a triploid. Likewise, a diploid gamete fertilized by another diploid gamete will produce a tetraploid. The equivalent of a nonreduced gamete will come about in meiosis if the parent cell is polyploid to begin with. For example, if a branch of a diploid plant is tetraploid, its flowers will produce diploid gametes. These gametes are not the result of a failure to reduce chromosome numbers meiotically but rather the result of successful meiotic reduction in a polyploid flower. The tetraploid tissue of the plant in this example can originate by the process of **somatic doubling** of diploid tissues.

The process of somatic doubling can come about spontaneously or be caused by any chemical or process that disrupts the normal sequence of a nuclear division. For example, colchicine will induce somatic doubling by inhibiting microtubule formation. This prevents the formation of a spindle and thus prevents the movement of the chromosomes during either mitosis or meiosis. The end result is a cell with double the chromosome number. Other chemicals, temperature shock, and physical shock can produce the same effect.

Allopolyploidy.   Allopolyploidy comes about by cross-fertilization between two species. The resulting offspring have the sum of the reduced chromosome number of each parent species. If the chromosome set of each is distinctly different, these new organisms have difficulty in meiosis because no two chromosomes are sufficiently homologous. Every chromosome will form a univalent (unpaired) figure. They will separate independently during meiosis, producing aneuploid gametes. However, if, say, an organism can survive by vegetative growth until somatic doubling takes place in a gamete precursor cell ($2n \rightarrow 4n$), or alternatively, if the zygote was formed by two unreduced gametes ($2n + 2n$), the resulting offspring will be fully fertile because each chromosome has a pairing partner for meiosis. An example can be drawn from the work of a Russian geneticist, Karpechenko.

In 1928 Karpechenko worked with radishes (*Raphanus sativus*, $2n = 18$) and cabbage (*Brassica oleracea*, $2n = 18$). When these two plants are crossed, an F_1 results with $2n = 18$. This plant has characteristics intermediate between the two parental species (Figure 7–36) and is an allodiploid.

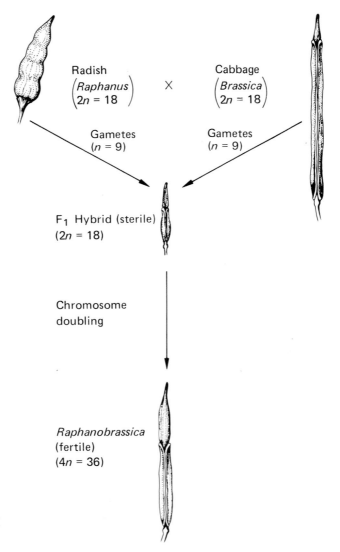

Radish
$\begin{pmatrix} Raphanus \\ 2n = 18 \end{pmatrix}$ $\times$ Cabbage $\begin{pmatrix} Brassica \\ 2n = 18 \end{pmatrix}$

Gametes ($n = 9$)   Gametes ($n = 9$)

F_1 Hybrid (sterile)
($2n = 18$)

Chromosome doubling

Raphanobrassica
(fertile)
($4n = 36$)

Figure 7–36.
Hybridization of Cabbage and Radish Showing the Fruiting Structures

If somatic doubling takes place, then the chromosome number is doubled to 36, and the plant becomes an allopolyploid (an allotetraploid of $4n$). Since each chromosome has a homologue, this allotetraploid is referred to as an **amphidiploid.** Without knowledge of its past history, this plant would simply be classified as a diploid with $2n = 36$. In this case, the new amphidiploid cannot successfully breed with either parent because the offspring are sterile triploids. It is, therefore, a new species and has been named *Raphanobrassica*. As far as an agricultural experiment, however, it was a failure. It had the root of a cabbage and the leaves of a radish!

Polyploidy in Plants versus Animals. Although polyploids in the animal kingdom are known, such as some lizards, fish, and invertebrates, polyploidy as a successful evolutionary strategy is primarily a plant phenomenon. There are several reasons for this difference between plants and animals. To begin with, many more animals than plants have chromosomal sex-determining mechanisms. These mechanisms are severely disrupted by polyploidy. For example, Bridges discovered a tetraploid female fruit fly. However, it has not been possible to produce a tetraploid male. The tetraploid female's progeny were triploids and intersexes. Second, plants can generally avoid the meiotic problems of polyploidy longer than most animals. Some plants can exist vegetatively, waiting for the rare somatic doubling event that will produce an amphidiploid. Animal life spans are more precisely defined, allowing less time for a somatic doubling. And third, many plants that depend on the wind or insect pollinators to fertilize them have more of an opportunity for hybridization. Many animals have relatively elaborate courting rituals that tend to control mating behavior and restrict interspecific hybridization.

Polyploid Genetics

Simple Mendelian analyses, including mapping, can be done on even-numbered polyploids. These analyses are complicated because of the large number of homologous chromosomes that are involved, as well as the numerous combinations that can ensue. We will briefly look at autotetraploids.

A heterozygote of the *AAaa* type can produce three types of gametes: *AA, Aa,* and *aa.* Assume for a moment that the locus lies near the centromere and that pairing normally produces two bivalents from each quartet of the same homologous chromosome. In that case, gametes will be a random assortment of the four chromosomes into groups of two. We should thus get a ratio of 1*AA*:4*Aa*:1*aa* (Figure 7–37). If a plant like this is selfed, a homozygote *aaaa* will occur in 1/36 of the offspring; in 1/6 if testcrossed with *aaaa.*

The situation becomes more complicated if the locus is further from its centromere. When crossing over occurs between the locus and the centro-

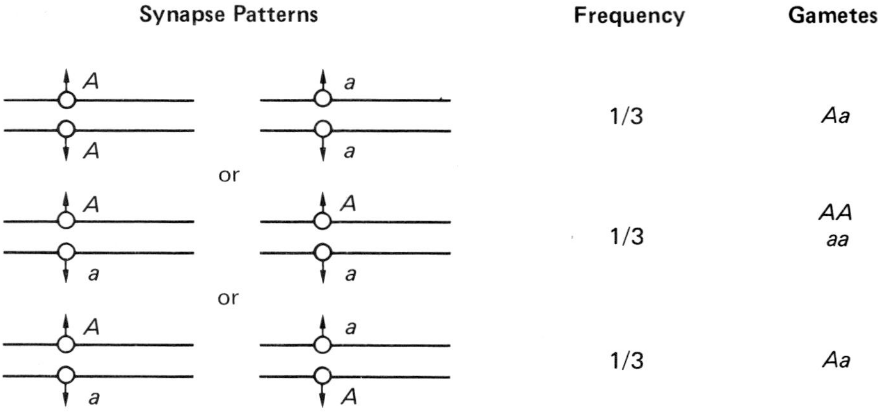

Synapse Patterns   **Frequency**   **Gametes**

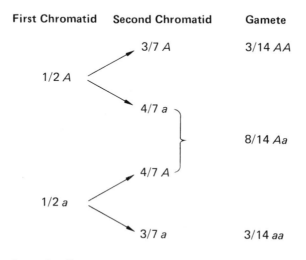

1/3   Aa

1/3   AA
           aa

1/3   Aa

Figure 7-37.
Segregation in a
Tetraploid
The A locus with *A*
and *a* alleles is
located near the
centromere.

Summing Gametes

1/3 *Aa* + 1/3(1/2 *AA* + 1/2 *aa*) + 1/3 *Aa* = 1 *AA* : 4 *Aa* : 1 *aa*

mere, the number of combinations is increased. That is, if the locus is far
enough from the centromere so that it is independent of it, we must consider
not just 4 chromosomes but 8 chromatids assorting in groups of 2. The
resulting ratio of gametes is 3*AA*:8*Aa*:3*aa* (Figure 7–38). In this case, the

First Chromatid   **Second Chromatid**   **Gamete**

                      3/7 A   3/14 AA

1/2 A

                      4/7 a

                                             8/14 Aa

                      4/7 A

1/2 a

                      3/7 a   3/14 aa

Figure 7-38.
Segregation in a
Tetraploid with the
A Locus More Than
50 Map Units from
Its Centromere
Eight Chromatids
(4*A* and 4*a*) are
taken and sorted in
groups of two to form
gametes.

Summing Gametes

3 *AA* : 8 *Aa* : 3 *aa*

probability of a homozygous *aaaa* in a self-cross is $(3/14)^2 = 9/196$ or approximately 1/22. We can see that the actual ratio will be somewhere between 1/36 and 1/22 depending on the distance of the locus from its centromere.

While polyploidy in the main is restricted to plants, it should be noted that many instances of polyploidy are known in animals. Triploid *Drosophila* are viable. Mammalian liver cells, cancer cells, and cells in tissue culture are often polyploid. Polyploidy has been used in agriculture for the production of "seedless" as well as "jumbo" varieties of crops. Seedless watermelon, for example, is a triploid. Its seeds are mostly sterile and do not develop. It is produced by growing seeds from the cross between a tetraploid variety and a diploid variety. Jumbo Macintosh apples are tetraploid.

CHAPTER SUMMARY

Variation can occur in the structure and the number of chromosomes an organism can have. When chromosomes break, the ends act in a "sticky" fashion; they tend to reunite with other broken ends. A single break can lead to the possibility of deletions or the formation of acentric or dicentric chromosomes. Dicentrics tend to go through breakage–fusion–bridge cycles, which result in duplications and deficiencies.

Two breaks in the same chromosome can yield deletions or inversions. Variegation position effects as well as new linkage arrangements can result. Inversion heterozygotes produce loop figures during synapsis, which can be either at meiosis or in polytene chromosomes. Inversion heterozygotes suppress crossovers and cause semisterility.

Reciprocal translocations can result from single breaks in nonhomologous chromosomes. These produce cross-shaped figures at synapsis and result in semisterility. *Oenothera,* the evening primrose, exhibits a balanced lethal system with many reciprocal translocations. The *Bar* eye phenotype of *Drosophila* is an example of a duplication that causes a position effect.

Changes in chromosome number can involve whole sets (euploidy) or single or partial sets (aneuploidy) of chromosomes. Aneuploidy usually results from nondisjunction or chromosome lagging. Several medical syndromes are caused by aneuploidy, such as the Down's, Turner's, Klinefelter's, and XYY syndromes.

Polyploidy leads to difficulties in sex determination, general chromosomal imbalance, and meiotic segregation. It has been more successful in plants than in animals because plants circumvent sex-determination problems by generally not having chromosomal mechanisms. Plants can also avoid meiotic problems by growing vegetatively. In both animals and plants, even-numbered polyploids do better than odd-numbered polyploids because there is a better chance of even-numbered polyploids producing balanced gametes during meiosis. Somatic doubling provides each chromosome in an auto- or allopolyploid with a homologue and thus makes tetrad formation at meiosis possible. New species have been formed by allopolyploidy followed by somatic doubling.

Polyploid genetics is a complex subject. Segregation patterns depend not only on the number of chromosomes but also on the distance of loci from their centromeres.

EXERCISES AND PROBLEMS

1. What kind of figure is observed in meiosis of a reciprocal translocation homozygote?

2. Can a deletion result in the formation of a variegation position effect?

3. Does crossover suppression occur in an inversion homozygote? Explain.

4. Which arrangements of chromosomal structural anomalies cause semisterility?

5. What are the consequences of single crossovers during tetrad formation in a reciprocal translocation heterozygote?

6. Is a tetraploid more likely to show irregularities in meiosis or mitosis? Explain. What about these processes in a triploid?

7. How many chromosomes would a human tetraploid have? How many chromosomes would a human monosomic have?

8. Do autopolyploids or allopolyploids experience more difficulties during meiosis? Do amphidiploids have more or less trouble than auto- or allopolyploids?

9. If a diploid species of $2n = 16$ hybridizes with one of $2n = 12$ and the resulting hybrid doubles its numbers to produce an allotetraploid (amphidiploid), how many chromosomes will it have?

10. If nondisjunction of the sex chromosomes occurs in a female at the second meiotic division, what types of eggs will arise?

11. In terms of acentrics, dicentrics, duplications, and deficiencies, give the

gametic complement when a three-strand, double crossover occurs within a paracentric inversion loop.

12. In studying a new sample of fruit flies, a geneticist noted a position effect, semisterility, and the nonlinkage of previously linked genes. What probably caused this and what cytological evidence would strengthen your theory?

13. In a second sample of flies, the geneticist found a position effect and semisterility. The linkage groups were correct, but the order was changed and crossing over was suppressed. What probably caused this and what cytological evidence would strengthen your theory?

14. Diagram the results of alternate segregation for a three-strand, double crossover between a centromere and the cross center in a reciprocal translocation heterozygote.

15. Label the arms of the 14 metacentric chromosomes of a race of *Oenothera* ($2n = 14$) that forms a ring and 4 bivalents during meiosis.

SUGGESTIONS FOR FURTHER READING

Brewster, T., and P. Gerald. 1978. Chromosome disorders associated with mental retardation. *Pediatr. Ann.* 7:82–89.

Cleland, R. 1972. Oenothera: *Cytogenetics and Evolution.* New York: Academic Press.

Futch, D. 1966. A study of speciation in South Pacific populations of *Drosophila ananassae. Univ. Texas Stud. Genet.,* no. 6615:79–120.

Hook, E. 1976. Estimates of maternal age-specific risks of Down's-syndrome birth in women aged 34–41. *Lancet* 2:33–34.

Schmid, W. 1977. Cytogenetical problems in prenatal diagnosis. *Hereditas* 86:37–44.

Stebbins, G. 1971. *Chromosomal Evolution in Higher Plants.* Reading, Mass.: Addison-Wesley.

Sturtevant, A. 1925. The effects of unequal crossing over at the *Bar* locus in *Drosophila. Genetics* 10:117–147.

Swanson, C. 1957. *Cytology and Cytogenetics.* Englewood Cliffs, N.J.: Prentice-Hall.

Valentine, G. 1975. *Chromosome Disorders: An Introduction for Clinicians,* 3rd ed. Philadelphia: Lippincott.

White, M. J. D. 1973. *Animal Cytology and Evolution,* 3rd ed. New York: Cambridge University Press.

8 Quantitative Inheritance

Earlier, when we talked of genetic traits, we were usually discussing traits controlled by single genes. Their inheritance pattern led to simple ratios. However, many traits, including many of economic importance—such as yields of milk, corn, and beef—exhibit what is called continuous variation (Figure 8–1).

In Mendel's peas, while there was some variation in height, all plants could be scored as either tall or dwarf; there was no overlap. We can set up the same kind of cross using ear length in corn. In Mendel's peas all of the F_1 were tall. In a cross between long and short ears of corn, all of the F_1 ears will be intermediate in length. When both these F_1s are self-fertilized, the results are again different. In the F_2 generation Mendel obtained exactly the same height categories (tall and dwarf) as in the parental generation, only the ratio was different—3:1. In corn, however, ears of every length, from the shortest to the longest, will be found in the F_2; there will be no discrete categories. Ear length in corn exhibits a type of variation in which virtually every phenotypic (length) category is found in the F_2. A genetically controlled trait exhibiting this type of variation is controlled by many loci. In this chapter we will study this type of variation by looking at traits that are controlled by progressively more loci. We will then turn to the concept of *heritability,* which is used as a statistical tool to evaluate the genetic control of traits determined by many loci.

TRAITS CONTROLLED BY MANY LOCI

Let us begin by considering grain color in wheat. When a particular strain of wheat having red grain is crossed with another strain having white grain, all the F_1 are intermediate in color. When these are selfed, the F_2 ratio is 1 red : 2 intermediate : 1 white (Figure 8–2). This is simple inheritance involving

261

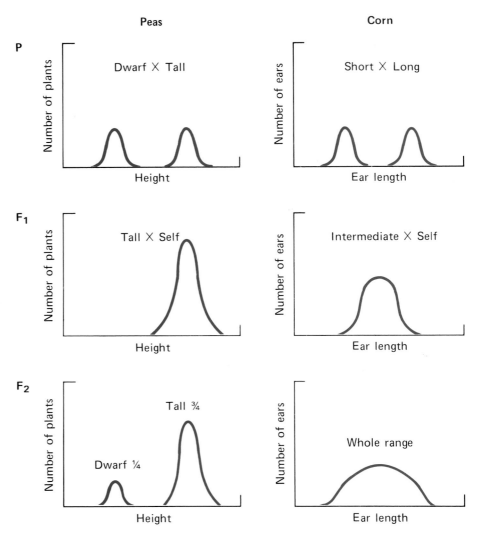

Figure 8-1.
Comparison of
Continuous
Variation (Ear
Length in Corn) with
Discontinuous
Variation (Height in
Peas)

one locus with two alleles. The white allele, *a,* produces no pigment (which results in "white"); the red allele, *A,* produces red pigment. The heterozygote, *Aa,* is intermediate and shows incomplete dominance. When this monohybrid is self-fertilized, the typical 1:2:1 ratio results.

Two-Locus Control

Now let us examine the same kind of cross using two other stocks of red and white wheat. Here, when the resulting intermediate (medium red) F_1 are

P	Red	X	White
	AA		*aa*

F₁ → **F₁**   Intermediate color

Aa

F₂   Red : Intermediate : White

AA   *Aa*   *aa*

Figure 8-2.
Cross Involving the
Grain Color of
Wheat

1 : 2 : 1

selfed, five color classes emerge in a ratio of 1 dark red : 4 medium dark red : 6 medium red : 4 light red : 1 white. This cross is shown in Figure 8–3. The ratios, in sixteenths, come from self-fertilizing a dihybrid of two unlinked loci, each segregating two alleles. In this case both loci affect the same trait in the same way. In Figure 8–3 each capital letter represents an allele that produces one unit of color and each lowercase letter represents an allele that produces no color. Thus a genotype of *AaBb* has two units of color just as a genotype of *AAbb* or *aaBB*. All three produce the same intermediate grain color in wheat. Recall that a cross such as this results in nine genotypes in a ratio of 1:2:1:2:4:2:1:2:1. If these classes are grouped according to numbers

P   Red   X   White

AABB   *aabb*

F₁   Medium red

AaBb

F₂   Dark red : Medium dark red : Medium red : Light red : White

Dark red	Medium dark red	Medium red	Light red	White
AABB	*AaBB*	*AaBb*	*Aabb*	*aabb*
	AABb	*AAbb*	*aaBb*	
		aaBB		

Figure 8-3.
Another Cross
Involving Grain
Color in Wheat

1 : 4 : 6 : 4 : 1

of color-producing alleles as in Figure 8–3, the 1:4:6:4:1 ratio found in the binomial distribution appears.

Three-Locus Control

In yet another cross of this nature, Nillson-Ehle in 1909 crossed two wheat strains, one red and the other white, that yielded all intermediate plants in the F_1 generation. When these intermediate plants were self-fertilized, at least seven color classes, from red to white, were distinguishable in a ratio of 1:6:15:20:15:6:1 (Figure 8–4). This is explained by assuming that three loci are assorting independently, each with two alleles, such that one allele produces a unit of color while the other allele does not. We then see, assuming we can distinguish them, seven color classes, from red to white in this ratio of 1:6:15:20:15:6:1. This is a ratio in sixty-fourths, directly from the 8×8 (trihybrid) Punnett square, and comes from grouping genotypes in accordance with the number of color-producing alleles that they contain. Again, the ratio is one that is generated in a binomial distribution.

Multilocus Control

From here we need not go on to an example with four loci and then one with five, and so on. We have enough information to perceive generalities. It should not be hard to see how discrete loci can generate a continuous distribution (Figure 8–5). Theoretically, it should be possible to distinguish different color classes down to the level of the eye's ability to perceive differences in wavelengths of light. In fact, we rapidly lose the ability to distinguish separate color classes because the variation within each class is such that before long the color classes begin to overlap. Thus, as with three loci, a color somewhat lighter than medium dark red may belong to the medium-dark-red class with three color alleles, or it may belong to the medium-red class (only two color alleles). The variation within each class is due to environmental interaction—that is, two organisms with the same genotype may not necessarily be identical in phenotype because nutrition, physiological state, and many environmental variables will influence the phenotype. Figure 8–6 shows that the environment can obscure genotypes even in a one-locus, two-allele system. That is, a 17-cm height could result from either aa or Aa (Figure 8–6, column 3) in the F_2 when there is a good deal of environmental variation. In the figure's other two cases, there would not even be a 17-cm height. Systems such as we are considering, where each allele contributes a small unit to the phenotype, are easily influenced by the

P   Red   X   White

        AABBCC   *aabbcc*

F_1 Intermediate color   X   Self

        AaBbCc

	ABC	*ABc*	*AbC*	*aBC*	*Abc*	*aBc*	*abC*	*abc*
ABC	6	5	5	5	4	4	4	3
ABc	5	4	4	4	3	3	3	2
AbC	5	4	4	4	3	3	3	2
aBC	5	4	4	4	3	3	3	2
Abc	4	3	3	3	2	2	2	1
aBc	4	3	3	3	2	2	2	1
abC	4	3	3	3	2	2	2	1
abc	3	2	2	2	1	1	1	0

Phenotype   Red ——————▶ White

Number of Dominants   6 : 5 : 4 : 3 : 2 : 1 : 0

Ratio   1 : 6 : 15 : 20 : 15 : 6 : 1

Figure 8-4. One of Nillson-Ehle's Crosses Involving Three Loci Controlling Grain Color in Wheat
Within the Punnett square, only the number of color-producing alleles are shown to emphasize color production.

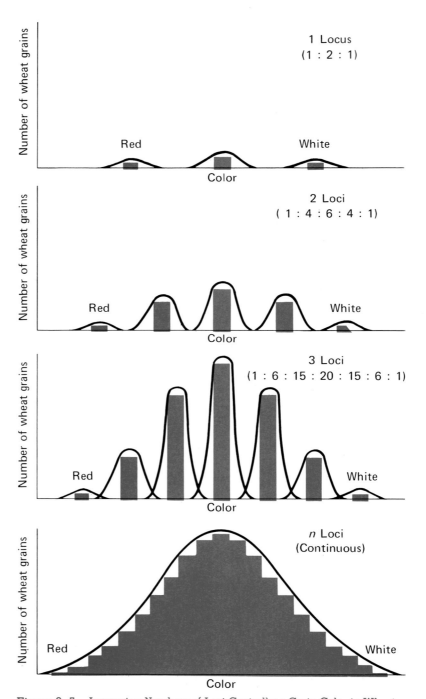

Figure 8-5. Increasing Numbers of Loci Controlling Grain Color in Wheat
Each locus segregating two alleles and each affecting the same trait will eventually generate a continuous distribution in the F_2 generation.

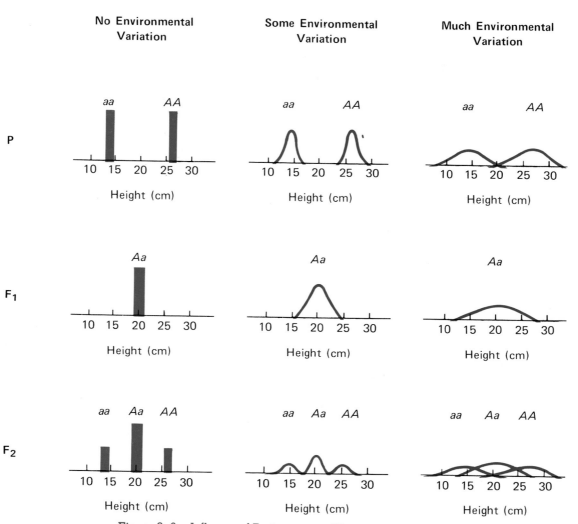

Figure 8-6. Influence of Environment on Phenotypic Distribution

environment and cause the distribution of phenotypes in Figure 8–5 to approach the bell-shaped curve at the bottom.

Thus, phenotypes that are determined by alleles that contribute identical units to the phenotype will approach a continuous distribution in the F_2 generation. This type of trait is said to exhibit **continuous, quantitative,** or **metrical variation.** The inheritance pattern is called **polygenic** or **quantitative.** The system is termed an **additive model** because each allele adds a certain amount to the phenotype.

From the three wheat examples just discussed, we can generalize to systems with more than three polygenic loci, each segregating two alleles. From Table 8–1 we can determine the distributions of genotypes and phenotypes to be expected from any number of loci segregating two alleles in an additive model. If it is possible to separate the various phenotypic classes, this table is useful when we seek to estimate how many loci are involved in producing a trait. For example, when a strain of heavy mice was crossed with a lighter strain, the F_1 were of intermediate weight. When these were interbred, a continuous distribution of adult weights was obtained in the F_2. Since only about 1 mouse in 250 was as heavy as the heavy parent stock, we could guess that if an additive model holds, then four loci are segregating. This is because we expect $(1/4)^n$ to be as extreme as either parent. One in 250 is roughly $(1/4)^4 = 1/256$.

TABLE 8-1. Generalities from an Additive Model of Polygenic Inheritance

	1 Locus	2 Loci	3 Loci	n Loci
Number of gamete types produced by an F_1 multihybrid	2 (A, a)	4 (AB, Ab, aB, ab)	8 (ABC, ABc, AbC, Abc, aBC, aBc, abC, abc)	2^n
Number of different F_2 genotypes	3 (AA, Aa, aa)	9 (AABB, AABb, AAbb, AaBB, AaBb, Aabb, aaBB, aaBb, aabb)	27 (AABBCC, AABBCc, AABBcc, AABbCC, AABbCc, AABbcc, AAbbCC, AAbbCc, AAbbcc, AaBBCC, AaBBCc, AaBBcc, AaBbCC, AaBbCc, AaBbcc, AabbCC, AabbCc, Aabbcc, aaBBCC, aaBBCc, aaBBcc, aaBbCC, aaBbCc, aaBbcc, aabbCC, aabbCc, aabbcc)	3^n
Number of different F_2 phenotypes	3	5	7	$2n + 1$
Denominator for multihybrid F_2 ratio*	4	16	64	4^n
Number of F_2 as extreme as one parent or the other	1/4 (AA or aa)	1/16 (AABB or aabb)	1/64 (AABBCC or aabbcc)	$1/4^n$
Distribution pattern of F_2 phenotypes	1:2:1	1:4:6:4:1	1:6:15:20:15:6:1	$(A + a)^{2n}$

*Number of boxes in F_2 Punnett square

Location of Polygenes

The fact that traits with a continuous variation can be controlled by genes dispersed over the whole genome was shown by James Crow, who studied DDT resistance in *Drosophila*. A DDT-resistant strain of flies was created by growing them on increasing quantities of the insecticide. Crow then systematically tested each chromosome in resistant flies for the amount of resistance it conferred. Susceptible flies were mated to resistant flies and the sons from this cross were back-crossed. Offspring of this cross were scored for the particular resistant chromosomes that they contained (each chromosome had a visible marker) and were then tested for their resistance to DDT. Sons were used in the backcross because there is no crossing over in sons. Thus the resistant and susceptible chromosomes are passed on by the sons intact. Crow's results are shown in Figure 8–7. As we can see, each chromosome has the potential to increase the fly's resistance to DDT. Thus, each chromosome contains loci (polygenes) that contribute to the phenotype of this additive trait.

James F. Crow
(1916–)
Courtesy of Dr.
James F. Crow

Significance of Polygenic Inheritance

The concept of additive traits is of great importance to genetic theory because it demonstrates that Mendelian rules of inheritance can explain traits that have a continuous distribution—that is, Mendel's rules for discrete characteristics also hold for polygenic traits. Additive traits are also of practical interest. Many agricultural products, both plant and animal, follow polygenic inheritance, including milk production and fruit and vegetable yield. In addition, many human traits, such as height and IQ, appear to be polygenic.

Historically, the study of quantitative traits began before the rediscovery of Mendel's work at the turn of the century. In fact, there was a great debate among biologists in the early part of the century as to whether "Mendelians" were correct or whether the "biometricians" were correct in regard to the rules of inheritance. Biometricians used statistical techniques to study traits characterized by continuous variation and claimed that single discrete genes were not responsible for the observed inheritance patterns. Mendelians claimed that inheritance was of discrete traits controlled by discrete "genes." Eventually the Mendelians were proven correct, but the biometricians' tools were the only ones valid for studying quantitative traits.

The biometric school was founded by Galton and Pearson, who showed that many quantitative traits, such as height, were inherited. They invented the statistical tools of correlation and regression analysis in order to study the inheritance of traits that fall into smooth distributions.

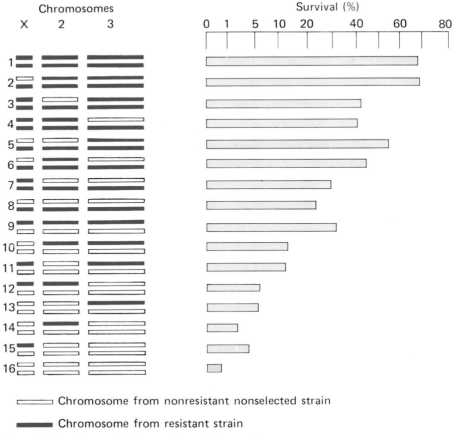

Chromosome from nonresistant nonselected strain

Chromosome from resistant strain

Figure 8-7. Survival of *Drosophila* with Varying Numbers and Arrangements of Resistant and Susceptible Chromosomes

Source: James F. Crow, "Genetics of Insect Resistance to Chemicals." Reproduced, with permission, from the *Annual Review of Entomology*, Volume 2. © 1957 by Annual Reviews, Inc.

POPULATION STATISTICS

A distribution can be described in several ways. One way is the formula for the shape of the curve formed by the frequencies within the distribution. A more functional description of a distribution is a statement about where its center is and how much variation occurs around this center, or **mean** (Figure 8–8). As can be seen from the figure, the center, or mean, is not itself enough to describe the distribution. Variation about this mean will determine the actual shape of the curve. (We will confine our discussion to symmetrical,

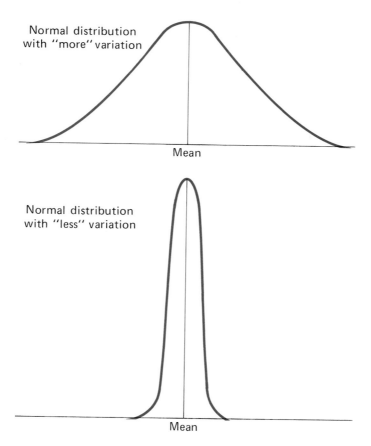

Normal distribution with "more" variation

Mean

Normal distribution with "less" variation

Figure 8-8.
Two Normal
Distributions (Bell-
Shaped Curves) with
the Same Mean

Mean

bell-shaped curves called **normal distributions.** Many distributions approach a normal distribution.)

Mean, Variance, and Standard Deviation

The center, or mean, of a set of numbers is the arithmetic average of the set of data and is defined as

$$\bar{x} = \frac{\Sigma x}{n}$$

(8.1)

where $\bar{x}$ = the mean
Σx = the summation of all values
n = the number of values summed

TABLE 8-2. Hypothetical Data Set of Ear Lengths (x) Obtained When Corn Is Grown from an Ear of Length 11 cm

x	$x - \bar{x}$	$(x - \bar{x})^2$
7	−4.12	16.97
8	−3.12	9.73
9	−2.12	4.49
9	−2.12	4.49
10	−1.12	1.25
10	−1.12	1.25
10	−1.12	1.25
10	−1.12	1.25
10	−1.12	1.25
10	−1.12	1.25
11	−0.12	0.01
11	−0.12	0.01
11	−0.12	0.01
11	−0.12	0.01
11	−0.12	0.01
11	−0.12	0.01
12	0.88	0.77
12	0.88	0.77
12	0.88	0.77
13	1.88	3.53
13	1.88	3.53
13	1.88	3.53
14	2.88	8.29
14	2.88	8.29
16	4.88	23.81

$\Sigma x = 278$ $\qquad \Sigma(x - \bar{x})^2 = 96.53$

$$n = 25$$

$$\bar{x} = \frac{\Sigma x}{n} = \frac{278}{25} = 11.12$$

$$s^2 = V = \frac{\Sigma(x - \bar{x})^2}{n - 1} = \frac{96.53}{24} = 4.02$$

$$s = \sqrt{s^2} = \sqrt{4.02} = 2.0$$

The data are graphed in Figure 8–9.

In Table 8–2 the mean is calculated for the distribution shown in Figure 8–9. The variation about the mean is calculated as the average squared deviation from the mean:

$$s^2 = V = \frac{\Sigma(x - \bar{x})^2}{n - 1} \tag{8.2}$$

This value (V) is called the estimated population **variance.** Observe that the flatter the distribution is, the greater the variance will be.

The ear lengths measured for the data in Table 8–2 are a sample of all ear lengths in the theoretically infinite population of that variety of corn. Statisticians call the sample values **statistics** (and use letters from the Roman alphabet) whereas they call the population values **parameters** (and use Greek letters). The sample is used as an estimate of the true value for the population. Thus, in the variance formula **8.2,** the sample value, V or s^2, is an estimate of the population variance, σ^2. When sample values are used to estimate parameters, one degree of freedom is lost for each parameter estimated. Thus, to determine the sample variance, we divide not by the sample size, but by the degrees of freedom (as defined in Chapter 3). The variance for the entire population (assuming we know the population mean μ) would be calculated by dividing by n. The sample variance is calculated in Table 8–2.

The variance has several interesting properties, not the least of which is the fact that it is additive—that is, if we can determine how much a given variable contributes to the total variance, we can subtract that amount of variance from the total and the remainder is caused by whatever other variables affect the trait. This property makes the variance extremely important in quantitative genetic theory.

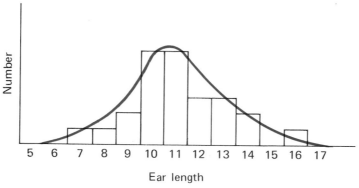

Figure 8-9. Normal Distribution of Ear Lengths in Corn
Data are given in Table 8–2.

The **standard deviation** is a measure of variation of a distribution. It is the square root of the variance:

$$s = \sqrt{V} = \sqrt{s^2} \qquad (8.3)$$

In a normal distribution approximately 67% of the area of the curve lies within one standard deviation on either side of the mean, 96% lies within two standard deviations, and 99% lies within three standard deviations (Figure 8–10). Thus, for the data in Table 8–2, about 2/3 of the population would have ear lengths between 9.12 and 13.12 cm (the mean $\pm$ 1 standard deviation).

One final measure of variation about the mean is the **standard error of the mean** ($s_{\bar{x}}$ or s.e.):

$$\text{s.e.} = s_{\bar{x}} = \frac{s}{\sqrt{n}}$$

The standard error (of the mean) is the standard deviation of the mean if the mean itself were considered a sample from a population of means. We would expect the variation among a population of means to be less than among individual values—and it is. Data are often summarized as "the mean $\pm$ s.e." In our example of Table 8–2, s.e. = $2.0/\sqrt{25}$ = 2.0/5.0 = 0.4. We can summarize the data set of Table 8–2 as 11.1 $\pm$ 0.4 (mean $\pm$ s.e.).

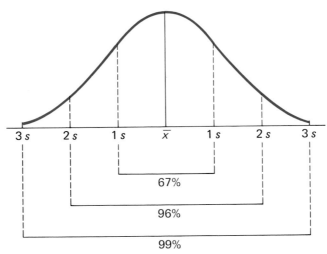

Figure 8–10. Area under the Bell-Shaped Curve
The abscissa is in units of standard deviation (s) around the mean ($\bar{x}$).

Covariance, Correlation, and Regression

It is often desirable to know if there is a relationship between two given characteristics in a series of individuals. For example, is there a relationship between height of a plant and its weight, or between scholastic aptitude and grades? If one increases, does the other also? An example is given in Table 8–3; the same data set is graphed in Figure 8–11 (which is referred to as a *scatter plot*). There does appear to be a relation between the two variables. With increasing aptitude (x axis), there is an increase in grade point (y axis). We can determine how closely the two variables are related by calculating a **correlation coefficient**—an index that goes from -1 to $+1$ depending on

TABLE 8–3. The Relationship between Two Variables, x (aptitude test score) and y (grade-point average)

x	y	x	y	x	y	x	y	x	y
14	4.0	11	2.9	10	2.9	10	1.4	8	2.6
14	3.4	11	2.8	10	2.8	9	2.8	8	2.4
14	3.2	11	2.7	10	2.7	9	2.7	8	2.3
13	3.7	11	2.6	10	2.6	9	2.6	8	1.8
13	2.7	11	2.5	10	2.2	9	2.4	8	1.4
12	2.7	11	2.4	10	2.1	9	2.1	8	1.1
12	2.4	11	2.2	10	1.9	9	1.7	8	0.9
12	2.2	11	2.0	10	1.8	9	1.5	8	0.8
12	2.1	11	1.9	10	1.7	9	1.1	7	1.7
11	3.5	10	3.2	10	1.6	9	0.9	7	0.8

$$\Sigma x = 505 \qquad n = 50 \qquad \Sigma y = 112.4$$

$$\bar{x} = \frac{\Sigma x}{n} = 10.1 \qquad \bar{y} = \frac{\Sigma y}{n} = 2.25$$

$$s_x^2 = \frac{\Sigma(x - \bar{x})^2}{n - 1} = 150.5 \qquad s_y^2 = \frac{\Sigma(y - \bar{y})^2}{n - 1} = 27.88$$

$$s_x = \sqrt{s_x^2} = 12.27 \qquad s_y = \sqrt{s_y^2} = 5.28$$

$$\text{cov}(x,y) = \frac{\Sigma(x - \bar{x})(y - \bar{y})}{n - 1} = 42.76$$

$$r = \frac{\text{cov}(x,y)}{s_x s_y} = \frac{42.76}{(12.27)(5.28)} = 0.66$$

Source: From *Elementary Statistical Methods in Psychology and Education*, 2nd ed., by Paul Blommers and Robert A. Forsyth. Copyright © 1977 by Houghton Mifflin Company. Used by permission.

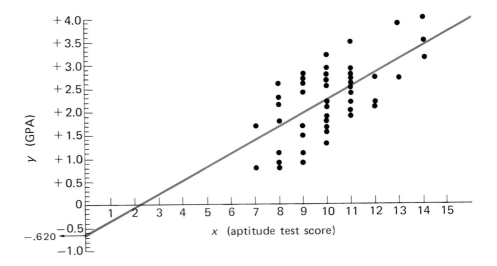

$y = a + bx$

$b = \dfrac{cov(x,y)}{s_x^2} = \dfrac{42.76}{150.5} = .284$

$a = \bar{y} - b\bar{x} = 2.5 - (.284)(10.1) = -.620$

$y = -0.620 + 0.284x$

Figure 8-11.   Relationship between Two Variables, Grade Point Average (GPA) and Aptitude Test Score
The raw data are given in Table 8–3.

Source: From *Elementary Statistical Methods in Psychology and Education,* 2nd ed., by Paul Blommers and Robert A. Forsyth. Copyright © 1977 by Houghton Mifflin Company. Used with permission.

the degree of relationship between the variables. If there is no relation (if the variables are independent), then the correlation coefficient will be zero. If there is perfect correlation, where an increase in one variable is associated with a proportional increase in the other, the coefficient will be +1.0. If an increase in one is associated with a proportional decrease in the other, the coefficient will be −1.0 (Figure 8–12). The formula for the correlation coefficient (r) is

$$r = \dfrac{\text{covariance of } x \text{ and } y}{s_x \cdot s_y} \tag{8.4}$$

where s_x and s_y are the standard deviations of x and y respectively.

To calculate the correlation coefficient, we need to define and calculate

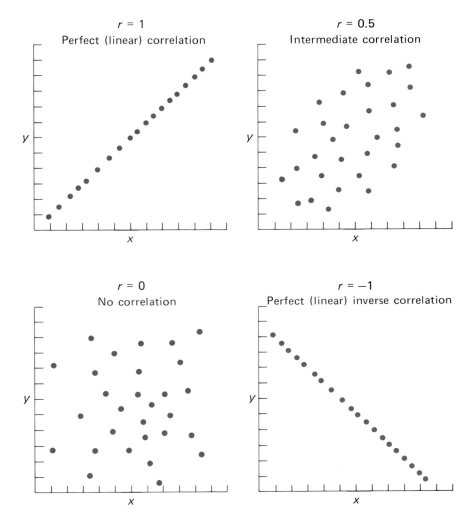

Figure 8-12.
Plots Showing
Varying Degrees of
Correlation within
Data Sets

the **covariance** of the two variables, cov(x,y). The covariance is analogous to the variance but involves the simultaneous deviations from the means of both the x and y variables:

$$\text{cov}(x,y) = \frac{\Sigma(x - \bar{x})(y - \bar{y})}{n - 1} \tag{8.5}$$

The analogy between variance and covariance can be seen by comparing equations **8.5** and **8.2.** The covariance, variances, and standard deviations are derived in Table 8–3, where the correlation coefficient is calculated to be $r = 0.66$. (There are computation formulas available that substantially cut down on the difficulty of calculating these statistics. If a computer is used, only the individual data points need be entered—the computer does all the computations.)

The correlation coefficient indicates the degree of relatedness between the two variables. The assumption throughout the analysis has been that the two variables are changing together—one is not dependent on the other. Many experiments, however, deal with a situation where we assume that one variable is dependent on the other (cause and effect). For example, we may ask: What is the relationship of *Drosophila* DDT-resistance to an increased number of DDT-resistant alleles? With more of these alleles (Figure 8–7), the DDT resistance of the flies should increase. Number of DDT-resistant alleles is the independent variable and resistance of the flies is the dependent variable. Going back to Figure 8–11, we could make the assumption that grade-point average is dependent on scholastic aptitude. If this were so, an analysis called regression analysis could be used. This analysis allows us to predict a grade-point average (y variable) given a particular aptitude (x variable). (It is important to note here that once a cause-and-effect relationship is assumed, a correlation analysis is not valid. Correlation or regression may be used on a given data set, but not both. Regression analysis assumes a cause-and-effect relationship.)

The formula for the straight-line relationship (regression line) between the two variables is $y = a + bx$, where b is the slope of the line ($\Delta y / \Delta x$) and a is the y-intercept of the line (see Figure 8–11). To define any line we need only to calculate the slope, b, and the intercept, a.

$$b = \frac{\text{cov}(x,y)}{s_x^2} \qquad (8.6)$$

$$a = \bar{y} - b\bar{x} \qquad (8.7)$$

Thus equipped, if a cause-and-effect relationship does exist between the two variables, we can predict any y value given an x value. We now continue our examination of the genetics of quantitative traits.

POLYGENIC INHERITANCE IN BEANS

In 1909 Johannsen, who studied bean weight of the dwarf bean plant *(Phaseolus vulgaris)*, demonstrated that polygenic traits are controlled by many genes. The parent population was made up of seeds (beans) with a continuous distribution of weights. Johannsen divided this parental group into classes according to weight, planted them, self-fertilized them, and weighed the F_1 beans. He found that the parents with the heaviest beans produced the progeny with the heaviest beans and the parents with the lightest beans produced the progeny with the lightest beans (Table 8–4). There was a significant correlation coefficient between parent and progeny bean weight

TABLE 8-4. Johannsen's Findings of Relationship between Bean Weights of Parents and Their Progeny

Weight of Parent Beans	Weight of Progeny Beans (centigrams)																n	Mean ± s.e.
	15	20	25	30	35	40	45	50	55	60	65	70	75	80	85	90		
65–75				2	3	16	37	71	104	105	75	45	19	12	3	2	494	58.47 ± 0.43
55–65			1	9	14	51	79	103	127	102	66	34	12	6	5		609	54.37 ± 0.41
45–55			4	20	37	101	204	281	234	120	76	34	17	3	1		1138	51.45 ± 0.27
35–45	5	6	11	36	139	278	498	584	372	213	69	20	4	3			2238	48.62 ± 0.18
25–35		2	13	37	58	133	189	195	115	71	20	2					835	46.83 ± 0.30
15–25			1	3	12	29	61	38	25	11							180	46.53 ± 0.52
Totals	5	8	30	107	263	608	1068	1278	977	622	306	135	52	24	9	2	5491	50.39 ± 0.13

($r = 0.34 \pm 0.01$). He continued this work by choosing 19 particular lines from various points on the original distribution and selfing each successive generation for the next several years. After a few generations, the means and variances stabilized within each line. That is, when Johannsen chose, within each line, parent plants with heavier-than-average or lighter-than-average seeds, the offspring had the parental mean with the parental variance for seed size. For example, after six years of selfing of the 19 lines, plants with both the lightest average bean weights (24 cg) and plants with the heaviest average bean weights (47 cg) produced offspring with average bean weights of 37 cg. By selfing the plants each generation, Johannsen had made them more and more homozygous, thus lowering the number of segregating polygenes. Thus, the lines became homozygous for certain of the polygenes and any variation in bean weight was then caused only by the environment. Johannsen thus showed that quantitative traits were controlled by many segregating loci.

Regression to the Mean

It is regularly found in polygenic traits that the offspring of extremes tend toward the population mean. Table 8–4 illustrates this phenomenon—known as **regression to the mean.** Although parents with heavy beans had progeny with heavy beans and parents with light beans had progeny with light beans, the heavy progeny beans were not as heavy as the seeds of their parents nor were the light progeny beans as light as their parents' seeds. The parents with the heaviest beans (65–75 cg) had progeny with beans weighing 58.47 cg on average (between the parent and the mean of the population). The parents with the lightest bean weights (15–25 cg) had progeny with beans weighing 46.53 cg on average (between the parent and the population mean). Figure 8–13 illustrates this regression.

Regression to the mean is primarily caused by dominance, epistasis, and environmental interactions. Figure 8–14 shows how dominance can explain the regression toward the mean of height extremes. For example, the *A, B,*

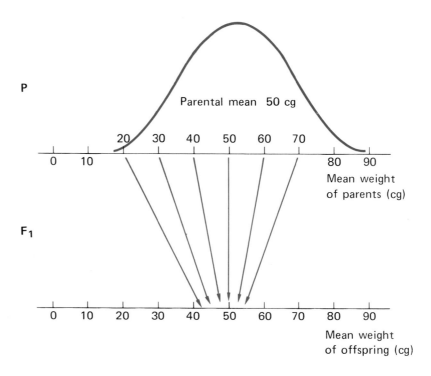

P

Parental mean 50 cg

Mean weight
of parents (cg)

F₁

Mean weight
of offspring (cg)

Figure 8-13.
Regression to the
Mean of the Data of
Table 8-4

and *C* alleles in beans are dominant and, as a homozygote dominant or a heterozygote, each adds 2 inches of height above a base height of 12 inches. The *a, b,* and *c* alleles are recessive and add nothing above the base height of the organisms. A trihybrid F_1 is created by mating homozygous tall (18 in.) with homozygous short (12 in.). If the F_1 trihybrids are selfed, they will produce an F_2 with a mean of 16.50 inches. (An effect of dominance is to **skew** the normal distribution curve toward the dominant end.) If only the tall F_2 are used as parents in a selection experiment, they will have a mean height of 18 inches. However, they are a heterogeneous lot and some will be heterozygous and produce recessive homozygous offspring. This is the key to regression to the mean. The recessive homozygotes (for 1, 2, or all 3 loci) will bring the mean down to 17.80 from 18 inches—regression toward the mean. A similar example could be constructed for the other end of the spectrum.

Epistasis is the hardest-to-measure cause of regression to the mean and is often neglected. The environment can cause regression to the mean when a less-than-perfect environment is present for the highest extreme or when an improved environment is present for the lowest extreme. Regression to the mean is thus a function of dominance, epistasis, and environmental interaction and is a complexity above the simple additive model—reality is rarely a follower of the simple additive model.

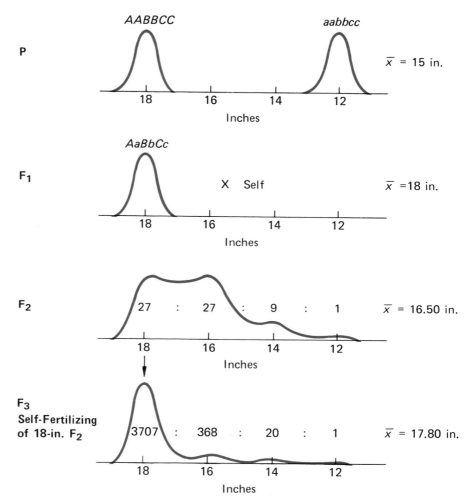

Figure 8-14. Explanation for Regression to the Mean Based on Dominance
A, B, C are dominant; each adds 2 inches height above a base of 12 inches. *a, b, c* are recessive and do not contribute to the height above a base of 12 inches.

Selection Experiments

At this point let us go into somewhat more detail about selection experiments, which are experiments where only certain of the organisms are chosen to be the parents of the next generation. In Johannsen's work, in most plant and animal breeding programs, and in the study of quantitative genetics with fruit flies, selection experiments are done. Plant and animal breeders select the most desirable individuals as parents in order to improve their stock. Laboratory geneticists select specific characteristics for study in order to understand the mechanics of genetic control.

Figure 8–15 illustrates a case in point. *Drosophila* were tested in a 15-choice maze for geotactic response (up or down). The flies with the highest score were chosen as parents for the "high" line (positive geotaxis) and the flies with the lowest score were chosen as parents for the "low" line (negative geotaxis). The same selection was made for each generation. As time progressed, the two lines diverged quite significantly. This tells us that there is a large genetic component to the response; the experimenters are successfully amassing more of the "up" alleles in the high line and the "down" alleles in the low line. Several other points emerge from this graph. First, the high and low responses are slightly different, or asymmetrical—a usual phenomenon—that is, the high line responds more quickly, levels out more quickly, and tends toward the original state much more slowly after selection is relaxed. (The relaxation of selection occurs when the parents of the next generation are a random sample of the adults rather than the extremes for geotactic scores.) The low line responds more slowly and erratically. It was still responding when selection was relaxed. In addition, it returned toward the original state more quickly when selection was relaxed.

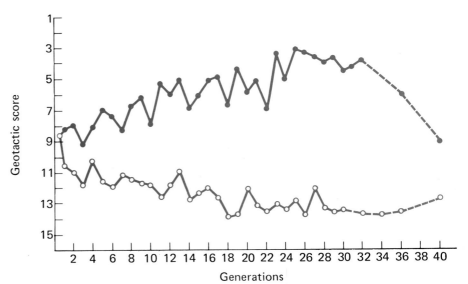

Figure 8–15.   Selection for Geotaxis
The dotted line represents relaxed selection.

Source: From T. Dobzhansky and B. Spassky, "Artificial and Natural Selection for Two Behavioral Traits in *Drosophila pseudoobscura*," *Proceedings. The National Academy of Sciences, USA* 62 (1969):75–80.

Several things are indicated by the information in Figure 8–15. First, the high line becomes more homozygous more quickly. This is shown by its very slight response when selection is relaxed. It has exhausted a good deal of its variability for the polygenes responsible for the geotaxis. The low line, however, seems to have not used up as much of its genetic variability because it was still responding when selection was relaxed and the relaxation caused the mean score to rapidly increase toward the original mean. The response to selection is one way that plant and animal breeders can predict future response.

HERITABILITY

Plant and animal breeders seek to improve their yields of milk, beef, and crops to the greatest degree they can. They must choose the parents of the next generation of cattle and crops on the basis of that generation's yields; consequently, they must perform selection experiments. This requires some measure of genetic control over the trait for which they are breeding so that they can predict how the trait will respond when they practice selection. Breeders run into two economic problems. They cannot pick only the very best to be the next generations' parents because (1) they cannot afford to decrease the size of a crop by using only a very few select parents and (2) they must avoid **inbreeding depression,** which occurs when plants are self-fertilized or animals are bred with close relatives for many generations. After frequent inbreeding, too much homozygosity occurs, and many genes that are slightly or partially deleterious begin to show themselves, depressing vigor and yield. (More on inbreeding will be presented in the last section of the book.) Thus breeders must use a wider-than-ideal range of parents, and therefore they need some index of predictability so that they will know at what point selection will be ineffective.

Breeders often calculate a **heritability** estimate, a value that will predict what direction their selection will take. Heritability is defined in the following equation:

$$Y_O = \overline{Y} + H(Y_P - \overline{Y}) \tag{8.8}$$

where Y_O = offspring yield
$\overline{Y}$ = mean yield of the population
H = heritability
Y_P = parental yield
This equation can be rearranged to solve for the heritability:

$$H = \frac{Y_O - \overline{Y}}{Y_P - \overline{Y}} = \frac{\text{gain}}{\text{selection differential}} \tag{8.9}$$

Restated in words, this equation says that the heritability is the gain due to selection divided by the amount of selection that has been practiced. $Y_0 - \overline{Y}$ is the improvement over the population average due to $Y_P - \overline{Y}$, which is the amount of difference between the parents and the population average (Figure 8–16). If there is no gain ($Y_0 = \overline{Y}$), then the heritability will be zero, and breeders will know that no matter how much selection they practice, they will not improve their crops and might as well not waste their time. Since this value is calculated after the breeding has been done, it is referred to as a **realized heritability.** Some typical values for realized heritabilities are shown in Table 8–5.

The following example may help to clarify the calculation of realized heritability. The number of bristles on the sternopleurite, an abdominal plate in *Drosophila,* is under polygenic control. In a population of flies, the mean bristle number was 6.4. Three pairs of flies were used as parents; they had a mean of 7.2 bristles. Their offspring had a mean of 6.6 bristles. Hence, $Y_0 = 6.6$, $\overline{Y} = 6.4$, and $Y_P = 7.2$. Then, dividing the gain by the selection differential—that is, substituting in equation **8.9**—gives us

$$H = \frac{6.6 - 6.4}{7.2 - 6.4} = \frac{0.2}{0.8}$$

$$= 0.25$$

If both a low line and a high line were begun and if both were carried over several generations, the heritability would be measured by the final difference in means of the high and low lines (gain) divided by the cumulative selection differentials summed for both the high and low lines.

Quantitative geneticists treat the realized heritability as an estimate of the **true heritability.** True heritability is actually defined two different ways: as *heritability in the narrow sense* and *heritability in the broad sense.* We will define these on the basis of partitioning of the variance of the quantitative character under study.

Partitioning of the Variance

Given that the variance of a distribution has genetic and environmental causes and given that the variance is additive, we can construct the following formula:

$$V_{Ph} = V_G + V_E \tag{8.10}$$

where V_{Ph} = total phenotypic variance
V_G = variance due to the genotype
V_E = environmental part of the variance

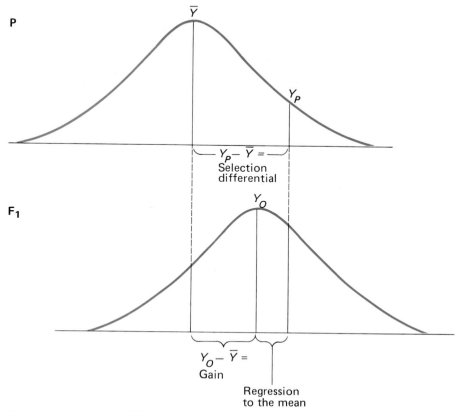

Figure 8–16. Realized Heritability Is the Selection Gain Divided by the Selection Differential

TABLE 8–5. Some Realized Heritabilities

Animal	Trait	Heritability
Cattle	Birth weight	.49
	Milk yield	.30
Poultry	Body weight	.31
	Egg production	.30
	Egg weight	.60
Swine	Birth weight	.06
	Growth rate	.30
	Litter size	.15
Sheep	Wool length	.55
	Fleece weight	.40

Throughout the rest of this discussion we will stay with this model. A more complex variance model can be constructed if there are interactions between variables. For example, if one genotype responded better in one soil condition than in another soil condition, there would be an environmental–genotype interaction that would require a separate variance term (V_{GE}).

The variance due to the genotype can be further broken down according to the effects of additive polygenes (V_A), dominance (V_D), and epistasis (V_I) to give us the final formula of

$$V_{Ph} = V_A + V_D + V_I + V_E \qquad (8.11)$$

We can now define the two commonly used, and often confused, measures of heritability. Heritability in the narrow sense is

$$H_N = \frac{V_A}{V_{Ph}} \qquad (8.12)$$

This heritability is the proportion of the total phenotypic variance that is caused by additive genetic effects. It is of greatest interest to plant and animal breeders because it predicts how much of a response will occur under selection. Heritability in the narrow sense is a measure of how much of the phenotype is inherited.

Heritability in the broad sense is

$$H_B = \frac{V_G}{V_{Ph}} \qquad (8.13)$$

This heritability is the proportion of the total phenotypic variance that is caused by all genetic factors, not just additive factors. It measures the extent to which individual differences in a population are caused by genetic differences. It is the measure most often used by psychologists. We will be concerned primarily with H_N, heritability in the narrow sense.

Measurement of Heritability

Heritability can be directly measured by estimating the components of variance. This statistical tool is called an analysis of variance. We will look at several ways of estimating the various components of the total phenotypic variance, which itself is directly measured from the distribution of the trait in the population (Tables 8–2 and 8–3).

The environmental component of variance is theoretically estimated by comparing identical genotypes in a series of random environments. If the genotypes are identical, then the additive, dominance, and interaction variances are zero and all that is left is the environmental variance. For example, Robertson determined the variance components for the length of the thorax in *Drosophila*. The total variance in a genetically heterogeneous population was 0.366. He then looked at the variance in flies that were genetically homogeneous. These were from isolated lines that had been inbred in the laboratory over many generations to become virtually homozygous. He studied the F_1 in several different matings of inbred lines and found the variance in thorax length to be 0.186. By subtraction, the total genetic variance is 0.180. From this we can calculate heritability in the broad sense as

$$H_B = \frac{V_G}{V_{Ph}} = \frac{0.180}{0.366} = 0.49$$

In order to calculate a heritability in the narrow sense, it is necessary to extract the components of the genetic variance, V_G.

Several methods exist to sort out the additive from the dominant and epistatic portions of the variance. The methods mostly rely on correlations between relatives. That is, the expected amount of genetic similarity between certain relatives can be compared to the actual similarity. The expected amount of genetic similarity is the proportion of genes shared; it is a known quantity for any form of relatedness. For example, parents and offspring have half their genes in common. Since heritability in the narrow sense measures the additive component of the genetic variance (that is, the portion of the variance that is due to inherited genes), a direct measure of heritability in the narrow sense comes from the relation of observed and expected correlations between relatives. We can thus define

$$H_N = \frac{r_{obs}}{r_{exp}} \qquad\qquad \textbf{(8.14)}$$

where r_{obs} is the observed correlation between the relatives and r_{exp} is the expected correlation. The expected correlation is simply the proportion of the genes in common.

For example, in humans, finger-ridge counts (fingerprints, Figure 8–17) have a very high heritability; there is very little environmental interference in the embryonic development of the ridges. Table 8–6 shows the correlations and the heritabilities estimated from each of the types of relationships. Monozygotic twins are from the same egg, which divides into two embryos at a very early stage. They have an identical genotype. Dizygotic twins result from the simultaneous fertilization of two eggs. They have the same relationship to each other as siblings. The data therefore suggest that human finger ridges are almost completely controlled by additive genes with a negligible input of environmental and dominance variation. Few human traits are this simple (Table 8–7).

This brief discussion should make it clear that the components of the total variance can be measured. For a given quantitative trait, the total variance can be measured directly. If identical genotypes can be used, then the environmental component of variance can be discovered. By correlation of

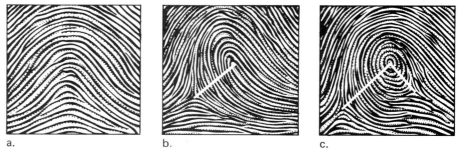

Figure 8-17. The Three Basic Fingerprint Patterns
Ridges are counted where they intersect the line connecting a triradius with a loop or whorl center. (a) An arch; there is no triradius; the ridge count is zero. (b) A loop; 13 ridges. (c) A whorl; there are two triradii and counts of 17 and 8 (the higher one is routinely used).

Source: Redrawn with permission from Sarah B. Holt, "Quantitative Genetics of Finger-Print Patterns," *British Medical Bulletin* 17 (1961).

various relatives, it is possible to measure heritability in the narrow sense directly. If heritability is known and if the total phenotypic variance is known, then all that is left, assuming no interaction, is the dominance and epistatic components. In practice the epistatic components are usually neglected. Thus, operationally, all that is left is the dominance variance, obtained by subtracting the additive from the total variance. In actual practice, plant and animal breeders use very sophisticated statistical techniques of covariance and variance analysis—techniques that are beyond our scope here.

TABLE 8-6. Correlations between Relatives, and Heritabilities, for Finger-Ridge Counts

Relationship	r_{obs}	r_{exp}	H_N
Mother–child	.48	.50	.96
Father–child	.49	.50	.98
Siblings	.50	.50	1.0
Dizygotic twins	.49	.50	.98
Monozygotic twins	.95	1.00	.95

Source: Sarah B. Holt, "Quantitative Genetics of Finger-Print Patterns," *British Medical Bulletin* 17 (1961). Reprinted courtesy of the British Medical Bulletin and Dr. Sarah B. Holt.

TABLE 8-7. Some Estimates of Heritabilities for Human Traits and Disorders

Trait	Heritability
Schizophrenia	.85
Diabetes mellitus	
Early onset	.35
Late onset	.70
Asthma	.80
Cleft lip	.76
Heart disease, congenital	.35
Peptic ulcer	.37
Depression	.45
Stature*	1.00^+

*A heritability higher than 1 can be obtained when the correlation among relatives is higher than expected. This is usually the result of dominant alleles.

QUANTITATIVE INHERITANCE IN HUMANS

As with most human studies, the study of heritability in humans is limited by a lack of certain types of information. We cannot develop pure lines, nor can we manipulate humans into various kinds of environments or do selection experiments. However, there are certain kinds of information available that allow some estimation of heritability.

Skin Color

Skin color is a quantitative human trait for which a simple analysis can be done on naturally occurring matings. Certain groups of people have black skin while other groups do not. These groups breed true and when intermarriage occurs the F_1 are intermediate in skin color. When F_1 individuals intermarry and produce offspring, the skin color of the F_2 is, on the average, about the same as the F_1 but with more variation (Figure 8–18). The data are consistent with a model of four loci, each segregating two alleles—one black, one white—with no dominance.

Twin Studies

In humans the use of twin studies has been helpful in determining the heritability of quantitative traits. Some monozygotic twins (MZ) are known that

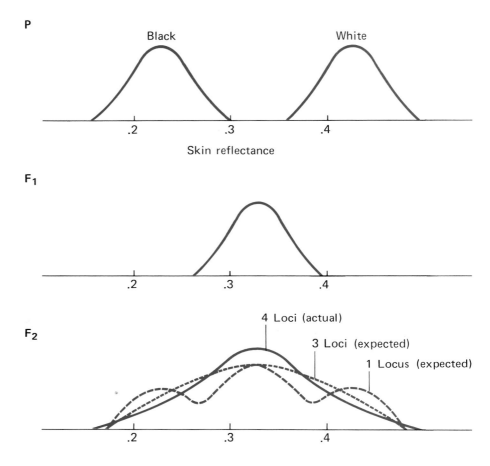

Figure 8-18.
Inheritance of Skin
Color in Humans

have been reared apart. The same is true for dizygotic twins (DZ) and siblings. Twin studies provide some information about environmental influence on many quantitative traits as well as estimates of the heritabilities. Figure 8–19 shows the range of data available for the quantitative trait intelligence. It shows that while there is some environmental effect (one-egg twins vary in their correlation between 0.75 to 0.85, depending on whether they were raised apart or together), there is in general a high correlation between relatives (parent–child and siblings including DZ twins all have mean correlation coefficients near the expected 0.5). However, see the discussion of Burt's research. We will return to the topic of intelligence in a moment.

Another way of looking at quantitative traits is by the **concordance** of twins. Concordance means that if one twin has the trait, the other does also. Discordance is the case where one has the trait and the other does not. Table 8–8 shows some values. High concordance is another indicator of the heritability of a trait. Concordance for measles susceptibility and handedness demonstrate the high environmental influence on some traits.

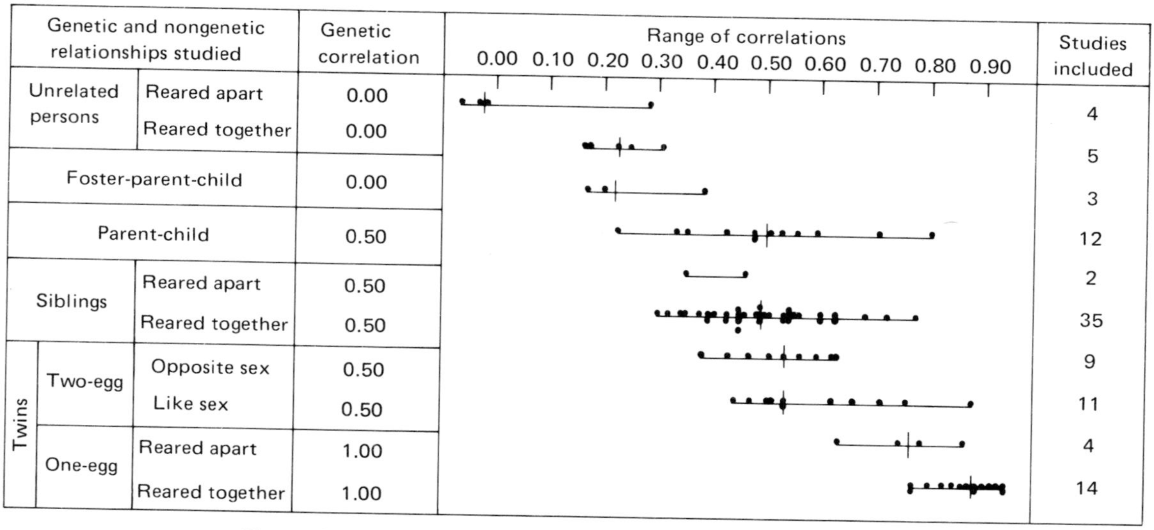

Genetic and nongenetic relationships studied		Genetic correlation	Range of correlations 0.00 0.10 0.20 0.30 0.40 0.50 0.60 0.70 0.80 0.90	Studies included
Unrelated persons	Reared apart	0.00		4
	Reared together	0.00		5
Foster-parent-child		0.00		3
Parent-child		0.50		12
Siblings	Reared apart	0.50		2
	Reared together	0.50		35
Twins — Two-egg	Opposite sex	0.50		9
	Like sex	0.50		11
Twins — One-egg	Reared apart	1.00		4
	Reared together	1.00		14

Figure 8-19. Correlations from 52 Studies of Intelligence Test Scores
Dots are correlation coefficients and vertical bars are median (middle) values.

Source: Erlenmeyer-Kimling and L. F. Jarvik, "Genetics and Intelligence: A Review," *Science* 142 (December 1963). Copyright 1963 by the American Association for the Advancement of Science. Reprinted by permission.

IQ

Let us return for a moment to the topic of intelligence. This subject has unfortunately taken a controversial turn in recent times. It has been reported that the mean IQ difference between whites and blacks in the United States is about 15 points (Figure 8–20), where blacks average one standard deviation lower than whites. Since the measured heritability of IQ is relatively high, it has been suggested that this discrepancy is evidence that blacks are genetically "inferior" to whites. That view is not supported by the evidence for several very important reasons.

First, there is disagreement over the validity of the IQ measurement. While it purportedly measures "intelligence," psychologists strongly disagree as to what it actually does measure. Some tend to think it measures only a thin slice of a personality, while others, mainly educators, tend to see it as an important predictor of achievement in the current educational system. They view it as a good measure of some inherent quality called intelligence. Given that there is this universal quality called intelligence, there seems to be good agreement that the current IQ tests are very much oriented to middle-class white Americans. This is inherent in the kinds of questions these tests ask and the manner in which they ask them. For example, other kinds of tests have been experimented with that have different cultural orientations. Tests oriented to ghetto life will give an advantage to ghetto children

TABLE 8-8. Concordance of Traits between Identical and Fraternal Twins

	Identical (MZ) Twins (percent)	Fraternal (DZ) Twins (percent)
Hair color	89	22
Eye color	99.6	28
Blood pressure	63	36
Handedness (left or right)	79	77
Measles	95	87
Clubfoot	23	2
Tuberculosis	53	22
Mammary cancer	6	3
Schizophrenia	80	13
Down's syndrome	89	7
Spina bifida	72	33

over nonghetto, or an advantage to northern blacks over southern whites. Thus the Stanford-Binet IQ test, which is the basis for most of this controversy, is not necessarily a valid test for all persons of all ethnic and cultural backgrounds. It is not clear at the present time that it measures what it claims to.

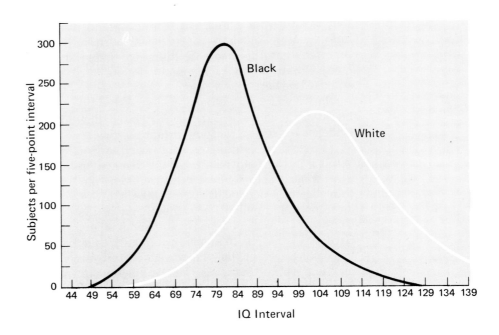

Figure 8-20. Reported IQ Distribution of American Whites and Blacks

Second, the observed differences in mean IQ between blacks and whites is a population phenomenon—it does not describe an individual. Eleven percent of blacks score above the white mean of 100, and 18% of whites score below the black mean of 85. Hence, a population mean does not describe every individual in that population.

Third, the measure of heritability is being misinterpreted. Human geneticists tend to use heritability in the broad sense (V_G/V_{Ph}) to make statements about the genetic control of a trait. The heritability of IQ usually is about 0.4, which has been taken to mean that 40% of the difference between races is due to genetics. This is a misinterpretation of heritability. The value means that 40% of the variation within the white population is due to genetic differences among people. For almost all other traits measured, it has been shown that variation between races is much less than variation within a race. To actually determine the genetic component in the difference between races takes data that we do not have. Heritability measures variation within a population at a given time and it is incorrect to use it to point out differences between races.

A last point is that there is an enormous environmental component to the heritability estimate. Remember that we estimated heritability on the basis of the observed and expected correlation between parents and offspring. This technique assumes independence in the environments between parents and offspring, which is, of course, an unrealistic assumption for humans. Those with environments of high quality tend to pass those environments on to their children and likewise for environments of poorer quality. The same difference in IQ between whites and blacks tends also to hold for the quality of the environment. If environments were made equal, we could not predict what, if any, would be the difference in IQ distributions.

CHAPTER SUMMARY

Some genetically controlled traits yield offspring whose phenotypes fall along a gradient rather than into discrete categories. This type of variation is referred to as quantitative, continuous, or metrical. The genetic control of this variation is referred to as polygenic variation. As long as the number of loci is small, and offspring fall into recognizable classes, it is possible to analyze the genetic control of variable phenotypes with standard methods.

When offspring phenotypes fall into continuous distributions, the methods of analysis change. We must learn how to describe a distribution, using means, variances, and standard deviations. Then, we must learn how to describe the relationship between two variables—we need to calculate covariances and correlation coefficients.

Equipped with these tools, we can turn to analyzing the genetic control of a continuous trait. The heritability estimate tells us how much of the variation in the distribution of a trait can be attributed to genetic causes. Her-

SCIENTIFIC FRAUD: THE CASE AGAINST SIR CYRIL BURT

In Chapter 1 we examined the criticisms leveled at Mendel that he "cheated" in the gathering and analysis of his data. Our conclusion was that although he could have cheated, he probably did not. He did not have to cheat to come up with the results he obtained. Not all scientists hold up under such criticism: There are known cases of fraud within the establishment. One of the more recent additions to the list of likely frauds is Cyril Burt, a noted and respected British psychologist, who died in 1971. In 1946 he was knighted, the first psychologist to be so honored.

Burt's research was concerned with the measurement and inheritance of IQ. He studied the relationship of IQs among parents and their offspring and between identical twins reared together and apart. His results, supporting the genetic determination of IQ, and his data have been widely cited. Starting in about 1976, a groundswell of criticism of Burt's work arose. He has been accused of fabricating his data.

Anomalies have been found in some of Burt's most widely used data. In one case, the correlation coefficients of intelligence test results with monozygotic twins reared apart did not change as the sample sizes increased. A value of 0.771 was reported by Burt from 1955 through 1966 as the sample size increased from 21 pairs through 53 pairs. The lack of change in the correlation coefficient is highly suspect. In 1978, D. Dorfman published an extensive reanalysis of Burt's data dealing with the IQs of children and their fathers classified according to working class. The data are an exceptionally good fit to Burt's theories about how the data should be distributed.

For example, Burt supported the theory that the children's IQs could be predicted from their fathers' as half way between their fathers' IQs and 100—that is, if a father's IQ were 130, then his child's IQ would be 115. In addition, he suggested

itability can be calculated in the narrow sense, being the amount of variance due to additive loci. Heritability in the broad sense considers all the genetic variance, including dominance and epistasis. In practice, heritability can be calculated as realized heritability—gain divided by selection differential. Some idea of human heritability can be obtained from correlations among relatives, concordance and discordance between twins, and studies where monozygotic twins are reared apart.

In this chapter we also show that polygenes controlling DDT resistance are located on all chromosomes in *Drosophila;* that regression to the mean is usually a function of dominance; and that IQ differences among races are not an accurate reflection of the relative intelligence of individuals.

EXERCISES AND PROBLEMS

1. A geneticist wished to know if the number of egg follicles produced by a chicken was inherited. As a first step in his experiments, he wished to determine if the number of eggs laid could be used to predict the number

that both distributions (fathers' and children's) would fit a normal distribution with mean of 100 and standard deviation of 15. In fact, the fit of Burt's data to these predictions is almost perfect and at variance with Burt's own earlier assessment (as well as what is generally known) that the IQ distributions are skewed.

Dorfman's analysis leaves virtually no doubt that Burt decided on the results, fixed the totals on his tables, and then filled in the data in the tables to make them fit the totals. Burt is evasive throughout about methods.

It has also been suggested that Burt fabricated the existence of colleagues who collaborated with him. In particular, two women, a Miss Howard and a Miss Conway, were coauthors of several of his works. There is some evidence that Howard actually existed, but Conway almost certainly did not. Even if they both did exist, it seems certain that they did not do the work Burt said they did. In his diary Burt wrote in one place that he was preparing "Howard's reply" to a particular criticism.

The final blow against Burt seems to have come from a 1979 authorized biography by Leslie Hearnshaw, who began the project as an admirer of Burt and ended by confirming Burt's deception and fabrication. Hearnshaw analyzed Burt's diaries, which have exceptional detail. Hearnshaw concluded that Burt probably fabricated all new data from 1950 onward and possibly as early as 1939.

Why did Burt do this? Hearnshaw concluded that Burt invented people and data to save face and to boost his own ego. He used mythical people to expound his views. Fabricated data allowed him to support his previous contentions and to give the false impression that he was actively carrying out research. Perhaps he thought that he could so dominate his contemporaries that he would never be found out.

How much of this goes on in science? In general, we do not know, but we assume that fraud is minimal because any work published by scientists can immediately come under the most careful scrutiny by peers. This factor by itself probably discourages most dishonesty because a scientist knows that his or her work will be repeated by others. Although frauds are occasionally discovered, there seems to be enough checks and balances in the system to restrict unethical behavior.

of follicles. If this were true, he could then avoid the sacrifice of chickens. He obtained the following data from 14 sacrificed chickens:

Chicken Number	Eggs Laid	Ovulated Follicles
1	39	37
2	29	34
3	46	52
4	28	26
5	31	32
6	25	25
7	49	55
8	57	65
9	51	44
10	21	25
11	42	45
12	38	26
13	34	29
14	47	30

Calculate a correlation coefficient. Graph the data and then calculate the slope and y-intercept of the regression line. Draw the regression line on the same graph.

2. A variety of squash has fruits that weigh about 5 pounds. In a second variety the average weight is 2 pounds. When the two varieties are crossed, the F_1 yield fruit with an average weight of 3½ pounds. When 2 of these are crossed, a range of fruit weights is found. Of 200 offspring, 3 produce fruits weighing about 5 pounds and 3 produce fruits about 2 pounds in weight. How many allele pairs are involved in the weight difference between the varieties, and how much does each effective gene contribute to the difference?

3. In rabbit variety 1 the length of the ear averages 4 inches. In a second variety it is 2 inches. Hybrids between the varieties average 3 inches in ear length. When these hybrids are crossed to each other, there is much greater variation in ear length. Of 500 F_2 animals, 2 have ears about 4 inches in length and 2 have ears about 2 inches long. How many allele pairs are involved, and how much does each effective gene seem to contribute to the length of the ear? What do the distributions of P_1, F_1 and F_2 probably look like?

4. Assume that height in humans depends on four pairs of alleles. How can two persons of moderate height produce children who are much taller than they are? Assume that environment is exerting a negligible effect. How can very short parents produce children closer to the mean in height?

5. How are polygenes different from traditional Mendelian genes?

6. Outstanding athletic ability is often found in several members of a family. Devise a study to determine to what extent athletic ability is inherited.

7. Variations in stature are almost entirely due to heredity. Yet average height has increased substantially since the Middle Ages, and the increase in height of children of immigrants to the United States, as compared with height of the immigrants themselves, is especially noteworthy. How can these observations be reconciled?

8. Would you expect good nutrition to increase or decrease the heritability of height?

9. Two adult plants of a particular species have extreme phenotypes for a quantitative character: 1 foot tall and 5 feet tall. If you had only one

uniformly lighted greenhouse, how would you determine whether plant height is environmentally or genetically caused? If genetically caused, how would you attempt to determine the number of allele pairs that may be involved in this trait?

10. If skin color is caused by additive genes, can matings between individuals with intermediate-colored skin produce lighter-skinned offspring? Can such matings produce dark-skinned offspring? Can matings between individuals with light skin produce dark-skinned offspring?

11. The tabulated data from Emerson and East (The inheritance of quantitative characters in maize. *Univ. Nebraska Agric. Exp. Sta. Bull.*, no. 2, 1913) show the results of crosses between two varieties of corn and their F_2. Provide an explanation for these data in terms of number of allele pairs. Do all the genes involved affect weight additively? Why or why not?

	5	6	7	8	9	10	11	12	13	14	15	16	17	18	19	20	21
							Ear Length in Corn (cm)										
Variety P_{60}	4	21	24	8													
Variety P_{54}									3	11	12	15	26	15	10	7	2
F_1					1	12	12	14	17	9	4						
F_2 ($F_1 \times F_1$)			1	10	19	26	47	73	68	68	39	25	15	9	1		

12. In *Drosophila* a marker strain exists containing dominant lethals on both chromosome 2 homologues, both chromosome 3 homologues, and both chromosome 4 homologues. These lethals are within inversions so there is no crossing over. The strain thus remains perpetually heterozygous for all three loci and thus all three chromosomes. The markers are: chromosome 2, Curly and Plum (*Cy/Pm*); chromosome 3, Hairless and Stubble (*H/S*); and chromosome 4, Cell and Minute(4) [*Ce/M(4)*]. With this strain, which allows you to follow particular chromosomes by the presence or absence of markers, construct crosses to give the strains used by Crow (Figure 8–7) to determine the location of polygenes for DDT resistance.

13. The components of variance for two characters of *D. melanogaster* are shown in the table [data from A. Robertson, Optimum group size in progeny testing . . . , *Biometrics* 13(1957):442–450]. Estimate the dominance and epistatic components and calculate heritabilities in the narrow and broad sense.

Variance Components	Thorax Length	Eggs Laid in 4 Days
V_{Ph}	100	100
V_A	43	18
V_E	51	38
$V_D + V_I$	?	?

14. One behavioral measurement in rats is termed "emotionality" (it is a defecation rate). The accompanying figure [data modified from Broadhurst, *Experiments in Personality*, vol. 1, (London: Eysenck, 1960).] shows mean emotionality scores over five generations in high and low selection lines. In the final generation the parental mean was 4 for the high line and 0.2 for the low line. The cumulative selection differential is five for each line.

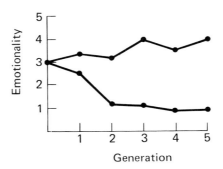

Calculate realized heritability by interpolating from the graph. Calculate separate heritabilities using only the high line or the low line. Do these differ? Why? Why was the response to selection asymmetrical?

15. The accompanying table [data from Ehrman and Parsons, *The Genetics of Behavior* (Sunderland, Mass.: Sinauer Associates, 1976), p. 121] gives heights in centimeters of 11 pairs of brothers and sisters. Calculate a correlation coefficient and a heritability. Is this realized heritability, heritability in the broad sense, or heritability in the narrow sense?

Pair	Brother	Sister	Pair	Brother	Sister
1	180	175	7	178	165
2	173	162	8	186	163
3	168	165	9	183	168
4	170	160	10	165	160
5	178	165	11	168	157
6	180	157			

How can environmental factors influence this heritability value?

16. Data were gathered during a selection experiment for 6-week body weight in mice. Graph these data and calculate a realized heritability.

Generation	High Line $\overline{Y}$	High Line Y_P	High Line Y_O	Low Line $\overline{Y}$	Low Line Y_P	Low Line Y_O
0	21			21		
1		24	22		18	20
2		24	23		18	20
3		26	23		18	20
4		26	24		16	19
5		26	23		16	18

SUGGESTIONS FOR FURTHER READING

Allen, G. 1965. Twin research: problems and prospects. *Prog. Med. Genet.* 4:242–269.

Blommers, P., and R. A. Forsyth. 1977. *Elementary Statistical Methods in Psychology and Education,* 2nd ed. Boston: Houghton Mifflin.

Cavalli-Sforza, L. L., and W. F. Bodmer. 1971. *The Genetics of Human Populations.* San Francisco: Freeman.

Crow, J. 1957. Genetics of insect resistance to chemicals. *Annu. Rev. Entomol.* 2:227–246.

Dorfman, D. 1978. The Cyril Burt question: new findings. *Science* 201:1177–1186.

Falconer, D. D. 1960. *Introduction to Quantitative Genetics.* New York: Ronald Press.

Feldman, M., and R. Lewontin. 1975. The heritability hang-up. *Science* 190:1163–1168.

Harrison, G. A., and J. J. T. Owen, 1964. Studies on the inheritance of human skin color. *Ann. Human Genet.* 28:27–37

Hearnshaw, L. 1979. *Cyril Burt, Psychologist.* Ithaca, N.Y.: Cornell University Press.

Holt, S. B. 1961. Inheritance of dermal ridge patterns. In L. S. Penrose, ed., *Recent Advances in Human Genetics.* London: J. and A. Churchill. Pp. 101–119.

Mather, K., and J. L. Jinks. 1971. *Biometrical Genetics,* 2nd ed. Ithaca, N.Y.: Cornell University Press.

Stern, C. 1970. Model estimates of the number of gene pairs involved in pigmentation variability of the Negro-American. *Human Hered.* 20:165–168.

Wright, S. 1934. The results of crosses between inbred strains of guinea pigs differing in number of digits. *Genetics* 19:537–551.

PART II

MOLECULAR GENETICS

9 Chemistry of the Gene

In 1953 James D. Watson and Francis H. C. Crick published a two-page paper in the journal *Nature,* entitled "Molecular Structure of Nucleic Acids." It began as follows: "We wish to suggest a structure for the salt of deoxyribose nucleic acid (D.N.A.). This structure has novel features which are of considerable biological interest." This paper is a milestone in the modern era of molecular genetics—the authors' model of DNA was correct. (James Watson, Francis Crick, and x-ray crystallographer Maurice Wilkins won the Nobel prize for their work on DNA). Once the structure of the genetic material had been determined, an understanding of its functioning and its method of replication followed quickly.

PROPERTIES OF GENETIC MATERIAL

This discussion will begin with a look at the properties that a genetic material must have. To be a genetic material, DNA (and in some cases ribonucleic acid, RNA) must fulfill the requirements of function—including replication—and location.

Function

The development, growth, and functioning of a cell are controlled by the proteins within it. All the metabolic processes of a cell are dependent on the enzymes it contains. Thus the nature of a cell can be controlled by controlling the protein synthesis within that cell. The genetic material controls the presence and effective amounts of a cell's enzymes and thereby specifies which proteins a cell will have. For example, in humans only a cell genetically directed to be an erythrocyte synthesizes hemoglobin in large quantities; no other cell does that.

MOLECULAR STRUCTURE OF NUCLEIC ACIDS: A STRUCTURE FOR DEOXYRIBOSE NUCLEIC ACID

We wish to suggest a structure for the salt of deoxyribose nucleic acid (D.N.A.). This structure has novel features which are of considerable biological interest.

A structure for nucleic acid has already been proposed by Pauling and Corey.[1] They kindly made their manuscript available to us in advance of publication. Their model consists of three intertwined chains, with the phosphates near the fibre axis, and the bases on the outside. In our opinion, this structure is unsatisfactory for two reasons: (1) We believe that the material which gives the X-ray diagrams is the salt, not the free acid. Without the acidic hydrogen atoms it is not clear what forces would hold the structure together, especially as the negatively charged phosphates near the axis will repel each other. (2) Some of the van der Waals distances appear to be too small.

Another three-chain structure has also been suggested by Fraser (in the press). In his model the phosphates are on the outside and the bases on the inside, linked together by hydrogen bonds. This structure as described is rather ill-defined, and for this reason we shall not comment on it.

We wish to put forward a radically different structure for the salt of deoxyribose nucleic acid. This structure has two helical chains each coiled round the same axis (see diagram). We have made the usual chemical assumptions, namely, that each chain consists of phosphate diester groups joining β-D-deoxyribofuranose residues with 3′,5′ linkages. The two chains (but not their bases) are related by a dyad perpendicular to the fibre axis. Both chains follow right-handed helices, but owing to the dyad the sequences of the atoms in the two chains run in opposite directions. Each chain loosely resembles Furberg's[2] model No. 1; that is, the bases are on the inside of the helix and the phosphates on the outside. The configuration of

This figure is purely diagrammatic. The two ribbons symbolize the two phosphate-sugar chains, and the horizontal rods the pairs of bases holding the chains together. The vertical line marks the fibre axis.

the sugar and the atoms near it is close to Furberg's "standard configuration," the sugar being roughly perpendicular to the attached base. There is a residue on each chain every 3.4 A in the z-direction. We have assumed an angle of 36° between adjacent residues in the same chain, so that the structure repeats after 10 residues on each chain, that is, after 34 A. The distance of a phosphorus atom from the fibre axis is 10 A. As the phosphates are on the outside, cations have easy access to them.

The structure is an open one, and its water content is rather high. At lower water contents we would expect the bases to tilt so that the structure could become more compact.

The novel feature of the structure is the manner in which the two chains are held together by the purine and pyrimidine bases. The planes of the bases are perpendicular to the fibre axis. They are joined together in pairs, a single base from one chain being hydrogen-bonded to a single base from the other chain, so that the two lie side by side with identical z-co-ordinates. One of the pair must be a purine and the other a pyrimidine for bonding to occur. The hydrogen bonds are made as follows: purine position 1 to pyrimidine position 1; purine position 6 to pyrimidine position 6.

If it is assumed that the bases only occur in the structure in the most plausible tautomeric forms (that is, with the keto rather than the enol configurations) it is found that only specific pairs of bases can bond together. These pairs are: adenine (purine) with thymine (pyrimidine), and guanine (purine) with cytosine (pyrimidine).

In other words, if an adenine forms one member of a pair, on either chain, then on these assumptions the other member must by thymine; similarly for guanine and cytosine. The sequence of bases on a single chain does not appear to be restricted in any way. However, if only specific pairs of bases can be formed, it follows that if the sequence of bases on one chain is given, then the sequence on the other chain is automatically determined.

It has been found experimentally[3,4] that the ratio of the amounts of adenine to thymine, and the ratio of guanine to cytosine, are always very close to unity for deoxyribose nucleic acid.

It is probably impossible to build this structure with a ribose sugar in place of the deoxyribose, as the extra oxygen atom would make too close a van der Waals contact.

The previously published X-ray data[5,6] on deoxyribose nucleic acid are insufficient for a rigorous test of our structure. So far as we can tell, it is roughly compatible with the experimental data, but it must be regarded as unproved until it has been checked against more exact results. Some of these are given in the following communications. We were not aware of the details of the results presented there when we devised our structure, which rests mainly though not entirely on published experimental data and stereochemical arguments.

It has not escaped our notice that the specific pairing we have postulated immediately suggests a possible copying mechanism for the genetic material.

Full details of the structure, including the conditions assumed in building it, together with a set of co-ordinates for the atoms, will be published elsewhere.

We are much indebted to Dr. Jerry Donohue for constant advice and criticism, especially on interatomic distances. We have also been stimulated by a knowledge of the general nature of the unpublished experimental results and ideas of Dr. M. H. F. Wilkins, Dr. R. E. Franklin and their co-workers at King's College, London. One of us (J. D. W.) has been aided by a fellowship from the National Foundation for Infantile Paralysis.

J. D. Watson
F. H. C. Crick
Medical Research Council Unit for the Study of the
Molecular Structure of Biological Systems,
Cavendish Laboratory, Cambridge.
April 2.

[1] Pauling, L., and Corey, R. B., *Nature*, 171, 346 (1953); *Proc. U.S. Nat. Acad. Sci.*, 39, 84 (1953).
[2] Furberg, S., *Acta Chem. Scand.*, 6, 634 (1952).
[3] Chargaff, E., for references see Zamenhof, S., Brawerman, G., and Chargaff, E., *Biochim. et Biophys. Acta*, 9, 402 (1952).
[4] Wyatt, G. R., *J. Gen. Physiol.*, 36, 201 (1952).
[5] Astbury, W. T., Symp. Soc. Exp. Biol. 1, Nucleic Acid, 66 (Camb. Univ. Press, 1947).
[6] Wilkins, M. H. F., and Randall, J. T., *Biochim. et Biophys. Acta*, 10, 192 (1953).

J. D. Watson and F. H. C. Crick, "Molecular Structure of Nucleic Acids: A Structure for Deoxyribose Nucleic Acid." *Nature*, 171, no. 4356 (1953): 737–738. Reprinted by permission of the authors and publisher.

James D. Watson
(1928–   )
Courtesy of Dr.
James D. Watson.

Francis Crick
(1916–   )
Reproduced by
permission of Herb
Weitman,
Washington
University, St. Louis,
Mo.

Maurice H. F.
Wilkins (1916–   )
Courtesy of Dr.
Maurice H. F.
Wilkens.

Enzymes.   At this point we need to review some basic information regarding enzymes. An enzyme is a protein, or proteinaceous substance, that acts as a catalyst of a specific metabolic process without itself being markedly altered by the reaction. Most reactions that are catalyzed by enzymes could occur anyway, but only under conditions too extreme for them to take place within living systems. For example, many oxidations occur naturally at high temperatures. Enzymes accomplish these reactions by lowering what is called the *activation energy* of a particular reaction. We are in an early stage of understanding the physical chemistry of how this comes about.

Most metabolic processes, such as the biosynthesis or degradation of molecules, occur in pathways where each step in the pathway is facilitated by an enzyme. For example, the metabolic pathway for the conversion of threonine into isoleucine is shown in Figure 9–1. Threonine and isoleucine are 2 of the 20 amino acids of which all proteins are made—all living things require both these amino acids. Each reaction product in the pathway is altered by an enzyme that converts it to the next product. The enzyme threonine dehydratase, for example, converts threonine into α-ketobutyric acid. Enzymes are composed of folded polymers of amino acids, and the sequence of these amino acids determines the final structure of the enzyme. The genetic material determines the sequence of the amino acids.

It is the three-dimensional structure of enzymes that permits them to perform their function. An enzyme combines with its substrate or substrates (the molecules it works on) at a part of the enzyme called the **active site** (Figure 9–2). The substrates "fit" into the active site, which has a shape that allows only the specific substrates to enter. This view of the way that an enzyme "recognizes" its substrates is called the lock-and-key model of enzyme functioning. When the substrates are in their proper position in the active site of the enzyme, the particular reaction that the enzyme catalyzes takes place. The reaction products then separate from the enzyme and leave it free to repeat the process. Enzymes can work at phenomenal speeds. Some can catalyze as many as a million reactions per minute.

Not all of the cell's proteins function as catalysts. Some are structural proteins, such as keratin, the main component of hair. Other proteins are regulatory when they function to control the rate of production of other enzymes. Still others are involved in various other functions; albumens, for example, help regulate the osmotic pressure of blood.

Replication.   The genetic material must replicate itself precisely so that every daughter cell receives an exact copy. Only thus can humans continue to produce humans or can liver cells, when they divide, continue to produce other liver cells. DNA and RNA are the only biological molecules that serve as direct models for the synthesis of new copies of themselves. In their 1953 paper Watson and Crick had already worked out the replication process based on the structure of DNA.

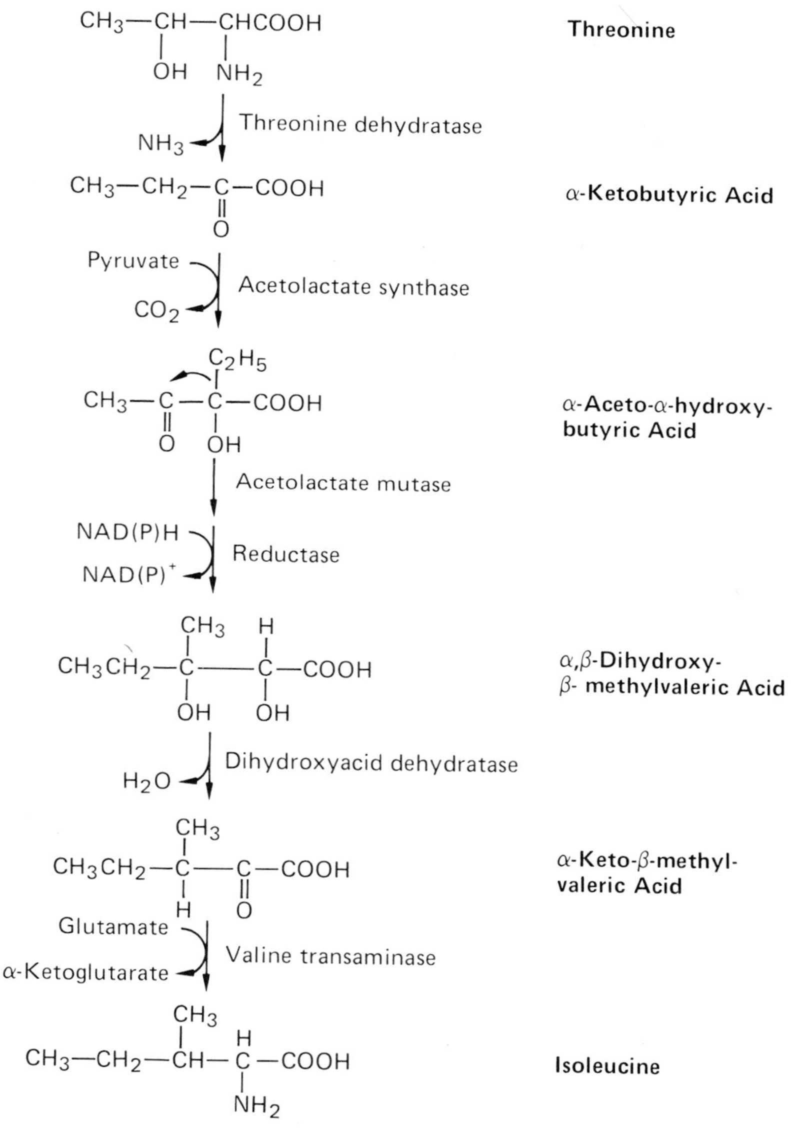

$$CH_3-CH-CHCOOH$$
$$||$$
$$OHNH_2$$

Threonine

Threonine dehydratase

NH_3

$$CH_3-CH_2-C-COOH$$
$$||$$
$$O$$

α-Ketobutyric Acid

Pyruvate

Acetolactate synthase

CO_2

$$C_2H_5$$
$$|$$
$$CH_3-C-C-COOH$$
$$|||$$
$$OOH$$

α-Aceto-α-hydroxy-butyric Acid

Acetolactate mutase

NAD(P)H

Reductase

NAD(P)$^+$

$$CH_3H$$
$$||$$
$$CH_3CH_2-C-----C-COOH$$
$$||$$
$$OHOH$$

α,β-Dihydroxy-β- methylvaleric Acid

Dihydroxyacid dehydratase

H_2O

$$CH_3$$
$$|$$
$$CH_3CH_2-C-C-COOH$$
$$|||$$
$$HO$$

α-Keto-β-methyl-valeric Acid

Glutamate

Valine transaminase

α-Ketoglutarate

$$CH_3$$
$$|H$$
$$CH_3-CH_2-CH-C-COOH$$
$$|$$
$$NH_2$$

Isoleucine

Figure 9–1.
Metabolic Pathway
of Conversion of the
Amino Acid
Threonine into
Isoleucine

Location

It has been known since the turn of the century that genes, the discrete functional units of genetic material, are located in chromosomes within the nuclei of cells. The behavior of chromosomes during the cellular division stages of mitosis and meiosis mimics the behavior of genes. Thus, our putative genetic material must be a part of the chromosome that lies within the nucleus of

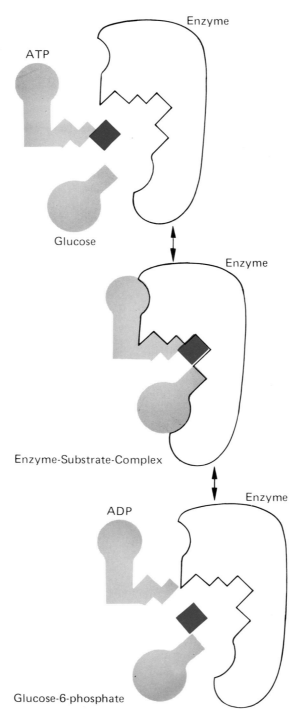

ATP

Glucose

Enzyme

Enzyme

Enzyme-Substrate-Complex

Enzyme

ADP

Glucose-6-phosphate

Figure 9 – 2.
Active site of an
Enzyme Recognizing
a Specific Substance
In this case, ATP
plus glucose is
converted into ADP
and glucose-6-
phosphate by the
enzyme hexokinase.

eukaryotic cells (cells with nuclei). The Watson and Crick model in 1953 ended a period when DNA was thought to be the genetic material but its structure had yet to be determined. First proof that the genetic material is deoxyribonucleic acid (DNA) was provided in 1944 by Oswald Avery. Prior to this time, of the many cell components that might possibly be the genetic material, proteins, because they have the necessary molecular complexity, were considered the most probable candidates. In the light of available chemical techniques, biologists erroneously believed that the DNA molecule consisted of a regularly repeated block of its constituent four nucleotides (the **tetranucleotide hypothesis**). Although such a structure is far too simple to perform as the genetic material, numerous experiments during the 1940s and early 1950s nevertheless confirmed that DNA was indeed the genetic material. In 1953 Watson and Crick solved the structural dilemma with their model of DNA.

TRANSFORMATION

In 1928 F. Griffith reported that heat-killed bacteria of one type could "transform" living bacteria of a different type. Griffith demonstrated this **transformation** using the bacterium *Diplococcus pneumoniae,* one strain of which did not produce a polysaccharide capsule around the bacterial cell and formed rough colonies on a solid medium (Figure 9–3). This R strain did not have a pathological effect on mice. Another strain (S), because of its polysaccharide coat, produced smooth colonies and caused a fatal bacteremia (bacterial infection) in mice. Bacteria of the rough strain were destroyed by

Figure 9–3.
Petri Plate with Smooth and Rough Colonies of *Diplococcus pneumoniae*
R (rough) strain colonies on the left and S (smooth) colonies on the right on the same blood agar. magnification 3.5X

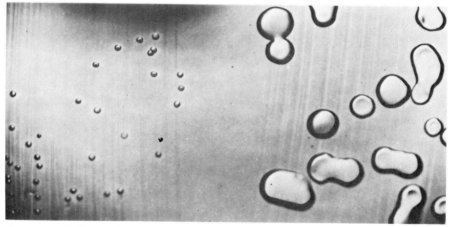

Source: O. T. Avery, C. M. MacLeod, and M. McCarty, ''Studies on the Chemical Nature of the Substance Inducing Transformation of Pneumococcal Types,'' *Journal of Experimental Medicine* 79 (1944):137–158. Reproduced by permission.

the immune system of the mice; the virulent smooth-strain bacteria survived attack of the immune system because they were protected by their polysaccharide coating.

DNA as the Transforming Agent

Griffith found that neither heat-killed S-type nor normal R-type cells, by themselves, caused bacteremia in mice. However, if a mixture of R-type and heat-killed S-type cells was injected into mice, they developed a bacteremia identical to that caused by injection of normal S-type cells (Figure 9–4). Thus, something in the heat-killed S cells transformed the R-type bacteria into S type. This work was continued by Oswald Avery and two of his associates, C. MacLeod and M. McCarty. In 1944 they established the nature of the transforming substance.

Avery and his colleagues did their work **in vitro** (literally, in glass), using colony morphology on culture media rather than bacteremia in mice as evidence of transformation. They found that deoxyribose nucleic acid (DNA) was the agent responsible for *transforming* R-type bacteria into S-type bacteria. They ruled out proteins, carbohydrates, and fatty acids both by their extraction procedure and by demonstrating that the only enzymes that destroyed the transforming ability were enzymes that destroyed DNA. This was the first proof that DNA was the genetic material.

The Hershey and Chase Experiments

Valuable information about the nature of the genetic material has also been obtained from viruses. Of particular value are the studies of viruses that attack bacteria (bacteriophages or phages). Since phages consist of only a nucleic acid surrounded by protein, they lent themselves nicely to the problem of determining whether protein or the nucleic acid is the genetic material.

The viral research of A. D. Hershey and M. Chase, published in 1952, supported the notion that DNA is the genetic material and helped to establish the nature of the viral infection process. Since all nucleic acids contain phosphorous while no proteins contain phosphorous, and since most proteins contain sulfur while no nucleic acids contain sulfur, Hershey and Chase designed an experiment using radioactive isotopes of phosphorous and sulfur to keep track of the viral proteins and nucleic acids during the infection process. They then proceeded to identify the material injected into the cell by phage attached to the bacterial wall. They used the bacteriophage named T2 and the bacterium *Escherichia coli*. The bacteria were labeled by growing them on culture medium containing the radioactive isotopes ^{35}S or ^{32}P. The phage were labeled by growing them on labeled bacteria.

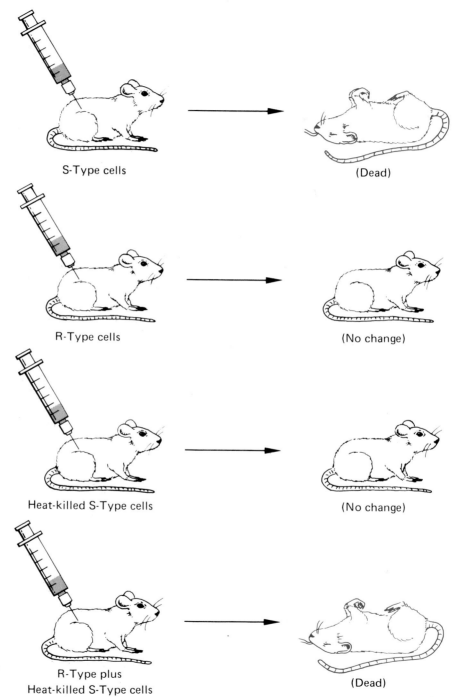

S-Type cells

(Dead)

R-Type cells

(No change)

Heat-killed S-Type cells

(No change)

Figure 9-4.
Griffith's Experiment
with the
Pneumococcus

R-Type plus
Heat-killed S-Type cells

(Dead)

When ^{35}S-labeled phage were mixed with unlabeled *E. coli,* it was found that the ^{35}S label, for the most part, stayed on the outside of the bacteria. When ^{32}P-labeled phage were mixed with *E. coli,* it was found that the ^{32}P label entered the bacterial cells and that the next generation of phage burst from the infected cells carrying a significant amount of the ^{32}P label. Hershey and Chase thus demonstrated that the outer protein coat of a phage does not enter the bacterium it infects, while the phage's inner material, consisting of DNA, does enter the infected bacterial cell (Figure 9–5). Since the DNA is responsible for the production of the new phage during the infection process, the DNA, not the protein, is the genetic material.

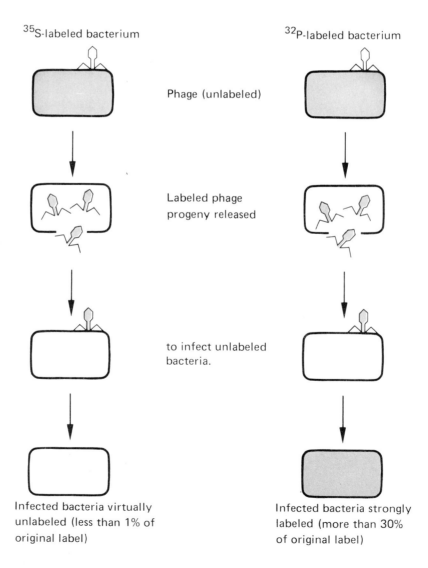

^{35}S-labeled bacterium

^{32}P-labeled bacterium

Phage (unlabeled)

Labeled phage progeny released

to infect unlabeled bacteria.

Infected bacteria virtually unlabeled (less than 1% of original label)

Infected bacteria strongly labeled (more than 30% of original label)

Figure 9 – 5.
The Hershey and
Chase Experiments
Using ^{32}P-Labeled
and ^{35}S-Labeled
Bacteriophage T2
The nucleic acid label (^{32}P) enters the bacterium during infection. The protein label (^{35}S) does not.

RNA as Genetic Material

The tobacco mosaic virus, TMV, that infects tobacco plants consists only of RNA (ribonucleic acid) and protein. The RNA, about 6400 nucleotides long, is packaged within a rod-like structure formed by over 2000 copies of a single protein (Figure 9–6). No DNA is present in TMV virions. Several classical experiments by H. Fraenkel-Conrat and others have demonstrated that RNA is the genetic material. In 1955 Fraenkel-Conrat and Williams showed that a virus can be separated, in vitro, into its component parts and reconstituted as a viable virus. This led to experiments by Fraenkel-Conrat and Singer, who reconstituted TMV with parts from different strains (Figure 9–6). For example, they combined the RNA from the common TMV with the protein from the masked (M) strain of TMV. They then made the reciprocal combination of common-type protein and M-type RNA. In both cases the TMV produced during the process of infection with the new combination was of the type associated with the RNA, not with the protein. Thus, it was the nucleic acid (RNA in this case) that was shown to be the genetic material. In later experiments pure TMV RNA was rubbed into plant leaves. Normal infection and a new generation of TMV resulted.

CHEMISTRY OF NUCLEIC ACIDS

Having identified the genetic material as the nucleic acid DNA (or RNA), we should examine the chemical structure of these molecules. Their structure will tell us a good deal about how they function.

Nucleic acids are made by joining **nucleotides** in a repetitive way. Nucleotides are made of three subunits: phosphate, sugar, and a nitrogenous base (Table 9–1). When incorporated into a nucleic acid, a nucleotide contains one each of the three components. But when free in the cell pool, nucleotides usually occur as triphosphates. The energy held in the extra phosphates is used to synthesize the polymer.

The structures of these various nucleic acid components are shown in Figure 9–7. The carbons of the ribose and deoxyribose sugars are numbered $1'$ to $5'$. The sugars differ only in the presence (ribose: in RNA) or absence

TABLE 9 - 1.   Components of Nucleic Acids

	Phosphate	Sugar	Base	
			Purines	Pyrimidines
DNA	Present	Deoxyribose	Guanine	Cytosine
			Adenine	Thymine
RNA	Present	Ribose	Guanine	Cytosine
			Adenine	Uracil

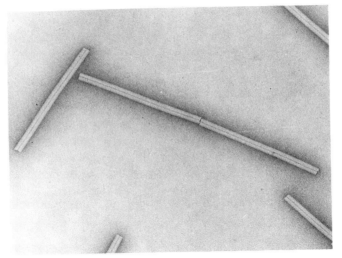

a.

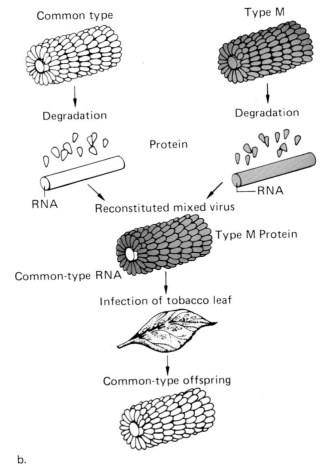

Common type

Type M

Degradation

Degradation

Protein

RNA

RNA

Reconstituted mixed virus

Type M Protein

Common-type RNA

Infection of tobacco leaf

Common-type offspring

Figure 9 – 6.
(a) Electron Micrograph of TMV. magnification 300,000X (b)Reconstitution Experiment of Fraenkel-Conrat and Singer Inheritance is controlled by the nucleic acid RNA.

Source: Reproduced courtesy of Dr. John T. Finch.

b.

314

PHOSPHATE

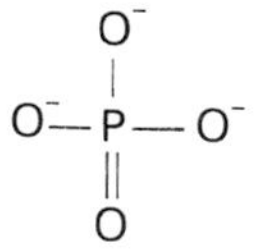

SUGAR

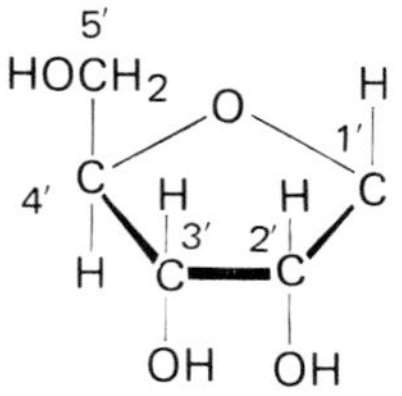

Ribose 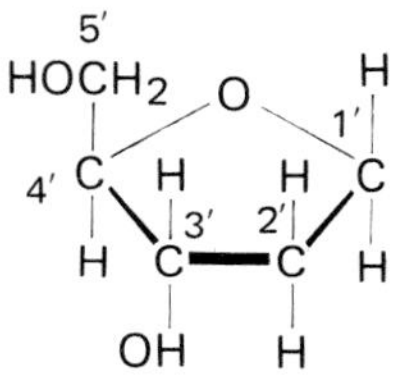   Deoxyribose

BASE   Purines

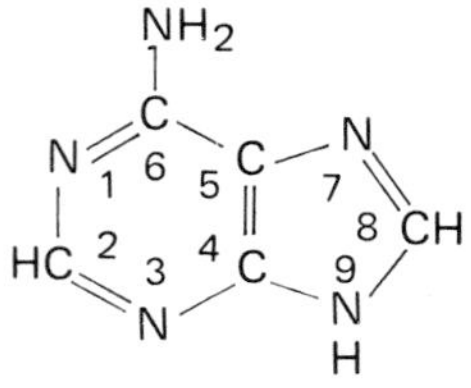

Adenine (6-aminopurine) 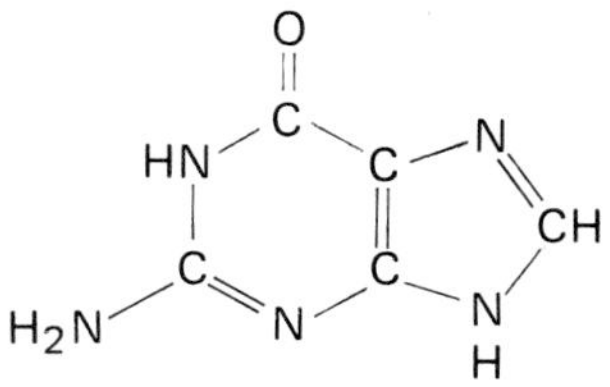   Guanine (2-amino-
                                  6-hydroxypurine)

                    Pyrimidines

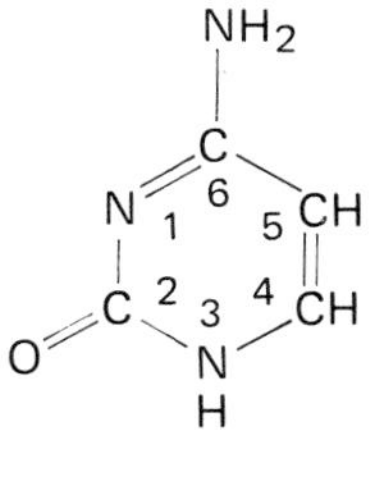

Cytosine   Thymine   Uracil
(2-hydroxy-   (2,6-dihydroxy-   (2,6-dihydroxy-
6-aminopyrimidine)   5-methylpyrimidine) pyrimidine)

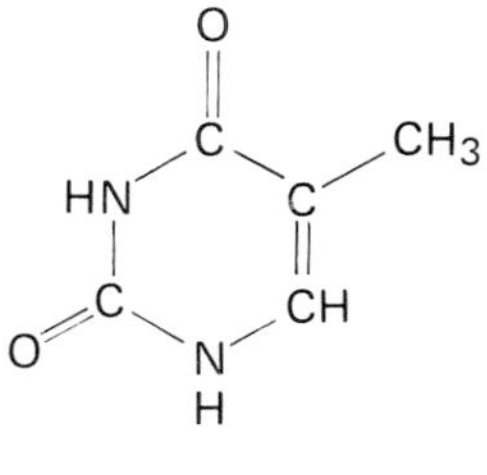

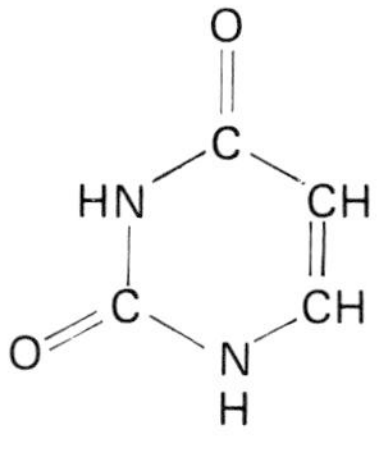

Figure 9–7.   Components of Nucleic Acids

(deoxyribose: in DNA) of an oxygen in the 2′ position. DNA and RNA both have four types of bases (two **purines** and two **pyrimidines**) in their nucleotide chains. Both molecules have the purines **adenine** and **guanine** and the pyrimidine **cytosine.** DNA has the pyrimidine **thymine;** RNA has the pyrimidine **uracil.** Thus, three of the nitrogenous bases are found in both DNA and RNA, whereas thymine is unique to DNA and uracil is unique to RNA.

A nucleotide is formed in the cell by attachment of a base to the 1′ carbon of the sugar and attachment of phosphate to the 5′ carbon of the same sugar (Figure 9–8); it takes its name from the base (Table 9–2). Nucleotides are linked together (polymerized) by the formation of a bond between the

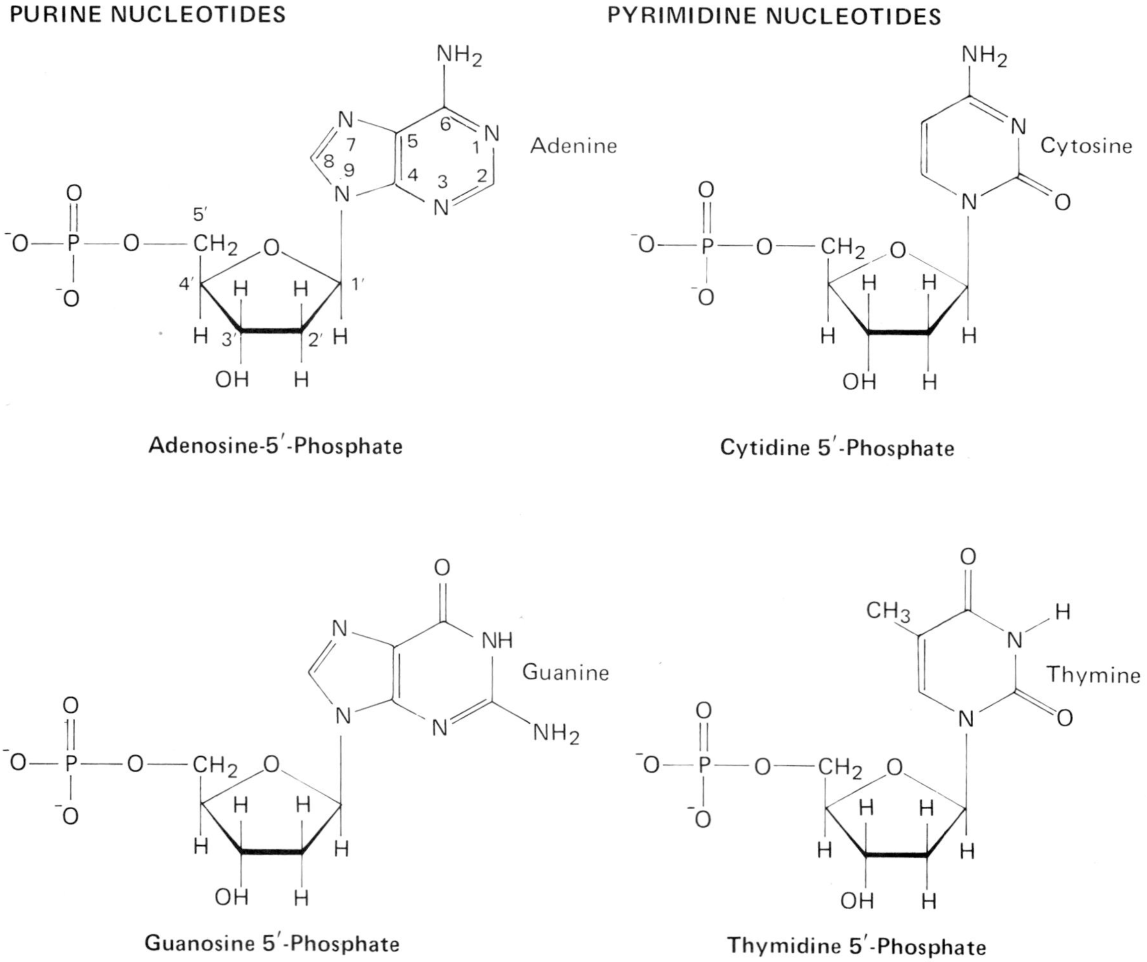

PURINE NUCLEOTIDES   **PYRIMIDINE NUCLEOTIDES**

Adenosine-5′-Phosphate   Cytidine 5′-Phosphate

Guanosine 5′-Phosphate   Thymidine 5′-Phosphate

Figure 9–8.   Structure of the Four Deoxyribose Nucleotides

TABLE 9-2. Nucleotide Nomenclature

Type of Nucleotide	Base	Nucleotide	Symbol
Deoxyribose	Guanine	Deoxyribose guanosine monophosphate (deoxyguanylate)	dGMP
	Adenine	Deoxyribose adenosine monophosphate (deoxyadenylate)	dAMP
	Cytosine	Deoxyribose cytidine monophosphate (deoxycytidylate)	dCMP
	Thymine	Deoxyribose thymidine monophosphate (deoxythymidylate)	dTMP
Ribose	Guanine	Guanosine monophosphate (guanylate)	GMP
	Adenine	Adenosine monophosphate (adenylate)	AMP
	Cytosine	Cytidine monophosphate (cytidylate)	CMP
	Uracil	Uridine monophosphate (uridylate)	UMP

phosphate of one nucleotide and the 3′ carbon of an adjacent nucleotide. Very long strings of nucleotides can be polymerized by this **phosphodiester bonding** (Figure 9–9).

Biologically Active Structure

Although the identity of the nucleotides that polymerized to form a strand of DNA or RNA was known, the actual structure of these nucleic acids when they function as the genetic material remained unknown until 1953. The general feeling was that the biologically active structure of DNA was more complex than a single string of nucleotides linked together by phosphodiester bonds—and that several interacting strands were involved. In 1953 Linus Pauling, a Nobel laureate who had discovered the α-helix structure of proteins, was investigating a three-stranded structure for the genetic material, while Watson and Crick decided that a two-stranded structure was more in line with available evidence.

DNA X-Ray Crystallography. Meanwhile, Maurice Wilkins, Rosalind Franklin, and their colleagues were using **x-ray crystallography** to analyze the structure of DNA. The molecules in a crystal are arranged in an orderly fashion, and, when a beam of x rays is passed through the crystal, the beam will be scattered. The pattern of the scatter can be recorded with a photographic plate. Analysis of such scatter photographs will reveal structural details of the molecules in the crystal (Figure 9–10).

The x-ray scatter patterns of DNA indicated that the molecule was a spiral or helix. With this information Watson and Crick began making molecular models.

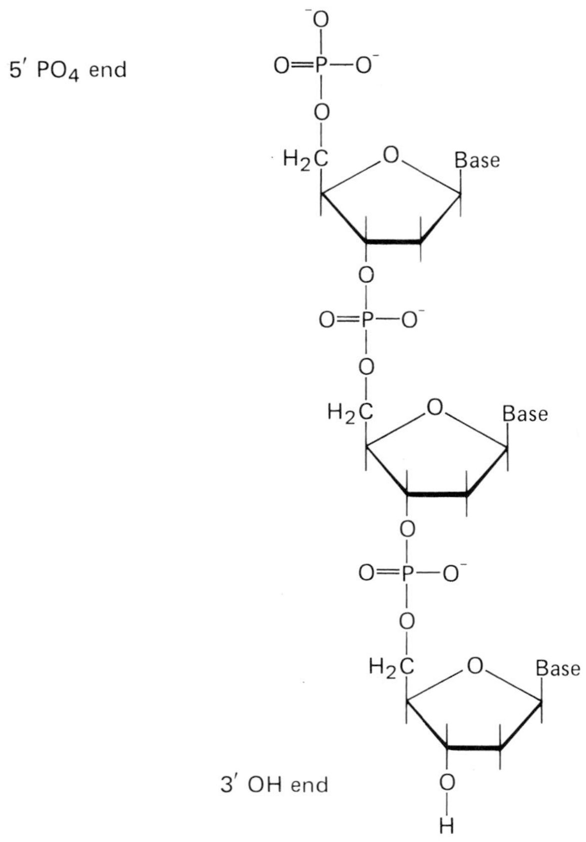

5′ PO₄ end

Figure 9–9.
Polymerization of
Adjacent
Nucleotides to Form
a Sugar-Phosphate
Strand

3′ OH end

The Watson-Crick Model. Watson and Crick found that a possible structure was one in which two helices coiled around one another (a **double helix**) with the sugar-phosphate backbones on the outside and the bases on the inside. This structure would fit the dimensions established for DNA by crystallography if the bases from the two strands were opposite each other and formed rungs in a helical ladder (Figure 9–11). The diameter of the helix could only be kept constant (about 20 Å units) if there were one purine and one pyrimidine base per rung. Two purines per rung would be too big and two pyrimidines would be too small.

After further experimentation with models of the bases, they found that the hydrogen bonding necessary to form the rungs of their helical ladder could occur only between certain base pairs. Hydrogen bonds are very weak bonds that involve the sharing of a hydrogen between two electronegative atoms, such as O and N. Hydrogen bonding can occur between thymine and adenine and between cytosine and guanine (Figure 9–12). There will be two hydrogen bonds between adenine and thymine and three between cytosine and guanine.

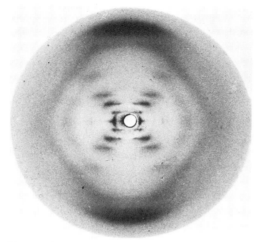

Figure 9-10.
Scatter Pattern of a
Beam of X Rays
Passed through
Crystalline DNA

Source: R. E. Franklin and R. Gosling, "Molecular Configuration in Sodium Thymonucleate," *Nature* 17 (1953):740–741. Reproduced by permission of the publisher and Professor M. H. F. Wilkins, the University of London King's College.

Erwin Chargaff
(1905–   )
Reproduced, with
permission, from the
Annual Review of
Biochemistry,
Volume 44. © 1975
by Annual Reviews
Inc.

Chargaff's Ratios. The structural effects of these constraints on hydrogen bonding explain the earlier results of E. Chargaff in his analysis of the base composition of DNA in various species (Table 9–3). Chargaff found that although the relative amount of a given nucleotide differs among species, there is an equivalent amount of adenine and thymine and an equivalent amount of guanine and cytosine in all the organisms studied. That is, in DNA there is a 1:1 correspondence between the purine and pyrimidine bases. (This is known as **Chargaff's rule**). Wherever adenine is on one strand of the DNA, thymine must be the corresponding base on the other strand; the same relationship holds for guanine and cytosine. The relation is called **complementarity.**

Another point about DNA structure relates to the fact that a **polarity** exists in the two strands. That is, one end of a DNA strand will have a 5'

TABLE 9-3. Percentage Base Composition of Some DNAs

Species	Adenine	Thymine	Guanine	Cytosine
Human (liver)	30.3	30.3	19.5	19.9
Mycobacterium tuberculosis	15.1	14.6	34.9	35.4
Sea urchin	32.8	32.1	17.7	18.4

Source: Data from Chargaff and Davidson, 1955.

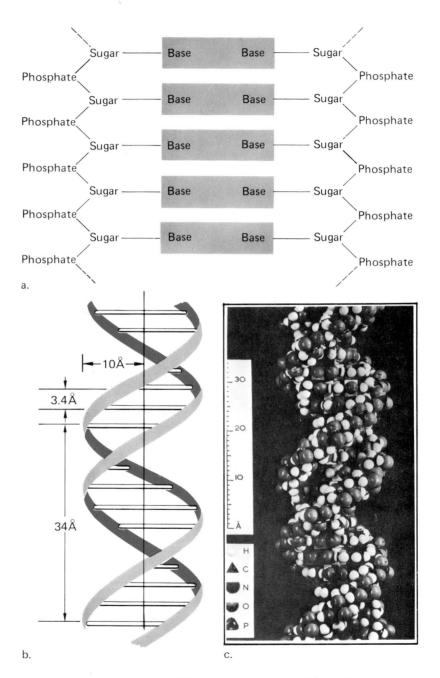

a.

b.

c.

Figure 9–11.
Double Helix
Structure of DNA
(a) Component parts
(b) Line drawing (c)
Space-filling model

Source: M. Feughelman et al., "Molecular Structure of DNA and Nucleoprotein," *Nature* 175 (1955):834–838. Reproduced by permission of the publisher and Professor M. H. F. Wilkins, the University of London King's College.

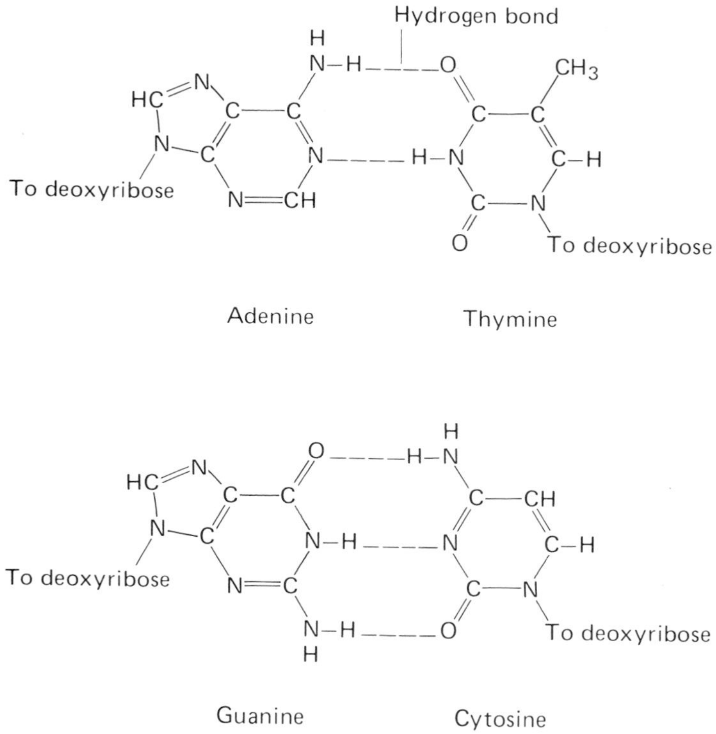

Figure 9–12. Hydrogen Bonding between the Nitrogenous Bases in DNA

phosphate and the other end will have a 3′ hydroxyl group. Watson and Crick found that an outcome of the hydrogen bonding was that the polarity of the two strands of DNA must run in opposite directions (Figure 9–13). This was later proved with a very elegant technique referred to as *nearest-neighbor analysis*. We will examine that technique later.

DNA Denaturation. Denaturation studies indicate that the hydrogen bonding in DNA occurs in the way suggested by Watson and Crick. Hydrogen bonds, while individually very weak, give structural stability to a molecule with large numbers of them. The hydrogen bonds can be broken and the DNA strands separated by heating the DNA molecule. A point is reached where the thermal agitation overcomes the hydrogen bonding and the molecule becomes **denatured** (or "melts"). It is logical that the more hydrogen bonds it contains, the higher the temperature needed to denature DNA. It follows that—since a G-C (guanine-cytosine) base pair has three hydrogen bonds to every two in an A-T (adenine-thymine) base pair—the higher the G-C content in a given molecule of DNA, the higher the temperature required to denature that DNA. This relationship is shown in Figure 9–14.

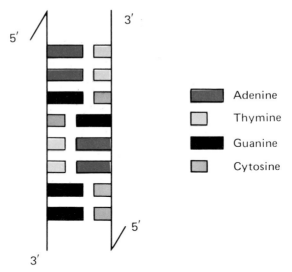

Figure 9-13. Polarity of the DNA Strands

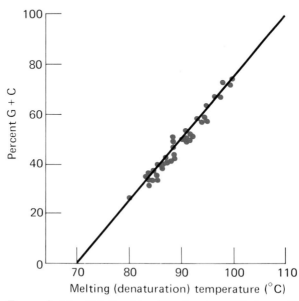

Figure 9-14. Relationship of the Number of Hydrogen Bonds (G-C content) and the Thermal Stability of DNA from Different Sources

322

Requirements of Genetic Material

Let us now return briefly to the requirements we have said a genetic material needs to meet: (1) control of protein synthesis, (2) self-replication, and (3) location in the nucleus (in organisms with nuclei). Does DNA meet these requirements?

Control of Protein Synthesis.   The next several chapters will examine the details of protein synthesis. At this point it is necessary only to observe that DNA possesses the complexity required to direct protein synthesis. Since there are only four bases in the DNA molecule with which to direct 20 amino acids, the relationship cannot be one base to one amino acid. The only way that four bases can direct 20 amino acids is to work in groups of sequential bases. Then, each of the 20 amino acids must be specified by a unique linear sequence of bases on the DNA strand. That is, these linear sequences act as a **genetic code**—one specific sequence for one specific amino acid.

Self-Replication.   Watson and Crick suggested in their 1953 paper how DNA might replicate itself. Their observation stemmed from the property of complementarity. Since the base sequence on one strand specifies the base sequence on the opposite strand, each strand could act as a complementary model for a new double helix. This could happen if the molecule simply "unzipped," allowing the inside of each strand to be a template for the new strand being synthesized (Figure 9–15).

Location.   The location requirement is that DNA must reside in the nucleus, where the genes occur on chromosomes. There is no problem here. In eukaryotes the chromosomes are made up of nucleic acids and proteins. In prokaryotes the chromosomes consist only of DNA.

Thus, DNA potentially fulfills all the requirements of a genetic material. Before going on to a discussion of DNA replication, let us examine the evidence in support of the contention that the polarity of the two strands of the double helix goes in opposite directions. This is the work involving nearest-neighbor analysis.

Arthur Kornberg
(1918–   )
Photograph by
Karsh.

Nearest-Neighbor Analysis

In 1961 J. Josse, A. Kaiser, and A. Kornberg published a technique known as **nearest-neighbor analysis.** By using a then newly discovered enzyme, the Kornberg enzyme (later to be DNA polymerase I), DNA could be synthesized in vitro. In any particular experiment one of the four nucleotides used in the synthesizing mixture contained radioactive phosphate. After the nucleotides had been incorporated into the DNA, the DNA was cleaved by means of an enzyme (spleen phosphodiesterase) that breaks the phospho-

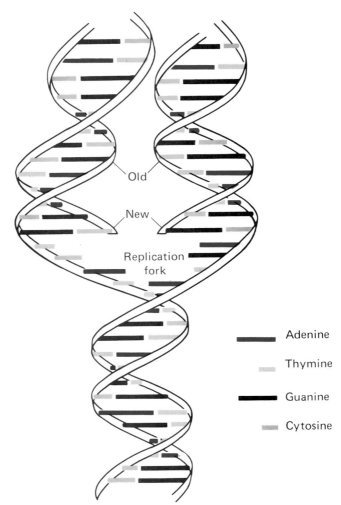

Figure 9-15.
A Possible Model for
DNA Replication

diester bond and leaves the radioactive phosphate on the 3′ carbon of the adjacent nucleotide (Figure 9–16). The distribution of the radioactivity was ascertained to determine nearest neighbors to the specific radioactively labeled nucleotide. For example, if radioactive phosphate is incorporated into adenosine triphosphate, we might find the recovered label associated with

Thymine	7%
Adenine	15%
Cytosine	38%
Guanine	40%

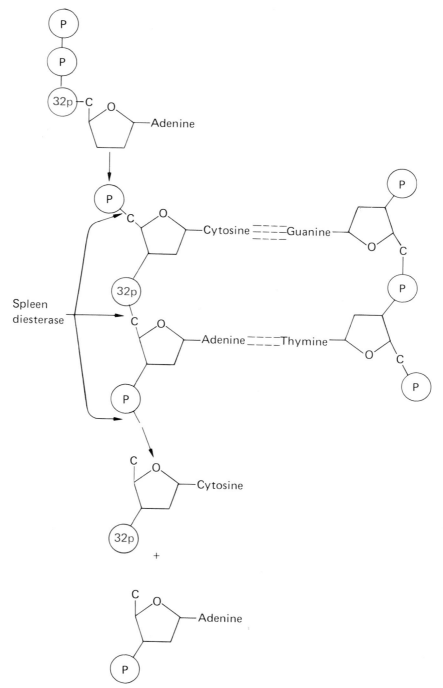

Figure 9-16. ^{32}P-Labeled Adenosine Triphosphate Incorporated into DNA Using the Kornberg Enzyme

The DNA is then cleaved using spleen diesterase. The labeled phosphoric acid is transferred to its nearest neighbor base, cytosine.

If the polarity is in the same direction, the transfer of label from C to T equals the amount of transfer of label from G to A.

If the polarity is in the opposite direction, the transfer of label from C to T equals the amount of transfer of label from A to G.

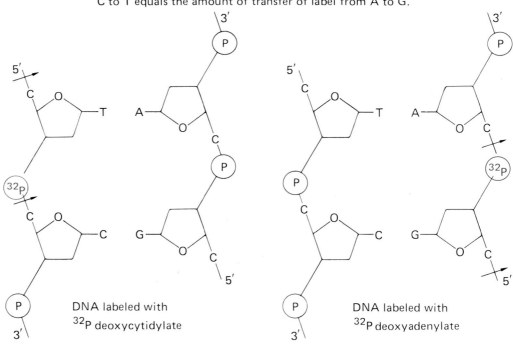

Figure 9–17. Alternative Predictions Arise in Nearest-Neighbor Analysis Depending on Whether or Not the Polarity of the Two Strands of the Double Helix Runs in Opposite Directions

If this experiment is repeated with each of the precursor nucleotides labeled, a table can be constructed of the frequency of nearest neighbors of all the nucleotides in DNA.

The nearest-neighbor analysis will give different results depending on whether the two strands of the double helix run in the same direction (parallel) or in opposite directions (antiparallel). These two alternatives are analyzed in Figure 9–17. The results of the entire experiment are shown in Table 9–4. The table shows that the transfer of radioactivity is the same from C to T as from A to G but not the same as from G to A. Thus it can be concluded that the strands run in opposite directions.

DNA REPLICATION

In their 1953 paper Watson and Crick suggested that the replication of the double helix could take place by unwinding of the DNA, so that each strand would form a new double helix by acting as a **template** for a newly synthesized strand. For example, when a double helix is unwound at an adenine-thymine base pairing, one unwound strand would carry A and the other would carry T. During replication, the A would pair with T on the new strand, giving rise to an A-T base pair and T would pair with A on the new strand, giving rise to an A-T base pair. Thus, one A-T base pair in one double helix would result in two base pairs—in two double helices. This process would repeat at every base pair in the double helix of the DNA molecule.

This process is called **semiconservative** replication because the double helix is not conserved in replication, although each strand is. It is not the only way in which replication could occur. The alternative methods of replication would be **conservative**—the whole original double helix acts as a template for a new one—and **dispersive**—some parts of the original double helix are conserved and some parts are semiconserved. In reality, the dispersive category includes any possibility other than conservative and semiconservative.

TABLE 9–4. The Results of Nearest-Neighbor Analysis in *Mycobacterium phlei*. The data are adjusted to the relative proportion of each base in the DNA.

	Location of Original Label			
Location of Eventual Label	A	T	G	C
T	.012	.026	.063	.061
A	.024	.031	.045	.064
C	.063	.045	.139	.090
G	.065	.060	.090	.122

Source: Data from Josse, Kaiser, and Kornberg, 1961.

DNA SEQUENCING

Major advances in molecular genetics over the past several years have come about through innovative technological developments. We are currently experiencing an information explosion centered around two new, and interrelated, techniques: *nucleic acid sequencing* and *recombinant DNA* (Chapter 12). Major questions about gene structure and function in both prokaryotes and eukaryotes are being answered with these techniques. Here we direct our attention to the newest techniques for determining the nucleotide sequences in DNA.

Sequencing has until recently been done more on RNA than DNA because several enzymes have been available to cleave RNA at particular points. This produces short, homogeneous segments of RNA that are relatively easy to sequence.

The same methodology has just now become available for DNA with the discovery of restriction endonucleases—DNA-degrading enzymes that cleave DNA at specific points (Chapter 12). By using different restriction endonucleases, it is now possible to obtain short, homogeneous segments of DNA.

Two methods of sequencing DNA are currently in use: the chemical method of Maxam and Gilbert and the "plus-and-minus" method of Sanger and Coulson. Here, we will describe the plus-and-minus method. The first complete genome sequenced was the single-stranded DNA virus φX174, which has 5375 nucleotides. We will use that as an example.

Within the viral protein coat is a single-stranded circle of DNA. When this DNA enters the

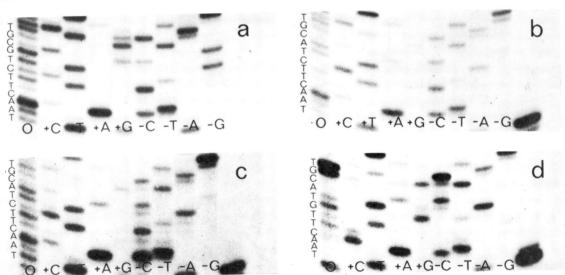

Figure 1. **Nucleotide sequences of the Same DNA Region in Four Different Mutants of** φX174.
The sequences are read directly from the banding patterns on the gels.

Source: Reprinted by permission of the publisher and the author from M. Smith et al., "DNA Sequence at the C Termini of the Overlapping Genes *A* and *B* in Bacteriophage φX174," *Nature* 265 (1977): 702–705.

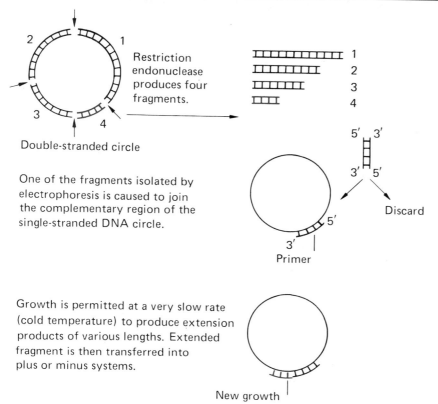

Restriction endonuclease produces four fragments.

Double-stranded circle

One of the fragments isolated by electrophoresis is caused to join the complementary region of the single-stranded DNA circle.

Discard

Primer

Growth is permitted at a very slow rate (cold temperature) to produce extension products of various lengths. Extended fragment is then transferred into plus or minus systems.

New growth

Figure 2. Initial Steps in the Plus-and-Minus Method of DNA Sequencing of Virus ϕX174

host cell, it is replicated to form a double-stranded circle. Therefore, if DNA is isolated either from the virus itself or from an infected cell, either single-stranded or double-stranded DNA can be obtained.

The basic concept of the sequencing method is that a small segment of DNA will be used as a primer for polymerization of DNA up to a specific nucleotide ("minus") or for degradation of DNA down to a specific nucleotide ("plus"). This will generate very small segments of DNA of various sizes all ending with a known nucleotide. These segments can be separated from each other by gel electrophoresis, a technique for separating molecules of different sizes. (See Chapter 4.) Nucleic acids of varying sizes will form bands such that the larger sequences will move shorter distances. A

step-ladder pattern is produced, with each rung representing a component of one less nucleotide (Figure 1). Let us return to ϕX174.

In order to produce heterogeneous DNA segments, we must grow DNA and for this we need a primer: a segment of DNA that is double stranded for part of the way and single stranded the rest of the way (Figure 9–25). Restriction endonucleases (see Chapter 12) are used to form small uniform segments of the viral DNA and these segments are isolated by electrophoresis. A particular segment is placed with the single-stranded viral DNA. It will form a double helix at its point of complementarity and thus form the primer (Figure 2). DNA polymerase is then added, along with the four nucleoside triphosphates, at a very low temperature in order to slow the reaction. Usually one of the

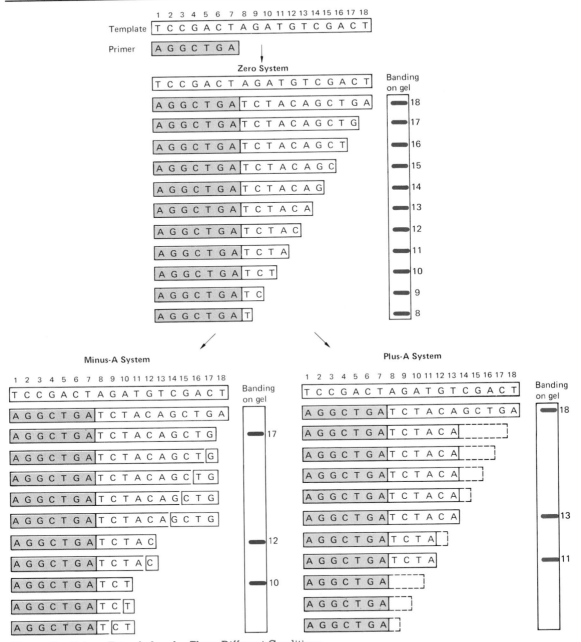

Figure 3. Primer Extended under Three Different Conditions

Source: From "The Nucleotide Sequence of a Viral DNA" by John C. Fiddes. Copyright © 1977 by Scientific American, Inc. All rights reserved. Reprinted by permission.

nucleoside triphosphates has a radioactive label for future autoradiography. The system is sampled frequently. The goal is a mixture that contains primer plus extension products representing every single growth step, where one step is the addition of one more nucleotide. This mixture is called the *zero* sample. Unused nucleotides are washed out and eight subsamples are separated; four of these are put into plus systems and four into minus systems.

The four minus systems are similar to the original growth mixtures that produced the zero sample, but each lacks one of the four nucleoside triphosphates ($-A$, $-T$, $-G$, $-C$). Thus growth in these systems will continue on each primed extension product until the place for the missing nucleotide is encountered. For example, the minus-A system will cause chain growth until a T is encountered on the template. Growth will then stop. Each extension product will be of a size such that the next addition should be an A (Figure 3). This is done for all four minus systems.

The four plus systems are similar to the minus except that they contain only one of the four nucleoside triphosphates. In the plus systems the DNA polymerase tends to degrade the newly synthesized DNA when it cannot synthesize new DNA. This degradation continues until the polymerase comes to the nucleotide that has been added in the system. For example, in the plus-A system, the polymerase will degrade the newly formed DNA

until it comes to the last A. At this point, degradation ceases because there are abundant adenosine triphosphates in the medium. The result is that the primed extension piece added to the plus system will have growth only to its last A (or whichever nucleotide for which the system is plus). (See Figure 3.) The end products of all eight experiments are then cleaved again by the exonuclease in order to isolate just the new growth product (Figure 4). Then the eight experiments plus the zero sample are run next to each other on the same gel.

The nucleotide sequence can be read directly off the gel (Figure 5). Each group of molecules travels down the gel a distance proportional to its length and shows as a band on the radiograph. The zero sample shows the number of possible positions and the plus samples show the nucleotides directly. The minus systems act as a check. For example, when there is a band in the minus-A system at position 12, there should be an A in position 13 of the plus-A system. The only possible difficulty is when there is a run of the same nucleotide. For example, if positions 10–15 were all G, there would be a band in the +G at position 15 and a band in the −G at position 9, but no bands in between. This would be read correctly however by noting the presence of bands in the zero sample. Upwards of 200 nucleotides can be read from a single gel. By sequencing each segment, the whole genome can be read, as will be shown in Chapter 12.

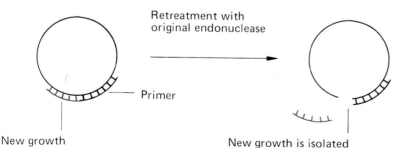

Retreatment with original endonuclease

Primer

New growth

New growth is isolated

Figure 4. After Extension Is Finished in Either Zero, Plus, or Minus Systems, New Growth is Isolated by Second Treatment with Restriction Endonuclease

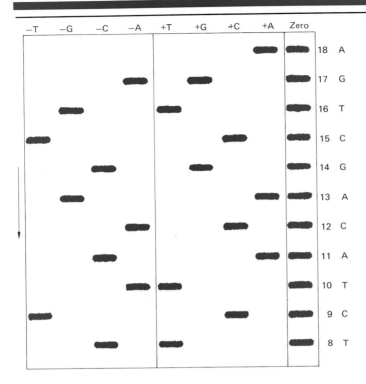

	−T	−G	−C	−A	+T	+G	+C	+A	Zero		
								▬	▬	18	A
				▬	▬				▬	17	G
		▬			▬				▬	16	T
	▬						▬		▬	15	C
			▬		▬				▬	14	G
		▬						▬	▬	13	A
				▬			▬		▬	12	C
			▬					▬	▬	11	A
				▬	▬				▬	10	T
	▬						▬		▬	9	C
			▬						▬	8	T

Figure 5. Schematic Representation of Autoradiograph Resulting from Procedures Illustrated in Figure 4

Source: From the "The Nucleotide Sequence of a Viral DNA" by John C. Fiddes. Copyright © 1977 by Scientific American, Inc. All rights reserved. Reprinted by permission.

The Meselson and Stahl Experiment

In 1958 M. Meselson and F. Stahl reported an elegant experiment designed to determine the mode of DNA replication. They grew *E. coli* in a medium containing a heavy isotope of nitrogen, ^{15}N. (The normal form of nitrogen is ^{14}N.) After growing for several generations on the ^{15}N medium, the DNA of *E. coli* is denser. The density of the strands can be determined using a technique known as **density-gradient centrifugation.** In this technique a cesium chloride (CsCl) solution is spun in an ultracentrifuge at high speed for several hours. Eventually, an equilibrium between centrifugal force and diffusion occurs, such that a density gradient is established in the tube. There will be a greater concentration of the CsCl at the bottom of the tube than at the top. If DNA (or any other substance) is added, it will concentrate and form a band in the tube at the point where its density is the same as that of the CsCl at that point in the density gradient. If there are several types of DNA with different densities, they will form several bands. The bands can be detected by observing the tubes with ultraviolet light at a wavelength of 260 nm, where nucleic acids absorb strongly. (One nanometer, nm, is 10 Å or 1/1000 of a micron, μ.)

After growing several generations of bacteria on ^{15}N medium, Meselson and Stahl transferred the bacteria to a medium containing only ^{14}N. The new DNA that was replicated in the ^{14}N medium was less dense than the DNA in the transferred bacteria. The results of the experiment are shown in Figure 9–18. This experiment shows that DNA replication is semiconservative.

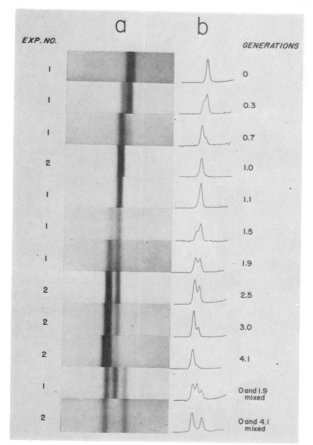

Figure 9–18. The Meselson and Stahl Experiment to Determine the Mode of Replication of DNA

Ultraviolet absorption photographs of bands resulting from density-gradient centrifugation of bacterial DNA (a) and densitometer tracings of the same bands (b).

Source: M. Meselson and F. Stahl, "The Replication of DNA in *Escherichia coli*," *Proceedings of the National Academy of Sciences USA* (1958)44:671–682. Reproduced by permission.

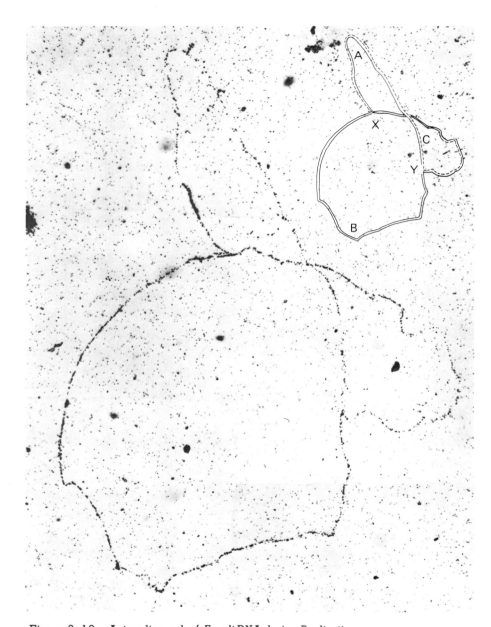

Figure 9–19. Autoradiograph of *E. coli* DNA during Replication
Inset diagram labels the three segments, A, B, C, created by the existence
of two forks, X and Y, in the DNA. Forks are created when circle opens
for replication. Length of the chromosome is about 1100 mμ.

Source: J. Cairns, "The Chromosome of *E. coli,*" *Cold Spring Harbor
Symposium on Quantitative Biology* 28 (1963):43–46. Copyright 1963.
Reproduced by permission.

If replication had been conservative, there would have been two bands after the first generation of replication—an original ^{15}N DNA and a new ^{14}N double helix. And, throughout the experiment, if the method of replication had been conservative, the original DNA would have continued to show as an ^{15}N band. This of course did not happen. If the method of replication had been dispersive, the result would have been various multiple-banded patterns, depending on the degree of dispersiveness. The results shown in Figure 9–18 are completely consistent with semiconservative replication.

Autoradiographic Demonstration of DNA Replication

The semiconservative method of replication was photographically verified by J. Cairns in 1963. Cairns used the technique of **autoradiography,** where radioactive atoms can be detected because, upon decay, they expose photographic film on which the silver grains can be counted to provide an estimate of the quantity of radioactive material present. Cairns grew *E. coli* in a medium containing radioactive thymine nucleotides. The radioactivity was in heavy, or tritiated, hydrogen (^{3}H). The DNA was then carefully extracted from the bacteria and autoradiographs were made. Figure 9–19 is an example. From this autoradiograph several points emerge. The first is the fact, known at the time, that the *E. coli* DNA occurs as a double-stranded circle. The second point is that the DNA is replicated while maintaining the integrity of the circle—that is, the circle is not broken in order to permit DNA replication; an intermediate **theta structure** is formed. Third, replication of the DNA seems to be occurring at one or two moving *Y-junctions* in the circle. This further supports the semiconservative mode of replication. The DNA is unwound at a point and replication proceeds in a semiconservative manner.

How the two Y-junctions move along the circle to the final step in which two circles are formed is diagrammed in Figure 9–20. The steps by them-

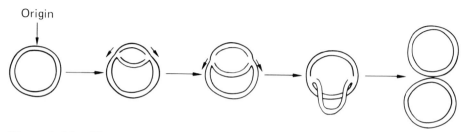

Origin

Figure 9–20.   Observable Stages in the DNA Replication of a Circular Chromosome, Assuming Bidirectional DNA Synthesis
The intermediate figures are called theta structures.

selves do not support either a unidirectional or a bidirectional mode of replication. But with autoradiography it is possible to determine whether new growth is occurring in only one or in both directions (Figure 9–21). By counting silver grains, Cairns found growth to be bidirectional. This has subsequently been verified by both autoradiography and genetic analysis.

In eukaryotes the DNA molecules (chromosomes) are larger than in prokaryotes and are not circular. In eukaryotes there are multiple sites of initiation of replication. Thus each eukaryotic chromosome is composed of several replicating units, or **replicons,** whereas the *E. coli* chromosome is composed of only one replicon. These replicating units form **bubbles** (or **eyes**) in the DNA (Figure 9–22).

Replication at the Y-Junction

Let us turn now to the processes that take place at the replicating junction in the DNA. Like all metabolic processes, DNA replication is controlled by enzymes. There are only three known enzymes that will polymerize nucleotides into a growing strand of DNA in *E. coli*. These enzymes are **DNA polymerase** I, II, and III. DNA polymerase I is the Kornberg enzyme used in the nearest-neighbor analysis and is most likely utilized in filling in small DNA segments during replication. At present, the precise role of polymerase II is not clear. Polymerase III is probably the primary active polymerase during normal DNA replication.

A Simple Model of Replication.   In the simplest model for DNA replication, new nucleotides would be added, according to the rules of complementarity, on both strands as the DNA opens up. But a problem exists, cre-

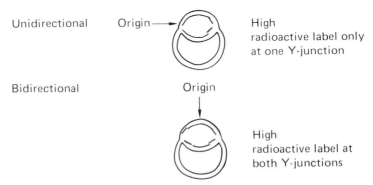

Figure 9–21.   Pattern of Radioactive Label Distinguishing Unidirectional from Bidirectional Mode of DNA Replication

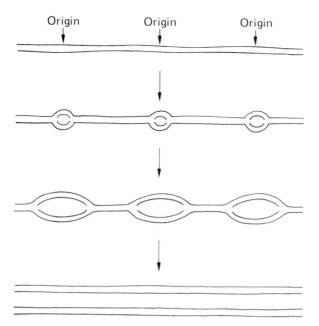

a.

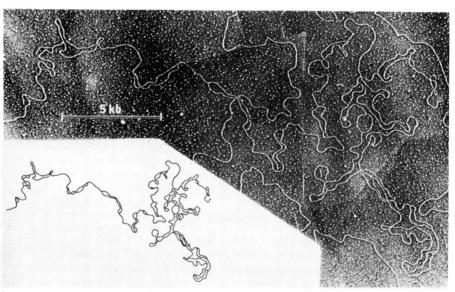

b.

Figure 9-22.   Replication Bubbles
(a) Formation of bubbles (eyes) in eukaryotic DNA because of multiple sites of origin of DNA synthesis (b) Electron micrograph of replicating *Drosophila* DNA showing these bubbles

Source: H. Kriegstein and D. Hogness, "Mechanism of DNA Replication in *Drosophila* Chromosomes: Structure of Replication Forks and Evidence for Bidirectionality," *Proceedings of the National Academy of Sciences USA* 71 (1974):135–139. Reproduced by permission.

ated by the polarity of DNA. The polarity of the two strands of a DNA double helix runs in opposite directions—that is, the strands are **antiparallel.** Where the molecule begins to unwind, one strand will have a terminal 3′ hydroxyl group and its unwinding complementary strand will have a terminal 5′ phosphate group. This would require one strand to replicate in the 5′→3′ direction and one strand to replicate in the 3′→5′ direction. But all the known polymerase enzymes add nucleotides only in the 5′→3′ direction. That is, the polymerase will, after assuring complementary base pairing, catalyze an ester bond between the 5′—PO_4 group of each new nucleotide and the 3′—OH carbon of the last nucleotide in the newly synthesized strand (Figure 9–23). The polymerase cannot effect the same bond with the 5′ phosphate of a nucleotide already in the DNA and the 3′ end of a new nucleotide. Thus, the simple model needs some revision, unless an undiscovered enzyme exists that can synthesize DNA by adding to the 5′—PO_4 end. Since most molecular biologists believe that a polymerase with that specificity will not be found, we are left with the need for another type of model.

Discontinuous DNA Replication. Evidence, such as Cairn's photos (Figure 9–19), suggests that replication is occurring simultaneously on both strands. Continuous replication is, of course, possible on the 3′→5′ template strand, which begins with the necessary 3′—OH **primer.** (Primer is DNA that is double stranded up to the point of the addition of new nucleotides and becomes a single-stranded template thereafter.) A **discontinuous** model of replication has been proposed for the complementary strand, which begins with an unusable 5′—PO_4 group. In this model the Y-junction opens to allow replication, which, as a rule, is **continuous** on the 3′→5′ template strand and discontinuous on the complementary 5′→3′ template strand, where it occurs in short segments *backwards,* away from the Y-junction (Figure 9–24). R. Okazaki has reported short, discontinuous segments of newly replicated DNA on one strand. These segments, called **Okazaki fragments,** range in length from about 1000 to 100,000 nucleotides.

Polymerase III can only initiate and carry out the replicating synthesis of complementary base nucleotides on a single strand from a double helix and only if there is a free 3′—OH primer end to which to bond a new nucleotide. These conditions are met by the 3′→5′ strand (Figure 9–23). But the second of these conditions, the presence of a 3′—OH primer, must be created de novo on the 5′→3′ strand before an Okazaki fragment can be synthesized. This is done by an RNA polymerase, which synthesizes a short RNA molecule complementary to the 5′→3′ template to act as primer (Figure 9–25). Polymerase III then begins adding DNA nucleotides and the RNA segment is degraded.

In order for discontinuous replication to be complete, an additional enzyme is required to seal the short sections (Okazaki fragments) polymerized backwards and hydrogen bonded to their 5′→3′ template. This enzyme,

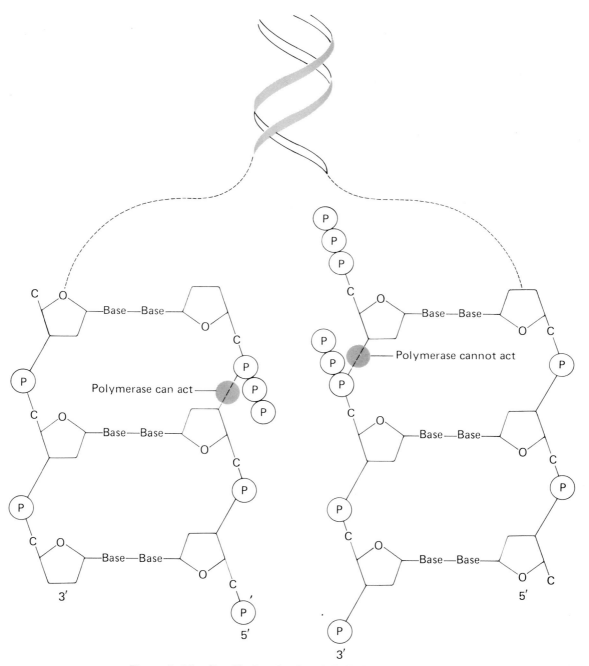

Figure 9-23.   New Nucleotides Can Only Be Added to DNA during
Replication in the 5′ → 3′ Direction

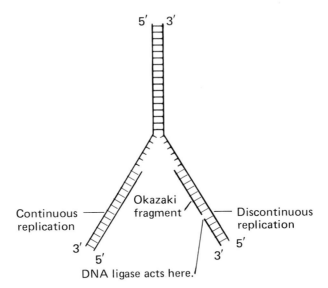

Figure 9-24.
Discontinuous Model
of DNA Replication

called **DNA ligase,** creates an ester bond between the last Okazaki fragment and the previously replicated DNA (Figure 9–24). The enzyme links adjacent nucleotides that are already incorporated into DNA.

Unwinding of DNA.   In the replication of a circular DNA, a topological problem occurs. If the double helix is opened at a point and unwound, ten-

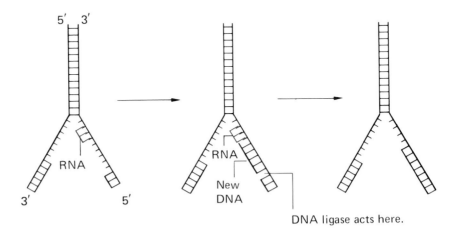

Figure 9-25.   Replication away from the Y-Junction
New DNA is added complementary to the 5′ → 3′ strand by first having an RNA segment synthesized. Then DNA polymerase can function and the RNA is subsequently removed.

sion builds up causing supercoiling of the DNA strands. (During replication, the rate of revolution is about 25,000 rpm.) Unless this tension is relieved, the entire circle cannot be replicated. It was originally believed that a swivel mechanism of some sort existed in the DNA molecule. Current evidence points to the action of an enzyme or enzymes (**nickase** or **swivelase**) that make nicks in one of the strands of the double helix near the point of replication. This cutting of the helix allows relaxation of the molecule and replication can continue. There also exist unwinding enzymes that open the DNA for replication at the replication Y-junction and unwinding proteins that attach to the single-stranded DNA and stabilize it prior to replication.

Rolling-Circle Model. There is a model other than the theta structure shown in Figure 9–20 for the mode by which circular DNA can be replicated. Evidence exists that a **rolling-circle** mode of replication holds for replication of several viral chromosomes (T4, ϕX174), as well as for chromosome transfer during bacterial conjugation (Chapter 6).

In the model for rolling-circle replication, a nick (actually, a break in one of the phosphodiester bonds) is made in one strand, creating a free 3'—OH end and a free 5'—PO_4 end (Figure 9–26). Synthesis of a new circular

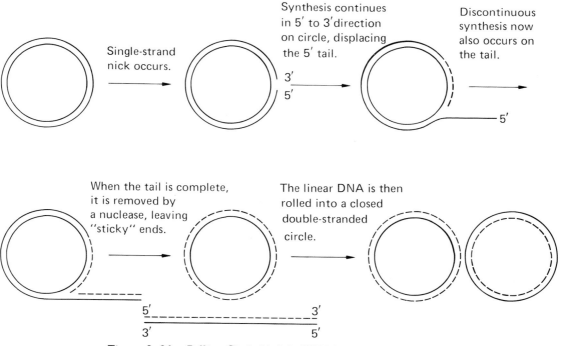

Figure 9–26. Rolling-Circle Model of DNA Replication

strand occurs by addition of nucleotides to the 3′ end, using the complementary intact strand as a template. As nucleotides are added to one end of the broken strand, the other end is displaced as a 5′—PO_4 tail. As replication of the circular template occurs, the 5′—PO_4 tail is replicated and the resulting double helix is severed from the double-helix circle by a DNA-degrading enzyme or **nuclease.** DNA ligase closes the replicated circular strand and joins the ends of the replicated tail into a circle. There are then two circular double helices. There is some evidence that the free 5′ tail becomes associated with the cell membrane of the host bacterium. While all the details are not in, there is strong evidence that this model holds in several systems.

DNA Polymerases as Exonucleases. A fact of interest is that all three polymerases not only can add new nucleotides to a growing strand in the 5′ →3′ direction but also can remove nucleotides in the opposite, 3′→5′ direction. This property is referred to as *exonuclease activity.* Enzymes that degrade nucleic acid are classified as **exonucleases** if they remove nucleotides from the end of a nucleotide strand or as **endonucleases** if they can break the sugar-phosphate backbone of a nucleotide strand. At first glance, exonuclease activity seems like an extremely curious property for a polymerase to have. Curious, unless we think about its ability to read for complementarity. After adding a nucleotide in the 5′→3′ direction, the polymerase can then check to be sure that there is proper complementarity. If the complementarity is improper, the polymerase can remove the incorrect nucleotide, put in the proper one, and continue on its way.

When an incorrect nucleotide is found and removed, the next nucleotide brought in, in the 5′→3′ direction, will have a triphosphate end available to provide the energy for its own incorporation. This fact might explain, in an evolutionary sense, why polymerization only occurs in the 5′→3′ direction (Figure 9–23). Consider what would happen if the polymerase were capable of adding nucleotides in the opposite direction. The energy for the diester bond would be coming from the triphosphate already attached in the growing 3′→5′ strand (Figure 9–23); then, if an error in complementarity were detected and the most recently added nucleotide were removed from the 3′→5′ strand by the polymerase, the last nucleotide in the double helix would no longer have a triphosphate available to provide energy for the diester bond of the next nucleotide brought in. Continued polymerization would thus require additional enzymatic steps to provide the energy for the process to continue. This could slow the process down considerably: The normal rate of incorporation of nucleotides is about 5000 per minute. This allows for the replication of the *E. coli* DNA in about 40 minutes.

CHAPTER SUMMARY

A genetic material must be able to direct protein synthesis, must replicate itself, and must be located in the chromosomes. Avery and his colleagues demonstrated that DNA was the genetic material when they showed that the transforming agent was DNA. Griffith had originally demonstrated the phenomenon of transformation of pneumococcus bacteria in mice. Hershey and Chase demonstrated that the DNA of bacteriophage T2 entered the bacterial cell. Fraenkel-Conrat demonstrated that in viruses without DNA (RNA viruses), such as tobacco mosaic virus, the RNA acted as the genetic material. Thus, by 1953, the evidence was strongly supportive of nucleic acids (DNA or, in its absence, RNA) as the genetic material.

Chargaff showed a 1:1 relationship of adenine (A) to thymine (T) and cytosine (C) to guanine (G) in DNA. Wilkins and his colleagues showed, by x-ray crystallography, that DNA was a helix of specific dimensions. Following these lines of evidence, Watson and Crick in 1953 suggested the double-helix model of the structure of DNA. In the Watson-Crick model, DNA is made up of two strands, running in opposite directions, with a sugar-phosphate backbone and bases facing inwards. Bases from the two strands form hydrogen bonds with each other with the restriction that only A and T and only G and C can hydrogen bond. This explains the relationships that Chargaff found among the bases. Melting temperatures of DNA also support this structural hypothesis because DNAs with higher G-C contents have higher melting, or denaturation, temperatures. (G-C base pairs have three hydrogen bonds versus only two in an A-T base pair.)

Nearest-neighbor analysis, using radioactive nucleotides, elegantly demonstrates that the two strands of DNA do, in fact, go in opposite directions.

DNA replicates by the unwinding of the double helix, with each strand subsequently acting as a template for a new double helix. This works because of complementarity: Only A-T, T-A, G-C, or C-G base pairs form hydrogen bonds within the structural constraints of the model. This model of replication is termed semiconservative. It was shown to be correct by Meselson and Stahl's experiment using heavy nitrogen. Autoradiographs of replicating DNA showed that replication proceeds bidirectionally from a point of origin. Prokaryotic chromosomes are circular with a single initiation point of replication. Eukaryotic DNA is linear with multiple initiation points of replication.

DNA polymerase enzymes add nucleotides only in the $5' \rightarrow 3'$ direction. This necessitates a complexity on the $5' \rightarrow 3'$ template strand. Replication must proceed in small segments working backwards from the Y-junction. Presumably, the $5' \rightarrow 3'$ restriction has to do with the ability of DNA polymerases to read for complementarity. Polymerase III is the active replicating enzyme and polymerase I is involved in DNA repair.

EXERCISES AND PROBLEMS

1. Diagram the results that Meselson and Stahl would have obtained (a) if DNA replication were conservative and (b) if it were dispersive.

2. Another suggested model of DNA replication is called a "knife-and-fork" model. In this model replication occurs in short segments at the Y-junction, but the polymerase starts 5'→3', crosses over the junction, and then comes back 5'→3'. What complication does this model introduce? What new enzymes would be needed?

3. Diagram the process of DNA replication that would take place if replication occurred in both directions.

4. If the tetranucleotide hypothesis were correct, could DNA be the genetic material?

5. Nucleic acids, proteins, carbohydrates, and fatty acids have been mentioned as potential genetic material. What other molecular moieties in the cell could function as the genetic material?

6. In what component parts do DNA and RNA differ?

7. What type of photo would J. Cairns have obtained if DNA replication was conservative? Dispersive?

8. Following is a section of a single strand of DNA. Supply a strand, by the rules of complementarity, that would turn this into a double helix.

A T T C T T G G C A T T C G C

9. Draw the structure of a short segment of DNA (three base pairs) at the molecular level and indicate the polarity of the strands.

10. In nearest-neighbor analyses, the transfers of radioactive label between particular bases must, of necessity, be equal to other particular transfers. For example, the amount of radioactive label transferred from adenine to cytosine (A to C) nucleotides equaled the amount transferred from guanine to thymine nucleotides (G to T). What transfer equaled a transfer of label from C to A? From A to A? From G to T? What would these predictions be if the polarity of both strands of DNA ran in the same direction? If any nucleotide could be added to DNA, at any point, in either direction, what predictions could be made about label transfers?

SUGGESTIONS FOR FURTHER READING

Avery, O., C. MacLeod, and M. McCarty. 1944. Studies on the chemical nature of the substance inducing transformation of *Pneumococcal* types. *J. Exp. Med.* 79:137–158.

Cairns, J. 1963. The chromosome of *E. coli. Cold Spr. Harb. Symp. Quant. Biol.* 28:43–46.

Chargaff, E. 1951. Structure and function of nucleic acids as cell constituents. *Fed. Proc., Fed. Am. Soc. Exp. Biol.* 10:654–659.

Chargaff, E., and J. Davidson, eds. 1955. *The Nucleic Acids.* New York: Academic Press.

Fraenkel-Conrat, H., and B. Singer. 1957. Virus reconstitution. II. Combination of protein and nucleic acid from different strains. *Biochim. Biophys. Acta.* 24:540–548.

Griffith, F. 1928. Significance of pneumococcal types. *J. Hygiene* 27:113–159.

Hershey, A., and M. Chase. 1952. Independent functions of viral protein and nucleic acid in growth of bacteriophage. *J. Gen. Physiol.* 36:39–56.

Josse, J., A. Kaiser, and A. Kornberg. 1961. Enzymatic synthesis of deoxyribonucleic acid. VII. Frequencies of nearest-neighbor base sequences in deoxyribonucleic acid. *J. Biol. Chem.* 236:864–875.

Kornberg, A. 1980. *DNA Replication.* San Francisco: Freeman.

Lewin, B. 1974. *Gene Expression, 1: Bacterial Genomes.* New York: Wiley.

Meselson, M., and F. Stahl. 1958. The replication of DNA in *Escherichia coli. Proc. Nat. Acad. Sci. U.S.* 44:671–682.

Stent, G., and R. Calendar. 1978. *Molecular Genetics: An Introductory Narrative,* 2nd ed. San Francisco: Freeman.

Watson, J. 1968. *The Double Helix.* New York: Signet. (A lively account of the discovery of the DNA structure.)

Watson, J. 1976. *Molecular Biology of the Gene,* 3rd ed. Menlo Park, Calif.: Benjamin.

Watson, J., and F. Crick. 1953. Molecular structure of nucleic acids. *Nature* 171:737–738.

10 Gene Expression: Transcription

All living things synthesize proteins. In fact, the kinds of proteins that a cell synthesizes determine the kind of cell or organism that it is. Hence the genetic material must control the types of proteins that a cell contains. Proteins are made up of long strings of amino acids joined together by peptide bonds (Figure 10–1). Each kind of protein is determined by the sequence of its amino acids, a sequence put together from a choice of only 20 naturally occurring amino acids. The amino acid sequence is specified by the sequence of nucleotides in DNA or RNA. In all prokaryotes, eukaryotes, and DNA viruses, the gene is a sequence of nucleotides in DNA that holds the information for the sequencing of amino acids in a polypeptide (a protein or one of its subunits if it is made of subunits). RNA serves as an intermediary between DNA and proteins. (In RNA viruses the RNA may serve as a template for the eventual synthesis of DNA or the RNA may serve as genetic material without DNA ever being formed. We consider these cases at the end of the chapter. In addition, some RNAs function directly.)

Originally, the direction of information transfer was termed the **central dogma:** DNA transferred information to an intermediate RNA, which then directly controlled protein synthesis (Figure 10–2). DNA also controls its own replication. The process whereby RNA is synthesized from a DNA template is called **transcription:** The DNA information is rewritten, but in basically the same language of nucleotides. The process whereby RNA controls the synthesis of proteins is termed **translation** because information in the language of nucleotides is translated into information in the language of amino acids. In this chapter we concentrate on the events surrounding transcription. In the next several chapters we will focus on translation and the various mechanisms of control of both transcription and translation in viruses, prokaryotes, and eukaryotes.

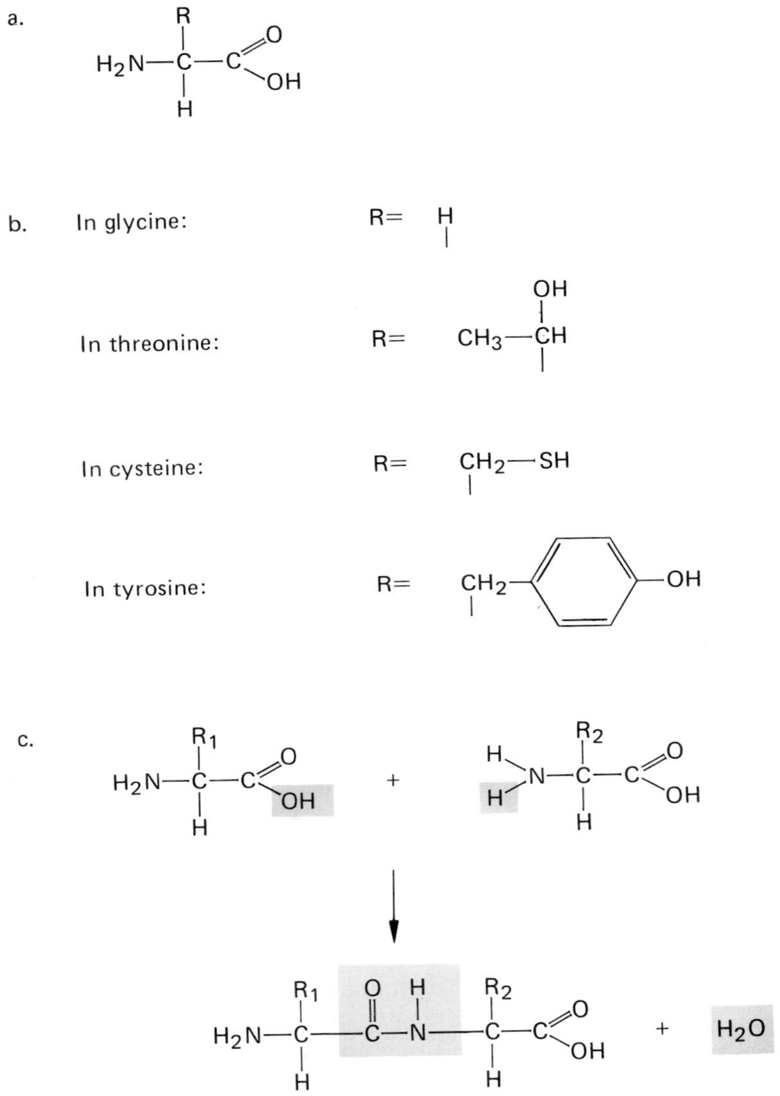

a.

b. In glycine: R= H

In threonine: R= CH_3—CH OH

In cysteine: R= CH_2—SH

In tyrosine: R= CH_2—〈benzene ring〉—OH

c.

Figure 10-1.
Amino Acids
(a) General form
(b) Complexity (see
Also Figure 11-1)
(c) Polymerization by
peptide bonding

TYPES OF RNA

In the process of protein synthesis, three different kinds of RNA serve in three unique roles. The first role is as a messenger (**mRNA**) to carry the sequencing information of DNA to particles in the cytoplasm known as **ribosomes,** where it will be translated. The second role of RNA is as a molecule (**tRNA**) to transfer the amino acids to the ribosomes, where protein

Figure 10-2.
Central Dogma

synthesis actually takes place. The third role of RNA is as part of the ribosome. This last type of RNA is called ribosomal RNA **(rRNA).** (Ribosomes will be described later in the chapter.) The general relationship of these three unique roles of RNA is diagrammed in Figure 10–3. By having an RNA messenger at the ribosome, the cell avoids the necessity of having DNA take a direct role in amino acid polymerization.

We know that DNA does not directly control protein synthesis because, in eukaryotes, translation occurs in the cytoplasm while DNA remains in the nucleus. In fact, the unicellular alga, *Acetabularia,* can grow (regenerate) a complex cap structure even in the absence of its nucleus (Figure 10–4). In addition, we have known for a long time that the genetic intermediate in prokaryotes and eukaryotes is RNA because the cytoplasmic RNA concentration increases with increasing protein synthesis and the cytoplasmic RNAs carry nucleotide sequences complementary to the cell's DNA.

TRANSCRIPTION

DNA-RNA Complementarity

What proof have we that the RNA found in the cytoplasm is a template copy of the DNA in the nucleus? Two lines of evidence are important here. First,

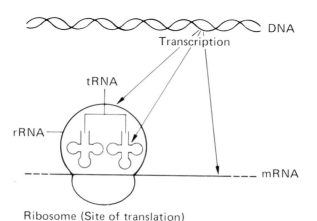

Figure 10-3.
Relationship among
the Three Types of
RNA during Protein
Synthesis

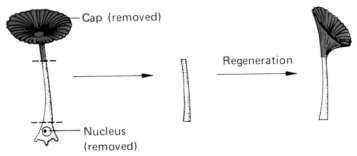

Figure 10-4. Regeneration in the Giant Unicellular Alga,
Acetabularia
The cell is several centimeters long.

it has been shown that the RNAs produced by various organisms have base
ratios very similar to the base ratios in their DNA (Table 10–1). The second
line of evidence comes from experiments by Hall, Spiegelman, and others
with **DNA–RNA hybridization.** DNA is denatured by heating, which
causes the two strands of the double helix to separate. When the solution is
cooled, a certain proportion of the DNA strands will rejoin and rewind. That
is, complementary strands "find" each other and reform double helices.
When RNA is added to the solution of denatured DNA and the solution is
slowly cooled, some of the RNA will form double helices with the DNA if
those fragments of RNA are complementary to a section of the DNA (Figure
10–5). The existence of complementarity between DNA and RNA is a per-
suasive indication that DNA acts as a template on which complementary
RNA is made.

Several interesting experiments have involved DNA–RNA hybridization. For example, RNA extracted from *Escherichia coli* before and after
infection by phage T2 has been tested to see if it hybridizes with the DNA
of the T2 phage or with the DNA of the *E. coli* cell. Interestingly enough, it
is found that the RNA in the *E. coli* cell hybridizes with the *E. coli* DNA
before infection but with the T2 DNA after infection. Thus, it is apparent

TABLE 10-1. Correspondence of
Base Ratios between DNA and RNA of
the Same Species

	RNA % G + C	DNA % G + C
E. coli	52	51
T2	35	35
Calf*	40	43

*Thymus gland cells.

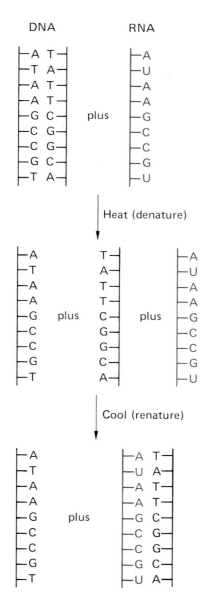

Figure 10-5.
DNA-RNA
Hybridization
This hybridization
can occur between
DNA and
complementary
RNA.

that when the phage attacks the *E. coli* cell, it starts to manufacture RNA
complementary to its own DNA.

Having reached the conclusion that protein synthesis is directed by
RNA that is transcribed (synthesized) from a DNA template, we need
answers to two questions. Is this RNA single or double stranded? Is is synthesized (transcribed) from one or both strands of the parent DNA? For the

most part, RNA does not exist as a double helix. It has the ability to form double helical sections when complementary parts come into apposition, but its general form is not as a regular double helix. The simplest, and most convincing, evidence for this is that the complementary bases are not found in corresponding proportions. That is, in RNA, uracil does not usually occur in the same quantity as adenine, and guanine does not usually occur in the same quantity as cytosine (Table 10–2).

The answer to the second question is that RNA is not copied from both strands of any given segment of the DNA double helix. Consider the problem of having a sequence of nucleotides (a gene) specify a sequence of amino acids to form a protein in the cell. Now impose the restriction that the complementary sequence must also specify a sequence of amino acids that acts as a functional enzyme, or structural protein, in the cell. Since most enzymes are about 300 amino acids long, the virtual impossibility of this task is obvious. It is therefore assumed a priori that, for any particular gene—that is, in any particular segment of DNA—the sequence on only one strand (the "sense" strand) is transcribed and its complementary sequence is not. There is considerable evidence to support this assumption.

The most impressive evidence that only one DNA strand is transcribed comes from work done with phage SP8, which attacks *Bacillus subtilis*. This phage has the interesting property of having a great disparity in the purine-to-pyrimidine ratio of the two strands of its DNA. The disparity is significant enough so that the two strands can be separated on the basis of their densities. After denaturation and separation of the two strands, DNA–RNA hybridization can be carried out using one of the two strands and the RNA produced after the virus infects the bacterium. Marmur and his colleagues found that hybridization occurred only between the RNA and the heavier of the two DNA strands. Thus, only the heavy strand acted as a template for the production of RNA during the infection process.

That only one strand of DNA serves as a transcription template for RNA has been verified for several other phages. However, when we get to larger viruses and cells, we find that both strands are used for transcription, but only one strand is used as a template for any one gene. This was clearly shown in phage T4 of *E. coli*, where certain RNAs hybridize with one strand while other RNAs hybridize with the other strand. Let us now look at the transcription process itself and then proceed to examine the three types of RNA in detail.

TABLE 10–2. Base Ratios in RNA (percentage)

	Adenine	Uracil	Guanine	Cytosine
E. coli	24	22	32	22
Euglena	26	19	31	24
Polio virus	30	25	25	20

RNA Polymerase

In prokaryotes, RNA transcription is controlled by a polymerase enzyme—
RNA polymerase (technically, DNA-dependent RNA polymerase). Using
DNA as a template, this enzyme polymerizes RNA triphosphate nucleotides.
In eukaryotes there are three RNA polymerases. But since much less is
known about them, we will focus our attention on prokaryotes. The complete
RNA polymerase enzyme—the **holoenzyme**—is composed of 5 subunits
(Figure 10–6). The core enzyme is made up of 4 parts: 2 alpha subunits, 1
beta subunit, and 1 beta-prime subunit. This core enzyme actually carries
out the polymerization process. The fifth part, the **sigma factor,** is a subunit that is primarily involved in the recognition of transcription start signals on the DNA. It can disassociate from the core enzyme.

When DNA polymerase is functioning properly, it will completely replicate the strands of DNA that it is using for templates. RNA polymerase,
however, must only use small segments (genes or small groups of genes) of
the DNA as templates for transcription. A given cell does not need all of its
enzymes functioning to the same extent all the time. (Nor, as we have noted,
is it feasible for the complementary strand of a gene sequence to be transcribed.) If there were no control of protein synthesis, all the cells of a higher
organism would be identical, and a bacterial cell would be producing all of
its proteins all the time. Since some enzymes depend on substrates not present all the time and since some reactions in a cell occur less frequently than
others, the cell—be it a bacterium or a human liver cell—needs to control
its protein synthesis to some extent. One of the most efficient ways for a cell
to exert the necessary control of protein synthesis is to perform transcription
in a selective manner. Transcription of unneeded enzymes is a wasteful procedure. Therefore, the RNA polymerase concerned with the transcription
process must be selective.

The mechanisms of transcriptional control need to be examined in two
ways. First, we need to understand how the beginnings and ends of tran-

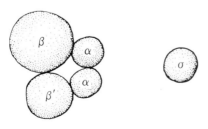

Core enzyme   Sigma factor

Figure 10–6.
Diagrammatic
Structure of RNA
Polymerase

Holoenzyme

scribable sections (a single gene or a series of adjacent genes) are demarcated. Second, we need to understand how the cell can selectively repress transcription of certain of these transcribable sections. The latter question is covered in Chapter 12.

If RNA polymerase is to act as the agent of transcriptional control, it must be able to recognize both the beginnings and the ends of genes (or gene groups) on the DNA double helix in order to initiate and terminate transcription. It must also be able to recognize the beginning of a gene to avoid transcribing the DNA strand that is not informational. RNA polymerase must be able to associate with the DNA double helix at the beginning of a gene and be oriented to transcribe only the "sense" strand of the double helix. This is accomplished because RNA polymerase recognizes certain *start* and *stop signals* in the DNA strands.

Initiation and Termination Signals for Transcription

The start signal is a sequence of nucleotides, known as the **promoter,** that is recognized by the holoenzyme. Without the sigma factor, RNA polymerase would transcribe randomly along the DNA (as it does in vitro). Thus, the sigma factor is involved in promoter recognition. Since the sigma factor itself does not bind to DNA and since sigma induces a conformational change in the core enzyme when the two are bound, the inference is that sigma changes the structure of the core enzyme such that the core enzyme recognizes the promoter. Shortly after the initiation of transcription, sigma disassociates from the core enzyme.

Promoters. Recent sequencing of DNAs has shown that promoter regions, known sometimes as **Pribnow boxes** after one of their discoverers, consist of relatively invariant sequences of 7 nucleotides located shortly before the first base transcribed. Most of these nucleotides are adenines and thymines (Figure 10–7); so the region is primarily held together by only two hydrogen bonds per base pair. Presumably, local DNA unwinding is a consequence of transcription, and presumably, RNA polymerase itself causes this local unwinding. Furthermore, when the polymerase is bound at the promoter region (Figure 10–7), it is in position to begin polymerization 5 to 7 nucleotides down, rather than at the next base after the promoter. The difference in the sequences of the various promoters may reflect a difference in affinity for RNA polymerase and, thus, may be a form of transcriptional control. All other things being equal, the genes with the more efficient ability to bind RNA polymerase will be transcribed more often.

Transcription, like DNA replication, proceeds always in the $5' \rightarrow 3'$ direction of the RNA. However, unlike DNA polymerase, RNA polymerase does not seem to **proofread**—that is, verify the complementarity of the hydro-

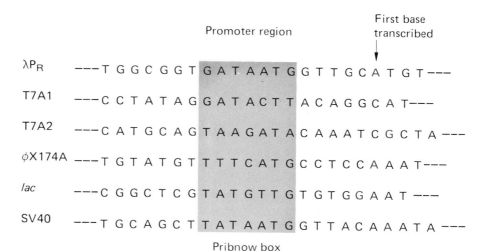

Figure 10-7. Nucleotide Sequences of the Promoter Region and the First Base Transcribed on Several Different DNA Sequences
Lambda (λ), T7, and φX174 are phages. *Lac* is an *E. coli* gene and SV40 is an animal virus.

gen bonding. This is not a serious deficiency, since mRNA is short-lived and many copies are made from each gene. Therefore, an occasional mistake does not produce permanent or overwhelming damage. If a particular segment of RNA is not functional, a new one will soon be made. It is probably more important to make RNA quickly than to proofread each RNA made.

Terminators.   Transcription continues until a stop signal, or **terminator sequence,** is perceived by the RNA polymerase. These terminator signals are strings of G-C base pairs followed by strings of A-T base pairs. They signal the polymerase by the alternation of a tight sequence, containing three hydrogen bonds per base pair, with a loose sequence, containing two hydrogen bonds per base pair. For termination to be specific, a protein known as **rho** is required. Rho binds to the developing RNA strand. Its exact role is not clear, but one current model has it traveling along the newly synthesized RNA strand at a speed comparable to that of the transcription process itself. A slowdown of the transcription process, such as would take place at a "tight" sequence of G-C base pairs, allows the rho factor to "catch up" to the polymerase. The catching up would result in the termination of transcription.

An overview of the transcription process is shown in Figure 10–8. The central part of the final piece of RNA transcribed contains the complement of a gene. That is, the information of a gene, coded in the sequence of nucleotides in the DNA, has been transferred in the process of transcription to a complementary sequence of nucleotides in the RNA. This RNA transcript

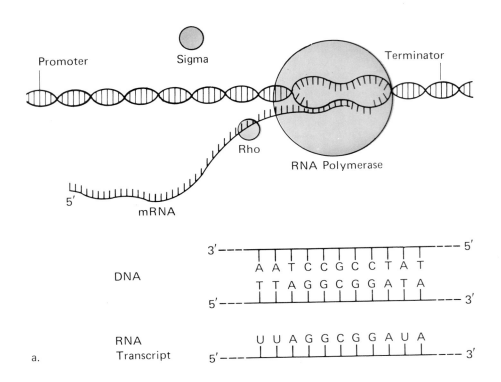

Promoter   Sigma   Terminator

RNA Polymerase

Rho

mRNA

5'

DNA

3'———|———|———|———|———|———|———|———|———|——— 5'
        A   A   T   C   C   G   C   C   T   A   T
        T   T   A   G   G   C   G   G   A   T   A
5'———|———|———|———|———|———|———|———|———|——— 3'

RNA
Transcript

        U   U   A   G   G   C   G   G   A   U   A
5'———|———|———|———|———|———|———|———|———|——— 3'

a.

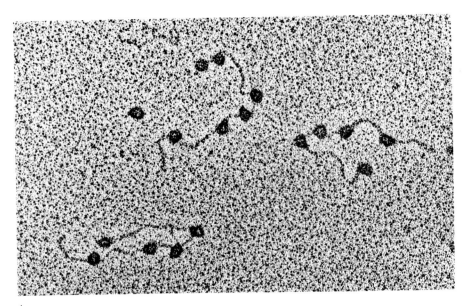

b.

contains a complement of the "sense" strand of the DNA of a gene and thus justifies its being termed a messenger. However, the entire transcript is more than a gene (or gene sequence): There are sequences of nucleotides before and after the segment that will be translated into amino acids (the gene itself). The translatable segment, or gene complement, always begins with a three-base sequence, AUG, which is known as an initiator sequence, and ends with one of three base sequences, UAA, UAG, or UGA, known as nonsense sequences. The discussion of these signals will be expanded in Chapter 11.

From the start of transcription to the translation initiation sequence (AUG) is a length of RNA referred to as a **leader.** From the nonsense sequence (UAA, UAG, or UGA) to the last nucleotide transcribed is the **trailer.** These sequences play a role in recognition and structural stability of the mRNA at the ribosome during the process of translation. A complete RNA transcript is diagrammed in Figure 10–9. More will be said about its parts later in this chapter and the next. Now, we turn our attention to ribosomal, transfer, and messenger RNA.

RIBOSOMES AND RIBOSOMAL RNA

Ribosomes are organelles in the cell, composed of proteins and RNA (rRNA), where protein synthesis occurs. In a rapidly growing *E. coli* cell, ribosomes can make up as much as 25% of the mass of the cell. Ribosomes, as well as other small particles and molecules, have their sizes measured in units that describe their rate of sedimentation in sucrose, using *density-gradient centrifugation.* This technique is used because it gives information on size (due to speed of sedimentation) while simultaneously isolating many of the molecules under study. Isolation by centrifugation is a relatively gentle isolation technique; the isolated molecules still retain their biological properties and can then be further used for experimentation. Since the physical chemist Svedberg developed ultracentrifugation in the 1920s, the unit of sedimentation is given his name.

In sucrose density-gradient centrifugation, the gradient is created by layering on decreasingly concentrated sucrose solutions. In cesium chloride

Figure 10–8. Transcription Overview and RNA Polymerase Molecules
(a) RNA polymerase is transcribing a gene. The sigma factor is seen disassociated near the promoter. The rho factor is shown on the newly formed RNA. (b) RNA polymerase molecules from *E. coli* bound to several promoter sites on T7 DNA. magnification 200,000X

Source: H. W. Fisher and R. C. Williams, "Electron Microscope Visualization of Nucleic Acids and of Their Complexes with Proteins," *Ann. Rev. Biochem.* 48 (1979):649–679. Reproduced with permission, from the *Annual Review of Biochemistry,* Volume 48. © 1979 by Annual Reviews Inc.

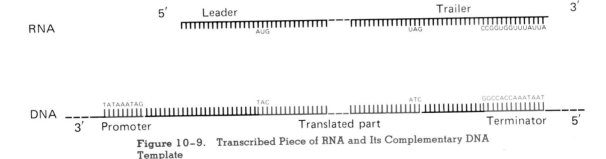

Figure 10-9.   Transcribed Piece of RNA and Its Complementary DNA Template

density-gradient centrifugation, mentioned in the previous chapter, the gradient develops during centrifugation. The sucrose centrifugation is stopped after a fixed time whereas in the cesium chloride technique the system spins until equilibrium is reached; the sucrose method tends to be more rapid. Samples can be isolated from a sucrose gradient by punching a hole in the bottom of the tube and collecting the drops in sequentially numbered containers. The first (lowest numbered) containers will have the heaviest molecules.

Ribosome Subunits

Ribosomes in all organisms are made of two subunits of unequal size. In *E. coli* the sedimentation value of the larger one is 50S (50 **Svedberg units**) and of the smaller one 30S. Together they sediment at about 70S. The eukaryotes have ribosomes of 80S. Most of our discussion will be confined to the well-studied ribosomes of *E. coli*.

Each ribosomal subunit is made up of one or two pieces of rRNA and a fixed number of proteins. The 30S subunit has 21 proteins and a 16S section of rRNA and the 50S subunit has 34 proteins and 2 pieces of rRNA—one 23S and one 5S section (Figure 10–10). Currently, the role of most of these proteins and the rRNAs is unknown but is being investigated. It is known, for example, that one of the 50S proteins is the enzyme peptidyl transferase, which forms the peptide bond between amino acids during protein synthesis. Additionally, the 3′ end of the 16S rRNA in the 30S subunit forms complementary hydrogen bonds with the 5′ leader of mRNA to help align it properly at the start of translation.

The Nucleolus in Eukaryotes

In *E. coli* all three rRNA segments are transcribed as a single long piece of RNA that is then subsequently cleaved and modified to form the final three

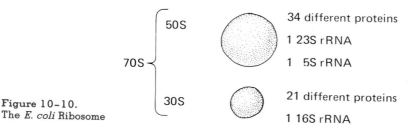

Figure 10-10.
The *E. coli* Ribosome

pieces (16S, 23S, and 5S) of rRNA. There appear to be about 5 to 10 copies of this rRNA gene in each chromosome of *E. coli*. In eukaryotes the two larger rRNA sections (18S and 28S) are transcribed as part of the same piece of RNA. The eukaryotic cells, however, have many copies of these rRNA genes, depending on the species. For example, fruit flies, *Drosophila mela-nogaster,* have about 130 copies of the gene on which the two larger segments of rRNA are transcribed. These genes occur in tandem on the sex (X and Y) chromosomes and are known as the **nucleolus organizer** (Chapter 2). The smallest rRNA subunit is also produced from a duplicated gene, but at a different point in the genome. For example, in *D. melanogaster* the 5S sub-unit is produced on chromosome 2. At the nucleolus organizer the nucleolus forms; this is the familiar dark blob found in the nuclei of eukaryotes. The nucleolus is the place of assembly of ribosomes. The various ribosomal pro-teins that have been manufactured in the cytoplasm migrate to the nucleus and eventually to the nucleolus where, with the final forms of the rRNAs, they are assembled into ribosomes.

An interesting development in the study of the nucleolus organizer is the discovery that between each repeat of the large rRNA gene is a region of **spacer DNA** that is not transcribed. This is shown in Figure 10–11 and diagrammed in Figure 10–12. In the electron micrograph of Figure 10–11, the polarity of transcription is indicated by the fact that the RNA is long at one end of the gene and short at the other end with a uniform gradation between. Notice that many RNA polymerases are transcribing each gene at the same time. The regions between the transcribed DNA segments are the spacer DNA regions.

TRANSFER RNA

During protein synthesis (Figure 10–3), the individual amino acids attached to particular tRNAs are taken to the site of protein synthesis, the ribosome. A messenger RNA will be at the ribosome and the code of information from the gene (DNA) will reside in this messenger. The code is read in sequences of three nucleotides, called **codons.** Each codon, with a few exceptions, spec-ifies an amino acid. The nucleotides of the codon on mRNA are complemen-

Figure 10-11. Transcription in the Nucleolus of *Triturus*
The polarity of the process as well as the spacer DNA is shown. magnification 18,000X

Source: O. L. Miller, Jr. and R. R. Beatty, "Visualization of Nuclear Genes," *Science* 164 (May 23, 1969):955–957. Copyright 1969 by the American Association for the Advancement of Science. Reproduced by permission.

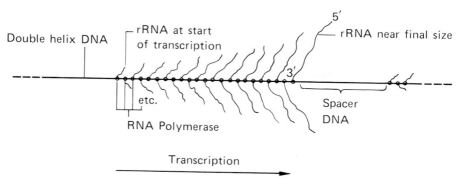

Figure 10-12. Details of the Transcription of the Large rRNA Gene Shown in Figure 10-11

tary to and pair with a sequence of three bases—the **anticodon**—on tRNA. Each different tRNA carries a specific amino acid. Thus, the specificity of the genetic code resides with the tRNA (Figure 10-13).

The correct amino acid is attached to the correct tRNA by one of a group of enzymes called **aminoacyl-tRNA synthetases.** There is one specific aminoacyl synthetase for every amino acid. However, there is more than one tRNA for almost every amino acid. This is due to the fact that there are more codons than there are amino acids. A triplet code will produce $4 \times 4 \times 4 = 64$ different combinations. The next chapter will amplify discussion of the code. Here it is sufficient to say that about 50 of the possible 64 tRNAs have been isolated. R. W. Holley and his colleagues were the first to discover the nucleotide sequence of tRNA; they published the structure of the tRNA for alanine in yeast in 1964. This structure is shown in Figure 10-14. The average tRNA is about 80 nucleotides long.

Similarities of All tRNAs

There are several unusual properties of transfer RNA. For one, all the different tRNAs of a cell have virtually the same general shape. (When purified, the heterogeneous mixture of all of a cell's tRNAs can form very regular crystals.) The regularity of the shape of tRNAs makes sense. During the process of protein synthesis, two tRNAs will occur next to each other on the ribosome and a peptide bond will be formed between their two amino acids. Thus, both (and therefore any two) tRNAs must have the same general dimensions, as well as similar structures, for recognition and manipulation at the ribosome.

An obvious feature of the tRNA in Figure 10-14 is that it has **unusual bases.** This tRNA is originally transcribed from the chromosomal DNA as a molecule about 50% longer than the final 80 nucleotides. The transcription

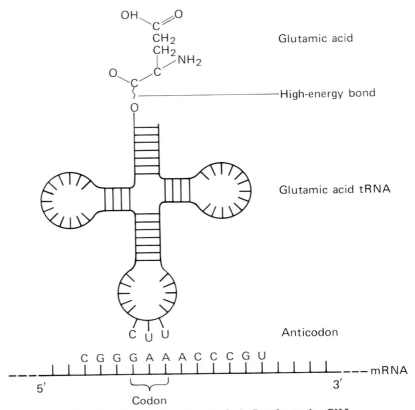

Figure 10-13. Specificity of the Genetic Code Resides in the tRNA where a Particular Anticodon Is Associated with a Particular Amino Acid

process is completely regular; it does not involve unusual bases. The original piece of RNA is degraded to the final 80 nucleotides, which are then enzymatically modified, primarily by the addition of methyl groups to the bases already in the RNA (Figure 10–15). Presumably, these unusual bases disrupt normal base pairing and are in part responsible for the loops formed by unbonded bases (Figure 10–14).

tRNA Loops

It is believed that the first loop on the 3′ side is involved in binding to the ribosome, the center loop is the anticodon loop, and the loop on the 5′ side is associated with recognition by the aminoacyl synthetases. Every tRNA so far examined starts with an adenine-cytosine-cytosine sequence on the 3′ end to which the amino acid is attached. The ribosome-binding loop on all has

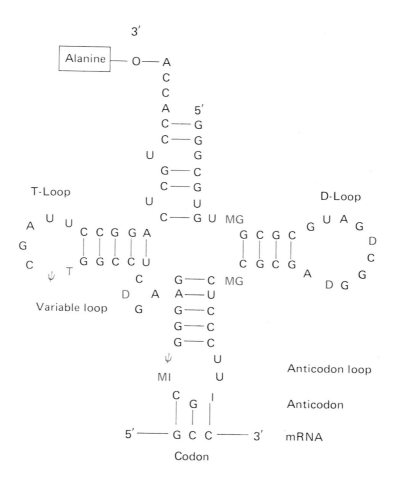

Figure 10–14.
Structure and
Sequence of Alanine
tRNA of yeast

Unusual Bases

ψ	Pseudouridine
I	Inosine
D	Dihydrouridine
T	Ribothymidine
MG	Methylguanosine
MI	Methylinosine

Source: Data from Holley et al. (1965).

the T-Ψ-C-G sequence. The anticodon on all is bounded by uracil on the 5′ side and a purine on the 3′ side. Thus there is a good deal of general similarity among all the tRNAs. This is consistent with the fact that they all enter into protein synthesis in the same way.

The actual shape of the functional tRNA in the cell is not an open cloverleaf as shown in Figure 10–14. Rather, there is helical twisting of the whole

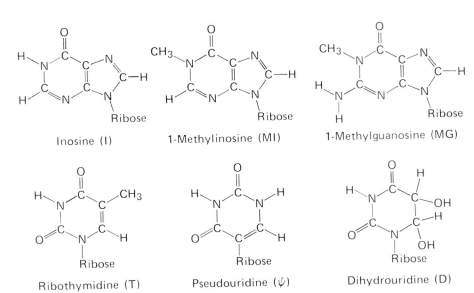

Figure 10–15.
Structures of the
Rare (Unusual)
Nucleotides Found
in Alanine tRNA of
Yeast

Inosine (I)   1-Methylinosine (MI)   1-Methylguanosine (MG)

Ribothymidine (T)   Pseudouridine (ψ)   Dihydrouridine (D)

molecule. A three-dimensional structure is shown in Figure 10–16, where the bases are numbered sequentially from the 5′ end.

MESSENGER RNA

After the discovery that RNA synthesis is correlated with protein synthesis and that in eukaryotes protein synthesis takes place in the cytoplasm rather than the nucleus, it was assumed that RNA was the messenger that carried the information from DNA into the cytoplasm. However, it was not known which form of RNA was the messenger. At first, after the discovery of ribosomal RNA, it was thought that the messenger was the rRNA; a mechanism was envisioned whereby RNAs associated with the ribosomes carried the messages for protein sequences. Each ribosome would be specific for a given protein. However, in 1959 it was found that no new ribosomes were synthesized when the phage T2 attacked *E. coli*. Since phages produce proteins only after infection of host cells and no new ribosomes were synthesized, the ribosomal RNA could not be the messenger.

The technique of DNA–RNA hybridization was used to demonstrate that another species of RNA with messenger properties existed in the cytoplasm. These experiments were described earlier in this chapter to show that RNA is made as a template copy of DNA. For example, *E. coli* RNA hybridizes with *E. coli* DNA. However, after the T2 phage attack, the RNA from within the *E. coli* cell (being neither tRNA nor rRNA) hybridizes with the T2 DNA. This suggests that shortly after infection the phage, using its own DNA as a template, transcribes mRNA, which then controls the protein synthesis in the host *E. coli* cell.

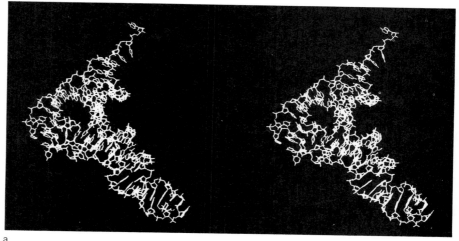

a.

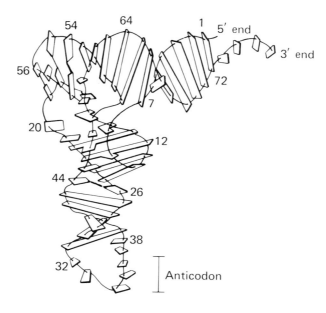

b.

Figure 10-16. Stereo (a) and Polar (b) Views of Yeast Phenylalanine tRNA

To make the stereo view three dimensional, fuse the images by "crossing" your eyes.

Source: J. L. Sussman and S. H. Kim, "Three Dimensional Structure of a Transfer RNA in Two Crystal Forms," *Science* 192 (May 28, 1976): 853–858. Copyright 1976 by the American Association for the Advancement of Science. Reproduced by permission.

Coupling of Transcription and Translation in Prokaryotes

In *E. coli,* translation into a protein of the newly transcribed mRNA can take place before the transcription process is complete. The mRNA is transcribed in the 5′→3′ direction, and it is at the 5′ end that translation begins. We can thus picture ribosomes attaching to mRNA as it is being transcribed from the *E. coli* chromosome. This has been beautifully shown in electron micrographs (Figure 10–17). In eukaryotes, however, mRNA is synthesized in the nucleus, but protein synthesis takes place in the cytoplasm. We do not have this regional division of labor in *E. coli* because the bacterium has no nucleus.

Eukaryotic Transcripts

In eukaryotes the mRNAs are much more complex than they are in prokaryotes. As in prokaryotes, the translatable part of the mRNA (gene complement) is surrounded by nontranslated segments that have been transcribed (Figure 10–9). However, there are also nontranslated segments that are added at both ends after the process of transcription is completed. (A sequence of methyl groups, referred to as a **cap,** is added to the 5′ end; a sequence of adenosine nucleotides, termed a **poly-A-tail** is added at the 3′ end.) Often the nontranslated parts form double helices (Figure 10–18),

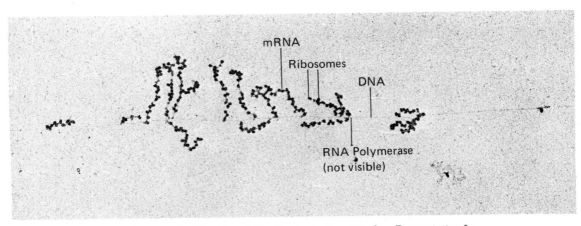

Figure 10–17. Translation Begins in *E. coli* before Transcription Is Finished
magnification 71,000X

Source: O. L. Miller, Jr. and B. R. Beatty, "Portrait of a Gene," *Journal of Cellular Physiology* 74, suppl. 1 (1969):225–232. Reproduced by permission.

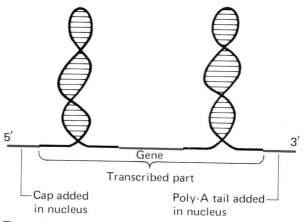

Figure 10–18. Eukaryotic mRNA

which are important in recognition of the mRNA by the ribosome and protection of the mRNA from degradation. Other modifications take place in
the nucleus. Sequences within the gene complement may be removed (see
next chapter). Before transportation into the cytoplasm, the mRNA is packaged with protein. Thus, unlike *E. coli* mRNA, the eukaryotic mRNA is a
highly modified, protein-associated molecule by the time it reaches the cytoplasm where translation will take place. The original transcribed RNA is
referred to as **heterogeneous nuclear RNA** and the changes are known
as **posttranscriptional modifications.**

We earlier considered a rough definition of a gene as that length of DNA
that codes for one protein. Although technically incorrect (we will correct it
later), this serves as a good working definition. However, we have just
encountered an inconsistency: Both tRNA and rRNA are coded for by genes;
yet neither is eventually translated into a protein. Both the tRNAs and
rRNAs function as final products. Thus tRNA and rRNA are the major
exceptions to the general rule that a gene codes for a protein.

NEW INFORMATION ABOUT THE CENTRAL DOGMA

The central dogma describes genetic information as flowing from DNA to
RNA to protein, with a DNA-to-DNA loop thrown in for self-replication.
Until relatively recently, the central dogma was believed to be inviolable, but
more recent discoveries require that it be modified. In Figure 10–19 two new
arrows are added to the original central dogma (Figure 10–2) to indicate
modifications.

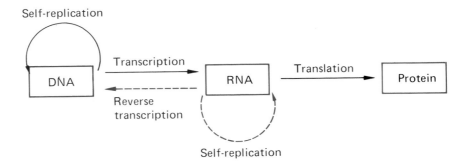

Figure 10–19.
Modified Central
Dogma

Reverse Transcription

First, the return arrow from RNA to DNA indicates that RNA can be used as a template for the synthesis of DNA. All RNA tumor viruses, such as Rous sarcoma virus, have the information for the translation of an RNA-dependent DNA polymerase (often referred to as **reverse transcriptase**) that synthesizes a DNA strand complementary to the viral RNA. Howard Temin of the University of Wisconsin and David Baltimore of MIT received Nobel prizes for their discovery of this polymerase enzyme, the functioning of which is especially important since it is a step in the process of infection of a normal cell and its transformation into a cancerous cell. When the viral RNA enters a cell, it translates the reverse transcriptase, which is a normal component of all RNA tumor viruses. The enzyme synthesizes a DNA–RNA double helix, which then converts into a DNA–DNA double helix that can integrate into the host chromosome. After integration, the DNA transcribes copies of the viral RNA, which is both translated and packaged into new viral particles that are released from the cell to repeat the process.

RNA Self-Replication

The second modification to the central dogma is the loop at RNA. This loop indicates that RNA can act as a template for its own replication, a process observed in a small but interesting class of phages. These **RNA phages,** such as R17, f2, MS2, and Qβ, are the simplest phages known. Recently, the complete nucleotide sequence has been worked out for one of them, MS2. (See Chapter 9 for details of nucleotide sequencing.) This phage contains about 3500 nucleotides and codes for only 3 proteins: a coat protein, an attachment protein, and a subunit of the enzyme **RNA replicase** (Figure 10–20). The coat protein is needed 180 times per new phage. The attachment protein is needed in only one copy per phage and is responsible for the attachment of the phage and subsequent penetration into the host. The

Figure 10-20.   RNA "Chromosome" of MS2

RNA replicase subunit combines with three of the cell's proteins to form the enzyme (RNA replicase) that allows the single-stranded RNA of the phage to replicate itself.

Since the new protein needed to construct the RNA replicase enzyme must be synthesized before the phage can replicate its own RNA, it is necessary for the phage RNA to first act as a messenger when it infects the cell. Thus we have the interesting situation of genetically directed protein synthesis without the process of transcription ever taking place. The viral genetic material, its RNA, is first used as a messenger in the process of translation and then used as a template for RNA replication. Even simpler than these RNA viruses are **viroids,** bare RNA particles that are plant pathogens. They do not even have a protein coat!

The next chapter continues this discussion of the protein synthesis process by describing the process of translation, in which the information in messenger RNA is turned into the sequence of amino acids in a protein.

CHAPTER SUMMARY

The central dogma is a description of the direction of information transfer between DNA, RNA, and protein. The previous chapter described the DNA self-replication loop. This chapter has described the transcription process, whereby DNA acts as a template for the production of RNA. Messenger RNA (mRNA) carries a complementary copy of the gene (a length of DNA that codes for a protein) to the ribosome, where protein synthesis actually takes place. Transfer RNAs (tRNAs) transport the amino acid building blocks to the ribosome. Complementarity between the mRNA codon and the tRNA anticodon establishes the amino acid sequence in the synthesized protein specified by the gene. Ribosomal RNA (rRNA) is also involved in this process of gene-directed protein synthesis.

Intracellular RNA is single stranded. At any one gene, RNA is transcribed from only one strand of the DNA double helix. The transcribing enzyme is RNA polymerase. The core enzyme, when associated with the sigma factor, recognizes the transcription start signal, the promoter. Termination of transcription involves the rho factor and a sequence on the DNA called the terminator. When RNA polymerase encounters the terminator, the rho factor, in some as yet to be determined fashion, takes part in the termination of transcription and release of the transcript from the DNA template.

The ribosome is made of two subunits, each having protein and RNA components. There are approximately 50 different tRNAs. They are charged with their particular amino acids by enzymes called aminoacyl synthetases. Each tRNA has about 80 nucleotides, including several "unusual" bases. All tRNAs have similar structures and dimensions.

Messenger RNAs are transcribed with a leader before and a trailer after the translatable part, or gene. In prokaryotes translation begins before transcription is completed. In eukaryotes these processes are completely uncoupled: Transcription is nuclear and translation is cytoplasmic. In eukaryotes mRNA is modified after transcription. A cap and tail are added and association with protein takes place prior to transport into the cytoplasm.

The study of several RNA viruses has shown that RNA can act as a template to replicate itself and also to synthesize a template DNA. These two processes add new directions of information transfer in the central dogma.

EXERCISES AND PROBLEMS

1. How could DNA–DNA or DNA–RNA hybridization be used as a tool to construct a phylogenetic tree (evolutionary tree) of organisms?

2. Given that RNA polymerase does not proofread, do you expect high or low levels of error in the transcription process? Think about this from an evolutionary point of view.

3. What are the transcription start and stop signals? How are they recognized? Can a transcription unit include more than one translation unit (gene)?

4. Diagram a moment in time at a ribosome. Show complementarity among the three RNA varieties.

5. Make a diagram of Figure 10–17 in order to clarify precisely what it shows.

6. What would the effect be on transcription if a cell had no sigma factor? No rho factor?

7. Draw a double helical section of DNA containing transcription start and stop information, as well as translation start and stop information. Give the base sequence of the mRNA transcript.

SUGGESTIONS FOR FURTHER READING

Crick, F. 1970. Central dogma of molecular biology. *Nature* 227:561–563.

Hall, B. D., and S. Spiegelman. 1961. Sequence complementarity of T2-DNA and T2-specific RNA. *Proc. Nat. Acad. Sci. U.S.* 47:137–146.

Holley, R. W., et al. 1965. Structure of a ribonucleic acid. *Science* 147:1462–1465.

Lewin, B. 1980. *Gene Expression, 2: Eukaryotic Chromosomes,* 2nd ed. New York: Wiley.

Marmur, J., et al. 1963. Specificity of the complementary RNA formed by *B. subtilis* infected with bacteriophage SP8. *Cold Spr. Harb. Symp. Quant. Biol.* 28:191–199.

Miller, O. L., Jr., et al. 1970. Electron microscopic visualization of transcription. *Cold Spr. Harb. Symp. Quant. Biol.* 35:505–512.

RNA Polymerase. 1976. Cold Spring Harbor, N.Y.: Cold Spring Harbor Laboratory.

Rosenberg, M., and D. Court. 1979. Regulatory sequences involved in the promotion and termination of RNA transcription. *Ann. Rev. Genet.* 13:319–353.

Stent, G., and R. Calendar. 1978. *Molecular Genetics.* San Francisco: Freeman.

Transcription of Genetic Material. 1970. Cold Spring Harbor, N.Y.: Cold Spring Harbor Laboratory.

Watson, J. 1976. *Molecular Biology of the Gene.* Menlo Park, Calif.: Benjamin.

Gunther S. Stent (1924–   ) Courtesy of Dr. Gunther S. Stent

11 Gene Expression: Translation

In order to synthesize a functional protein, basically all that needs to be done is to have the amino acids polymerized (Figures 11–1 and 11–2). The sequence of polymerized amino acids is termed the **primary structure.** This structure will spontaneously fold itself into the **secondary structure,** which is the flat or helical configuration of the polypeptide backbone. Further folding and formation of disulfide bridges between closely aligned cysteines create the **tertiary structure.** Many proteins in the active state are composed of several subunits that together are then referred to as the **quaternary structure** of the protein. The structures of some proteins are shown in Figures 11–3, 11–4, and 11–5.

From information about the sequencing of proteins, it is apparent that the role of the genetic material during the process of translation must be to specify the primary structures of the proteins to be synthesized. The average protein is about 300 to 500 amino acids long. Thus, since the genetic code works in triplets, the average translated part of a gene is about 900 to 1500 nucleotides, or base pairs, long.

INFORMATION TRANSFER

Before proceeding to the details of the translation process, a quick summarizing sketch of the process may be helpful (Figure 11–6). The ribosome with its rRNA and proteins is the site of protein synthesis. The information from the gene is in the form of mRNA, and the amino acids are carried to the ribosome attached to tRNAs. We will begin to look at the translation process by looking at the tRNAs.

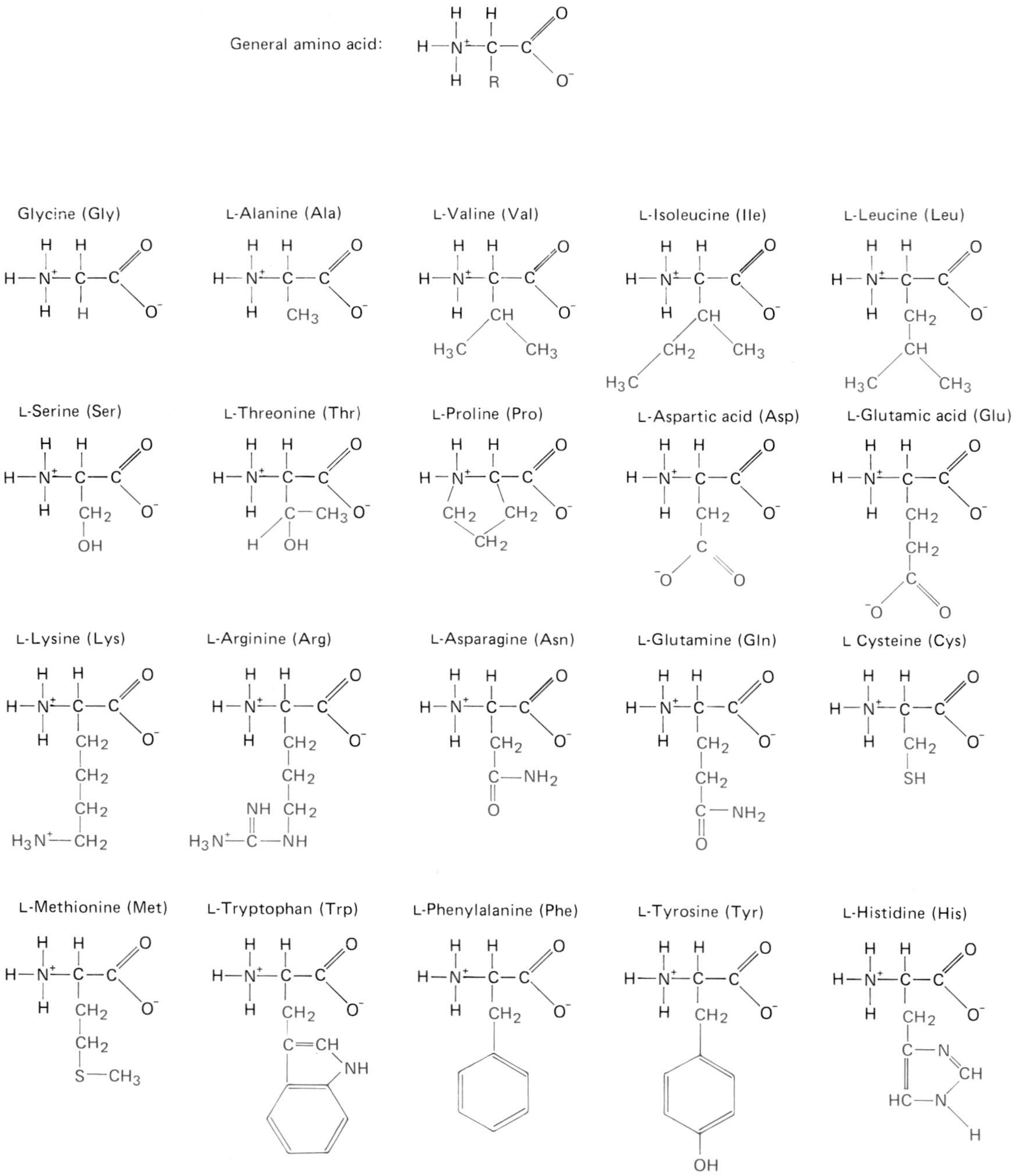

Figure 11 –1. The Twenty Amino Acids Found in Proteins
At physiological pH, the amino acids usually exist as ions.

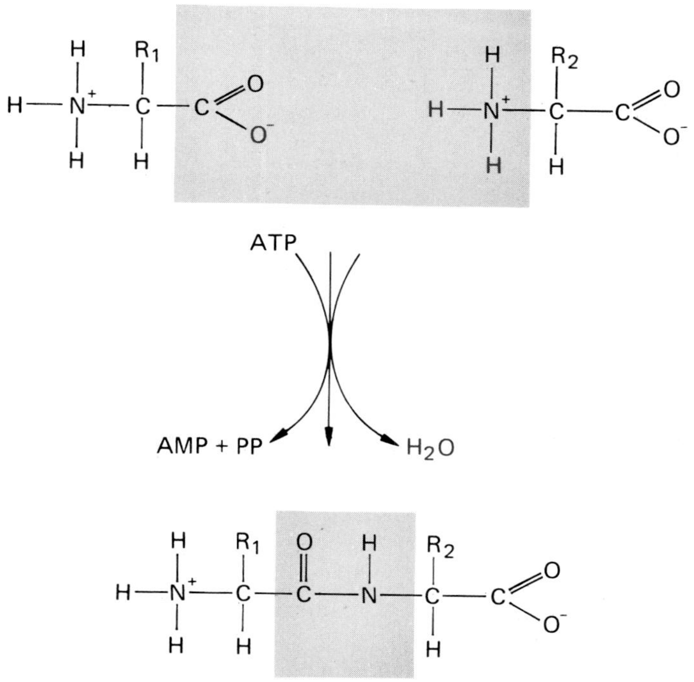

Figure 11–2.
Formation of a
Peptide Bond
between Two Amino
Acids

Transfer RNA

Attachment of Amino Acid to tRNA. The function of transfer RNA is to ensure that each amino acid incorporated into a protein corresponds to a particular codon on the messenger. The tRNA serves this function by its structure—it has an anticodon at one end and an amino acid attachment site at the other end. The "correct" amino acid—that is, the amino acid corresponding to the anticodon—is attached to the tRNA by enzymes known as **aminoacyl-tRNA synthetases** (for example, arginyl-tRNA synthetase, leucyl-tRNA synthetase, and so on). (Recall from the previous chapter the cloverleaf structure of tRNA where there is an anticodon loop as well as a loop recognized by an aminoacyl-tRNA synthetase.)

An aminoacyl-tRNA synthetase joins a specific amino acid to its "proper" tRNA in a two-stage reaction that takes place on the surface of the enzyme. In the first stage the amino acid is "activated" with ATP. In the second stage of the reaction, the amino acid is attached with a high-energy bond to the 3' end of the tRNA (Figure 11–7). Thus, during the process of protein synthesis, the energy for the formation of the peptide bond will be present where it is needed.

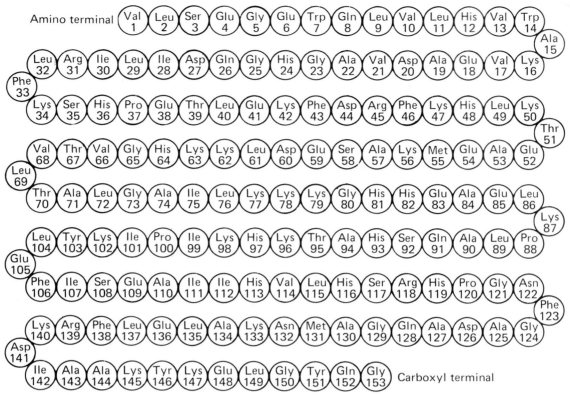

Figure 11-3. Primary Structure of the Protein Myoglobin (from Sperm Whales)

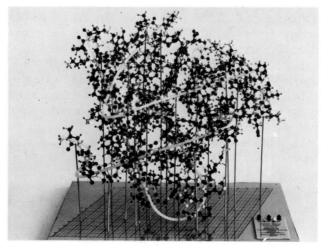

Figure 11-4.
The 3-D Structure of
Myoglobin

Source: Reproduced courtesy of Dr. John Kendrew.

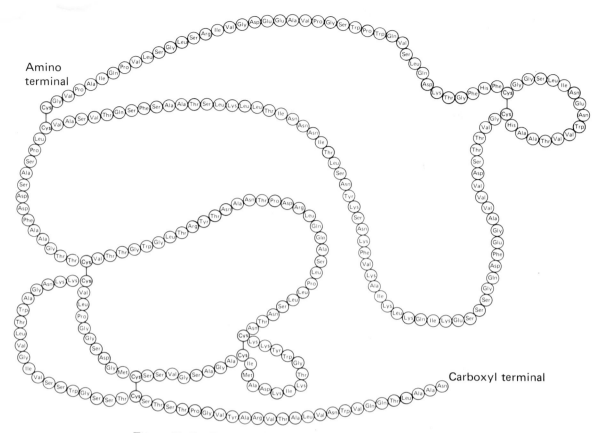

Figure 11-5. Position of Disulfide Bridges in the Protein Chymotrypsinogen

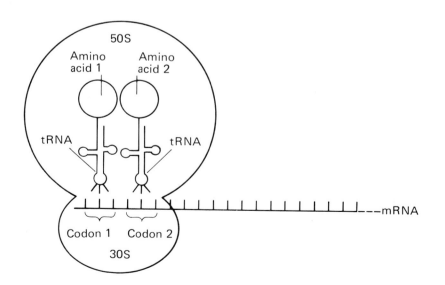

Figure 11-6.
Overview of the Translation Process at the Ribosome

1. Amino acid (aa) + ATP (Adenosine—(P)(P)(P)) ⟶ aa ∿ (P)—Adenosine + (P)(P)

or

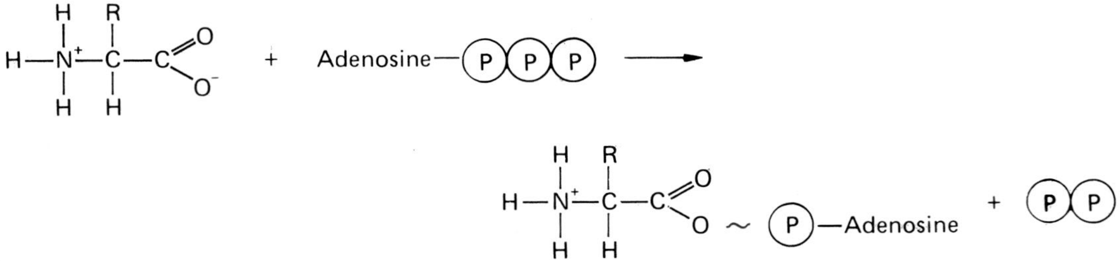

2. aa ∿ (P)—Adenosine + tRNA ⟶ aa ∿ tRNA + Adenosine + (P)

or

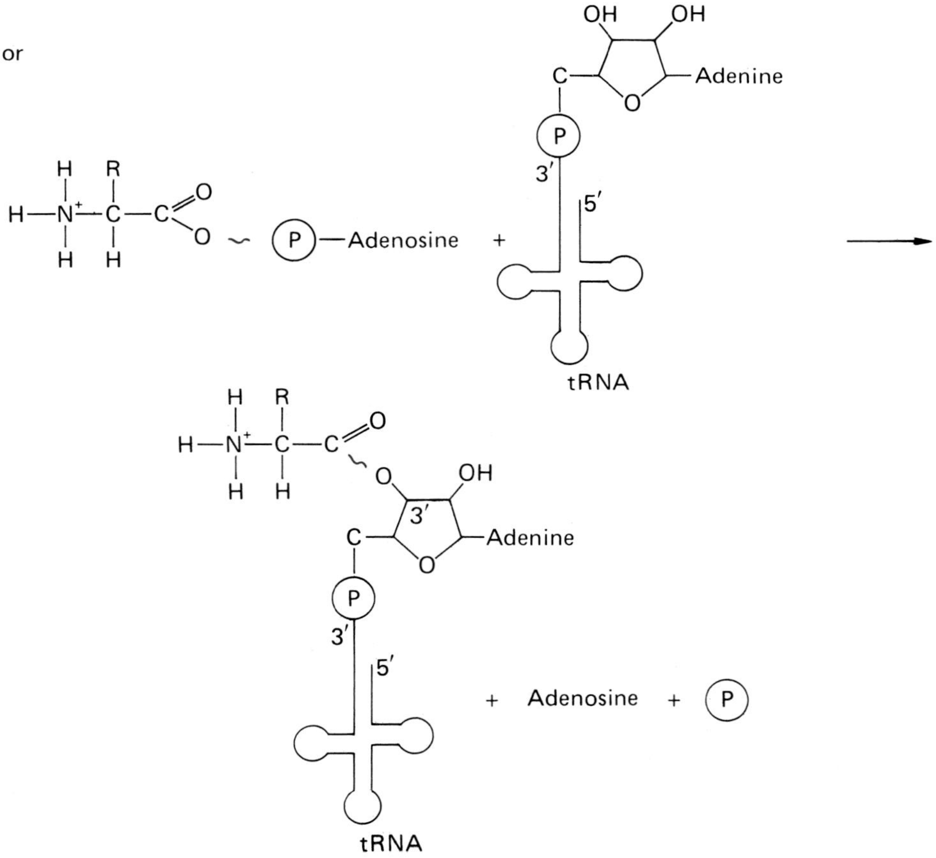

Moiety Numbers.   There must be a minimum of 20 tRNAs and 20 aminoacyl-tRNA synthetases because there are 20 amino acids. There are 64 possible codons in the genetic code. (Four nucleotide bases in groups of three = 4 × 4 × 4 = 64.) Three of these codons are used to terminate translation. Thus, there is an upper limit of 61 tRNAs because there are 61 different nonterminator codons. The precise number is between the minimum and the maximum: 50 tRNAs are known so far. The number 50 can be explained by the phenomenon of **wobble** occurring in the third position of the anticodon. We will examine this phenomenon in the section on the genetic code. The tRNAs are designated tRNA^{Leu}, tRNA^{His}, and so on.

The number of aminoacyl-tRNA synthetases can theoretically correspond either to the number of tRNAs or to the number of amino acids or to some number in between. Actually, the number is 20: one enzyme for each of the 20 amino acids. Thus, a particular enzyme recognizes a particular amino acid and all the tRNAs that code for this amino acid. The aminoacyl-tRNA synthetases recognize one of the loops of the tRNA.

Recognition of the Aminoacyl-tRNA during Protein Synthesis. The preceding discussion implies a scheme whereby, during protein synthesis, an amino acid is incorporated into a protein because the amino acid is on the proper tRNA. However, the amino acid could be recognized directly. A simple experiment could determine whether the amino acid or the tRNA is recognized. If the amino acid of a particular amino acid–tRNA complex were chemically altered to a "new" form and if the amino acid–tRNA complex is recognized through the tRNA, then the "new" amino acid should be incorporated into the newly synthesized protein in place of the "old" amino acid. If, however, the amino acid itself is recognized, then the "new" amino acid would not be incorporated and only the correct "old" amino acid would appear in the synthesized protein (Figure 11–8). We can thus test whether the amino acid itself or the tRNA is recognized by the protein-synthesizing machinery.

This experiment can readily be done using Raney nickel, a catalytic form of nickel that will change cysteine to alanine by removal of the SH group (see Figure 11–1). In 1962 Chapeville and colleagues isolated tRNA that had cysteine attached. They converted the cysteine to alanine by the catalytic nickel process. When the tRNAs were used in protein synthesis, alanine was incorporated where cysteine should have been . The experiment adequately demonstrated alternative 1 of Figure 11–8—it is the tRNA that is recognized during protein synthesis. The synthetase puts a specific amino acid on a specific tRNA and then during protein synthesis information on the tRNA, not on the amino acid, is checked.

Figure 11 –7.   Two-Step Process of Attachment of a Specific Amino Acid to a Specfic tRNA on an Amino-Acyl Synthetase
High-energy bonds are indicated by ⌐

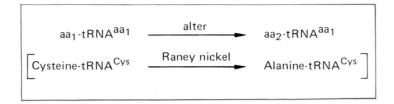

Alternative 1: Of the amino acid—tRNA complex, the tRNA is recognized at the ribosome; aa_2 (alanine) is incorporated where aa_1 (cysteine) should have been incorporated.

Alternative 2: Of the amino acid—tRNA complex, the amino acid is recognized at the ribosome; both amino acids (cysteine and alanine) are incorporated properly.

Figure 11–8.   Experiment to Determine Whether the Amino Acid or the tRNA Is Recognized at the Ribosome during Protein Synthesis

Initiation Complex

As with many processes that have a repetitive nature, once the process of protein synthesis has begun, it runs very smoothly. For most of the process an amino acid is added to the growing peptide at each cycle. There is, however, complexity in the initiation of protein synthesis and in its termination.

It is especially important to realize the need of getting the translation process started precisely. Remember that the genetic code is translated in groups of three nucleotides. If the reading of the messenger RNA begins one base too early or too late, the reading frame is shifted so that every codon is read incorrectly (Figure 11–9). The protein produced, if any, will bear no structural or functional resemblance to the protein coded for. To avoid a protein synthesis error, a relatively elaborate initiation of protein synthesis at the ribosome is required.

Role of N-Formyl Methionine.   Several interesting facts exist regarding the amino acid methionine and its transfer RNA. It has been found that the synthesis of every protein in *Escherichia coli* begins with N-formyl methionine (Figure 11–10). (In eukaryotes the initiation methionine does not have an N-formyl group.) However, none of the proteins in *E. coli* contains N-formyl methionine, and many of these proteins do not even have methionine as the first amino acid. Obviously, before a protein becomes functional, the initial amino acid is modified or removed.

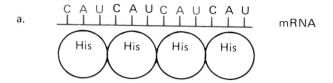

a.

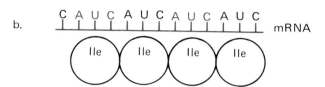

b.

Figure 11–9.
(a) Normal Reading
of the Messenger
RNA (b) Shift in the
Reading Frame of
the Messenger RNA

Methionine, with a codon of 5′AUG3′, known as the **initiation codon,** has two tRNAs with the same anticodon (3′UAC5′) but with different structures (Figure 11–11). One of these tRNAs (tRNA^{fMet}) serves as part of the initiation complex (see ensuing discussion) and prior to initiation will have its methionine chemically modified to N-formyl methionine (fMet). The other tRNA will not have its methionine modified (tRNA^{Met}). It will be used by the translation machinery to put methionine into proteins in all but the first position. The cell thus has a mechanism to make use of methionine in the normal way as well as to use a modified form of it to initiate protein synthesis. The initiator tRNA cannot be used later in protein synthesis because N-formyl methionine cannot undergo peptide bonding; its N-terminal end is blocked by the formyl group.

30S Ribosomal Subunit.   In prokaryotes the subunits of the ribosome (30S and 50S in *E. coli*) are usually dissociated from each other when not involved in translation. The **initiation complex** consists of the following components: the 30S subunit of the ribosome, the mRNA, the N-formyl methionine tRNA^{fMet}, and three **initiation factors (IF1, IF2,** and **IF3).** These components interact in a series of steps that culminate in the initiation complex. The interaction is not yet understood in its entirety. It is known that IF3 binds to the 30S ribosomal subunit. Then, the mRNA is

Figure 11–10.
Structure of N-
Formyl Methionine

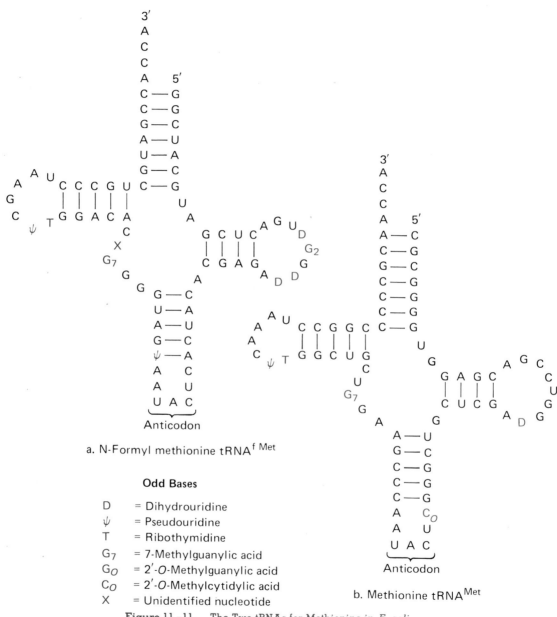

a. N-Formyl methionine tRNA$^{f\ Met}$

Odd Bases

D	= Dihydrouridine
ψ	= Pseudouridine
T	= Ribothymidine
G$_7$	= 7-Methylguanylic acid
G$_O$	= 2'-O-Methylguanylic acid
C$_O$	= 2'-O-Methylcytidylic acid
X	= Unidentified nucleotide

b. Methionine tRNA^{Met}

Figure 11 –11. The Two tRNAs for Methionine in *E. coli*
(a) The initiator tRNA; (b) The interior tRNA

bound to the 30S-IF3 complex. (Presumably, complementarity between the 5' end of the mRNA and the 3' end of the 16S rRNA comes into play.) Meanwhile, a complex has formed with the IF2, the N-formyl methionine tRNA, and GTP. These two complexes are then bound together with the IF1 playing a role that has not yet been clearly elucidated (Figure 11–12).

Aminoacyl and Peptidyl Sites in the Ribosome.   The functional 70S ribosome (which has not yet formed) consists of both a 30S and 50S subunit. When these two subunits bind together, two sites come into being at the

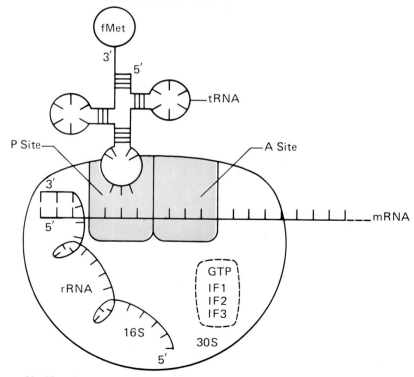

Figure 11–12.   Formation of the Initiation Complex

junction of the two subunits. These two sites, or cavities in the ribosome, are referred to as the (aminoacyl) **A site** and the (peptidyl) **P site** (Figure 11–13). Each site will contain a tRNA just prior to the formation of the peptide bond: The P site will contain the tRNA with the growing peptide chain (peptidyl-tRNA); the A site will contain a new tRNA with its single amino acid (aminoacyl-tRNA). Although there is evidence that the P site is more on the 30S subunit and the A site is more on the 50S subunit, both sites are on both subunits.

When the initiation complex of Figure 11–12 is formed, the initiation fMet tRNA (tRNA^{fMet}) is placed directly into the P site, where it should be so that when translation begins, the next tRNA can go into the A site and a peptide bond can be formed. Before translation can occur, however, the 50S subunit associates with the initiation complex and the 70S ribosome is formed. In the process GTP is hydrolyzed to GDP + P and the three initiation factors are released (Figure 11–14). Usually the hydrolysis of a nucleotide triphosphate in a cell is to provide the energy in the phosphate bonds for a metabolic process. In translation the hydrolysis is apparently to change the shape of the GTP so that it and the initiation factors can be released from the ribosome after the 70S particle has been formed. Thus hydrolysis of GTP in translation is for conformational change rather than covalent bond formation.

Elongation and Translocation

Elongation.   The next step is to position the second tRNA, which is specified by the codon at the A site. This step requires the correct tRNA, a GTP,

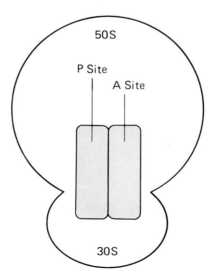

Figure 11 –13.
A Site and P Site of
the 70S Ribosome

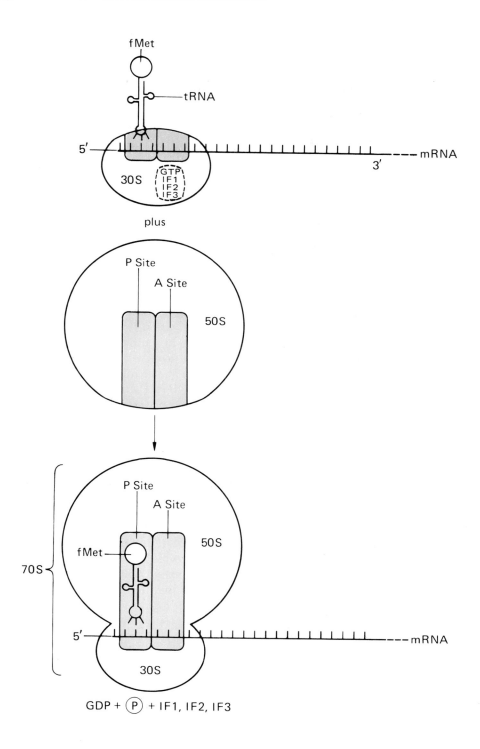

Figure 11–14.
Formation of the
Functional 70S
Ribosome

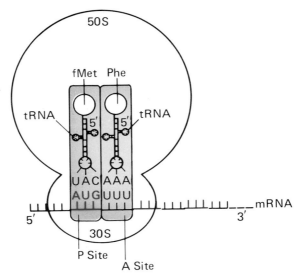

**Figure 11–15.
Ribosome with Two
tRNAs Attached**
In this case the
second codon is for
the amino acid
phenylalanine.

and (in *E. coli*) two proteins called **elongation factors (EF-Ts** and **EF-Tu).** (In eukaryotes the elongation factors are referred to as *EF-1* and *EF-2.* Their mode of action is slightly different.) This second tRNA is positioned in the A site of the ribosome so as to form hydrogen bonds between its anticodon and the second codon on the mRNA. The bonding is then followed by hydrolysis of the GTP to GDP + P and release of the elongation factors and the GDP. The elongation factors may serve as proofreaders, assuring that the anticodon of the tRNA added in the A site is indeed complementary to the codon of the mRNA at this site. Again, the hydrolysis may be for the purpose of changing the shape of the GTP so that GTP and EF-Ts and EF-Tu can depart from the ribosome after the second tRNA is in place. Figure 11–15 shows the ribosome at the end of this step.

Peptide Bond Formation. The two amino acids on the two tRNAs are now in position for formation of a peptide bond between them: Both amino acids should be in apposition to an enzyme, called **peptidyl transferase,** in the 50S subunit. This enzyme is an integral part of the 50S subunit (one of its 34 proteins) and catalyzes a bond transfer from the tRNA-bound carboxyl end of the N-formyl methionine and the amino end of the second amino acid (phenylalanine in Figure 11–15). Every subsequent peptide bond made is identical, regardless of the amino acids involved. The energy used is contained in the high-energy ester bond between the transfer RNA in the P site and its amino acid. The peptide bond formation is diagrammed in Figure 11–16. Immediately after the formation of the peptide bond, the tRNA with the dipeptide will be in the A site and depleted tRNA will be in the P site. The next step in the process is **translocation** of the ribosome in relation to the mRNA.

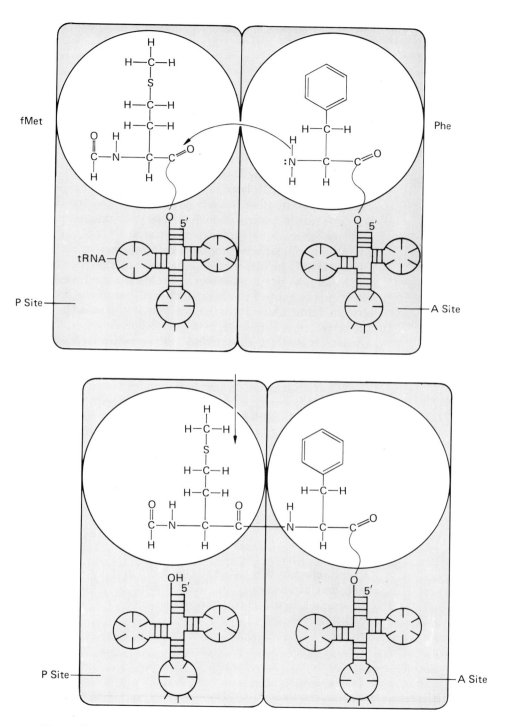

Figure 11–16. Peptide Bond Formation at the Ribosome between N-Formyl Methionine and Phenylalaline

Translocation.   The translocation process requires another GTP and a protein called *elongation factor G,* or **EF-G.** This elongation factor has also been called **translocase.** The process of translocation results in the movement of the ribosome in relation to the mRNA such that the tRNA with the polypeptide in the A site is moved with the mRNA and its codon into the P site. When this movement occurs, the depleted tRNA is ejected from the P site and the empty A site becomes available to the next tRNA. Translocation is completed with the hydrolysis of GTP to GDP + P and the release of EF-G. Again, the hydrolysis is probably for the purpose of a conformational change that will allow the elongation factor and the GTP to be released.

Presumably, during translocation the ribosome itself undergoes conformational changes to allow the one-codon shift of the tRNA and mRNA. A hybrid site model was proposed by Bretscher in 1968 and is illustrated in Figure 11–17. In this model there are two alternative routes that the ribosome can take to translocate; each route has a different intermediary hybrid ribosome. In route 1 the hybrid site consists of the 50S A site and the 30S P site. In route 2 the hybrid is the 50S P site and the 30S A site. Which of these two routes, if either, is correct is currently not known. Route 1 has the advantage that the next tRNA will be recognized during formation of the hybrid. That is, halfway through the process the next codon is in the A site on the 30S subunit.

That it is the tRNA that is transferred from site to site, with the mRNA passively following, has been demonstrated with a mutant tRNA that has an anticodon four nucleotides long. When this tRNA is used in protein synthesis, the translocation process moves the mRNA four nucleotides rather than the expected three.

When the first translocation process is complete, the situation is again as diagrammed in the bottom of Figure 11–14, except that instead of N-formyl methionine tRNA, the P site contains the second tRNA (tRNA^{Phe}) with a dipeptide attached to it. The process of elongation is then repeated, with a third tRNA coming into the A site. The process is repetitive from here to the end. Figure 11–18 is a summary diagram. During the repetitive aspect of protein synthesis, two GTPs are hydrolyzed per peptide bond: one GTP in the attachment of a tRNA to the A site and one GTP in translocation of the ribosome after the peptide bond is formed. In addition every tRNA has had an amino acid attached at the expense of the hydrolysis of an ATP to AMP + PP.

Termination

Nonsense Codons.   Termination of protein synthesis occurs when one of three **nonsense codons** appears in the A site of the ribosome. These codons

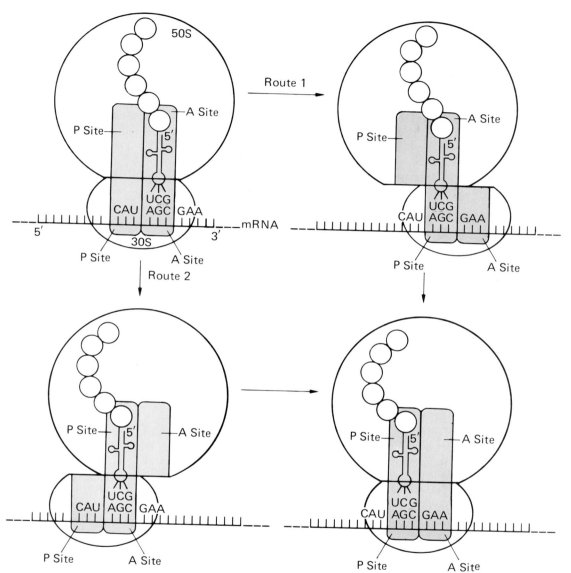

Figure 11–17.   Hybrid Site Model for Translocation of the Ribosome

Source: After Bretscher (1968).

are UAG (sometimes referred to as *amber*), UAA *(ocher),* and UGA *(opal).*
("Amber" is the English translation of the name Bernstein, a graduate stu-
dent who took part in the discovery of UAG in R. H. Epstein's lab at Caltech.
"Ocher" and "opal" are tongue-in-cheek extensions of the first label.) Three

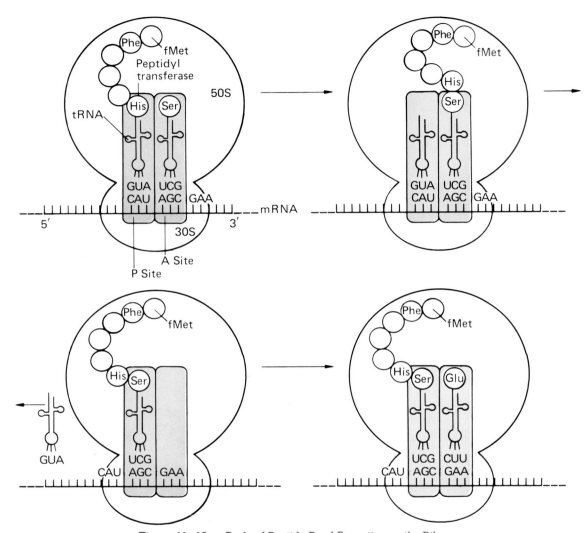

Figure 11-18.   Cycle of Peptide Bond Formation on the Ribosome

proteins called **release factors** and the hydrolysis of a GTP molecule to GDP + P are involved in termination.

When the nonsense codon appears in the A site on the ribosome, it is recognized by a release factor, either *RF-1* or *RF-2,* that enters the A site. The release factors differentially recognize the three nonsense codons. RF-1 recognizes UAA and UAG; RF-2 recognizes UAA and UGA. The result is a blocking of further chain elongation, as shown in Figure 11–19.

A third release factor, *RF-3,* then hydrolyzes the linkage of the terminal

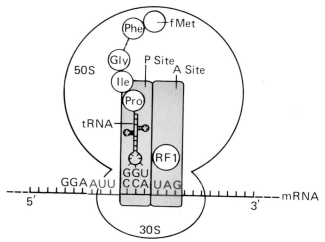

Figure 11 –19. Chain Termination
A nonsense codon in the A site is recognized by a release factor.

amino acid and its tRNA in the P site. The completed polypeptide is released and so is the depleted tRNA. Then, with the hydrolysis of a GTP, the mRNA is released, the releasing factors dissociate, and the ribosome dissociates into its component parts. Dissociation is aided by one of the original initiation factors, IF3, which rebinds to the 30S subunit and thus causes dissociation of the 70S ribosome.

Rate and Cost of Translation. The ribosomal subunits are now ready to reinitiate protein synthesis with another mRNA. The speed of the protein-synthesizing process is about 20 peptide bonds per second. Discounting the time for initiation and termination, an average protein of 300 amino acids is synthesized in about 15 seconds. (As mentioned before, the released protein will form its final structure spontaneously or will be modified with the aid of other enzymes.) The energy cost is 4 high-energy phosphate bonds per peptide bond (2 from an ATP during tRNA charging and 2 from GTP hydrolysis during tRNA binding at the A site and translocation) or about 1200 high-energy bonds per protein. This cost is very high—about 90 percent of the energy production of *E. coli* goes into protein synthesis. A high energy cost is presumably the price that a living system has to pay for the speed and accuracy of the synthesis of its proteins.

Coupling of Transcription and Translation. In prokaryotes such as *E. coli*, where no nuclear envelope exists, translation can begin before transcription is completed. Figure 11–20 shows a section of an *E. coli* chromosome in which transcription has not been completed. On the DNA molecule

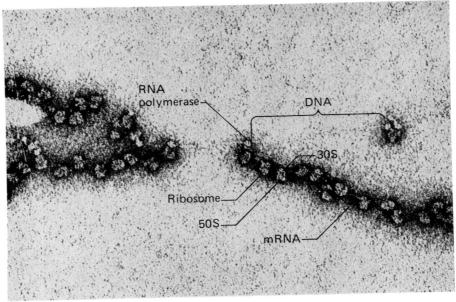

Source: Reproduced courtesy of Dr. Barbara Hamkalo.

Figure 11-20.
A Polysome (i.e.,
Multiple Ribosomes
on the Same mRNA)
Each ribosome is
approximately 250 Å
units across.

an RNA polymerase can be seen transcribing a gene. The transcribed mRNA can be seen, still attached, extending away from the DNA. Attached to the mRNA are several ribosomes, all acting on the same mRNA. Since translation starts at the same end of the messenger (5′) as transcription, shortly after transcription begins, an initiation complex can be formed and translocation can also begin. As translation proceeds along the messenger, the 5′ end of the messenger will again become exposed and a new initiation complex can be formed. The length of the protein is a function of the number of nucleotides translated. The occurrence of several ribosomes translating the same messenger is referred to as a *polyribosome,* or simply a **polysome.** Presumably, if we could see the peptides being translated by the ribosomes, we would see the equivalent of Figure 11–21.

ANTIBIOTICS

Antibiotics are substances that are produced by living organisms and that are toxic to other organisms. Antibiotics are of interest to us for two reasons. First, they have been extremely important in fighting human diseases and the diseases of farm animals. Second, they are a very useful tool for analyzing many of the steps of protein synthesis. Antibiotics impede the process of protein synthesis in a variety of ways and selectively poison bacteria. The effectiveness of antibiotics usually derives from the metabolic differences

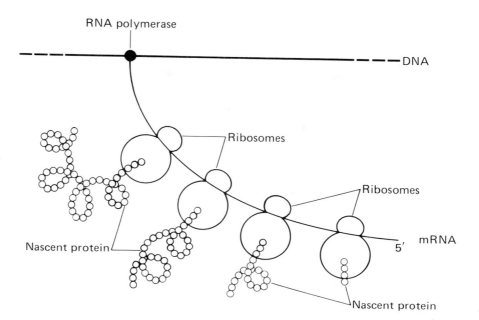

Figure 11 –21.
Protein Synthesis at
a Polysome

between bacteria and eukaryotes. For example, an antibiotic that will block a bacterial ribosome without affecting a human ribosome will be an excellent antibiotic.

Puromycin

Puromycin works by resembling the 3' end of an aminoacyl-tRNA (Figure 11–22). It is bound to the A site of the bacterial ribosome, where peptidyl transferase will transfer a bond from the existing peptide attached to the tRNA in the P site to puromycin. The peptide chain is then prematurely released and protein synthesis at the ribosome is terminated.

Streptomycin

Streptomycin binds to one of the proteins (protein S12) of the 30S subunit of the prokaryotic ribosome and inhibits initiation of protein synthesis (Figure 11–23). It also causes misreading of codons if chain initiation has already begun, presumably by altering the conformation of the ribosome so that the tRNAs are less firmly bound to it. There are bacterial mutants that are streptomycin resistant as well as mutants that are streptomycin dependent (they cannot survive without the antibiotic). Both mutants have altered 30S subunits, specifically altered protein S12.

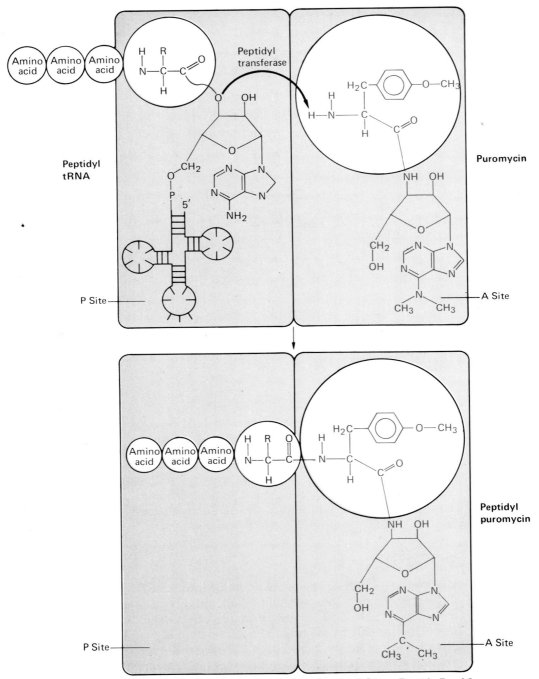

Figure 11-22. Puromycin Is Bound to the A Site, a Peptide Bond Is Formed, and Chain Termination Occurs

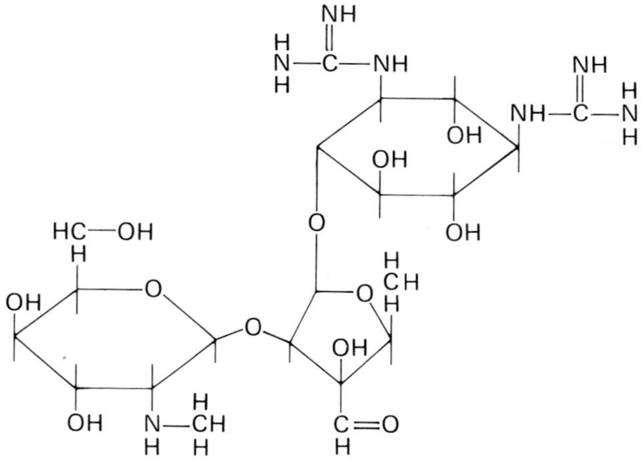

Figure 11 –23. Structure of Streptomyocin

Tetracycline

Tetracycline blocks protein synthesis by preventing the aminoacyl-tRNA from binding to the A site on the ribosome (Figure 11–24).

Chloramphenicol

Chloramphenicol blocks protein synthesis by binding to the 50S subunit of the prokaryotic ribosome, where it blocks the peptidyl transfer reaction (Figure 11–25). Chloramphenicol does not affect the eukaryotic ribosome. However, chloramphenicol, as well as several other antibiotics, is used cautiously because the mitochondria within eukaryotic cells have their own genomes as well as their own ribosomes. And, since the mitochondrial ribosomes are very similar to prokaryotic ribosomes, some of the antibiotics that affect prokaryotic ribosomes also affect the eukaryotic mitochondrion. The implications of the similarity of bacteria and the mitochondria will be discussed in the chapter on extrachromosomal inheritance.

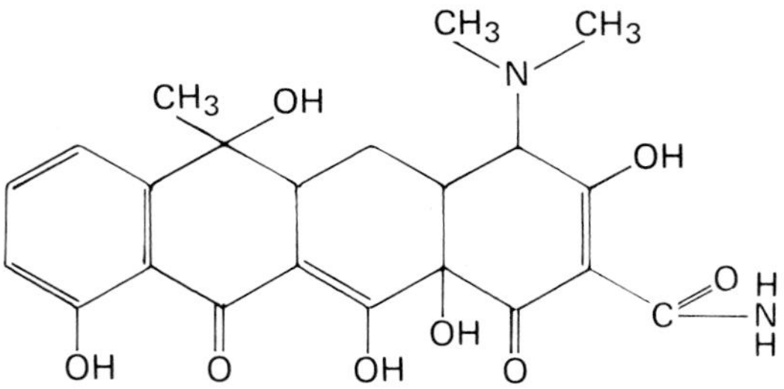

Figure 11 –24. Structure of Tetracycline

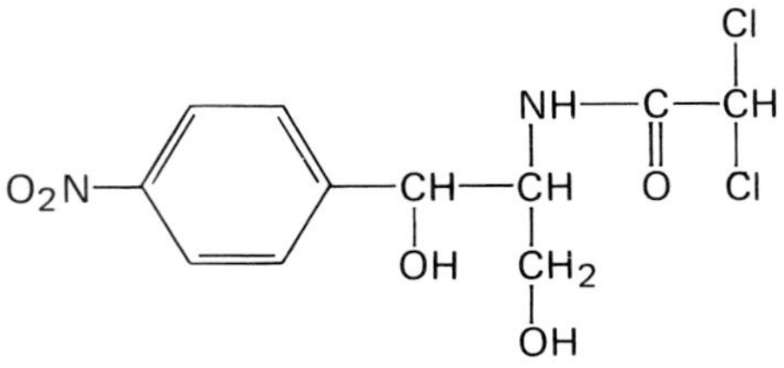

Figure 11-25.
Structure of
Chloramphenicol

THE GENETIC CODE

Researchers in the mid-1950s assumed that the genetic code would be found to consist of simple sequences of nucleotides specifying particular amino acids. They sought answers to questions such as: Is the code overlapping? Is there punctuation between code words? How many letters make up a code word (codon)? Logic and genetic experiments supplied some of the answers, but it was the rapidly improving techniques of biochemistry that provided a final solution to the genetic language.

Triplet Nature of the Code

Several lines of evidence had suggested that the nature of the code was triplet: Three bases in mRNA would specifiy one amino acid. Since there are 20 amino acids, there must be at least one code word for each. If the code contained only one base (represented in discussions of the code by one letter), it would only be able to specify 4 amino acids since there are only 4 different bases in DNA (or mRNA). A couplet code would have $4 \times 4 = 16$ 2-base (2-letter) words, or codons, which is still not enough for 20 amino acids. A triplet code would allow for $4 \times 4 \times 4 = 64$ codons, which is more than enough to specify 20 amino acids.

The concept of a triplet code is supported by the length of the DNA in some viruses. For example, the tobacco necrosis satellite virus contains an RNA molecule that has about 1200 nucleotides. This RNA molecule must code for 3 coat-protein subunits that are about 400 amino acids long. Assuming a continuously read code with no punctuation, 1200/400 gives 3 bases per amino acid.

Evidence for the Triplet Nature of the Code. The triplet-code concept was reinforced by the experimental manipulation of mutants, or altered genes. Francis Crick and his colleagues did most of the work, in which a chemical mutagen, acridine dye, was used to cause inactivation of the *rIIB* gene of the bacteriophage T4. Acridine inactivates the gene by either adding a nucleotide to the gene (DNA) or deleting a nucleotide from it (Chapter 14). The *rII* gene controls the plaque morphology of this bacteriophage growing on *E. coli* cells. Rapid-lysis mutants produce large plaques. The wild-type form of the gene, *rII*+, results in normal plaque morphology. Exactly what the protein product of this gene does is not important to us now. All that need concern us here is that the *rII* gene can be manipulated by causing the addition or deletion of a single nucleotide.

Figure 11–26 shows the consequences of adding or deleting nucleotides in the DNA codons in terms of the reading sequence of the mRNA. From the point of addition or deletion onward, there is a **frameshift** where all codons are misread. If a deletion is combined with an addition to produce a double-mutant gene, the frameshift occurs only in the region between the two effects. If this region is small enough or does not contain coding for vital amino acids, the function of the gene may be partially restored. Two deletions or two insertions combined will not restore the reading frame. However, Crick and his colleagues found that the combination of 3 additions or 3 deletions also restored gene function. This finding led to the conclusion that the genetic code was a triplet code because a triplet code would be put back into reading frame by 3 additions or 3 deletions (Figure 11–27).

Figure 11–26.
Frameshift Mutants in a Gene Result from Addition or Deletion of Single Nucleotides in the DNA
Note the mRNAs that result from transcription of these frameshift mutants. Asterisk (*) indicates deletion or insertion.

Normal
(CAG repeat)

First A deleted
(AGC repeat)

A inserted after third A of normal
(GCA repeat)

Deletion and insertion combined
(return to CAG repeat)

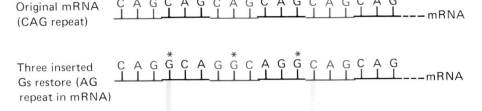

Figure 11 –27.
Restoration of Codon
Reading by Three
Additions
(Insertions)
Asterisk (*) indicates
insertion.

It should be noted that Crick's group was aware of the possibility that the codon is a multiple of 3, where each addition or deletion is of an even number of bases—that, for example, the codon could be 6 bases long and each addition or deletion could involve two bases. However, the length of the tobacco necrosis satellite virus and the final, actual working out of the code dispelled any minor doubts.

Overlap and Punctuation in the Code.   The questions still remained as to whether or not the code was overlapping or had punctuation (Figure 11–28). Several logical arguments favored a no-punctuation, nonoverlapping model. An overlapping code would be subject to two restrictions. First, a change in one base (a mutation) would affect more than one codon and thus affect more than one amino acid. But, studies of amino acid sequences in various mutants almost always showed that only one amino acid was changed and thus argued against overlap. Second, there are certain restrictions as to which codons could occur next to each other in an overlapping code; therefore, restrictions would also be placed on which amino acids occurred next to each other in proteins. For example, the amino acid coded by UUU could never be adjacent to the amino acid coded by AAA because the overlap codons UUA and UAA would always insert other amino acids between them. Overlap, then, appeared not to be the case since, in fact, every amino acid appears next to every other amino acid in one protein or another.

Punctuation between codons was also tentatively ruled out, in view of the length of the mRNA in the tobacco necrosis satellite virus, which has just about enough codons to specify its coat protein with no room left for a punctuating base or bases between each codon. (However see the information in the next chapter on ϕX174 where a given length of genetic material codes for an unusually large number of proteins. Its situation appears to be an oddity rather than the rule.)

Breaking the Code

Given that the genetic code is a code of nonoverlapping triplets, the 64 codons still had to be worked out. For example, which amino acid is specified

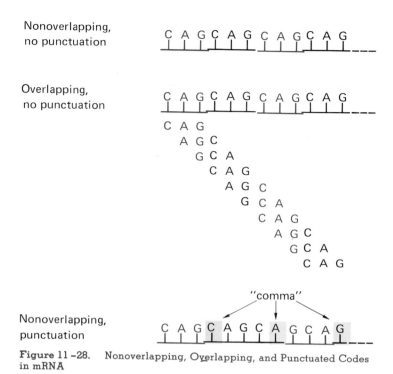

Figure 11–28. Nonoverlapping, Overlapping, and Punctuated Codes in mRNA

by ACC? The work was done in two stages. In the first stage the laboratories of Nirenberg and of Ochoa made long artificial mRNAs and determined which amino acids these mRNAs incorporated into protein. In the second stage specific triplet RNA sequences were synthesized. The amino acid–tRNA complex that was bound by each sequence was then determined.

Synthetic mRNAs. The ability to synthesize long-chain mRNAs resulted from the discovery, by Grunberg-Manago and Ochoa in 1955, of the enzyme **polynucleotide phosphorylase,** which will join diphosphate nucleotides into long-chain, single-stranded polynucleotides. Unlike a polymerase, polynucleotide phosphorylase does not need a primer on which to act. This enzyme is found in all bacteria. (Its main function in the cell is probably the reverse of the use made of it here: to degrade mRNA. If so, it would be important in keeping the life span of bacterial mRNAs short.) In 1961 Nirenberg and Matthei added artificially formed RNA polynucleotides of known composition to an *E. coli* ribosome system and looked for the incorporation of amino acids into proteins (Figure 11–29). They found that when uridine diphosphates are made into a messenger (poly-U) by the enzyme polynucleotide phosphorylase, the amino acid phenylalanine is incorporated as a poly-

Severo Ochoa (1905 –) Courtesy of Dr. Severo Ochoa

Figure 11–29.
Incorporation of
Amino Acids into
Proteins Using an
Artificial Messenger,
Poly-U

From *E. coli*

Ribosomes
Aminoacyl-t RNAs
Mg^+
GTP   + Poly-U ——→ Polymerization of
Initiation, translocation,   phenylalanine
  and release factors

peptide in the system of Figure 11–29. Thus the first code word established was UUU for phenylalanine. The work was continued by Nirenberg and his associates and by Ochoa and his associates. They found that AAA is the code word for lysine, CCC is the code word for proline, and GGG is the code word for glycine.

They then made synthetic mRNAs by using mixtures of the various diphosphate nucleotides in known proportions. An example is given in Table 11–1. From their experiment it is possible to determine the letters (bases) used in many of the code words, but not their specific order. For example, cysteine, leucine, and valine are all coded by two U's and a G, but the experiment cannot sort out the order of these letters, or bases, (5′UUG3′, 5′UGU3′, or 5′GUU3′) for any one of them. Determining the order required a step in sophistication—that is, of being able to synthesize known trinucleotides.

Synthetic Codons.   When trinucleotides of known composition could be manufactured, Nirenberg and Leder in 1964 developed a "binding assay." They found that isolated *E. coli* ribosomes, in the presence of high-molarity magnesium chloride, could bind trinucleotides as if they were messengers. Also bound would be the tRNA that carried the anticodon complementary to the trinucleotide. It was thus possible, using radioactive amino acids, to determine which trinucleotide coded for a particular amino acid. A given

TABLE 11–1.   Structure of Artificial mRNA Made by Randomly Assembling Uracil- and Guanine-Containing Ribose Diphosphate Nucleotides with a Ratio of 5U:1G

Codon	Frequency of Occurrence
UUU	$(5/6)^3 = 0.58$
UUG	$(5/6)^2(1/6) = 0.12$
UGU	$(5/6)^2(1/6) = 0.12$
GUU	$(5/6)^2(1/6) = 0.12$
UGG	$(5/6)(1/6)^2 = 0.02$
GUG	$(5/6)(1/6)^2 = 0.02$
GGU	$(5/6)(1/6)^2 = 0.02$
GGG	$(1/6)^3 = 0.005$

trinucleotide was mixed with all the components necessary to form a trinucleotide–ribosome–aminoacyl-tRNA complex, including one amino acid that was radioactively labeled. The reaction mixture was passed over a filter that would allow everything except the trinucleotide–ribosome–aminoacyl-tRNA complex to pass through. If the radioactivity passed through the filter, the experiment was repeated with another labeled amino acid. When the radioactive amino acid appeared on the filter, they knew that the amino acid corresponded to the trinucleotide codon. In a short period of time, all of the codons were deciphered (Table 11–2).

Wobble Hypothesis

The genetic code is a **degenerate code.** That is, a given amino acid may have more than one codon. As can be seen from Table 11–2, 8 of the 16 boxes contain just one amino acid per box. Therefore, for these 8 amino acids the codon need only be read in the first two positions because the same amino acid will be represented regardless of the third base of the codon. Six of the boxes are split in half in such a way that the codons are differentiated by the presence of a purine or a pyrimidine in the third base. For example, CAU and CAC both code for histidine; in both, the third base, U (uracil) or C

TABLE 11 –2. The Genetic Code

First Position (5′ end)	Second Position				Third Position (3′ end)
	U	C	A	G	
U	Phe	Ser	Tyr	Cys	U
	Phe	Ser	Tyr	Cys	C
	Leu	Ser	stop	stop	A
	Leu	Ser	stop	Trp	G
C	Leu	Pro	His	Arg	U
	Leu	Pro	His	Arg	C
	Leu	Pro	Gln	Arg	A
	Leu	Pro	Gln	Arg	G
A	Ile	Thr	Asn	Ser	U
	Ile	Thr	Asn	Ser	C
	Ile	Thr	Lys	Arg	A
	Met (start)	Thr	Lys	Arg	G
G	Val	Ala	Asp	Gly	U
	Val	Ala	Asp	Gly	C
	Val	Ala	Glu	Gly	A
	Val	Ala	Glu	Gly	G

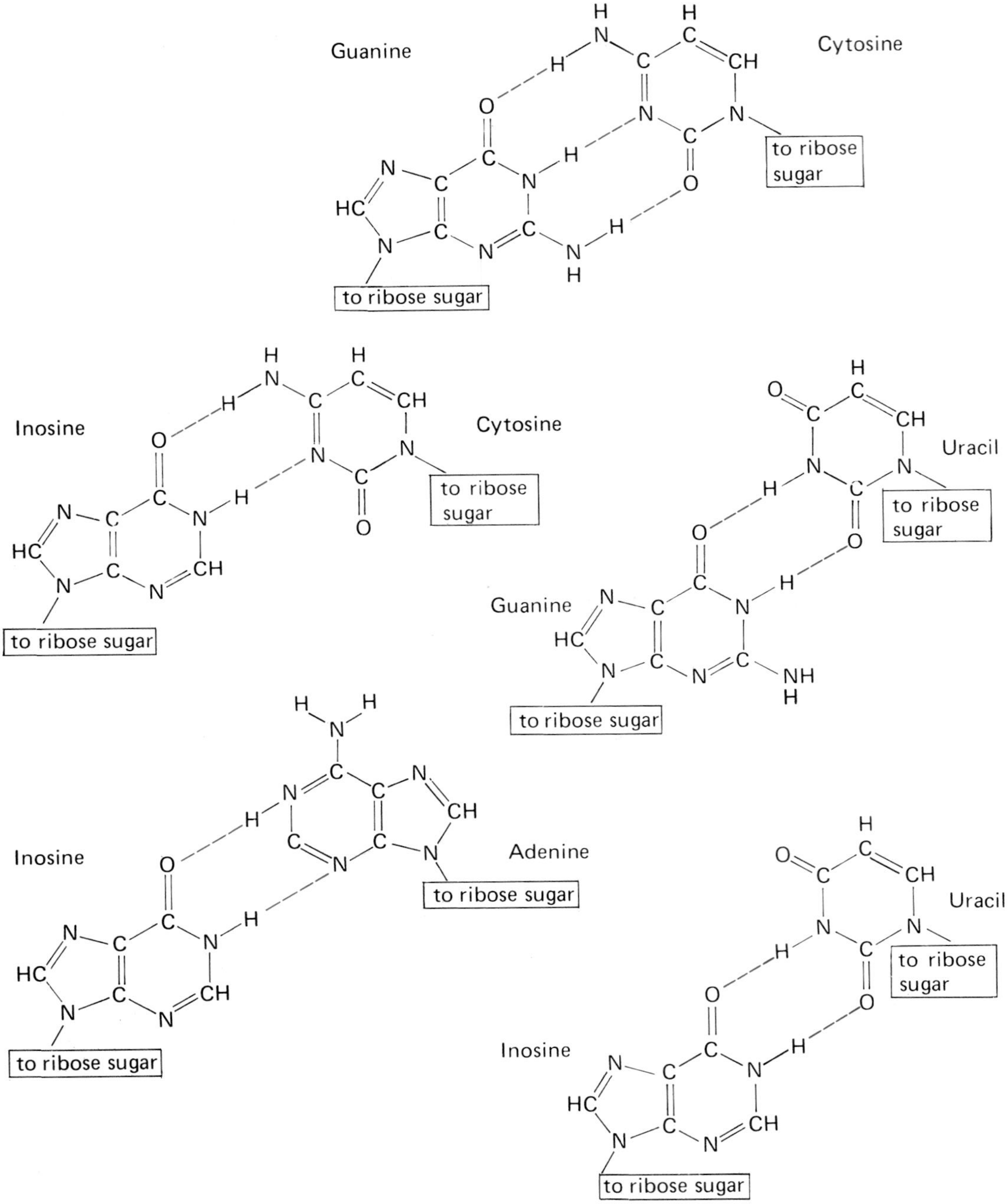

Figure 11–30.   Base Pairing Possibilities of Guanine and Inosine in the Third (3′) Position of a Codon

(cytosine), is a pyrimidine. The fact regarding the lesser importance of the third position in the code ties in with two facts about transfer RNAs. First, although there would seem to be a need for 62 tRNAs—since there are 61 codons specifying amino acids and an additional codon for initiation—there are actually only about 50 different tRNAs in an *E. coli* cell. Second, a rare base such as inosine can appear in the anticodon on the tRNA. These three facts lead to the belief that some kind of conservation of tRNAs is occurring and that rare bases may be involved.

Since the third position of the anticodon is not as constrained as the other two positions, a given base at the third position on the tRNA may be able to pair with any of several bases in the codon itself (Figure 11–30). Crick characterized the ability as *wobble*. Table 11–3 shows the possible pairings that would produce a tRNA system compatible with the known code. For example, if an isoleucine tRNA has the anticodon 3′UAI5′, it is compatible with the three codons for that amino acid (Table 11–2): 5′AUU3′, 5′AUC3′, and 5′AUA3′. That is, inosine in the third position of the anticodon can recognize U, C, or A in the codon and thus one tRNA will complement all three codons for isoleucine.

Universality of the Genetic Code

Until 1979 all experiments demonstrated that the genetic code is universal. That is, the codon dictionary (Table 11–2) is the same for *E. coli,* humans, African clawed toads, and all other species so far studied. The universality of the code has been demonstrated, for example, by taking the ribosomes and mRNA from rabbit reticulocytes and mixing them with the aminoacyl-tRNAs and other translational components from *E. coli.* Rabbit hemoglobin was synthesized.

In 1979 and 1980, however, discrepancies were noted in mitochondrial translation systems when genes for structural proteins were sequenced. (See Chapter 9. As we will discover in Chapter 16, mitochondria have DNA that codes for part of the translation apparatus and part of the enzyme apparatus

TABLE 11 –3. Pairing Combinations at the Third Anticodon Position

Base in tRNA (5′ end)	Base in mRNA (3′ end)
G	U *or* C
C	G
A	U
U	A *or* G
I	A, U, *or* C

AMINO ACID SEQUENCING

Protein sequencing techniques have been known since 1953 when Sanger worked out the complete sequence of the protein hormone insulin. The basic strategy is to purify the protein and then break it up into small peptides several different ways. These peptides are then sequenced. The whole protein sequence can then be determined by the overlap of the sequenced subunits.

A protein can be broken into peptide fragments by many different methods, including acid hydrolysis and alkaline hydrolysis. For the most part the methods used involve proteolytic enzymes (proteases) that hydrolyze the peptides at specific points. For example, *pepsin* preferentially hydro-

lyzes peptide bonds involving aromatic amino acids, methionine, and leucine. *Chymotrypsin* hydrolyzes peptide bonds involving carboxyl groups of aromatic amino acids. *Trypsin* hydrolyzes bonds involving the carboxyl groups of arginine and lysine.

The proteolytic digest is usually separated into a peptide map or peptide "fingerprint" by using a two-dimensional combination of paper chromatography and (or) electrophoresis. (Column chromatography, a technique that we will not consider, is also used.) In two-dimensional chromatography a sample is spotted onto a piece of paper and placed in a specific solvent system. After an allotted time

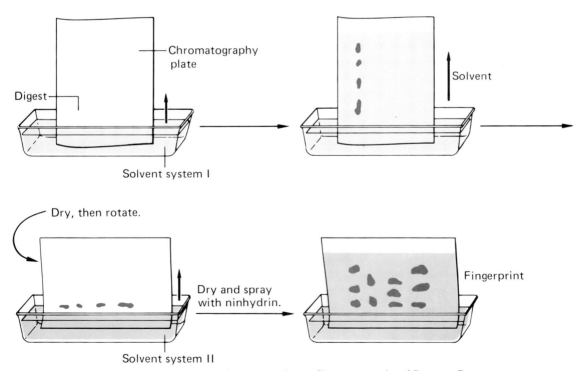

Figure 1.   Two-Dimensional Paper Chromatography of Protease Digest

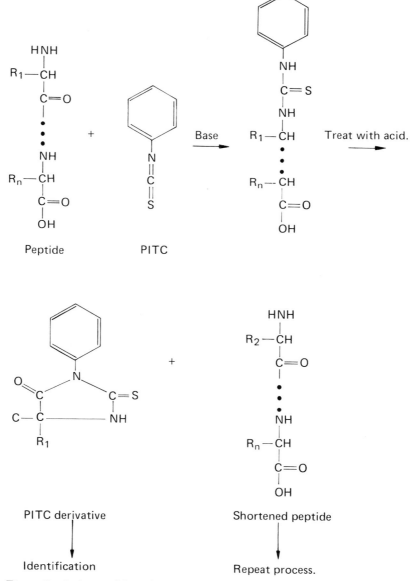

Figure 2. Isolation of Peptide Amino Acids for Purposes of Sequencing

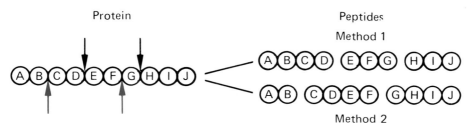

Figure 3. Sequenced Peptides Overlap in a Unique Way When a Polypeptide Is Digested Two Different Ways

the paper is dried, turned 90°, and placed in a second solvent system for an allotted time. In each solvent different peptides travel along the paper at different rates. The spots are then developed using ninhydrin, which reacts with the N-terminal amino acid and produces a colored product when heated.

The spots, which represent small peptides, can be cut out of a chromatogram that has not been sprayed with ninhydrin. These spots can then be sequenced by, for example, the Edman method whereby the peptide is sequentially degraded from the N-terminal end. Phenylisothiocyanate reacts with the amino end of the peptide. When acid is added, the N-terminal amino acid is removed as a derivative and can be identified. The process is then repeated until the whole peptide has been sequenced.

If the fingerprint pattern is worked out for two different digests of the same polypeptide, the unique sequence of the original polypeptide can be determined by overlap. In the illustration the letters A–J represent a polypeptide of 10 amino acids. A is known to be the first (N-terminal end) amino acid. The N-terminal end is known since the Edman method sequences peptides from this end.

We can thus summarize the methodology as follows:

1. A protein is purified. If it is made up of several subunits, these subunits are separated and purified. (If disulfide bridges exist within a peptide, they must be reduced. They are later determined by digestion, with the bridges intact, followed by resequencing.)
2. Different proteolytic enzymes are used on separate subsamples so that the protein is broken into different sets of peptide fragments.
3. Two-dimensional chromatography, electrophoresis and chromatography, or column chromatography are used to isolate the peptides.
4. The Edman method of sequentially removing amino acids from the N-terminal end is used to sequence each peptide.
5. The amino acid sequence from the N- to C-terminal ends of the protein is deduced from the overlap of sequences in peptide digests generated with different proteolytic enzymes.

for oxidative phosphorylation.) It was discovered that there are two deviations from universality in the reading of the code by mitochondrial tRNAs. First, fewer tRNAs are needed to read the code. Second, there are several instances where a codon is interpreted differently by the mitochondrial system than by the cellular system.

The minimal requirement of tRNAs, given that two methionine tRNAs are needed, is 32, according to Crick's wobble rules (Table 11–3). Unmixed families require two tRNAs and mixed families require one, two, or three tRNAs, depending on the family. (An unmixed family of codons consists of the four codons that specify a single amino acid. For example, the codon family GUX, where X is any of the four RNA bases, all code for valine. Mixed families code for two amino acids or an amino acid or acids and stop signals.)

In the mitochondrial coding system, apparently only 24 tRNAs are needed. The reduction in numbers is accomplished by having each unmixed family recognized by only one tRNA. Because mitochondrial tRNAs for unmixed families of codons have a U in the third (wobble) position of the anticodon, apparently, given the structure of the mitochondrial tRNAs, the U can pair with U, C, A, or G. Presumably, there has been evolutionary pressure to minimize the number of tRNA genes in the DNA of the mitochondrion. Reduction from 32 to 24 is a 25 percent savings.

It has also been found that yeast mitochondria read the CUX family as threonine rather than as leucine (Table 11–2) and the terminator UGA (opal) as tryptophan rather than as termination. However, in the reading of the CUX family, there appear to be differences among different groups of organisms. Human and *Neurospora* mitochondria appear to read the CUX codons as leucine, just as cellular systems do. Of the three groups so far analyzed, only yeast read the CUX family as threonine.

We can thus conclude that the genetic code seems to be universal among prokaryotes, eukaryotes, and viruses. Mitochondria, however, read the code slightly differently. Different wobble rules apply, and at least one terminator and one unmixed family of codons are read differently by mitochondria and cells. Also, the mitochondrial discrepancies are not universal among all types of mitochondria. Further work, involving the sequencing of more mitochondrial DNAs should elucidate the pattern of discrepancies among the mitochondria of diverse species.

Evolution of the Genetic Code

One theory suggests that the genetic code has wobble in it because it originally arose as a couplet code for a small number of amino acids that may have been in use 3.5 billion years ago. As new amino acids with useful properties became available, the code was modified by adding a third letter with less specificity. This theory is especially easy to envision if the code were originally punctuated with a base between each codon.

The fact that all 61 possible codons are used means that biological systems are protected to some extent against mutations of the kind where one nucleotide substitutes for another. For example, if, given a code of 5'CUX3', only U in the third position coded for leucine, any mutation in the third position would produce an unused codon (see Table 11–2) and would thus result in chain termination. In fact, however, any mutation in the third position of 5'CUU3' will produce another codon for the same amino acid. Wobble in the third position, therefore, ensures that less than half of the mutations causing base changes will actually change the amino acid called for by this codon.

There are also patterns in the code where the mutation of one codon to another will result in an amino acid of similar properties. There is a good chance that a functional protein will be produced by such a mutation. For example, all the codons with U as the middle letter are for amino acids that are hydrophobic (phenylalanine, leucine, isoleucine, methionine, and valine). Thus mutation in the first or third positions for any of these codons will still code a hydrophobic amino acid. The two negatively charged amino acids, aspartic acid and glutamic acid, both have codons that start with GA. The aromatic amino acids, phenylalanine, tyrosine, and tryptophan (Figure 11–1) all have codons that begin with uracil. Such patterns minimize the negative effects of mutations that change one codon into another by one base replacement.

This chapter completes the discussion of gene expression. The next two chapters are concerned with the control of gene expression in both prokaryotes and eukaryotes.

CHAPTER SUMMARY

During protein synthesis the translation apparatus at the ribosome recognizes the transfer RNA. A particular tRNA has an anticodon at one end and a specific amino acid at the other end. Through complementarity the anticodon recognizes a codon on the mRNA. The tRNAs are charged with the proper amino acid by aminoacyl-tRNA synthetase enzymes that incorporate the energy of ATP into amino acid–tRNA bonds. Hence, no outside source of energy is needed during translation.

An initiation complex forms at the start of translation. It consists of the mRNA, the 30S subunit of the ribosome, the initiator tRNA with N-formyl methionine, and the initiation factors IF1, IF2, and IF3. The 50S ribosomal subunit is then added and A and P sites form in the resulting 70S ribosome. The N-formyl methionine tRNA is in the P site. A GTP is hydrolyzed and the initiation factors are released.

A tRNA enters the A site, which requires the involvement of elongation factors EF-Ts and EF-Tu (in *E. coli*). Another GTP hydrolysis releases the elongation factors. Peptidyl transferase, a component of the 50S ribosome subunit, transfers the amino acid from the tRNA in the P site to the amino end of the amino acid in the tRNA in the A site. There then follows translocation of the ribosome.

With the help of elongation factor G (EF-G), the ribosome translocates in relation to the mRNA. The depleted tRNA is released from the P site and the tRNA with the growing peptide is moved into the P site. A GTP is hydrolyzed and EF-G is released. Elongation and translocation then continue until a nonsense codon enters the A site. With the aid of three release factors, RF-1, RF-2, and RF-3, the protein is released. The mRNA–ribosome complex then dissociates.

Antibiotics that interfere with translation are discussed. Puromycin, streptomycin, tetracycline, and chloramphenicol all act at the ribosome.

The genetic code was first assumed to be triplet on the basis of logical arguments regarding the minimum size of codons. Crick provided evidence that the code was triplet with his work on deletion and insertion mutants. The code was worked out first with the synthesis of long, artificial mRNAs and then with the synthesis of specific trinucleotide codons. The wobble hypothesis of Crick accounts for the fact that fewer than 61 tRNAs can read the entire genetic code. The reduction of tRNAs can happen because additional complementary base-pairings occur in the third position of the codon (3′ end of the codon, 5′ end of the anticodon).

The rule of universality of the genetic code has to be modified in light of findings regarding mitochondrial tRNAs, where only 24 are needed to read the code. In addition, the stop codon (UGA) is read as tryptophan in mitochondria; in yeast mitochondria the unmixed family, CUX, is translated as threonine instead of leucine. Some generalities of the code hold in both cells and mitochondria—including the structure of the code, which seems to mask a good deal of possible mutation.

EXERCISES AND PROBLEMS

1. Given the following piece of DNA that will be transcribed and then translated into a pentapeptide, give the base sequence for its mRNA. Give the anticodons on the tRNAs by making use of wobble rules. What amino acids are incorporated?

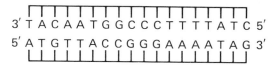

```
3′ T A C A A T G G C C C T T T T A T C 5′
5′ A T G T T A C C G G G A A A A T A G 3′
```

2. If DNA contained only the bases cytosine and guanine, how long would a code word have to be? How could we tell if this DNA were double stranded?

3. If an artificial mRNA contains 2 parts uracil to one of cytosine, which amino acids in which proportions should be incorporated into protein?

4. How could we determine whether the mutant that caused a strain to cease producing a given enzyme was due to a frameshift, a nonsense codon occurring within the gene, or a simple change of one amino acid to another?

5. Draw the details of translation at the ribosome for the mRNA produced in problem 1. Include in the diagram the ribosomal sites, the tRNAs, and the various nonribosomal proteins involved.

6. Redraw one stage in the preceding diagram given that the results of the Chapeville study indicated that the aminoacylated tRNA, not the amino acid itself, was recognized during translation.

7. What evolutionary implications can be seen in the fact that a blocked amino acid is used to initiate protein synthesis (in prokaryotes)?

8. Name the similarities and differences among the three nonsense codons. Draw their theoretical tRNAs by using the wobble rules. Is there an apparent potential difficulty with the tryptophan codon?

9. Other than the antibiotics named in the chapter, suggest 5 "theoretical" antibiotics that could interfere with the prokaryotic translation process.

10. What would be proved or disproved if an organism were discovered that did not follow the codon dictionary? Do we expect organisms from Mars (if they exist) to use our codon dictionary?

SUGGESTIONS FOR FURTHER READING

Bonitz, S. et al. 1980. Codon recognition rules in yeast mitochondria. *Proc. Nat. Acad. Sci., USA* 77:3167–3170.

Bretscher, M. S. 1968. Translocation in protein synthesis in a hybrid structure model. *Nature* 218:675–677.

Chapeville, F., et al. 1962. On the role of soluble ribonucleic acid in coding for amino acids. *Proc. Nat. Acad. Sci., USA* 48:1086–1092.

Crick, F. H. C. 1958. On protein synthesis. *Symp. Soc. Exp. Biol.* 12:138–163.

Crick, F. H. C. 1966. Codon–anticodon pairing: The wobble hypothesis. *J. Mol. Biol.* 19:548–555.

Crick, F. H. C., et al. 1961. General nature of the genetic code for proteins. *Nature* 192:1227–1232.

The Genetic Code. 1966. *Cold Spring Harb. Symp. Quant. Biol.* 31.

Grunberg-Manago, M. 1963. Polynucleotide phosphorylase. *Progr. Nucleic Acid Res.* 1:93–133.

Lewin, B. 1974. *Gene Expression, 1: Bacterial Genomes.* New York: Wiley.

Nirenberg, M. W., and P. Leder. 1964. RNA code-words and protein synthesis: The effect of trinucleotides upon the binding of tRNA to ribosomes. *Science* 145:1399–1407.

Nirenberg, M. W., and J. H. Matthei. 1961. The dependence of cell-free protein synthesis in *E. coli* upon naturally occurring or synthetic polyribonucleotides. *Proc. Nat. Acad. Sci., USA* 47:1588–1602.

Nomura, M.; A. Tissieres; and P. Lengyel, eds. 1974. *Ribosomes.* Cold Spring Harbor, N.Y.: Cold Spring Harbor Laboratory.

Rich, A., and U. L. RajBhandary. 1976. Transfer RNA: molecular structure, sequence, and properties. *Ann. Rev. Biochem.* 45:805–860.

Stent, G. S., and R. Calendar. 1978. *Molecular Genetics,* 2nd ed. San Francisco: Freeman.

Watson, J. D. 1963. The involvement of RNA in the synthesis of proteins. *Science* 140:17–26.

Watson, J. D. 1976. *Molecular Biology of the Gene,* 3rd ed. Menlo Park, Calif.: Benjamin.

12 Gene Expression: Control in Prokaryotes and Phages

Control of gene expression has several meanings. First, control is responsible for the fact that in bacteria not all types of proteins of the cell are present all the time. Second, control is responsible for the fact that human erythrocytes are the only human cells that contain large quantities of hemoglobin, although all human cells have the hemoglobin genes. Third, control is responsible for the extremely precise process in zygotic development of higher organisms whereby genes are turned on and off in sequence. Finally, lack of control is responsible for the diseases classified under the term *cancer* that result from cells whose growth control has gone awry.

In order to deal with cancer and various other diseases that result from control errors, it is necessary to understand the processes that control gene expression. Although at the present moment the great bulk of our knowledge concerns the simpler organisms—the viruses and bacteria—recent studies in nucleotide sequencing (Chapter 9) and recombinant DNA (this chapter) are paving the way for a new understanding of gene expression in eukaryotes. This chapter is devoted to control processes in prokaryotes; the next chapter will examine control processes in eukaryotes.

THE OPERON MODEL

In the process leading from a sequence of nucleotides in DNA to a protein, there are many places where control can be exerted. In general, control of gene expression can take place at the transcriptional level, at the translational level, or at the functional level where the synthesized proteins can be activated or inactivated.

One of the best understood control mechanisms is that of transcriptional control, in which the production of mRNA is regulated according to need. *Escherichia coli* mRNAs are short-lived in vivo—they are degraded enzymatically within about 2 minutes—and a complete turnover (that is, degradation and resynthesis) in the cell's mRNA occurs rapidly and continually. This rapid turnover is a prime prerequisite for transcriptional control.

Not all of the proteins *E. coli* can produce are needed in all circumstances in the same quantity. For example, the enzymes involved in particular synthetic pathways are in low concentration or are absent when an adequate quantity of the end product of the pathway exists in the cell. That is, when an abundance of histidine is encountered in the environment or if the cell is overproducing histidine, the cell stops the manufacture of histidine until a need again arises. A system of enzyme synthesis, wherein synthesis is repressed and production of the end product stops when this end product is not needed, is called a **repressible** system. Repressible systems are repressed by the appearance in the cell of an excess of the end product of a synthetic (anabolic) pathway.

Furthermore, some of the metabolites that the cell breaks down for energy and as carbon sources—for example, lactose—may not always be present in the cell's environment. If a given metabolite is not present, enzymes for its breakdown will not be used and producing these enzymes would be wasteful. When the cell produces enzymes for the degradation of a particular carbon source only when this carbon source is present in the environment, the enzyme system is known as an **inducible** system. Inducible enzymes are produced when the environment includes a carbon source that the cell can break down (catabolic pathway). The best studied inducible system is the *lac* operon in *E. coli*. Since the term *operon* refers to the control mechanism, its definition will follow a description of the mechanism.

lac OPERON (INDUCIBLE SYSTEM)

Lactose Metabolism

Lactose is a β-galactoside disaccharide that can be used for energy and as a carbon source by *E. coli* after it is broken down into glucose and galactose. The enzyme that performs the breakdown is **β-galactosidase** (Figure 12–1). This enzyme is inducible. There is virtually no β-galactosidase in a wild-type *E. coli* cell not growing on lactose (the inducer). However, within minutes after adding lactose (as the only carbon source) to the growth medium, this enzyme appears. When the synthesis of β-galactosidase (encoded by the *lacZ* gene) is induced, the production of two additional enzymes is also induced: **β-galactoside permease** (encoded by the *lacY* gene) and **β-thiogalactoside acetyltransferase** (encoded by the *lacA* gene). The permease is involved in concentrating lactose in the cell. The transferase (or

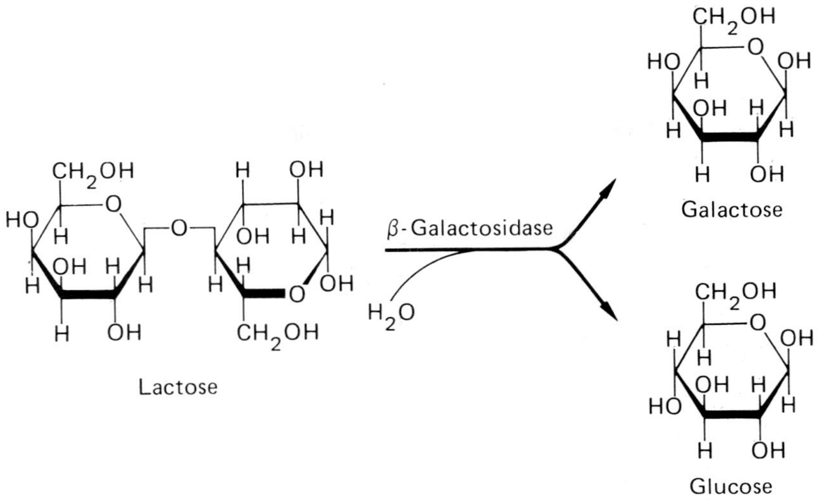

Figure 12–1.
β-Galactosidase
Hydrolytically
Cleaving Lactose
into Glucose and
Galactose

transacetylase) is presumably involved in the further metabolism of lactose in the cell. Its precise in vivo function is not presently known.

Regulator Gene

Not only are the three *lac* genes (*z*, *y*, *a*) induced together, but they are adjacent to one another in the *E. coli* chromosome; they are, in fact, transcribed on a single mRNA (Figure 12–2). Induction involves the product of another gene, called the **regulator gene,** or *i* gene (*lacI*). Although the regulator gene is located adjacent to the three *lac* genes, it is a totally independent transcriptional entity. The regulator specifies a protein called a **repressor** that, when synthesized, interferes with the transcription of the genes involved in lactose metabolism.

Operator

In order for the regulator gene to have an effect, there must be a control element located near the beginning of the β-galactosidase *(lacZ)* gene. This

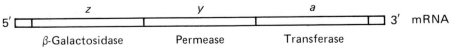

Figure 12–2.   The *lac* Operon Transcribed as a Multigenic
(Polycistronic) mRNA
z, *y*, and *a* indicate the *lacZ*, *lacY*, and *lacA* loci.

Figure 12–3.   The *lac* Operon and Its Regulator Gene (*i*)
The *o* stands for operator and the *p* for promoter. Both the operon and
the regulator gene have their own promoter.

control element is a region referred to as the **operator,** or operator site (Figure 12–3). The operator site is a sequence of DNA that is recognized by the product of the regulator gene, the repressor. When the repressor is bound to the operator, transcription of the operon is prevented. The nucleotide sequence of the *lac* operator is known and is shown in Figure 12–4. Note that here the concept of the promoter is expanded to be the region of DNA that not only is recognized by RNA polymerase but also has other controlling functions related to the initiation of transcription. In the absence of the inducer lactose, the repressor binds to the operator site and prevents transcription by physically preventing RNA polymerase from binding to the DNA sequence it recognizes in the promoter (Figure 12–5). With transcription prevented, none of the products specified by the *lac* genes will be present in the cell. We can now define an **operon** as a sequence of adjacent genes all under the transcriptional control of the same operator.

Induction of the *lac* Operon

The regulatory system just described has the sequence *p-o-z-y-a* (Figure 12–3), where *p* through *a* constitute the operon. The regulator (*i* gene) produces a repressor that binds to the operator (*o* site) and thus prevents RNA polymerase from binding to the promoter (*p*). Since RNA polymerase cannot initiate transcription at the promoter, the genes *lacZ*, *lacY*, and *lacA* (*z*, *y*, and *a* genes) are not transcribed (Figure 12–5); they are, so to speak, "turned

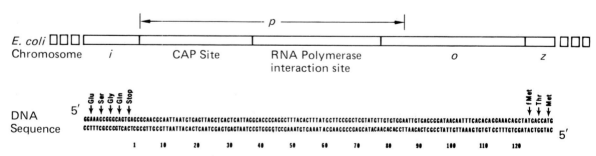

Figure 12–4.   The *lac* Operon Promoter and Operator Regions

Source: R. C. Dickson et al., "Genetic Regulation: The Lac Control Region, *Science* 187 (10 January 1975):27–35. Copyright 1975 by the American Association for the Advancement of Science. Reprinted by permission.

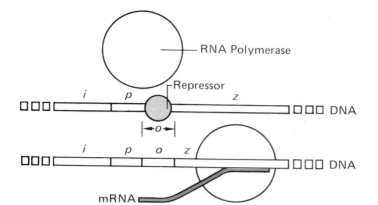

Figure 12 –5.
The Repressor
By binding to the operator, the repressor prevents RNA polymerase from binding to the promotor and transcribing the *lac* operon.

off." This system can only be useful to the cell if the genes are "turned on" when lactose enters the medium. Lactose induces the expression of the *lac* genes by inactivating the repressor. Actually, allolactose is the inducer. A few molecules of β-galactosidase are present in an uninduced cell. They convert lactose to allolactose (Figure 12–6), which then acts as the inducer. (Most experimental work uses the lactose analogue isopropylthiogalactoside, a molecule that does not require either a permease or a functional β-galactosidase in order to function.)

The repressor is an **allosteric protein.** That is, when it binds with a first molecule, the shape of the protein is changed and so, therefore, is its ability to react with a second molecule. Here, the first molecule is the inducer allolactose and the second molecule is the operator (Figure 12–7). Thus, until lactose begins to diffuse into the cell, the repressor is bound to the operator and thereby prevents transcription. When a repressor binds to an inducer molecule, the ability of the repressor to bind to the operator is greatly reduced. The repressor then dissociates from the operator because it has only been bound by very weak forces: No covalent bonds are involved. After the repressor releases from the operator, RNA polymerase can begin transcrip-

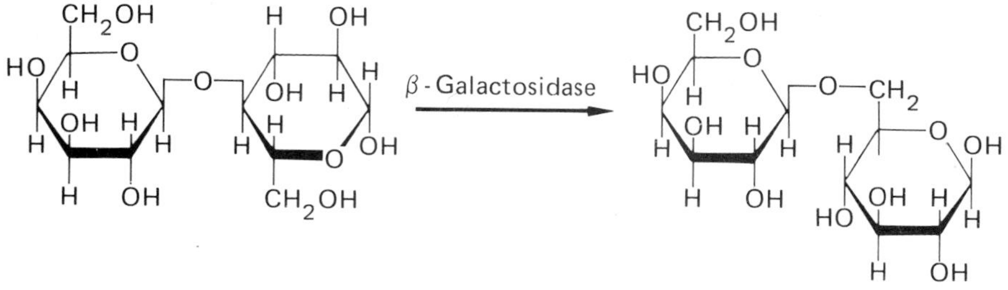

Figure 12 –6.
Allolactose, the *lac* Operon Inducer, Formed from Lactose by β-Galactosidase

Lactose

β - Galactosidase

Allolactose

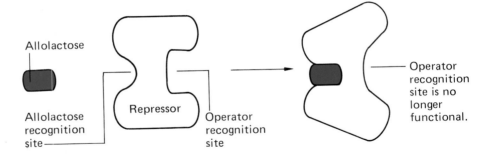

Figure 12-7.
Allosteric Property of the *lac* Repressor Protein

Allolactose

Allolactose recognition site

Repressor

Operator recognition site

Operator recognition site is no longer functional.

tion. The three *lac* operon genes are then transcribed and subsequently translated into their respective proteins.

This system of control is very efficient. The presence of the lactose molecule causes the genes of the *lac* operon, which act to break down the lactose, to be transcribed. After the lactose is metabolized, the repressor returns to its original shape and can again bind to the operator. The system is once again "turned off." The details of this system were worked out by François Jacob and Jacques Monod, who subsequently won the Nobel prize for their efforts.

lac OPERON MUTANTS

Merozygote Formation

Discovery and verification of this system came about through the use of mutant forms and partial diploids of the *lac* operon. The structural (protein-specifying) genes of the *lac* operon, z, y, and a, all have known mutant forms where the particular enzyme does not perform its function. These mutant forms are designated z^-, y^-, and a^-. The genes responsible for normal forms of the enzymes are designated as z^+, y^+, and a^+. Partial diploids in *E. coli* can be developed through sexduction (discussed in Chapter 6).

Some strains of *E. coli* are known to have the *lac* operon incorporated into an F' factor. This factor sets the stage for genetic manipulation since F' strains with certain *lac* operon mutants can pass the F' particle into F⁻ strains that have other *lac* mutants. If the F' factor that is passed circularizes in the F⁻ cell, it will continue to replicate in tandem with the cell and for all intents and purposes the cell becomes a *lac* operon diploid (merozygote).

François Jacob
(1920–)
Courtesy of Dr. François Jacob

Constitutive Mutants

There are mutants in which the three *lac* operon genes are transcribed at all times—that is, are not turned off even in the absence of lactose. They are

Jacques Monod
(1910–1976)
Archives
Photographiques,
Musée Pasteur

known as **constitutive mutants.** Inspection of Figure 12–5 will show that constitutive production of the enzymes can come about in several ways. A defective repressor produced by a mutant regulator gene will not turn off the system nor will a mutant operator that is no longer recognized by the normal repressor. The regulator constitutive mutants are designated i^- and the operator constitutive mutants are designated o^c (Figures 12–8 and 12–9). Both types of mutants produce the same phenotype: constitutive expression of the three *lac* operon genes. It is possible to determine whether such mutants are regulator or operator mutants—for example, the exact location of a mutant on the bacterial chromosome can be determined by the standard mapping techniques discussed in Chapter 6. An alternate technique uses the Jacob and Monod model, which predicts different modes of action for the two types of mutants. In merozygotes a constitutive operator mutant will affect only the operon of which it is a part. Operator mutants are **cis** dominant and **trans** recessive. A constitutive *i*-gene mutant, since it works through an altered protein, will be recessive to a wild-type regulator gene, regardless of which operon (chromosomal or episomal) the mutant is contiguous with—constitutive regulator mutants are, therefore, *cis* and *trans* recessive. (If two mutants are on the same piece of DNA, they are in the *cis* configuration. If they are on different pieces of DNA, they are in the *trans* configuration.)

In Figure 12–8 the episome has a regulator constitutive mutant (i^-). The cell, however, will have the normal (inducible) phenotype because the

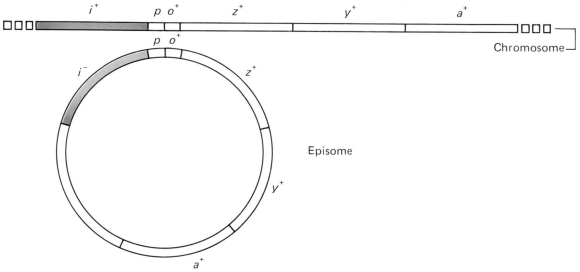

Figure 12 –8. The *lac* Operon Diploid of *E. coli* with a Regulator Constitutive Mutant

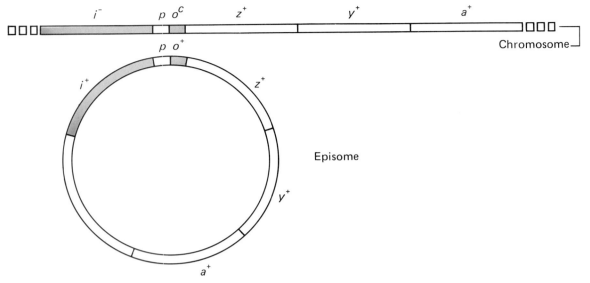

Figure 12 –9.   Another *E.coli lac* Operon Merozygote

chromosomal i^+ allele is dominant to the mutant: The i^+ regulates both the chromosomal and episomal operons. Hence both operons are inducible. In Figure 12–9, however, there are both regulator and operator constitutive mutants on the chromosome of the merozygote cell. Thus, the cell has the constitutive phenotype because the *lac* operon on the chromosome will be continually transcribed—regardless of the alleles of the regulator genes, the chromosomal operon has a *cis* dominant constitutive operator mutant.

Other *lac* Operon Control Mutants

Other mutants have also been discovered that support the Jacob and Monod operon model. A superrepressed mutant, i^s, has been isolated. This mutant represses the operon even in the presence of large quantities of the inducer. Thus, the repressor seems to have lost the ability to recognize the inducer. Basically, it is acting as a constant repressor rather than as an allosteric protein. Another mutant, i^Q, produces much more of the repressor than is normal and is presumably a mutant of the promoter region of the *i* gene.

In 1966 W. Gilbert and B. Müller-Hill isolated the *lac* repressor and ended all speculation as to the correctness of the model. The repressor is a tetramer with each subunit capable of binding one inducer molecule. At about the same time, M. Ptashne and his colleagues isolated the repressor for phage λ operons. (Control of gene expression in phage λ will be discussed at the end of this chapter.)

Walter Gilbert
(1932–   )
Photo: Rick Stafford

CATABOLITE REPRESSION

An interesting property of the *lac* operon and other operons that catabolize certain sugars—for example, arabinose and galactose—is that they are all repressed by the presence of glucose in the medium. That is, glucose is catabolized in preference to other sugars; the mechanism (**catabolite repression**) involves **cyclic AMP** (Figure 12–10). In eukaryotes cyclic AMP acts

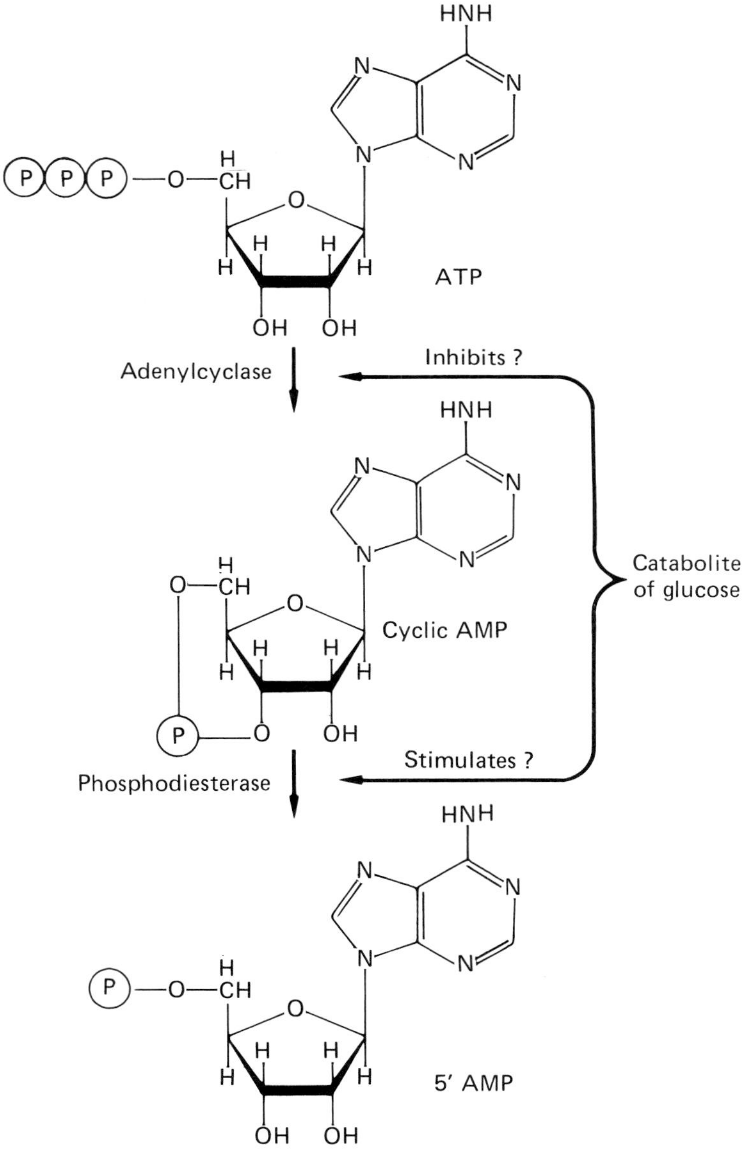

Figure 12 –10. Structure and Role of Cyclic AMP in Catabolite Repression
A catabolite of glucose lowers the quantity of cyclic AMP in the cell and inhibits transcription of many operons.

as a "second messenger," an intracellular messenger regulated by certain extracellular hormones (Chapter 13). Geneticists were surprised to discover cyclic AMP in *E. coli*, where it works in conjunction with another regulatory protein, the **catabolite activator protein (CAP).** First, cyclic AMP combines with CAP; then, the CAP–cyclic-AMP complex binds to a distal part of the promoter of the operon (the *lac* operon in Figure 12–4) and thereby apparently enhances the affinity of RNA polymerase for the promoter. Without the binding of the CAP–cyclic-AMP complex to the promoter, the transcription rate is very low. The addition of glucose to *E. coli* cells depresses the quantity of cyclic AMP by some unknown mechanism and thus lowers the CAP–cyclic-AMP level. The transcription rate of CAP–cyclic-AMP-dependent operons will, therefore, be reduced. The same reduction of transcription rates is noticed in mutant strains of *E. coli* when this part of the distal end of the promoter is deleted.

his OPERON (REPRESSIBLE SYSTEM)

The inducible operons are induced when the metabolite that is to be catabolized enters the cell. Anabolic operons function in a reverse manner: They are "turned off" (repressed) when their end product accumulates in excess of the needs of the cell. Transcription of repressible operons appears to be controlled by two entirely different mechanisms. The first mechanism follows the basic scheme of inducible operons. The second mechanism involves secondary structure in mRNA, which is controlled by translation of an attenuator region of the operon.

Histidine Synthesis

One of the best studied repressible operons is the histidine, or *his*, operon in *Salmonella typhimurium*, a close relative of *E. coli*. The *his* operon in *Salmonella* contains the 9 genes necessary for the enzymes that transform phosphoribosyl pyrophosphate into histidine (Figure 12–11).

Operator Control

In this repressible system the product of the *i* gene, the repressor, is inactive by itself: It does not recognize the operator sequence of the *his* operon. The repressor becomes active when it combines with the metabolite. The actual metabolite is not histidine itself but histidine bonded to its tRNA (histidinyl-tRNA). Thus, when there is an excess of histidinyl-tRNA, enough will be available to bind with and activate the repressor. The metabolite is

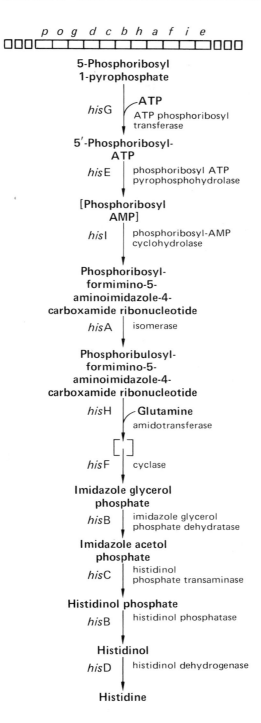

Figure 12–11.
Genes of the *his*
Operon in
Salmonella

referred to as the **corepressor.** The corepressor-repressor complex now recognizes the operator, binds to it, and prevents transcription by RNA polymerase. After the available histidine in the cell is used up, there will be a paucity of histidinyl-tRNAs and eventually the last one will detach from the repressor. The repressor will then diffuse from the *his* operator. The transcription process will no longer be blocked and can proceed normally (it is now **derepressed**). Transcription will continue until enough of the various enzymes have been translated to produce a sufficient quantity of histidine. Then there will again be an excess of histidinyl-tRNA. Some will be available to bind to the repressor and make a functional complex. Thus the system will be again shut off and the process will repeat itself, assuring that histidine is being synthesized when it is needed.

trp OPERON (REPRESSIBLE SYSTEM)

Attenuator-Controlled Operons

Only recently have details of the second control mechanism of repressible operons been elucidated, primarily by Yanofsky and his colleagues, who

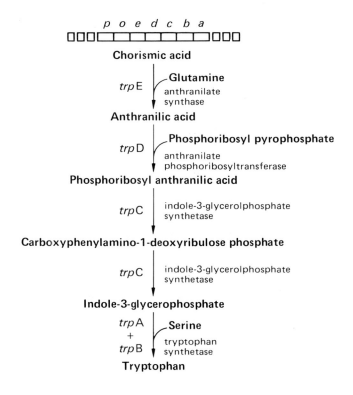

Figure 12 –12.
Genes of the
Tryptophan Operon
in *E. coli*

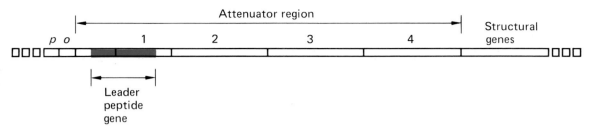

Figure 12–13.   Attenuator Region of the *trp* Operon
This region is transcribed into a leader transcript, which contains a leader peptide gene.

worked with the operon that transcribes the genes involved in the synthesis of tryptophan (*trp* operon) in *E. coli*. This type of operon control—that is, control by an **attenuator** region—has also been demonstrated for the leucine and histidine operons in *Salmonella*. These regulatory mechanisms may be the same for all operons involved in the synthesis of an amino acid.

Leader Transcript

The *trp* operon in *E. coli* contains the 5 genes necessary for synthesis of the enzymes that transform chorismic acid into tryptophan (Figure 12–12). In addition to the promoter and operator, there is an attenuator region in the *trp* operon between the operator and the first structural gene (Figure 12–13). The mRNA transcribed by the attenuator region, termed the **leader transcript,** has been sequenced, with two surprising and interesting facts emerging. First, the four subregions in Figure 12–13 are defined by the fact that they have base sequences that are complementary to each other such that three different **stem-loop structures** can form in the RNA (Figure 12–14). Depending on circumstances, regions 1–2 and 3–4 can form two stem-loop structures, or region 2–3 can form a stem and loop; the formation of other stem-loop structures is thus preempted. As we will see, the particular combination of stem-loop structures determines whether or not transcription will continue.

Leader Peptide Gene

The second interesting fact obtained by sequencing the leader transcript is that there is information for a small peptide from bases 27 to 68. The gene for this peptide is referred to as the **leader peptide gene.** It codes for 14 amino acids, of which two adjacent ones are tryptophan (Figure 12–15). These adjacent tryptophan codons are critically important in attenuator regulation. The presumed mechanism is as follows.

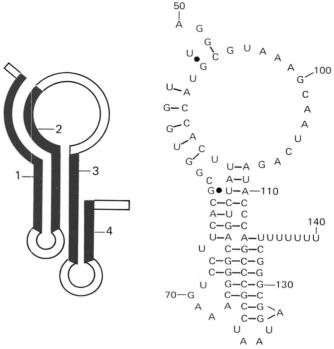

Figure 12 -14.
Nucleotide
Sequence of Part of
the Leader
Transcript of the *trp*
Attenuator Region
Region includes
bases 50 to 140. The
proposed secondary
structure (stem and
loops 1–2, 2–3, 3–4)
is shown.

Source: D. L. Oxender, G. Zurawski, and C. Yanofsky, "Attenuation in the *Escherichia coli* Tryptophan Operon: Role of RNA Secondary Structure Involving the Tryptophan Codon Region," *Proceedings of the National Academy of Sciences USA* 76 (1979):5524–5528. Reproduced by permission.

Excess Tryptophan. Assuming that the operator site is not blocked, transcription of the leader RNA will begin. As soon as the 5′ end of the leader peptide gene has been transcribed, a ribosome will attach and begin the process of translation of this gene. Depending on the levels of amino acids in the cell, three different outcomes of this translation process can take place. If the concentration of tryptophan in the cell is such that tryptophanyl-tRNAs exist, translation will proceed down the leader peptide gene. The

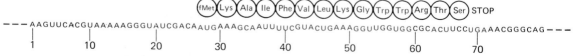

Figure 12 -15.   Nucleotide Sequence of the *trp* Leader Peptide Gene
Nucleotides 50–140 are shown in Figure 12–14.

Source: D. L. Oxender, G. Zurawski, and C. Yanofsky, "Attenuation in the *Escherichia coli* Tryptophan Operon: Role of RNA Secondary Structure Involving the Tryptophan Codon Region," *Proceedings of the National Academy of Sciences USA* 76 (1979):5524–5528. Reproduced by permission.

moving ribosome will overlap regions 1 and 2 of the transcript and allow the 3–4 stem and loop to form as shown in the configuration at the far left of Figure 12–16. This loop will cause transcription to be terminated. (In other operons this stem-loop structure is variously referred to as the **terminator** or **attenuator stem.**) Hence, when existing quantities of tryptophan are adequate for translation of the leader peptide, transcription is terminated.

Tryptophan Starvation. However, if there is a lowered quantity of tryptophanyl-tRNA, the ribosome will stall at the leader peptide as shown in the configuration in the middle part of Figure 12–16. The stalled ribosome will permit the 2–3 stem and loop to form, which precludes the formation of the terminator (3–4) stem and loop. In this configuration transcription is not terminated, so that, eventually, the whole operon is transcribed and translated, which will raise the level of tryptophan in the cell. (The 2–3 stem-loop structure has been referred to as a **preemptor stem** in the *leu* operon of *Salmonella.*)

General Starvation. A final configuration is possible, as shown on the far right in Figure 12–16. Here, no ribosome interferes with stem formation and, presumably, the 1–2 and 3–4 (terminator) stem-loops will form. This configuration will also terminate transcription because of the existence of the terminator stem. It is believed that this configuration will occur if the ribosome is stalled on the 5′ side of the *trp* codons, which will happen when the cell is starved for other amino acids. Presumably, it makes no sense to manufacture tryptophan when other amino acids are in short supply. Hence the cell can carefully bring up the levels of the various amino acids in the most efficient manner.

**Figure 12 – 16.
Model for
Attenuation in the *E.
coli trp* Operon**
The sphere is the
ribosome and the
strand is the leader
transcript of Figure
12–14.

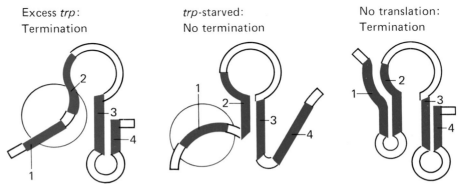

Excess *trp*:
Termination

trp-starved:
No termination

No translation:
Termination

Source: D. L. Oxender, G. Zurawski, and C. Yanofsky, "Attenuation in the *Escherichia coli* Tryptophan Operon: Role of RNA Secondary Structure Involving the Tryptophan Codon Region," *Proceedings of the National Academy of Sciences USA* 76 (1979):5524–5528. Reproduced by permission.

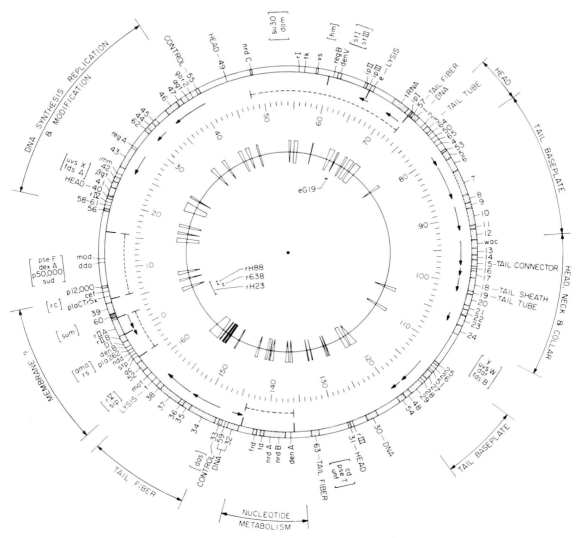

Figure 12 –17. Genetic Map of T4

Inner arrows indicate transcriptional units. Bracketed gene names
indicate loci whose exact position is not known. Outermost terms indicate
broad functional categories.

Source: W. B. Wood and H. R. Revel, "The Genome of Bacteriophage T-
4," *Bacteriological Review* 40 (1976):857. Reprinted by permission.

Redundant Controls

There are several reasons why the cell should have redundant mechanisms for control of amino acid operon expression. In the case of the *trp* operon, the cell is testing both the tryptophan levels (tryptophan is the corepressor) and the tryptophanyl-tRNA levels (in the attenuator control system). However, it is not precisely clear why the cell needs to "know" these levels. The attenuator system also allows the cell to regulate tryptophan synthesis on the basis of the shortage of other amino acids. For example, when there is a shortage of both lysine and tryptophan, operator control will allow transcription to begin, but attenuator control will terminate transcription because stem-loops 1–2 and 3–4 will form (Figure 12–16). More research will clarify our understanding of why the cell has these interesting and efficient mechanisms of operon control.

PHAGE OPERONS

When a phage infects a cell, it must express its genes in an orderly fashion because there are genes whose products are needed early in infection as well as genes whose products are not needed until late in infection. Early genes usually control phage DNA replication; late genes usually control phage coat proteins and the lysis of the bacterial cell. It is most efficient if a phage expresses the early genes first and the late genes last in the infection process. Also, during temperate life cycles, phages have the option either of entering into lysogeny with the cell or of lysing the cell, and here, too, control processes determine which path will be taken. One generalization that holds for most phages is that their genes are clustered into late and early operons, with different transcriptional control mechanisms for each.

Phage T4

For example, phage T4, with 73 genes (see Figure 12–17 and Chapter 6), has its transcription controlled by the nature of the RNA polymerase specificity factors. Early T4 genes have promoters whose specificity of recognition depends on the σ factor of the host. However, middle and late operons of T4 have promoters whose specificity is determined by other proteins that are synthesized during the early infection process. For example, late promoters require RNA polymerase plus the products of genes 33 and 55 of the T4 chromosome. Some proteins function both early and late and are specified by genes that have several promoters, with each promoter being recognized by a different specificity factor (Figure 12–18).

Early Gene

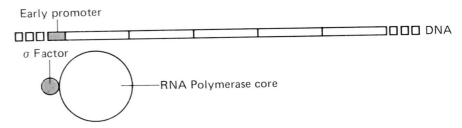

Late Gene

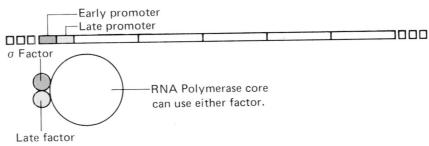

Early/Late Gene

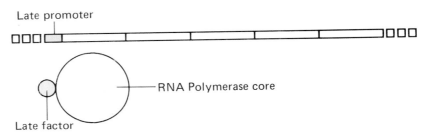

Figure 12–18.
Types of Promoters
in T4 Phage

Phage λ

λ Operons. Phage λ (Figure 6–26) is probably the best known phage. It exhibits a complex system of controls of both early and late operons as well as controls for the decision of lytic infection versus lysogenic integration. The genes of λ are grouped into three operons (early left, early right, and late) and a repressor region (Figure 12–19). The early left and early right operons contain the genes for DNA replication, recombination, and phage integration. The late operon contains the genes that control lysis of the host cell and phage head and tail proteins. The sequence of events following phage

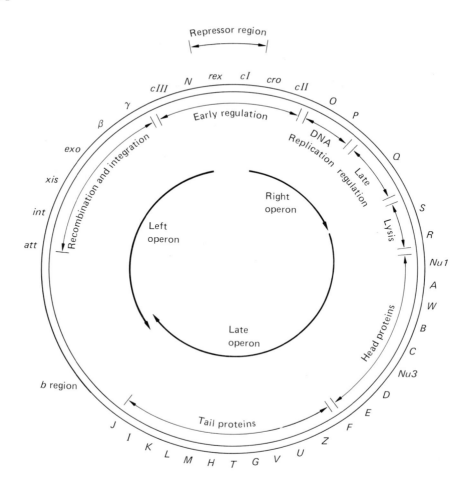

Figure 12–19. Genetic Map of Phage λ
The prophage begins and ends at *att*. The mature phage begins and ends at *Nu1*.

infection is relatively well known. It is less clear how the decision for lytic versus lysogenic response is determined.

Early and Late Transcription.   When the phage first infects an *E. coli* cell, transcription of the left and right operons begins. The *N* and *cro* genes are transcribed; transcription then stops on both operons and cannot continue until the protein product of the *N* gene is produced. This protein product interacts with RNA polymerase in such a way as to make the polymerase insensitive to the termination signals to the left of *N* and the right of *cro*. Transcription then continues along the entire left and right operons. The *Q* gene in the right operon has the same effect on the late operon as the *N* gene did on the two early operons: Without the *Q*-gene product, transcription of the late operon proceeds about 200 nucleotides and then terminates. The

RNA polymerase, when associated with the Q-gene product, becomes insensitive to the termination signal. Hence, in phage λ the left and right operons are transcribed first, and then the late operon is transcribed. Instead of control resting with transcription specificity factors, as in T4, control is with proteins that allow RNA polymerase to proceed past termination signals.

Repressor Transcription.   In order for the cI gene (the repressor) to be transcribed, the products of the cII and $cIII$ genes are needed. Together with RNA polymerase, the two protein products initiate transcription of cI (and rex) at a promoter termed p_{RE} (for *e*stablishment of *r*epression). We can now see how every gene on the λ chromosome can be transcribed. But to see the complex interaction of genes that determines whether the phage will show lytic or lysogenic responses, we need to look more closely at the early regulation region (Figure 12–20), where the cI gene is primarily concerned with lysogeny and cro (for *c*ontrol of *r*epression and *o*ther things—also known as tof for *t*urn *of*f) is concerned with the lytic pathway.

Maintenance of Repression.   Once cI is transcribed, it is translated into a protein called the *lambda repressor*, which interacts at the left and right operators, o_L and o_R. When these operators are blocked, transcription of the left and right operons (and therefore also the late operon) ceases. There are several ramifications of the repression. First, lysogeny still can be initiated because the left operon has been transcribed at the early stage of infection. Second, since cII and $cIII$ are no longer being synthesized, cI transcription from the p_{RE} promoter is stopped. However, cI can still be transcribed because there is a second promoter, p_{RM} (for *m*aintenance of *r*epression), that will allow low levels of transcription of the cI gene. The cI gene can further control its own concentration in the cell: When the right and left operators were sequenced, it was discovered that each has three sites of recognition of the repressor (Figure 12–21). On the right operator, for example, it was found that the rightmost site (o_{R3}) is most efficient at binding repressor—the right operon is thereby repressed and transcription of cI is some-

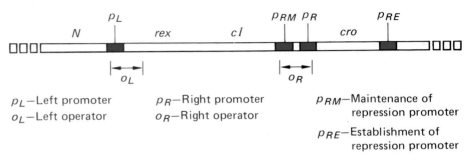

p_L—Left promoter p_R—Right promoter p_{RM}—Maintenance of
o_L—Left operator o_R—Right operator   repression promoter

                                                                   p_{RE}—Establishment of
                                                                         repression promoter

Figure 12 –20.   Early Regulation Region of Phage λ

Figure 12–21.
Region of the Right
Operator on the λ
Chromosome
The repressor
recognition sites
within the operator
are o_{R1}, o_{R2}, and o_{R3}.

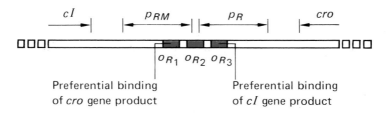

how enhanced (in a way similar to enhancement of transcription by binding at the CAP site in the *lac* operon). Excess repressor, when present, however, is also bound by the other two sites within o_R. The foregoing process results in the repression of the *cI* gene itself. Hence maintenance levels of *cI* can be kept within very narrow limits.

A third ramification of repression is the prevention of superinfection. That is, bacteria lysogenic for λ phage are protected from further infection by other λ phage because of the presence of repressor. (These bacteria are also protected from infection by T4 phage with *rII* mutants. The protection is controlled by the *rex*-gene product that is transcribed with the *cI* gene.)

The promoters for maintenance and establishment of repression differ markedly in their control of repressor gene expression. When p_{RE} is active, a very high level of repressor is present, while p_{RM} produces only a low level of repressor. The level of repressor is due to the length of the leader RNA transcribed on the 5′ side of the *cI* gene—p_{RE} transcribes a very long leader RNA and is very efficient at translation of the *cI* region; p_{RM} transcribes a very short leader (it is much closer to *cI*) and is very inefficient in translating *cI*.

Lysogenic versus Lytic Response.   So far, we have established a description of the lysogenic system. How then does λ turn toward the path of cell lysis? Here, control is exerted by the *cro*-gene product, a repressor that works at the left and right operators in exactly the reverse way that the *cI* repressor works. That is, using the right operator as an example, *cro*-gene product binds preferentially to the leftmost of the three sites within o_R and represses *cI* but enhances the production of *cro*. Hence there is a race between *cro* and *cI* to decide whether infection will lead to lysogeny or lysis. All the details of this race in the λ life cycle are not clearly understood.

OTHER FORMS OF TRANSCRIPTIONAL CONTROL

There are other ways to regulate the transcription of mRNA. One way is to control the efficiency of various processes. For example, we know that the promoter sequence of different genes in *E. coli* is different. Since the affinity for RNA polymerase is different for the different sequences, the rate of transcription of the genes will be different. The more efficient promoters will be transcribed at a greater rate than the less efficient promoters. An example is the promoter of the *i* gene of the *lac* operon. This promoter is for a constitutive gene that usually produces only about one mRNA per cell cycle. However, mutants of the promoter sequence are known that produce up to 50 mRNAs per cell cycle. Here, then, the transcriptional rate is controlled by the efficiency of the promoter in binding RNA polymerase.

TRANSLATIONAL CONTROL

Translational control takes on more of an importance in eukaryotes, and this importance will be discussed in the next chapter. In prokaryotes, translational control is of lesser importance than transcriptional control for two reasons. First, mRNAs are extremely unstable; they have a lifetime of only about 2 minutes. There is little room for controlling the rates of translation of existing mRNAs; they simply do not last long enough. Second, although there are some indications of translational control in prokaryotes, such control is inefficient—energy is wasted synthesizing mRNAs that will never be used.

Control can be exerted on a gene if the gene occurs distally from the operator in an operon. The genes that are transcribed last appear to be translated at a lower rate than the genes transcribed first. For example, the three *lac* operon genes are translated roughly in a ratio of 10:5:2. This ratio is due to the polarity of the translation process. That is, in prokaryotes, translation is directly tied to transcription; a messenger RNA can have ribosomes attached to it well before transcription is finished. Thus, genes at the 5′ end of the operon will be available for translation before genes at the 3′ end. Presumably, operon-linked enzymes needed in greater quantities should be at the proximal (5′) end of an operon.

Translational control can also be exerted by the efficiency with which the messenger is bound to the ribosome. This efficiency is related to some extent to the sequence of nucleotides at the 5′ end of the messenger RNA that is complementary to the 3′ end of the 16S rRNA segment in the ribosome. Different sequences will have different efficiencies of binding and therefore will be translated at different rates.

The redundancy in the genetic code can also play a part in translational control of some proteins since different tRNAs occur in the cell in different quantities. Genes with abundant protein products may have codons that specify the commoner tRNAs. Genes that code for proteins not needed in high abundance could have several codons specifying the rarer tRNAs (Table 12–1), which would slow down the process of translation of these genes.

A translational control mechanism called the **stringent response** conserves the production of tRNA and rRNA—and hence ribosome construction—under conditions of amino acid starvation. When the cell is deficient

TABLE 12-1. Codon Distribution in MS2, an RNA Virus

First Position	Second Position								Third Position
	U		C		A		G		
U	Phe	10	Ser	13	Tyr	8	Cys	7	U
	Phe	13	Ser	10	Tyr	13	Cys	4	C
	Leu	11	Ser	10	stop	1	stop	0	A
	Leu	4	Ser	13	stop	1	Trp	14	G
C	Leu	10	Pro	7	His	4	Arg	13	U
	Leu	14	Pro	3	His	4	Arg	11	C
	Leu	13	Pro	6	Gln	10	Arg	6	A
	Leu	6	Pro	5	Gln	16	Arg	4	G
A	Ile	8	Thr	14	Asn	11	Ser	4	U
	Ile	16	Thr	10	Asn	23	Ser	8	C
	Ile	7	Thr	8	Lys	12	Arg	8	A
	Met	15	Thr	5	Lys	17	Arg	6	G
G	Val	13	Ala	19	Asp	18	Gly	17	U
	Val	12	Ala	12	Asp	11	Gly	11	C
	Val	11	Ala	14	Glu	9	Gly	4	A
	Val	10	Ala	8	Glu	14	Gly	4	G

in one or more amino acids, there will be an accumulation of depleted tRNAs in the cell. Subsequently, protein synthesis will be stalled on many ribosomes. When a depleted tRNA appears in the A site of a ribosome, a protein called the **stringent factor** catalyzes the formation of two unusual nucleotides: guanosine-3'-diphosphate-5'-diphosphate (3'ppGpp5') and guanosine-3'-diphosphate-5'-triphosphate (3'ppGppp5') from guanosine diphosphate (3'Gpp5') and adenosine triphosphate (3'Appp5'). These unusual nucleotides accumulate in the cell with the concomitant cessation of the transcription of tRNA and rRNA. The exact mechanism of this process is not yet known. It is possible that ppGpp and ppGppp actually cause the cessation of transcription of the two RNA types, or possibly they are a consequence of the shutdown process. The gene for the production of the stringent factor is rel^+, for the **relaxed mutant** that neither accumulates ppGpp or ppGppp nor has the stringent response. Eukaryotes in general do not show the stringent response.

POST-TRANSLATIONAL CONTROL

Even after a gene has been transcribed and the mRNA translated, a cell can still exert some control over the functioning of the enzymes produced if the enzymes are allosteric proteins. We have discussed the activation and deac-

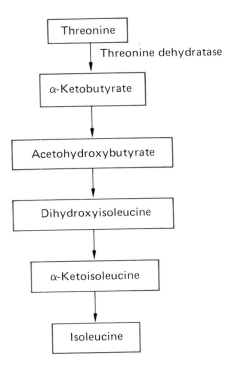

Figure 12-22.
Isoleucine Pathway
in *E. coli*

tivation of operon repressors owing to their allosteric properties. Similar effects occur with structural proteins. The need for post-translational control is apparent because of the relative longevity of proteins as compared to RNA. That is, when an operon is repressed, it no longer transcribes mRNA. However, the mRNA that was previously transcribed has been translated into protein, and this protein is still functioning. Thus, during the process of operon repression, it would be efficient for the cell to also prevent the activity of the translated protein.

An example of post-translational control occurs in the threonine-to-isoleucine pathway in *E. coli* (Figure 12–22). If an abundance of isoleucine is added to the cells, repression of mRNA transcription immediately occurs. Although the enzymes are still present, no more threonine is converted to isoleucine because isoleucine inhibits the enzyme threonine dehydratase (Figure 12–22), the first enzyme in the direct pathway to isoleucine. Without this enzyme none of the products of the pathway are formed, and therefore no synthesis of isoleucine will occur even though the other enzymes are unaffected.

This method of control is called **feedback inhibition** because some product of the pathway is the agent that turns off the functioning of the pathway. Threonine dehydratase is an allosteric enzyme. Its active site is responsible for the conversion of threonine to α-ketobutyrate. However, it also has a site that recognizes isoleucine. Recognition of isoleucine alters the shape of the enzyme and prevents its main active site from either recognizing threonine or acting on it (Figure 12–23). Thus, the cell not only has the ability to turn off transcription and therefore translation, it also can prevent the enzymes already translated from continuing to function.

A good deal is known about how regulation occurs in *E. coli* and many of its viruses. Much less is known about how eukaryotes control gene expression, the topic of the next chapter. So far, the greatest recent advances in the study of control of protein synthesis have been prompted by new and innovative biochemical techniques. One of the most important of these techniques is the sequencing of long segments of nucleic acid. It is now possible to actually read the genetic code and see precisely how the genetic information is organized. Knowledge of this organization and of the way that many of the controlling enzymes work is slowly but surely unlocking the secrets of molecular genetics. There are many surprises in store.

In the last section of this chapter, we look at the latest techniques of molecular genetic analysis: recombinant DNA techniques. These techniques allow the study of gene structure and function by isolating a foreign gene within an *E. coli* cell. These techniques are having major effects in eukaryote gene analysis and thus serve to tie this chapter to the next.

RECOMBINANT DNA

A new tool, in the form of gene cloning, has been added to the arsenal of the molecular geneticist. Gene cloning is the technique of taking a piece of DNA from any source and, regardless of its origin, inserting it into a suitable vehicle (usually plasmid or phage DNA; see later) and having this vehicle introduced into a suitable host cell (usually *E. coli*) where it can be replicated. Since cells carrying the recombinant DNA molecule can be grown in

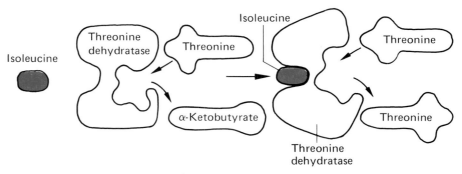

Figure 12–23.   View of Feedback Inhibition of Threonine Dehydratase, Which Normally Converts Threonine into α-Ketobutyrate

THE RECOMBINANT DNA DISPUTE

With the ability to manipulate genetic material—to remove it from one organism and put it into another—came the possibility of disaster. Particularly fearful was the chance of putting either carcinogenic viruses or other highly toxic genetic material into our common intestinal bacteria, *E. coli,* and having these bacteria escape into the environment. From the beginnings of recombinant DNA work in the earliest part of the 1970s, the issue has been debated hotly. Views have ranged from a desire to ban all recombinant DNA work to claims of nothing at all to worry about. It seems now that the latter view is correct. After a decade of research, there are no known cases of disease caused by an "Andromeda strain," and every reason exists to suspect that there will be no cases.

The recombinant DNA debate has had the advantages of bringing the issue to the public's attention and of forcing a concerted effort toward development of intelligent guidelines that will ensure minimal risk in the future. A series of guidelines for recombinant research has been put forward by NIH (The National Institutes of Health); some of them have, subsequently, been relaxed. These guidelines prescribe, among other things, the types of facilities and the mutant strains that must be used when doing the riskiest experiments—that is, where random pieces of mammalian DNA are placed into bacterial plasmids.

Most plasmid work is being done with bacteria or plasmids that are modified. A bacterium of the *E. coli* strain EK2 cannot survive in the human gut. It contains mutants that do not permit it to manufacture thymine and diaminopimelate. The lack of thymine-synthesizing ability is lethal because the cell cannot replicate its DNA. The diaminopimelate is a cell-wall constituent without which the cell bursts. The cell also carries mutants that make it extremely sensitive to destruction by bile salts. Thus, if by accident the cells were to escape, they would pose no threat. Less than 1 in 10^8 of these cells can survive the human gut for 24 hours.

The plasmids now used for recombinant research have been modified so that they cannot be transferred from one cell to the next. Again, if containment fails, neither the host cells nor their plasmids will survive.

Two other issues have surfaced in the recombinant DNA debate: the risk of "tampering with evolution" and the risk of "genetic engineering." Tampering with evolution is the charge directed at the transfer of genetic material between organisms that are evolutionarily unrelated (eukaryotes and prokaryotes). But, *E. coli* can pick up foreign DNA by the process of transformation. *E. coli* lives in the human gut. Cells from the human gut die and are sloughed off in the intestine. It seems inconceivable that *E. coli* has not in the past, over and over again, had the opportunity to incorporate vertebrate DNA into its own genome. That *E. coli* does not normally contain vertebrate genes would certainly seem to indicate that we are not really doing something totally new under the sun, but simply that the uptake of vertebrate DNA by *E. coli* is not a process that is threatening to humans.

Genetic engineering has raised the fear that some malevolent authority will seed the population with a bacteria containing the "evil" gene or the "subservient" gene. But, although risks can certainly be conjectured, our present knowledge is restricted to attempting to manipulate bacteria so as to uncover some of nature's secrets in order to cure certain cancerous or developmental disease states. This knowledge is far removed from understanding anything about behavior genetics, let alone having the ability to manipulate people's behavior by seeding them with the right gene. We are nowhere near the level of understanding of either the seeding techniques or of behavior. Philip Handler, president of the National Academy of Sciences, said in 1977: "Those who have inflamed the public imagination have raised fears that rest on no factual basis but their own science fiction."

Paul Berg (1926–   )
Courtesy of Dr. Paul
Berg

large numbers, the technique has been called *cloning*. Thus, it is possible to isolate large numbers of a particular gene, prokaryotic or eukaryotic, and study the structure, function, and regulation of this gene. Paul Berg of Stanford University won the 1980 Nobel Prize in chemistry for his pioneering work in gene cloning.

Restriction Endonucleases

Cloning techniques evolved with the discovery of specific **restriction endonucleases.** The 1978 Nobel Prizes in physiology and medicine were awarded to W. Arber, H. Smith, and D. Nathans for their pioneering work in the study of these restriction endonucleases, which are enzymes that serve to protect bacteria from phage infection by recognizing certain nucleotide sequences found on the attacking phage and then degrading the DNA with endonuclease activity. Two types of restriction endonucleases are known: type I and type II. Type I endonucleases recognize a specific sequence of DNA but then go on to cleave this DNA randomly. They have little value for recombinant work. Type II endonucleases, however, recognize specific sites and cleave at just these sites. The sites recognized by type II endonucleases are palindromes, sequences that read the same from either direction. For example, in Figure 12–24 the sequence CCGGATCCGG, recognized by endonuclease Bam I, is found 5′ to 3′ on both strands. After endonuclease

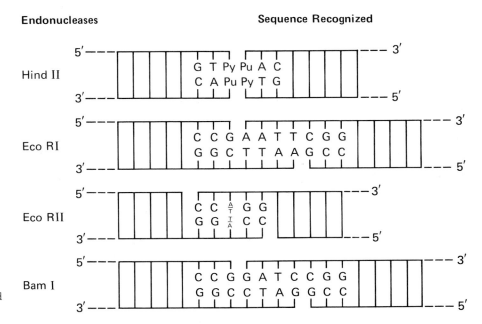

Figure 12 –24.
Sequences Cleaved
by Various
Restriction
Endonucleases
Py is pyrimidine and
Pu is purine.

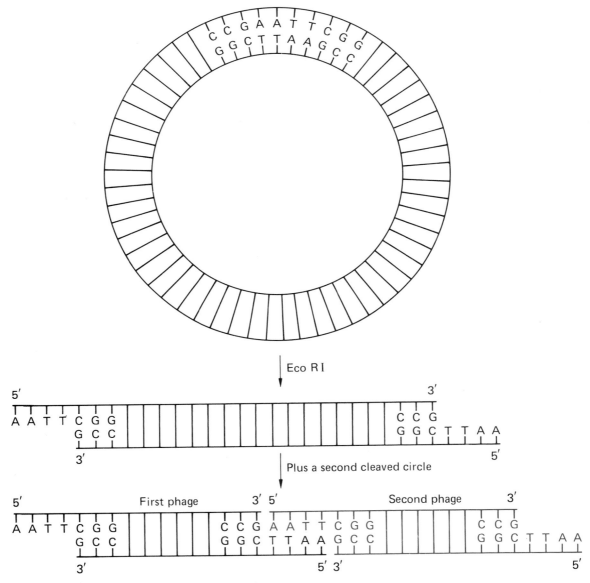

Ligase is needed to join the phages.

Figure 12–25. Circular Phage Chromosome with a Palindrome Recognized by Eco RI
The "loose ends" of cleaved DNA can join to recircularize the molecule or to unite two or more linear molecules of DNA cleaved by the same restriction endonuclease.

action the cell will degrade the cleaved ends of DNA with exonucleases. The cell is protected from the restriction enzymes either by the absence of the specific palindromes they recognize or by modification of the nucleotides—for example, by the addition of methyl groups—in the palindrome sequence.

Figure 12–25 shows how DNA cleaved by a specific restriction endonuclease can rejoin with itself and recircularize if it was only cleaved in one place or how different molecules with the same free-end patterns can fuse to form hybrid molecules. Only the action of a ligase is needed to make the molecule complete. At the present time, over 80 different restriction endonucleases have been purified and well over 100 are known.

Hybrid Plasmids

It is now feasible to join together in vitro in a systematic fashion DNAs from different species and from different superkingdoms (prokaryotes and eukaryotes). More significantly, one of the pieces of DNA involved in the hybridization can be a plasmid. (*Plasmid* refers to a piece of DNA that, like an episome, can replicate independently in a cell. *Episome* connotes the ability to integrate into the host chromosome, such as the F factor or λ phage; plasmids do not necessarily have this ability. The terms are sometimes erroneously used interchangeably.) The hybrid plasmid (Figure 12–26) can then be transferred into a host cell. A host bacterial cell can be made permeable to the DNA by addition of a dilute solution of calcium chloride.

A few conditions have to be met in order to succeed in the joining of DNAs from different species or different kingdoms. First, the vehicle plasmid must be cleaved at only one point by the endonuclease. If it is cleaved at more than one point, it will only be fragmented by the experiment. Second, the cleavage of this plasmid must occur in a nonessential region of the plasmid or the plasmid will be rendered ineffective by the process. Also, the **passenger** DNA that is being fused into the **vehicle** plasmid should be a functional unit—that is, it should be at least a complete gene. Other conditions also prevail. For example, it would be a good idea if the plasmid contained a drug-resistance locus so that selection could be done for this plasmid once it has been introduced into host cells. In the presence of drug, only cells that have the hybrid plasmid will survive. In fact, one of the most widely used vehicles is the colicin-producing plasmid (Col E1). Colicins are bacteriocidal substances produced by a number of bacterial strains. The most common plasmids fall into two groups: **Col plasmids** carry genes that control the synthesis of colicins; **R plasmids** carry genes that control resistance to various drugs. Col plasmids are useful as vehicles because the cells carrying them will kill other cells not carrying them. Another commonly used group of vehicles are derivatives of phage λ. They have been named **Charon phages** after the mythical boatman of the River Styx.

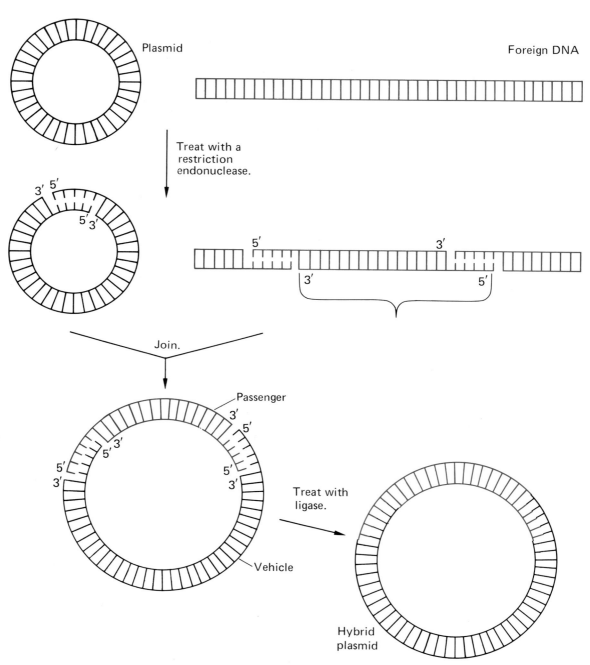

Figure 12–26. Formation of a Hybrid Plasmid
Restriction endonuclease is used to cleave both host and foreign DNA.

Poly-A/Poly-T Method of Cloning

A second method is used to produce hybrid plasmids when simple restriction endonuclease cleavages do not suffice—for example, when the endonuclease cut falls in the middle of a gene to be examined or renders the plasmid inactive. In the poly-A/poly-T technique, poly-A is added to the 5′ ends of the passenger and poly-T is added to the 5′ ends of the vehicle by treating the cleaved plasmids with deoxynucleoside transferase and ATP (poly-A addition) or TTP (poly-T addition). Discrepancies in the size of the overlap regions can be handled by repair enzymes that fill in gaps (Chapter 14). The process is illustrated in Figure 12–27. This second method is also useful for cloning randomly produced small segments of a genome that have been generated by methods other than endonuclease cleavage. For example, eukaryotic chromosome segments can be isolated by physical shearing of the DNA.

Heteroduplex Analysis

Other techniques are also used in recombinant work. In **heteroduplex analysis,** for example, a heteroduplex is formed when DNA is hybridized from two strands that do not have exactly the same sequences. Thus, regions are not joined by nucleotide base pairing, and single-stranded loops and substitution bubbles are formed. Heteroduplex analysis can show that there has been the expected insertion of foreign DNA into the plasmid since, by joining a normal plasmid with the **chimeric** (genetically mixed) one, the presence of a single-stranded loop can be verified. The method also allows the investigator to determine how large a sector is inserted and how much of the original plasmid still remains. *Heteroduplex mapping,* as the technique is called, has gained widespread use in analyzing recombinant plasmids. Cloned DNA can also be located within a vehicle by hybridizing the DNA with its messenger RNA. In Figure 12–28 a cloned rabbit β-globin gene is located by hybridization with rabbit β-globin mRNA.

Another technique that is used in recombinant work is **hybrid screening,** whereby the passenger DNA of a plasmid is determined to be a particular gene by hybridization with radioactively labeled complementary RNA. For example, the yeast genome can be fragmented and its pieces inserted into many plasmids. tRNA genes can be located by hybridizing each clone with radioactively labeled tRNA. Only clones with tRNA genes will retain the radioactive label by forming DNA-RNA hybrids (Figure 12–29).

It is apparent that genetic techniques have reached a point where recombinant vehicles with passengers of virtually any kind of DNA can be isolated with ease. At first, there were some fears as to the safety of the human race. With these fears dispelled, recombinant DNA work has become one of the most exciting and active areas in modern molecular genetics. It seems that each day new accomplishments are being reported. A few of the giant steps in genetic research are noted in the following discussion.

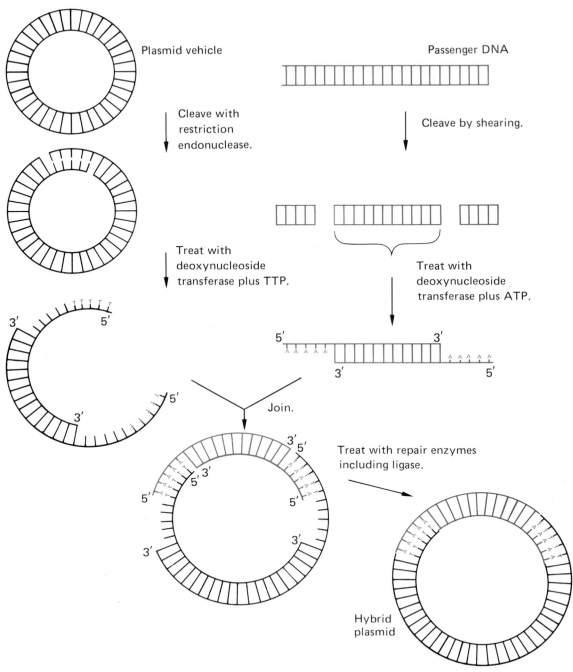

Figure 12 –27.   Poly-A/Poly-T Technique for Producing Hybrid Plasmids

444

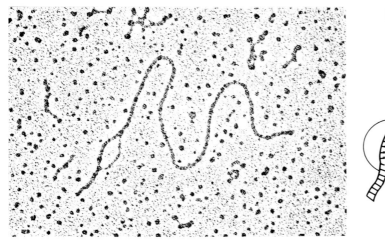

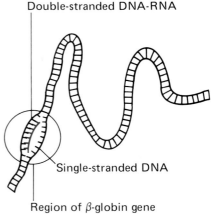

Double-stranded DNA-RNA

Single-stranded DNA

Region of β-globin gene

a.   b.

Figure 12 –28.   Electron Micrograph of a Linearized Plasmid
E. coli plasmid (PMB9), with cloned insert of the rabbit β-globin gene,
which is localized by hybridization with rabbit β-globin mRNA.
Magnification 16,000X

Source: Reproduced courtesy of Thomas R. Broker and Louise T. Chow.

Cloned Genes

Many genes from many sources have been cloned—the genes for human insulin, for example, and interferon, the protein that is the first cellular defense against viral disease. Interferon may prove to be extremely important as both an antiviral and an anticancer agent. Research on interferon has been hampered by the protein's prohibitive costs and limited supply. The hemoglobin gene has been cloned by using a method of isolating mRNA from erythrocytes and using this mRNA to manufacture DNA with reverse transcriptase. Although this cDNA (complementary DNA) has been cloned, this method does not allow study of the control processes in eukaryotic development because the transcribed mRNA does not have the transcriptional control information on it. Cloning to discover control mechanisms is possible and has been successful in several cases. For example, antigenic specificity on the flagellar proteins of *Salmonella* seems to be controlled by an inverted region in front of one of the genes.

The technique of cloning makes it possible to grow large quantities of both the passenger DNA and its gene product, which has eased the task of sequencing the genes or working with their protein products. Examples include DNA ligase, in which a 500-fold increase in production of the enzyme was achieved by cloning it. Cloning also opens the door to studies on the structure and function of the genes associated with organelle DNA, which

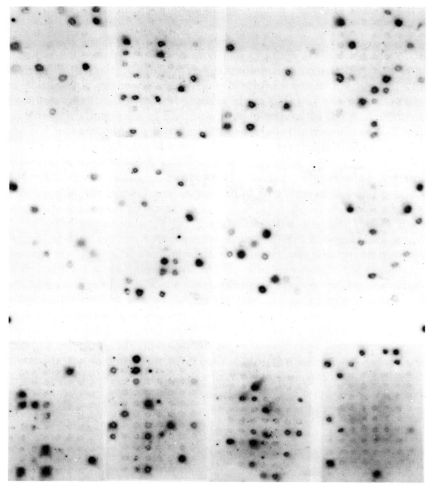

Figure 12–29.   Autoradiograph of 1,140 Clones Carrying Yeast DNA Fragments Obtained by Restriction Endonuclease Activity (*Hind*III) Clones were hybridized with radioactive yeast tRNA. Dark spots indicate clones carrying yeast tRNA genes.

Source: J. S. Beckmann, P. F. Johnson, and J. Abelson, "Cloning of Yeast Transfer RNA Genes in *Escherichia coli*," *Science* 196 (8 April 1977):205–208. Copyright 1977 by the American Association for the Advancement of Science. Reproduced by permission.

can be isolated from plasmids and mitochondria and then cloned. Viruses, such as SV40 (SV stands for Simian Virus), that induce tumors, can be partly cloned for the purpose of studying control mechanisms in tumor production. Thus, new techniques of recombinant DNA research should provide tremendous insights in molecular genetics during the next several years.

GENES WITHIN GENES

Recent work in molecular genetics has focused on sequencing the genetic material of bacterial viruses, the simplest "organisms." Complete sequencing has been accomplished with φX174, a virus that contains a single-stranded DNA circle within its protein capsule. Once injected into the host, the DNA is replicated to form a double helix that then proceeds in a normal viral fashion to replicate itself, manufacture its own coat proteins, lyse the cell, and escape. This virus has 9 genes; it is a small, 20-faced polyhedron with a small spike at each of its 12 vertices. It is this spike that attaches

φX174 to *E. coli.* The coat accounts for 1 protein and the spike accounts for 2. Thus 3 of the virus's 9 genes manufacture coat proteins. The location of the genes in φX174, obtained through mapping methods, is illustrated here.

The antisense strand (complementary to the transcribed strand) of φX174 DNA has 5387 nucleotides. The map (pp. 449–456) shows the translation start signals (ATG) for each gene, as well as their stop signals—the nonsense codons (TGA, TAA). The ribosome recognition signals precede the start of each gene. Transcription is

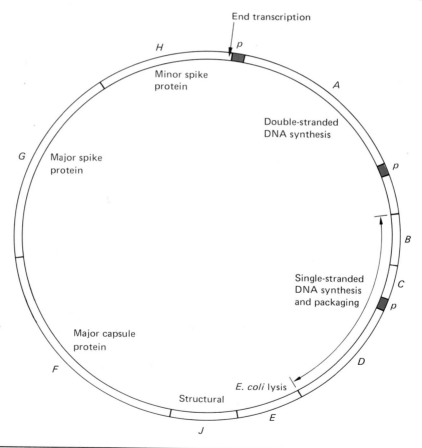

Figure 1.
Presumed Location of the 9 Genes of φX174 on Its Circular Chromosome

known to start in three places; the map shows the promoter sequences prior to the first nucleotides transcribed. In the middle of the start of transcription and translation for the A gene is a transcription stop signal for the previous transcript: TTTTTTA.

From the information obtained from the sequencing of MS2, an RNA virus, it was believed that there should be a nontranslated sequence between each gene for the purpose of control of one sort or another. However, careful perusal of the map provides several surprises. First, the ends of three genes overlap the beginnings of the next genes (A-C, C-D, and D-J); in the first two cases, the initiation codon is entirely within the end of the previous gene, *but,* read in a different frame of reference. In the sequence ATGA, the ATG is the initiation of the next gene whereas the TGA is the termination of the previous gene. In the D-J interface, one A is shared: TA<u>A</u>TG. It is the number 3 base of the termination codon and the number 1 base of the initiation codon. The surprises do not end here.

At first, with the sequence of nucleotides spread out in front of them, the researchers could not find the B and the E genes: these genes were missing! Upon careful analysis, however, with verification using mutated strains, they found that the B gene was entirely within the A gene and the E gene was entirely within the D gene. Their finding went against all theory. We are led to believe, from logical arguments, that genes cannot overlap. There would be too much of a constraint on function; the functional sequence of one gene would have to also be a functional sequence in the other. Similarly, there would be an evolutionary constraint involved. The genes would have to evolve together. But, here we have two cases in which genes do overlap. The explanation of overlapping genes is still tentative.

There are a large number of thymine bases in a φX174 virus. In the D gene particularly, many of the codons end with the letter T. The E gene is read on a shifted frame within D so that the terminal letter of D's codons is the middle letter of E's. A look at the genetic code (see Table 11–2) will show that the codons with T in the middle (E's codons) are mainly for hydrophobic amino acids. Thus, E is a protein with detergent properties. In fact, it serves precisely as a detergent; it is the protein responsible for dissolution of the outer cell wall of the host bacterium, a process that can be accomplished in vitro by a detergent. The properties of the E gene, then, are more the properties of a general detergent molecule than of a specific enzyme.

In the A-B case, there is an indication that the two genes were once autonomous. This indication is based on the patterns of the codons where A's tend to end in T prior to the overlap, but thereafter, in the region of overlap, B's codons end in T whereas A's do not. Presumably, a mutational event tagged the B material onto the end of the earlier, shorter A and improved its enzymatic ability. We can, however, only speculate.

The amazing arrangement of this viral DNA is one of extreme economy. The protein package is small; yet a minimum of 9 genes had to be packed into it. Thus there has been extreme evolutionary pressure to economize. Not only are beginnings and ends overlapping, but also genes occur within genes—an oddity that will probably not prove to be a general rule. Only time and the further sequencing of other phages will tell. However, for the first time, we have before us the entire map of a DNA virus and can see all of the sequences and controls present.

Frederick Sanger
(1918–   )
Courtesy of Dr.
Frederick Sanger

Figure 2. Nucleotide Sequence of Phage φX174, *cs*70 and the Amino Acid Sequences of the Proteins for Which It Codes
Letters in the left-hand margin indicate the protein whose amino acid sequence is shown in the corresponding line. Restriction enzyme recognition sites are indicated by underlining. The single-letter code for the enzymes is as follows: A—*Alu*I; F—*Hinf*I; H—*Hha*I; M—*Mbo*II; P—*Pst*I; Q—*Hph*I; R—*Hind*II; T—*Taq*I; Y—*Hap*II; Z—*Hae*III.

Source: With permission from F. Sanger et al., *Journal of Molecular Biology* 125 (1978):233–242. Copyright by Academic Press Inc. (London) Ltd.

```
CCGTCAGGATTGACACCCTCCCAATTGTATGTTTCATGCCTCCAAATCTTGGAGGCTTT
     3927      3937      3947      3957      3967      3977

        MET VAL ARG SER TYR TYR PRO SER GLU CYS HIS ALA ASP TYR PHE GLU ARG
A  TTTATGGTTCGTTCTTATTACCCTTCTGAATGTCACGCTGATTATTTTGACTTTGAGCGT
        3987      3997      4007      4017      4027      4037
   mRNA end
   T1/6

   ILE GLU ALA LEU LYS PRO ALA ILE GLU ALA CYS GLY ILE SER THR LEU SER GLN SER PRO
A  ATCGAGGCTCTTAAACCTGCTATTGAGGCTTGTGGCATTTCTACTCTTTCTCAATCCCCA
        4047      4057      4067      4077      4087      4097

   MET GLY PHE HIS LYS GLN MET ASP ASN ARG ILE LYS LEU LEU GLU GLU ILE LEU SER
A  ATGGCTGGCTTCCATAAGCAGATGGATAACCGCATCAAGCTCTTGGAAGAGATTCTGTCT
        4107      4117      4127      4137      4147      4157
                                     A7b/7a    M5/8       F5c/3

   PHE ARG MET GLN GLY VAL GLU PHE ASP ASN GLY ASP MET TYR VAL ASP GLY HIS LYS ALA
A  TTTCGTATGCAGGGCGTTGAGTTCGATAATGGTGATATGTATGTTGACGGCCATAAGGCT
        4167      4177      4187      4197      4207      4217
                            Q2/3a     R4/3      Z2/6b

   ALA SER ASP VAL ARG ASP GLU PHE VAL SER VAL THR GLU LYS LEU ALA
A  GCTTCTGACGTTCGTGATGAGTTTGTATCTGTTACTGAGAAGTTAATGGATGAATTGGCA
        4227      4237      4247      4257      4267      4277

   GLN CYS TYR ASN VAL LEU PRO GLN LEU ASP ILE ASN ASN THR ILE ASP HIS ARG PRO GLU
A  CAATGCTACAACTGTTCCGACTGCCCCAGTGATTATTAATAATACTACAGACGACCCCGAA
        4287      4297      4307      4317      4327      4337
              Origin of viral strand replication

   GLY ASP GLU LYS LYS TRP PHE LEU GLU ASN GLU LYS THR VAL THR GLN PHE CYS ARG LYS LEU
A  GGGGACGAAAAATGGTTTTAGAGAACGAGAAGACGGTTACGCAGTTTTGCCGCAAGCTG
        4347      4357      4367      4377      4387      4397
                            M8/6                          A7a/4

   ALA ALA GLU ARG PRO LEU LEU ASP ILE ARG ASP GLU TYR ASN TYR PRO LYS LYS GLY
A  GCTGCTGAACGCCCTCTTAAGGATATTCGCGATGAGTATAATTACCCCAAAAAGAAAGGT
        4407      4417      4427      4437      4447      4457

   ILE LYS ASP GLU CYS SER ARG LEU LEU GLU ALA SER THR MET LYS SER ARG ARG GLY PHE
A  ATTAAGGATGAGTGTTCAAGATTGCTGGAGGCCTCCACTATGAAATCGCGTAGAGGCTTT
        4467      4477      4487      4497      4507      4517
                            Z6b/6a

   ALA ILE GLN ARG LEU MET ASN ALA MET ARG GLN ALA HIS ALA ASP GLY TRP PHE ILE VAL
A  GCTATTCAGCGTTTGATGAATGCAATGCGACAGGCTCATGCTGATGGTTGGTTTATCGT
        4527      4537      4547      4557      4567      4577

   PHE ASP THR LEU THR LEU ALA ASP ASP ARG LEU GLU ALA PHE TYR ASP ASN PRO ASN ALA
A  TTTGACACTCTCACGTTGGCTGACGACCGATTAGAGGCGTTTTATGATAATCCCAATGCT
        4587      4597      4607      4617      4627      4637
```

```
A  LEU  ARG  ASP  TYR  PHE  ARG  ASP  ILE  GLY  ARG  MET  VAL  LEU  ALA  ALA  GLU  GLY  ARG  LYS  ALA
   TTG CGT GAC TAT TTT CGT GAT ATT GGT CGT GCT GCC GAG GGT CGC AAG GCT
        4647      4657      4667      4677      4687      4697
                    F3/5a

A  ASN  ASP  SER  HIS  ALA  ASP  CYS  TYR  GLN  TYR  PHE  VAL  SER  VAL  PRO  GLU  TYR  GLY  THR  ALA  ASN
   AAT GAT TCA CAC GCC GAC TGC TAT CAG TAT TTT GTG TCC GTG CCT GAG TAT GGT ACA GCT AAT
        4707      4717      4727      4737      4747      4757
                                        A4/11

A  GLY  ARG  LEU  HIS  PHE  HIS  ALA  VAL  HIS  PHE  MET  ARG  GLY  HIS  THR  LEU  PRO  GLY  SER  VAL  ASP
   GGC CGT CTC ATT TCA TTC CAT GCG GTC ACT TTA TGC GAG GCA CTT CCT ACA GGT AGC GTT GAC
        4767      4777      4787      4797      4807      4817
     Z6a/9   M6/1                                                    R3/8

A  PRO  ASN  PHE  GLY  ARG  ARG  VAL  ARG  ARG  GLN  ASN  SER  LEU  ASN  SER  GLN  ALA  GLN  ASN  THR  TRP
   CCT AAT TTT GGT CGT CGG GGT ACG CCA ATC GCC CAG TTA AAT ACG GTA AAA TAC GTG G
        4827      4837      4847      4857      4867      4877
                                        A11/10              Z9/1C

A  PRO  TYR  GLY  TYR  SER  MET  PRO  ILE  ALA
   CCT TAT GGT TAC AGT ATG CCC ATC GCC
        4887      4897
                   mRNA start

A  GLY  TRP  LEU  TRP  PRO  VAL  ASP  ALA  LYS  ALA  LYS  TYR  GLY  GLU  PRO  LEU  LYS  ALA  THR  SER  TYR  MET  ALA
   GGT TGG TTG TGG CCT GTT GAT GCT AAA GGT GAG CCG CTT AAA GCT ACC AGT TAT ATG GCT
     Z10/3    4947      4957      4967      4977      4987      4997
                                        Q3a/1

A  VAL  GLY  PHE  TYR  VAL  ALA  LYS  TYR  VAL  ASN  LYS  LYS  SER  ASP  MET  ASP  ALA  ALA  LYS  LYS
   GTT GGT TTC TAT GTG GCT AAA ATA CGT TAA CAA AAG TCA GAT ATG GAC GCT GCT AAA
        5007      5017      5027      5037      5047      5057
                                   R8/5

A  GLY  LEU  GLY  ALA  LYS  GLU  TRP  ASN  ASN  SER  LEU  SER  LEU  THR  LYS  LEU  SER  VAL  ALA  THR  LEU  PRO  LYS  GLU
B                               MET  GLU  GLN  LEU  THR  LYS  ASN  GLN  ALA  VAL  ALA  THR  SER  GLN  GLU
   GGT CTA GGA GGC TAA AGA AAT GGA ACA ACT ACT AAA AAC CAA GCT GTC GCT ACT TCC CAA G
        5067      5077      5087      5097      5107      5117
     A12b/15b                              A15b/17

A  LYS  LEU  PHE  ARG  ILE  ARG  MET  SER  ARG  ASN  PHE  GLY  MET  LYS  MET  LEU  THR  MET  THR  ASN  SER
B  ALA  VAL  GLN  ASN  GLN  ASN  GLU  ASP  ARG  LEU  ARG  ASP  GLU  ASN  ALA  HIS  ASN  ASP  LYS  ASN
   AAG CTG TTC AGA ATC AGA ATG AGC CGC AAC TTC GGG ATG AAA ATG CTC ACA ATG ACA AAT
   A17/12a         5127      5137      5147      5157      5167
                                        F5a/6

A  LEU  SER  THR  GLU  CYS  LEU  ILE  LYS  THR  LEU  TYR  ASP  GLY  TYR  ALA  ASP  THR  PRO  PHE  ASN
B  VAL  HIS  GLY  VAL  LEU  ASN  PRO  THR  ASN  TYR  GLN  GLU  ARG  ARG  ASP  VAL  ALA  VAL  GLN  PRO
   CTG TCC ACG GAG TGC TTA ATC AAG ACT TAC GAC GGT TAC GCC GAC GCC GTT CAA AC
        5187      5197      5207      5217      5227      5237
                                        A12a/5

A  GLN  ILE  LEU  LYS  GLN  ARG  ALA  LYS  GLU  MET  ARG  LEU  ARG  LEU  GLY  LYS  VAL  THR  VAL  CYS  SER
B  ASP  ILE  GLU  GLU  ALA  LYS  LYS  LYS  ARG  ASP  GLU  ILE  GLU  ALA  LEU  GLY  ALA  LYS  SER  TYR  CYS  SER
   CAG ATA TTG AAG CAG AAC GCA AAA AAG AGA GAT GAG AGA TTG AGG CTG GGA AAA GTT ACT GTA
        5247      5257      5267      5277      5287      5297
```

```
A   ALA ASP VAL THR THR ASN LEU LEU LYS PHE MET ARG ALA SER ASP
B   ARG ARG PHE GLY ALA    THR CYS ASP ASP LYS SER ALA GLN ILE TYR ALA ARG PHE
    G C C G A C G T T T G G C G G C G C A A C C T G T G A C G A C A A C C T C A A A T C T G C T C A A A T T T A T G C G C G C T T C G
      5307      5317      5327      5337      5347      5357
        H1/15         H15/8b                              T2/7

A   ILE LYS MET ILE GLY VAL SER ASN LEU GLN SER PHE ALA SER MET THR GLN LYS LEU
B   LYS ASN ASP TRP ARG ILE GLN PRO ALA GLU PHE TYR ARG ASP HIS ASP ALA GLU VAL ASN
    A T A A A A A T G A T T G G C G T A T C C A A C C T G C A A A G T T T T A T C G C T T C C A T G A C G C A G A A G T T A
      5367      5377      1         11        21        31
        5377                P1/1                        R5/7b

A   THR LEU SER ASP ILE SER ASP GLU SER LYS ASN TYR LEU ASP LYS ALA GLY ILE THR THR
K                           MET SER ARG LYS ILE ILE LEU LEU GLN GLU LEU THR LEU LEU
B   THR PHE GLY TYR PHE ***
    A C A C T T T C G G A T A T T T C T G A T G A G T C G A A A A A T T A T C T T G A T A A A G C A G G A A T T A C T A C T
      41        51        61        71        81        91
                      F6/9  T7/8              T9/10

A   ALA CYS LEU TYR ARG ILE LEU SER TRP THR ALA LEU LEU
K   LEU VAL TYR GLU LYS ASN ARG SER GLY ILE ALA GLU LYS ASN GLU LYS ILE ARG PRO ILE
C            ***          MET ARG LYS ASN GLU                      LEU ASP PRO SER
    G C T T G T T T A C G A A T T A A A T C G A A G T G G A C T G G A A G T G C T G G C G G A A A A T G A G A A A T T C G A C C T A T
      101       111       121       131       141       151
        H8b/4  A5/18 T10/4  T8/9               T9/10

K   LEU ALA GLN LEU GLU LYS LEU LEU LEU CYS ASP LEU SER PRO SER THR ASN ASP SER VAL
C   LEU ARG SER ARG SER TYR TYR PHE ALA THR PHE ARG HIS GLN LEU THR VAL ASN SER ILE LEU SER
    C C T T G G C T C A A C T G G A A G T G C T T A C T T T G C C A C C T T T G C G A C C T T T G C G A C C A T C A A C T A A C G A T T C T G T
      161       171       181       191       201       211
        A5/18 T10/4         F9/13

K   LYS ASN ***
C   LYS THR ASP ALA LEU ASP SER LEU ALA VAL ASP LYS ASP
    C A A A A A C T G A C G C G T T G G A T G A G G A G A A G T G G C T T A A T A T G C T T G G C A C G T T C G T C A A G G
      221       231       241       251       261       271
                F13/17                      F17/16a

C   TRP PHE ARG TYR GLU HIS SER HIS GLY ARG ASP SER HIS PHE VAL HIS MET LEU VAL ASP SER ILE LEU LEU LYS
    A C T G G T T T A G A T A T G A G T C A T T T G T T C A T G G T A G A G A T T C T C A T G G T A G A G A T T C T T G T T G A C A T T T T A A
      281       291       301       311       321       331
        F13/17                                 R7b/6c

D                                                                   MET ***
C   GLU ARG GLY GLY ARG ALA VAL GLN PRO LEU ILE GLY LYS LYS SER
    A A G A G C G T T A C T A C T T G G A T T A C T G A G T C C G A T G C T G T T C A A C C A C T A A T A G G T A A G A A A T C A T
      341       351       361       371       381       391
        F16b/1  F16a/16b                              M1/7
                  mRNA start

D   SER GLN VAL THR GLU GLN SER VAL ARG PHE GLN THR ALA LEU ALA SER ILE LYS LEU ILE LYS
    G A G T C A A G T T A C T G A A C A A T C C G T A C G T T T C C A G A C C G C T T T G G C C T C T A T T A A G C T C A T
      401       411       421       431       441       451
        F16b/1                      Z3/7          A6/1

D   GLN ALA SER ALA VAL LEU ASP ASP ASP LEU THR GLU PHE ASP ASP ASP PHE THR SER ASN LYS
    T C A G G C T T C T G C C G T T T T G G A T T T A A C C G A A G A T G A T T T C G A T T T T C T G A C G A G T A A C A A A
      461       471       481       491       501       511
                          M1/7          T4/5
```

D | VAL TRP ILE ALA THR ASP ARG SER ARG ALA ARG CYS VAL GLU ALA CYS VAL TYR GLY MET VAL
E | A GTTTGGATTGCTACTGACCGCTCTCGTCGCTCGCTTGGAGGCTTGCGTTTATGG
    521   531   541   551   561   571

D | THR LEU ASP PHE VAL GLY TYR PRO ARG PHE PRO ALA LEU LEU PRO VAL GLU GLY ALA VAL SER
E | TACGCTGGACTTTGTGGGATACCCTCGCTTCCTGCCCTGTTATTGCTGCCGT
    581   591   601   611   621   631

D | ILE ALA TYR TYR VAL HIS PRO VAL ASN ILE GLN THR ALA CYS ILE MET GLU GLY GLY ALA LEU
E | CATTGCTTATTATGTTCATCCCGTCAACATTCAAACGGCCTGTATCATGGAAGGCGGC
    641   651   661   671   681   691
    R6c/7a   Z7/5   H4/13

D | GLU PHE THR GLU ASN ILE ILE LEU LEU ASN GLY VAL GLU SER VAL LYS ALA GLU ASN CYS SER
E | TGAATTTACGGAAAACATTATTACTTAATGGCGTCGAGAGCGTAAAGCCGCTGAATTGTT
    701   711   721   731   741   751
    H13/11   T5/3   Y1/3

D | ALA PHE LEU ARG THR VAL ARG ALA GLN ALA GLU ASP THR ALA LEU THR VAL LEU PHE LEU THR GLU GLU GLN LYS LYS THR
E | CGCGTTTACCTTGCGTGTGTACGCGGCAGGAAAACACTGACGCGAGAAGAAA
    761   771   781   791   801   811
                                               M7/3

J | MET SER LYS GLY LYS SER
D | VAL ARG GLN LYS LEU ASN TYR VAL ARG ALA ALA GLU ALA GLU GLU ***
E | CGTGCGTCAAAAATTACGTGCGGAAGGAGTGATGTAAATGTCTAAAGGTAAAAAACGTTCT
    821   831   841   851   861   871
    H11/14

J | GLY ALA ARG PRO GLN PRO LEU ARG GLY THR LYS LYS GLY ALA
E | GGCGGCTCGCCCTGGTGCGCCGTCCGGTTGCGAGGTACTAAAGGCAAGCGTAAAGGCGCT
    881   891   901   911   921   931
                                               H14/12

J | ARG LEU TRP TYR VAL GLY GLY GLN GLN PHE ***
E | CGTCTTTGGTATGTAGGTGGTCAACAATTTAATTGCAGGGGCTTCGGCCCCTTACTTGA
    941   951   961   971   981   991
    R7a/6b   Z5/8   Minor mRNA end

F | MET SER ASN ILE GLN THR GLY ALA GLU ARG MET PRO HIS ASP LEU SER HIS
    GGATAAATTATGTCTAATATTCAAACTGGCGCCGAGCGTATGCCGCATGACCTTTCCCAT
    1001   1011   1021   1031   1041   1051
                         H12/10

F | LEU GLY PHE LEU ALA GLY GLN ILE GLY ARG LEU ILE THR ILE SER THR THR PRO VAL ILE
    CTTGGCTTCCTTGCTGGTCAGATTGGTCGTCTTATTACCATTTCAACTACTCCGGTTATC
    1061   1071   1081   1091   1101   1111
                                               Y3/2

F | ALA GLY ASP SER PHE GLU MET ASP ALA VAL GLY ALA LEU ARG LEU SER PRO LEU ARG ARG
    GCTGGCGACTCCTTCGAGGATGGCGCTCTCGGGCTCTCCGCTCCATTGCGTCGT
    1121   1131   1141   1151   1161   1171
    F1/14b   T3/1   H10/7

452

```
F  GLY LEU ALA ILE ASP SER THR VAL ASP ILE PHE THR PHE TYR VAL PRO HIS HIS ARG HIS VAL
   GGCTTGCTATTGACTCTACTGTAGACATTTTTACTTTTTATGTCCCTCATCGTCACGTT
        Z8/4    F14b/2     1181      1191      1201      1211      1221      1231

F  TYR GLY GLU GLN TRP ILE LYS MET PHE ASP GLY VAL ASN ALA THR PRO LEU PRO THR
   TATGGTGAACAGTGGATTAAGTTCATGAAGGTGTTAATGCCACTCCTCTCCCGACT
        Q1/3c     1241      1251      1261      1271      1281      1291

F  VAL ASN THR THR GLY TYR ILE ASP HIS ALA ALA PHE ILE GLY THR ILE ASN PRO ASP THR
   GTTAACACTACTGGTTATATTGACCATGCCGCTTTCTTGGCACGATTAACCCTGATACC
        R6b/1     1301      1311      1321      1331      1341      1351

F  ASN LYS ILE PRO LYS HIS LEU PHE GLN GLY TYR LEU ASN ILE TYR ASN ASN PHE LYS
   AATAAAATCCCTAAGCATTTGTTTCAGGGTTATTTGAATATCTATAACAACTATTTTAAA
        1361      1371      1381      1391      1401      1411

F  ALA PRO TRP MET PRO ASP THR ARG GLU ALA ASN PRO ASN LEU GLU ASN GLN ASP ASP ALA
   GCGCCGTGGATGCCCGACACCCGAGAGGCTAACCCTAATGAGCTTAATCAGAGATGATGCT
        H7/5      1421      1431      1441      1451      A1/12c    1461      1471

F  ARG TYR GLY PHE ARG CYS CYS HIS LEU LYS ASN ILE TRP THR ALA PRO LEU PRO GLU
   CGTTATGGTTTCCGTTGCTGCCATCTGAAAAACATTTGGACTGCTCCGCTTCCTGAG
        1481      1491      1501      1511      1521      1531

F  THR GLU LEU SER ARG GLN MET THR THR SER ILE ASP ILE MET GLY LEU GLN
   ACTGAGCTTTCTCGCCAAATGACGACTTCTACCATCGATATTGGGTCTGCAA
        A12c/13   1541      1551      1561      1571      1581      1591

F  ALA ALA TYR ALA ASN LEU HIS THR ASP GLN GLU ARG ASP TYR PHE MET GLN ARG TYR HIS
   GCTGCTTATGCTAATTTGCATACTGACCAAGAACGTGATTACTTCATGCAGCGTTACCAT
        A13/2     1601      1611      1621      1631      1641      1651

F  ASP VAL ILE SER SER PHE GLY GLY LYS THR SER GLY TYR ASP ALA ASP ASN ARG PRO LEU LEU
   GATGTTATTTCTTCTTTTGGAGGTAAAACCTCTGGCTATGACGCTGACAACGCTCCTTTACTT
        H5/9a     1661      1671      1681      1691      1701      1711

F  VAL MET ARG SER ASN LEU TRP ALA SER GLY VAL ASP GLY THR ASP GLN THR SER
   GTCATGCGCTCTAATCTCTGGGCATCTCTGGCTATGATGGTTGATGGTACCGATCAAACGTCG
        Z4/1      1721      1731      1741      1751      1761      1771

F  LEU GLY GLN PHE SER GLY VAL VAL ARG THR LYS HIS SER VAL ASP PRO ARG PHE PHE
   TTAGGCCAGTTTTCTGGTCGTGTTCAACAGACCTATAAACATTCTGTGCCGCGTTTCTTT
        Z4/1      1781      1791      1801      1811      1821      1831

F  VAL PRO GLU HIS GLY THR MET PHE THR LEU ALA LEU VAL ARG PHE PRO PRO THR ALA THR
   GTTCCTGAGCATGGCACTATGTTTACTCTTGCGCTTGTTCGTTTTCCGCCTACTGCGACT
        H9a/8a    1841      1851      1861      1871      1881      1891
```

453

F | LYS GLU ILE GLN TYR LEU ASN ALA LYS GLY ALA LEU THR TYR THR ASP ILE ALA GLY ASP
    AAAGAGATTCAGTACCTTAACGCTAAAGGTGCTTTGACTTATACCGATATTGCTGGCGAC
    1901   1911   1921   1931   1941   1951
    F2/11

F | PRO VAL LEU TYR GLY ASN LEU PRO PRO ARG GLU ILE SER MET LYS ASP VAL PHE ARG SER
    CCTGTTTTGTATGGCAACTTGCCGCGCGAAATTTCTATGAAGGATGTTTTCCGTTCT
    1961   1971   1981   1991   2001   2011

F | GLY ASP SER SER LYS LYS PHE LYS ILE ALA GLU GLY TRP TYR TYR ARG PRO ALA PRO SER
    GGTGATTCGTCTAAGAAGTTTAAGATTGCTGAGGGTCAGTGGTATCGTTATGCGCCTTCG
    2021   2031   2041   2051   2061   2071
    Q3c/6 F11/7   H8a/6

F | TYR VAL SER PRO ALA TYR HIS LEU GLU GLU GLY GLU PRO PHE ILE GLN PRO PRO SER
    TATGTTTCTCCTGCTTATCACCTTCTTGAAGGCTTCCCATTCATTCAGGAACCGCCGTTCT
    2081   2091   2101   2111   2121   2131
    Q6/5

F | GLY ASP LEU GLN GLU ARG VAL LEU ILE ARG HIS HIS ASP TYR ASP GLN GLN CYS PHE GLN SER
    GGTGATTTGGCAAGAGAACGCGTACTTATTCGCCACCATGATTATGACCAGTTCCAGTCC
    2141   2151   2161   2171   2181   2191
    Q5/3b

F | VAL GLN LEU LEU GLN TRP ASN SER GLN VAL LYS PHE ASN VAL THR ARG ASN LEU
    GTTCAGTTGTTGCAGTGGAATAGTCAGGTTAAATTTAATGTGACCGTTATCGCAATCTG
    2201   2211   2221   2231   2241   2251
    F7/5b

F | PRO THR THR ARG ASP SER ILE MET THR SER ***
    CCGACCACTCGCGATTCAATCATGACTTCGTGATAAAAGATTGAGTGTGAATAAAGAGATTGAGTGTGAAAACG
    2261   2271   2281   2291   2301   2311

     CCGAAGCGGGTAAAAATTTAATTTTGCCGTGAGGGTTGACCAAGCGACGGTAG
     2321   2331   2341   2351   2361   2371
     R1/9   H6/3

G | MET PHE GLN THR PHE ILE SER ARG HIS ASN SER ASN PHE
    GTTTTCTGCTTAGGAGTTAATCATGTTCAGACTTTCAGACTTATTTTCCGTCATAATTCAAAACT
    2381   2391   2401   2411   2421   2431

G | PHE SER ASP LYS LEU VAL LEU THR THR VAL THR PRO ALA SER VAL PRO VAL LEU GLN
    TTTTTCTGATAAGGCTGGTTCTCACTTCTGTTACTCCAGCTTCTTCGGCACCTGTTTTAC
    2441   2451   2461   2471   2481   2491
    A2/16   A16/15a   M4/10

G | THR PRO LYS LYS ALA THR LEU ILE SER SER THR LEU TYR PHE ASP SER SER LEU THR VAL ASN ALA GLY ASN
    AGACACCTAAAGCTACACATCACATCGTCAACTCGTTATATTTTGACGTTAATGCTGGTA
    2501   2511   2521   2531   2541   2551
    A15a/3   R9/10

G | GLY GLY PHE HIS LEU ILE GLN MET ASP THR SER VAL ASN ALA ALA ASN GLN VAL VAL
    ATGGGTTGGTTTTCTTCATTGCATTCATTCAGATGGATACATCTGTCAACGCGCTAATCAGGTTG
    2561   2571   2581   2591   2601   2611
    M10/9   R10/2

```
G   SER VAL GLY ALA ASP ILE ALA PHE ASP ALA ASP PRO LYS PHE PHE ALA CYS LEU VAL ARG
    TTTCTGTTGGTGCTGATATTGCTTTTGATGCCGACCCTAAATTTTTGCCTGTTTGGTTC
            2621      2631      2641      2651      2661      2671

G   PHE GLU SER SER SER VAL PRO THR THR LEU PRO THR ALA TYR ASP VAL TYR PRO LEU ASN
    GCTTTGAGTCTTCGGTTCCGGTTCCGACTACCCTCCGACTGCCTATGATGTTATCCTTTGA
            2681      2691      2701      2711      2721      2731
         F5b/8   M9/2

G   GLY ARG HIS ASP GLY TYR TYR THR VAL LYS ASP CYS VAL THR ILE ASP VAL LEU PRO
    ATGGTGCGCCATGATGGTTATTATACCGGTGTGACTTGTGACTGTGACGTCCTTC
            2741      2751      2761      2771      2781      2791

G   ARG THR PRO GLY ASN ASN VAL TYR VAL GLY PHE MET TRP SER ASN PHE THR ALA THR
    CCCGTACGCGGGCAATAACGTTTATGTTGGTTTCATGGTTGGTCTAACTTTACCGCTA
            2801      2811      2821      2831      2841      2851
         Y2/5

G   LYS CYS ARG GLY LEU VAL SER LEU ASN GLN VAL ILE LYS GLU ILE ILE CYS LEU GLN PRO
    CTAAATGCCGCGGATTGGTTCGCTGAATCAGGTTATTAAAGAGATTATTGTCTCCAGC
            2861      2871      2881      2891      2901      2911
                              F8/4

H   LEU LYS ***
G                   MET PHE GLY ALA ILE ALA GLY ILE LEU ALA
    CACTTAAGTGAGGTGATTATGTTTGGTGCTATTGCTGGCGGTATTGCTGGCGGTATTGCTGGC
            2921      2931      2941      2951      2961      2971
         Q3b/4

H   GLY GLY ALA MET SER LYS LEU PHE GLY GLY GLY MET ALA ALA SER GLY ILE GLN
    TGGTGGCGCCATGTCTAAATTGTTTGGAGGCGGTCAAAAAGCCGCCTCCGGTGGCATTCA
            2981      2991      3001      3011      3021      3031
         H3/2                                              Y5/4

H   GLY ASP VAL LEU ALA THR ASP ASN THR VAL GLY MET GLY ASP ALA GLY ILE LYS SER
    AGGTGATGTGCTTGCTACCGATAACAATACTGTAGGCATGGCTGATGGTTATTAAATC
            3041      3051      3061      3071      3081      3091
         Q4/7                              Q7/2

H   ALA ILE GLN GLY SER ASN VAL PRO ASN PRO ASP GLU ALA ALA PRO SER PHE VAL SER GLY
    TGCCATTCAAGGCTCTAATGTTCCTAACCCTGATGAGGCCGCCCCTAGTTTTGTTTCTGG
            3101      3111      3121      3131      3141      3151
         A3/9                              Z1/2

H   ALA MET ALA LYS ALA GLY LYS GLY LEU LEU GLU GLY THR LEU GLN ALA GLY THR SER ALA
    TGCTATGGCTAAAGCTGGTAAAGGACTTCTTGAAGGTACGTTGCAGGCTGGCACTTCTGC
            3161      3171      3181      3191      3201      3211

H   VAL SER ASP LYS LEU LEU ASP VAL GLY GLY LEU GLY LYS GLY LYS SER ALA ALA ASP LYS GLY
    CGTTTCTGATAAGTTGCTTGATTTGGTTGGTGGTGGACTTGGTGGCAAGTCTGCCGCTGATAAAGG
            3221      3231      3241      3251      3261      3271
```

```
H  LYS ASP THR ARG ASP TYR LEU ALA ALA ALA PHE PRO GLU LEU ASN ALA TRP GLU ARG ALA
   A A A G G A T A C T C G T G A T T A T C T T G C T G C T G C A T T T C C T G A G C T T A A T G C T T G G G G A G C G T G C
         3281      3291      3301      3311      3321      3331
                                        A9/12d

H  GLY ALA ASP ALA SER ALA GLY MET VAL ASP ALA GLY PHE GLU ASN GLN LYS ASN GLN LYS   GLU LEU
   T G G T G C T G A T G C T T C C T C T G C T G G T A T G G T T G A C G C C G G A T T T G A G A A T C A A A A G A A C C A G A A A A A G A G G C T
         3341      3351      3361      3371      3381      3391
                             R2/6a Y4/1        F4/14a              A12d/7c

H  THR LYS MET GLN ASP ASN LEU GLN ILE GLU ALA ASN GLN THR ALA ASN GLU GLN LYS
   T A C T A A A A T G C A A A C T G G A C A A T C A G A A A A T T G C C G A G A T G C A A A A T G C A A A A G A C T C A A A A A
         3401      3411      3421      3431      3441      3451
                                        F14a/12

H  GLU ILE ALA GLY ILE GLN SER ALA THR SER ARG GLN ASN ASP ASP VAL TYR ALA
   A G A G A T T G C T G G C A T T C A G T C G G C G A C T T C A C G C C A G A A T A C G A A A G A C C A G G T A T A T G C
         3461      3471      3481      3491      3501      3511

H  GLN ASN GLU MET GLU MET LEU LYS GLN TYR ARG ALA ASN SER VAL ALA SER ILE MET
   A C A A A A T G A G A T G A A T G A T G C T T A T C A G A A A G G A A G T G C T T C T G C G T T G C G T T C T A T T A T
         3521      3531      3541      3551      3561      3571
                                  F12/10

H  GLU ASN THR ASN LEU SER LYS GLN GLN GLN VAL SER GLU ILE MET ARG GLN MET LEU THR
   G G A A A A C A C C A A T C T T T C C A A G C A A C A G C A G G T T T C C G A G A T T A T G C G C C A A A T G C T T A C
         3581      3591      3601      3611      3621      3631
                                              H2/9b

H  GLN ALA GLN THR ALA GLY GLN TYR GLN THR ASN PHE THR ASP SER GLN ILE GLU LYS ILE GLU MET THR ARG LYS
   T C A A G C T C A A A C C G C T G G T C A G T A T C A G G G T C A G T A T T T T A C C A A T G A C C A A T C A A A A G A A A T C A A A G A A T G A C C G A A A A T C A A A A G A G A C T G C G C C A A
         3641      3651      3661      3671      3681      3691
        A7c/8                                                                   F10/15

H  VAL SER ALA GLU VAL ASP LEU VAL HIS GLN GLN THR GLN ASN GLN ARG TYR GLY SER SER
   G G T T A G T G C T G A G G T T G A C T T A G T T C A T C A G C A A A C G C A G A A T C A G C G G T A T G G C T C T T C
         3701      3711      3721      3731      3741      3751
               R6a/4                        F15/5c                           M2/5

H  HIS ILE GLY ALA THR ALA LYS ASP ILE SER ASN VAL VAL THR ASP ALA ALA SER GLY VAL
   T C A T A T T G G C G C T A C T G C A A A G G A T A T T T C T A A T G T C G T C A C T G A T G C T G C T T C T G G T G T
         3761      3771      3781      3791      3801      3811
        H9b/1

H  VAL ASP ILE PHE HIS GLY ILE ASP LYS ALA VAL ALA ASP THR ASP TRP ASN PHE TRP LYS
   G G T T G A T A T T T T T C A T G G T A T T G A T A A A G C T G T T G C C G A T A C T G G T T G G A A C A A T T T C T G G A A
         3821      3831      3841      3851      3861      3871
                            A8/14

H  ASP GLY LYS ALA ASP GLY ILE GLY SER ASN LEU SER ARG LYS ***
   A G A C G G T A A A G C T G A T G G T A T T G G C T C T A A T T T G T C T A G G A A A T A A
         3881      3891      3901      3911
        A14/7b
```

456

CHAPTER SUMMARY

Most bacterial genes are organized into operons, which can be either repressible or inducible. Inducible operons, such as *lac,* begin transcription when the metabolite, upon which the operon enzymes act, appears in the environment. The metabolite, or inducer, combines with the repressor, the product of the independent regulator gene, and renders it nonfunctional. In the absence of the inducer, the repressor binds to the operator, a segment between the promoter and the first gene of the operon. In place, the repressor blocks transcription. After combining with the inducer, the repressor is removed from the operator and transcription proceeds.

All operons responsible for the breakdown of sugars in *E. coli* are inducible. In the presence of glucose, other inducible sugar operons (such as the arabinose and galactose operons) are repressed, even if their sugar appears in the environment. The process is called *catabolite repression* and functions because the other operons have enhanced transcription in the presence of cyclic AMP and a catabolite activator protein (CAP). Glucose lowers the level of cyclic AMP in the cell and thus inhibits transcription of these other operons.

Repressible operons, such as the *his* operon in *Salmonella* and the *trp* operon in *E. coli,* have the same basic components as an inducible operon—polycistronic transcription controlled by an operator site between the promoter and the first structural gene. However, the repressor protein, controlled by an independent regulator gene, is only functional in blocking transcription after it has combined with the corepressor, which is the end product of the operon's pathway or some form of the end product (tryptophan in the *trp* operon, histidinyl-tRNA in the *his* operon). In addition, amino acid synthesizing operons also have an attenuator region. The ability of a ribosome to translate a leader peptide gene determines the secondary structure of the mRNA transcript. If the ribosome can translate the leader peptide gene, then there must be adequate quantities of the amino acid present. A terminator stem and loop forms in the mRNA and causes the termination of transcription.

Early and late promoters in phage are another form of transcriptional control. Control of gene expression in λ phage is well known. How the decision for lytic versus lysogenic response is determined is not clear.

Translational control can be exercised through a gene's position in an operon—genes at the 5′ end are transcribed most frequently—or through redundancy of the genetic code or by way of the stringent response. Post-translational control is primarily by feedback inhibition.

Recombinant DNA methods involve the insertion of a segment of DNA (passenger) inside a vehicle plasmid (F factor or λ phage) and the incorporation of the hybrid plasmid by a bacterium. These methods make it possible to obtain a large amount of experimental DNA and provide an opportunity for studying the control mechanisms of transcription.

EXERCISES AND PROBLEMS

1. Are the following *E. coli* cells constitutive or inducible for the z gene?
   a. $i^+o^+z^+y^+a^+$
   b. $i^-o^+z^+y^+a^+$
   c. $i^-o^cz^+y^+a^+$
   d. $i^+o^cz^+y^+a^+$
   e. $i^so^+z^+y^+a^+$
   f. $i^{Q+}z^+y^+a^+$

2. Given the following *lac* operon merozygotes, where the F′ component is given, determine which form or forms of the z gene (z^+, z^-) are transcribed either (i) with or (ii) without inducer present:
   a. $i^+o^+z^+y^+a^+/$F′ $i^+o^+z^+y^+a^+$
   b. $i^-o^+z^+y^+a^+/$F′ $i^+o^+z^-y^+a^+$
   c. $i^+o^+z^+y^+a^+/$F′ $i^-o^+z^-y^+a^+$
   d. $i^-o^cz^-y^+a^+/$F′ $i^+o^+z^+y^+a^+$
   e. $i^-o^+z^-y^+a^+/$F′ $i^+o^cz^+y^+a^+$

3. Construct a merozygote of the *his* operon in *Salmonella* with two forms of the first gene (G gene) in the operon. Describe the types of *cis* and *trans* effects that are possible, given mutants of any component of the operon, including mutants of the histidine tRNA gene. Can this repressible system work for any type of operon other than those controlling amino acid synthesis?

4. Describe the interaction of the attenuator and the operator control mechanism in the *trp* operon of *E. coli* under varying concentrations of tryptophan in the cell.

5. What is the fate of a λ phage entering an *E. coli* cell that contains quantities of λ repressor? What is the fate of the same phage entering an *E. coli* cell that contains quantities of *cro*-gene product?

6. Describe the fate of λ phages, with mutants in the following genes, during the infection process: *cI, cII, N, cro, att, Q*

7. What is the fate of λ phages, with mutants in the following areas, during the infection process: o_{R1}, o_{R3}, p_L, p_{RE}, p_{RM}, p_R

8. Prepare a table of the steps from transcription through translation to enzyme function and note all the points at which control can be exerted.

9. Outline the steps needed to get *E. coli* to produce human insulin.

10. Describe the role of cyclic AMP in transcriptional control in *E. coli*.

11. Operon systems exert negative control in the sense that they act through inhibition. The CAP system is a positive control system because it acts through enhancement of a process. Describe how an operon could work if it were dependent only upon positive control.

SUGGESTIONS FOR FURTHER READING

Abelson, J. 1977. Recombinant DNA: Examples of present day research. *Science* 196:159–160. (Followed by 24 articles of current research methods and results.)

Abelson, J., and E. Butz, eds. 1980. Recombinant DNA. *Science* 209: 1317–1438.

Beckwith, J., and D. Zipser. 1970. *The Lactose Operon.* Cold Spring Harbor, N.Y.: Cold Spring Harbor Laboratory.

Brenner, M., and B. Ames. 1971. The histidine operon and its regulation. In H. Vogel, ed., *Metabolic Regulation.* New York: Academic Press.

Davis, B. 1977. The recombinant DNA scenarios: Andromeda strain, chimera, and Golem. *Amer. Scientist* 65:547–555.

Dickson, R., et al. 1975. Genetic regulation: The *lac* control region. *Science* 187:27–35.

Fiddes, J. 1977. The nucleotide sequence of a viral DNA. *Sci. Amer.,* December, pp. 54–67.

Gemmill, R., et al. 1979. *Leu* operon of *Salmonella typhimurium* is controlled by an attenuation mechanism. *Proc. Nat. Acad. Sci., USA* 76:4941–4945.

Gilbert, W., and B. Müller-Hill. 1966. Isolation of the *Lac* repressor. *Proc. Nat. Acad. Sci., USA* 56:1891–1898.

Herskowitz, I. 1973. Control of gene expression in bacteriophage lambda. *Ann. Rev. Genet.* 7:289–324.

Jacob, F., and J. Monod. 1961. Genetic regulatory mechanisms in the synthesis of proteins. *J. Mol. Biol.* 3:318–356.

Johnston, H., et al. 1980. Model for regulation of the histidine operon of *Salmonella. Proc. Nat. Acad. Sci., USA* 77:508–512.

Lewin, B. 1974. *Gene Expression, 1: Bacterial Genomes.* New York: Wiley.

Lewin, B. 1977. *Gene Expression, 3: Plasmids and Phages.* New York, Wiley.

Jonathan R.
Beckwith (1935–   )
Courtesy of Dr.
Jonathan R. Beckwith

Oxender, D., G. Zurawski, and C. Yanofsky. 1979. Attenuation in the *Escherichia coli* tryptophan operon: Role of RNA secondary structure involving the tryptophan codon region. *Proc. Nat. Acad. Sci., USA* 76:5524–5528.

Ptashne, M., et al. 1976. Autoregulation and function of a repressor in bacteriophage lambda. *Science* 194:156–161.

Sanger, F., et al. 1978. The nucleotide sequence of bacteriophage ϕX174. *J. Mol. Biol.* 125:225–246.

13 Gene Expression: Control in Eukaryotes (Developmental Genetics)

Chapter 12 considered the control of gene expression in prokaryotic organisms and discussed the regulatory processes of prokaryotes and bacteriophages. Bacteriophages have two possible life cycles: lytic and lysogenic. When engaged in lysogeny, the genes required for phage replication are repressed; when virulent, the phage express all of their genes and produce more copies of the same virus. The DNA sequencing of ϕX174 and other phages has greatly increased our knowledge of the regulatory processes in phage.

Bacteria, compared with eukaryotes, are relatively uncomplicated. It is estimated that we already know more than half of the enzymes and processes of an *Escherichia coli* cell. Many control systems, such as operons, are well understood. In fact, it is now possible to imagine having sometime in the near future the *E. coli* genome mapped as has been done with the ϕX174 genome. It is not intended here to underplay the genetic complexity of prokaryotes but to set the stage for an appreciation of the complexity of eukaryotes. For the jump from the study of the biochemical control mechanisms of gene function in prokaryotes to the mechanisms in eukaryotes is a leap of the greatest magnitude.

EUKARYOTIC CELL

Eukaryotes and prokaryotes are the two superkingdoms of organisms. Prokaryotes lack a nuclear membrane and lack the process of meiosis. They include bacteria and blue-green algae (bacteria). Eukaryotes have nuclei, undergo mitosis, and include all higher organisms (protists, plants, fungi, and animals). The following examples show that eukaryotes are generally much more complex than prokaryotes:

461

1. *E. coli* contains approximately 4×10^{-12} mg of DNA. The haploid human genome contains nearly 1000 times as much DNA.

2. *E. coli*'s DNA is known to be a single, naked circle. Eukaryotic DNA is in the form of nucleoprotein. While it is now believed that each eukaryotic chromosome has a single linear DNA double helix and while great strides are being made in determining substructure, a complete picture of DNA and protein interaction has not yet emerged.

3. *E. coli* has very little internal structure. Eukaryotes have extensive lipid membrane systems, including the nuclear envelope itself.

4. *E. coli* is small. Eukaryotic cells are generally larger than prokaryotes.

5. The messenger RNA of *E. coli* is translated while it is being transcribed. Eukaryotic mRNA is modified within the nucleus at both ends by post-transcriptional additions. It is combined with proteins and then transported out of the nucleus for translation in the cytoplasm.

6. No mRNA isolated from eukaryotic cells, including the mRNA of animal viruses, has been found to be polycistronic (containing many genes). Most prokaryotic mRNAs are polycistronic.

7. Most *E. coli* genes are parts of inducible or repressible operons. There is very little evidence for operons or operon-type controls in eukaryotes.

8. *E. coli* exists as a simple, single cell. While some prokaryotes do have aggregation processes, sporulation processes, and a few other limited forms of differentiation, they are primarily one-celled organisms. The essence of eukaryotes, however, is differentiation. In humans a zygote gives rise to every other cell type in the body in an orderly, predictable manner.

In considering the details of the complexities encountered in the genetic control processes of eukaryotes, we will begin with the relationship of the nucleus and the cytoplasm during development and will then determine if development occurs as a result of loss or permanent change of chromosome material. Our focus will then shift to the structure of the eukaryotic chromosome and what is known about the manner in which the expression of its genes is controlled. The chapter closes with a look into three areas of developmental genetics: cancer, endocrinology, and immunology.

NUCLEAR-CYTOPLASMIC INTERACTIONS

Development is the orderly sequence of changes that produce increased complexity during the growth of an organism. The cytoplasm plays an important role in early development. It is, however, a product of the genome of the cell that produced it. Thus, the problem of development is not solved by discovering the roles of cytoplasmic ingredients. We are still left with the problem of how the genome is regulated to produce these ingredients.

Cap Formation in *Acetabularia*

An example of the role played by the cytoplasm in development is found in *Acetabularia*, which is a unique single-celled alga that grows to a size of several centimeters. *Acetabularia* is quite amenable to certain kinds of manipulation. Each individual consists of a stalk, a cap, and root-like rhizoids. When, in accordance with J. Hämmerling's classical experiments, the cap and the rhizoid with its nucleus are removed, the stalk eventually regenerates a new cap. The regeneration in the absence of the nucleus shows that the proximate control of cap development emanates from the cytoplasm. Further experimentation involves transplanting the nucleus (within the rhizoid) from one species onto the stalk of another species. Figure 13–1 shows experiments using two species with distinct caps.

If a stalk has its cap removed and has its rhizoid replaced by a rhizoid and nucleus from another species, the stalk will regenerate a new cap that is either similar to the original or intermediate between the cap types of the two species. If the cap is then removed again, the newly regenerated cap will

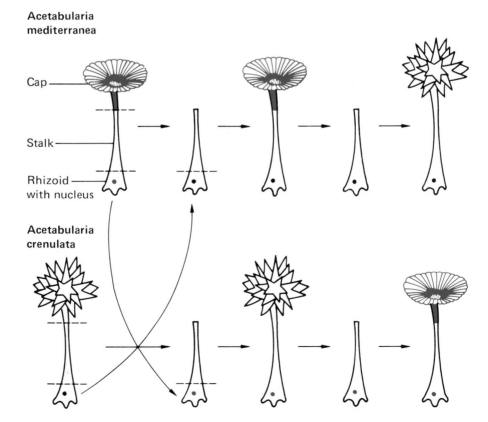

Figure 13–1. Nuclei in Rhizoids Transplanted between *Acetabularia* Species

be specific to the new nucleus. The experiment demonstrates that development is under the control of something in the cytoplasm that is relatively long lived (it takes several months for the cap to regenerate). However, the ultimate control is with the nucleus. Thus, while we can say that development is controlled by cytoplasmic substances, we are still left with explaining how the nucleus produced, or controlled the production of, these substances.

The prime candidate for cytoplasmic control of development would be messenger RNA. In fact, the ability of an anucleate stalk to regenerate the cap can be prevented by RNase. J. Brachet and others demonstrated that mRNA is produced by the nucleus and migrates to and accumulates in the apical tip of the alga where the cap will be formed. This mRNA is long lived and directs the protein synthesis that forms the cap. Thus, cytoplasmic control of development is in the form of mRNA, the synthesis of which was directed by the genome. There remain two basic questions concerning development: (1) How does the mRNA act to direct the production of a structure as complex as a cap and (2) what were the nuclear controls that directed that only certain mRNAs were to be transcribed? At present, we cannot give complete answers to these questions.

Gray Crescent in Amphibian Oocytes

Many other examples of cytoplasmic substances having direct effects on development are known. For example, the amphibian oocyte is a very large, complex cell. Shortly after the initiation of development, usually by fertilization (although pin-pricking and other stimuli are also effective), major reorganizations of cytoplasmic materials occur. Two examples from the amphibia will suffice to demonstrate these reorganizations: the gray crescent and the germ-cell determining factor.

The **gray crescent** of frogs and some salamanders is a cortical region that forms just after fertilization on the side of the egg opposite to the side penetrated by the sperm (Figure 13–2). It will mark the site of the dorsal lip of the opening (the **blastopore**) into the future gut. [The amphibian oocyte has already differentiated into pigmented (animal) and nonpigmented (vegetal) halves.] This gray crescent is extremely important in development.

Figure 13–2.
Formation of the Gray Crescent in an Amphibian Oocyte Just after Fertilization
The gray crescent will mark the site of the dorsal lip of the blastopore.

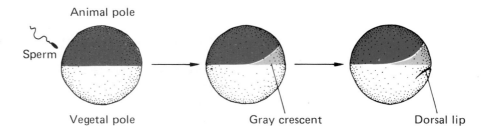

Animal pole

Sperm

Vegetal pole

Gray crescent

Dorsal lip

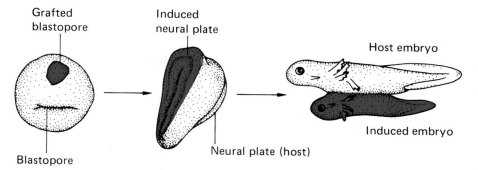

Figure 13–3.   Transplanting Part of a Gray Crescent to a Second Amphibian Oocyte to Induce the Formation of a Second Nervous System and Eventually a Second Embryo

If the region of the gray crescent is removed from one oocyte and implanted below the surface of another in a region where it does not belong, it will induce the formation of a nervous system at the site where it was transplanted (Figure 13–3). Note that the gray-crescent region is cortical egg material and contains no chromosomes or genetic material.

Recall that a developing embryo goes through several stages, the first of which is the **blastula**. The cells making up this blastula are called **blastomeres**. A blastula can be separated into two parts at the two-cell stage. In most cases, the first cleavage furrow will bisect the gray crescent so that the two separated blastomeres each contain gray-crescent material. However, occasionally the first cleavage furrow will have formed so that one of the two cells contains all of the gray crescent and the other cell, none of it. When this type of blastula is separated into two single cells, the developmental course of the two cells will be different (Figure 13–4). When the gray-crescent material is present, a frog blastomere develops into a perfectly normal tadpole. Without the gray-crescent material, there is no differentiation at all.

Germ-Cell Determination in
Amphibians

A second example of the developmental role of the cytoplasm in the amphibian oocyte is the germ-cell organizer material. Shortly after fertilization, a cytoplasmic material that will determine which cells become germ cells becomes localized at the bottom (vegetal pole) of the fertilized egg. No cells in the amphibian will be germ cells except cells that incorporate this localized material. Shining ultraviolet (UV) light on the vegetal pole will inactivate this material and result in an embryo without germ cells. RNase will also inactivate this material, indicating that it is probably mRNA.

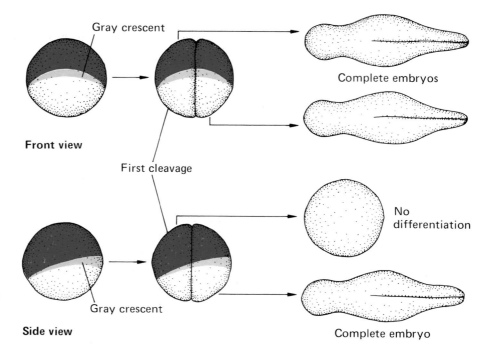

Figure 13 –4.
Development of an
Amphibian
Blastomere Depends
on the Presence of
Gray Crescent
Material

CHROMOSOME LOSS OR ALTERATION DURING DEVELOPMENT

The next question that could be asked about eukaryotic development is whether development proceeds by the sequential loss of genetic material from the chromosome. Karyotyping of tissues can, for the most part, demonstrate that no chromosome loss occurs—for example, the human pancreas cell has the same number of chromosomes as the spleen cell. Although there are some notable exceptions—liver cells tend to be polyploid, heart muscle is usually syncytial (multinucleate), and many cancer cells are aneuploid—eukaryotic development is not associated with chromosome changes.

That perhaps there are subtle changes in chromosomes during development that are not observable simply by karyotyping cells has been explored by the method of nuclear transplantation. In this method nuclei of differentiated cells are put into zygotes that are about to begin development. If the transplanted nuclei can support normal development, then it is demonstrated that development does not proceed by permanent chromosomal differentiation.

Briggs and King perfected a technique in frogs to illustrate this point. An egg can be activated to begin development and then enucleated to remove all of its original genetic material. As shown in Figure 13–5, a needle

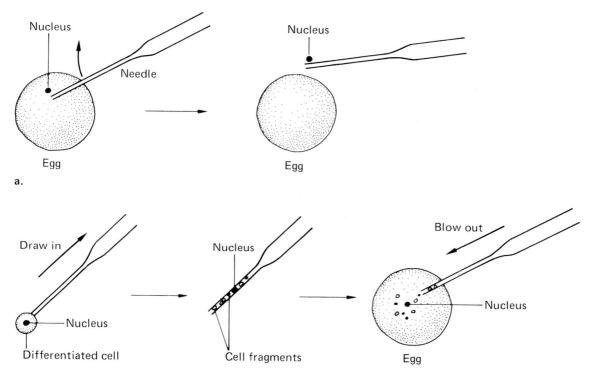

Figure 13 –5. Technique of Nuclear Transplantation

is used to activate development of the frog egg and then to remove the nucleus. Nuclei from tissues of more developed embryos can then be inserted into the enucleated egg. See Figure 13–5, which illustrates how a differentiated tissue cell is drawn into a pipette that has an inner diameter smaller than the cell. The cell is destroyed but the nucleus remains intact and is injected into the enucleated egg. The process produces an egg developing under the control of a nucleus from a known level of differentiation. When nuclei from late blastula cells are used, development proceeds normally. However, as nuclei from later stages of development are used, the proportion of cells that develop into normal embryos declines. In *Xenopus,* the African clawed toad, nuclei from the differentiated intestine will support normal growth. Steward and his colleagues have shown that single cells from the petiole or the root of a carrot, when grown in cell culture, will produce embryoids, embryo-like structures that will produce perfectly normal carrots when transplanted to soil.

Several conclusions can be drawn from these and similar studies. Cases such as the carrot and the clawed toad demonstrate that cell differentiation can and does take place without any permanent change in the genetic material. There is no loss or subtle change of chromosome parts to account for differentiation. The case of the frog, however, demonstrates that permanent changes can occur during development. While these changes may only be age-related phenomena whereby the nuclei from more differentiated tissue do not survive the experimental technique or do not undergo mitosis at a fast enough rate, we must be careful about oversimplifying our conclusions about development. Nevertheless, it does seem safe to suggest that we must look elsewhere than to simple loss of genetic material to account for organismal development; we must direct our attention to the structure of the eukaryotic chromosome.

EUKARYOTIC CHROMOSOME

DNA Arrangement

Evidence that the eukaryotic chromosome contains one double helix of DNA (is **uninemic**), which replicates semiconservatively, comes from several sources, the best of which is radioactive-labeling studies. If a eukaryote is allowed to undergo one DNA replication in the presence of tritiated (^{3}H) thymidine, it is to be expected that each of the daughter chromatids would contain a double helix with one unlabeled template strand and one labeled strand of newly synthesized bases. The configuration is expected on the basis

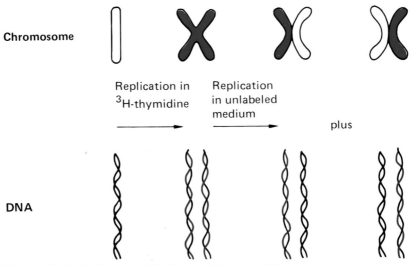

Figure 13–6.   Radioactive Labeling of a Uninemic Eukaryotic Chromosome Following Semiconservative Replication
Replication occurs first in the presence of ^{3}H-Thymidine and then in its absence. Color represents labeling.

of semiconservative replication, with each chromatid containing one double helix. A second round of replication, in the absence of ^{3}H-thymidine, should produce chromosomes where, prior to cell division, one chromatid would have unlabeled DNA and one would have labeled DNA (Figure 13–6).

The experiment has been done using the plant *Vicia faba,* which has a diploid number of $2n = 12$. The cells were grown in colchicine to prevent the segregation of chromosomes. The results were as predicted. Figure 13–7 shows the chromosomes after a division in nonlabeled media. As expected, one chromatid of every figure is labeled and one is not. Thus, the evidence is in complete concordance with the simple uninemic model of eukaryotic chromosome structure.

Nucleoprotein Composition

The constituency of calf thymus chromatin is given in Table 13–1. The **histones,** a group of arginine- and lysine-rich basic proteins, have been rela-

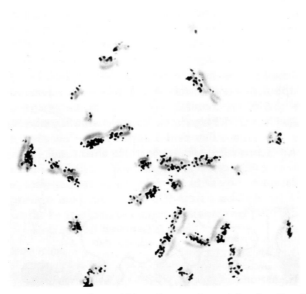

Figure 13 –7.   Second Metaphase in *Vicia faba* after One Replication in the Presence of ^{3}H-Thymidine
Cases where the label apparently switches from one chromatid to the other are caused by sister chromatid exchange.

Source: G. Marin and D. M. Prescott, "The Frequency of Sister Chromatid Exchanges Following Exposure to Varying Doses of H^3-Thymidine or X-ray," *Journal of Cell Biology* 21 (1964):159–167. Reproduced by permission.

TABLE 13-1.   The Constituency of Calf Thymus Chromatin

Constituent	Relative Weight[a]
DNA	100
Histones	114
Nonhistone proteins	33
RNA	1

[a] Weight relative to 100 units of DNA.

tively well characterized. At first, it was thought that the selective binding of these proteins to DNA might be the mechanism of transcriptional control. It is now known that histones are too homogeneous to act as selective control proteins. Table 13–2 provides details of histone composition. The sequence of one of the histones, H4, is shown in Figure 13–8.

Histones interact with DNA to form **nucleosomes** (Figure 13–9). A nucleosome consists of a section of DNA that is 166 base pairs long circled twice around an octamer of histones containing two each of histones H2A, H2B, H3, and H4. Histone H1 is not an integral part of the nucleosome but is associated with it (Figure 13–10). Since the length of DNA involved in a nucleosome is short and since the arrangement of histones is so regular, it is assumed that nucleosomes are nonspecific arrangements of the DNA. Nucleosomes, then, are a first-order packaging of DNA; they reduce its length and probably make the coiling and contraction required during mitosis and meiosis more efficient.

If the histones are removed from a chromosome, the DNA billows out and leaves a proteinaceous structure termed a **scaffold.** This scaffold structure is formed from **nonhistone proteins,** which are composed of 12 to 20 types of protein with very limited heterogeneity. Thus, the major nonhistone proteins are probably also involved in chromosome structure rather than in genetic control. However, the methods used to analyze the nonhistone proteins are, for the most part, incapable of separating proteins making up less than 1% of the chromatin protein. Thus, gene expression is most probably controlled by proteins that exist in very small quantities in the chromatin.

TABLE 13 –2.   Composition of Histones

Fraction	Class	Number of Amino Acids	Percentage of Basic Amino Acids
H1	Very lysine rich	213	30
H2A	Lysine, arginine rich	129	23
H2B	Moderately lysine rich	125	24
H3	Arginine rich	135	24
H4	Arginine, glycine rich	102	27

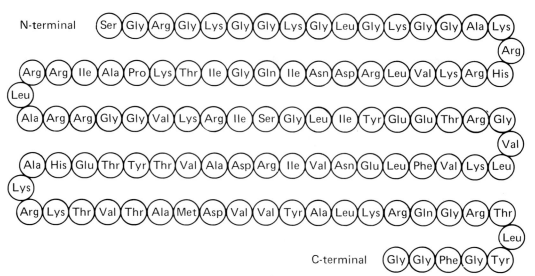

Figure 13 –8.   Amino Acid Sequence of Histone H4 from Calf Thymus Chromatin

The N-terminal serine is acetylated; the internal lysines may or may not be acetylated or methylated. Hence there is some variation in the molecules in respect to modification.

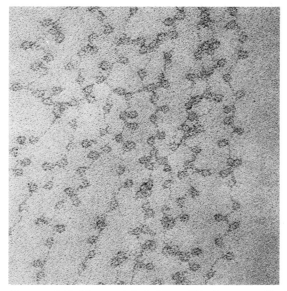

Figure 13 –9. Electron Micrograph of Chromatin Fibers Photo shows nucleosome structures (spheres) and connecting strands called *linkers.*

Source: D. E. Olins and A. L. Olins, "Nucleosomes: The Structural Quantum in Chromosomes," *American Scientist* 66 (November 1978):704–711. Reproduced by permission.

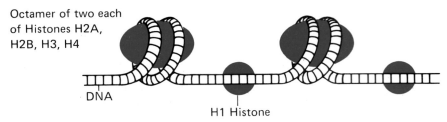

Octamer of two each
of Histones H2A,
H2B, H3, H4

DNA

H1 Histone

Figure 13-10.   Presumed Structure of Nucleosomes
Two loops of DNA interact with two copies each of histones H2A, H2B,
H3, and H4. Histone H1 is associated with the DNA strands (linkers)
between nucleosomes.

Many models have been developed to account for the higher-order
structure of the chromosome. Most of these models differ in the arrangement
of the fibers seen as the basic structural unit when chromosomes are examined carefully under high-power magnification (Figure 13–11). These fibers
are 200 to 500 Å wide, depending on the method of fixation. They can be
considered the basic 250 Å fiber of the chromosome and probably result from
the supercoiling of the nucleosomed double helix. On the basis of photos
such as Figure 13–11, and assuming one continuous fiber, DuPraw has suggested several models of chromosome structure. These models explain the
fact that no free ends of chromosome fibers are seen in electron micrographs
(Figure 13–12) and are also useful in explaining unique chromosome config-

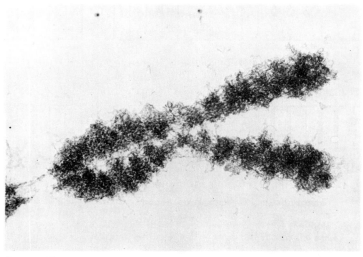

**Figure 13 -11.
Chinese Hamster
Chromosome**
magnification
20,000X

Source: Courtesy of Dr. Hans Ris.

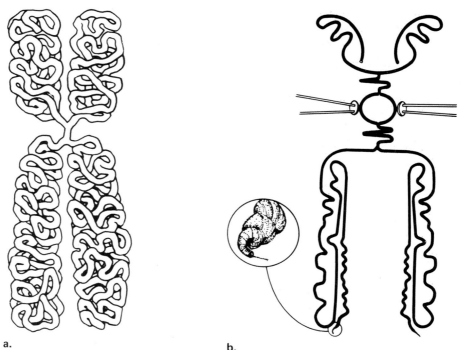

Figure 13–12.
Two Possible Models, Highly Diagrammatic, of the Configuration of the 250 Å Fiber within the Chromosome
In (b) the fiber can run up and down the chromosome arm several times.

a.   b.

Source: Reproduced by permission of Dr. E. J. DuPraw.

urations such as polytene chromosomes (with their associated puffs and Balbiani rings) and lampbrush chromosomes.

Polyteny, Puffs, and Balbiani Rings. The salivary glands, as well as some other tissues of *Drosophila* and other diptera, contain giant banded chromosomes (Figure 5–11). These chromosomes are the result of the synapsis of homologues followed by replication of the chromosomes without cell division (endomitosis). They represent more than 1000 copies of the same chromatid and appear as alternating dark bands and lighter interband regions. The structure of the polytene chromosome can be explained by such a diagram as is shown in Figure 13–13—dark bands are due to tight coiling of the 250 Å thread; light regions are due to looser coiling. The figure also shows how **chromosome puffs** would come about by the unfolding of fibers in regions of active transcription. **Balbiani rings** were originally defined as puffs specifically in *Chironomus,* whose polytene chromosomes were discovered by E. G. Balbiani in 1881. Currently, the term is used synonymously with all puffs, or at least the larger puffs, in all species with polytene chromosomes.

Methods specific for RNA—staining with toluidine blue or autoradiog-

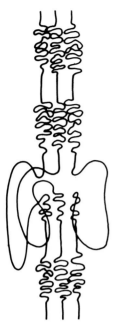

Figure 13 –13.
Polytene
Chromosome
Showing Parallel
Chromatids

raphy with ³H-uridine—have been used to demonstrate that there is active transcription going on in the puffs but not in neighboring regions of the polytene chromosomes. It has also been shown that mRNA isolated from cells with puffs will hybridize only to the puffed regions of the chromosomes. Thus, these regions of the DNA are complementary to the mRNA (Figure 13–14).

Puffs generally fall into four categories: (1) stage-specific; (2) tissue-specific (in dipteran larvae, tissues other than the salivary gland—such as the midgut and Malpighian tubules—have polytene chromosomes); (3) constitutive; and (4) environmentally induced. In *Drosophila* about 80% of the puffs are stage specific; in *Chironomus* only about 20% are. At the time of molt in insects, the hormone ecdysone is secreted by the prothoracic gland. At the same time, many puff patterns change (Figure 13–15). Similar changes in puff patterns can be induced by the injection of ecdysone. Hence, molting, a stage-specific developmental sequence, is related to a sequential transcription sequence in the chromosomes. We will return in a later section to look further at relationships of hormones and genes.

Lampbrush Chromosomes. Lampbrush chromosomes occur in amphibian oocytes and are so named because their looped-out configuration has the appearance of a lampbrush (Figure 13–16). The loops of the lampbrush chromosomes are covered by an RNA matrix and are undoubtedly the sites of active transcription. Presumably, the loops are unwindings of the single

Figure 13 –14.
Hybridization
between
Chironomus tentans
Salivary Gland
Chromosome and
Labeled RNA
Transcribed from
This Locus

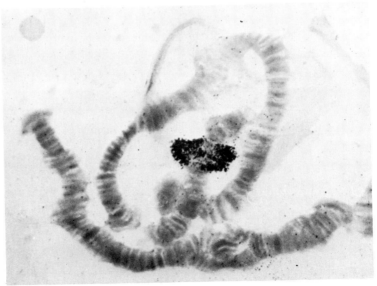

Source: Reprinted with permission from B. Lambert, "Repeated DNA Sequences in a Balbiani Ring," *Journal of Molecular Biology* 72 (1972):65–75. Copyright by Academic Press Inc. (London) Ltd.

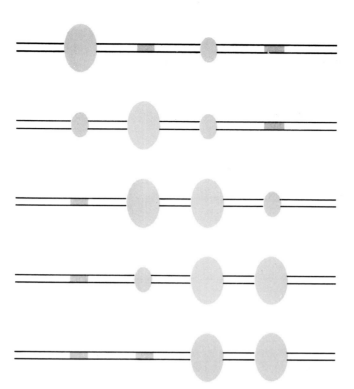

Figure 13–15.
Puff Pattern on a
Segment of a
Salivary Gland
Chromosome of
Chironomus tentans
during Molt

chromosome similar to the unwindings in the polytene chromosome shown
in Figure 13–13. Thus, under certain circumstances—such as in polytene
chromosome puffs and in lampbrush chromosomes—active transcription can
be seen. Since only certain bands puff at any moment in polytene chromo-

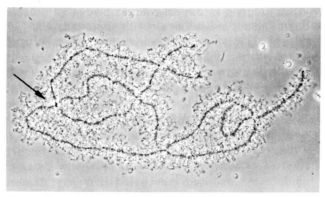

Figure 13–16.
Lampbrush
Chromosome of the
Newt, *Triturus*
viridescens
Centromeres are at
the arrow; the two
long homologues are
held together by
three chiasmata.
magnification 680X

Source: Joseph G. Gall, figure 2 in D. M. Prescott, ed., *Methods in Cell
Physiology*, vol. 2 (New York: Academic Press, 1966), p. 39. Reproduced
by permission.

somes and the loops of lampbrush chromosomes are of various sizes (with some regions not looped at all), we have evidence of specific transcription with no indication, so far, of the nature of the control.

Chromosome Banding

There are several chromosome-staining techniques that reveal consistent banding patterns. By means of these patterns, all of the human chromo-

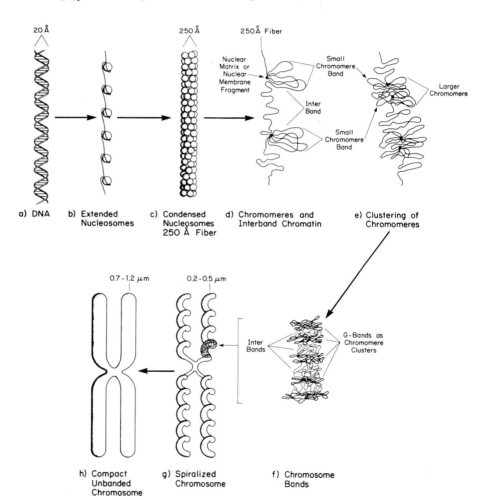

Figure 13-17. Model of Eukaryotic (Mammalian) Chromosome Structure

Source: David E. Comings, figure 1, p. 39, reproduced with permission from the *Annual Review of Genetics*, Volume 12. © 1978 by Annual Reviews Inc.

somes can be differentiated. Of possibly greater importance is the fact that these staining techniques have provided some insight into the structure of the chromosome. The techniques for staining the C, G, and R chromosome bands will serve as an illustration.

G bands are obtained with **Giemsa stain,** which is a complex of stains specific for the phosphate groups of DNA. Treatment of fixed chromatin with hot salts brings out the G bands (see Chapter 7). Giemsa stain enhances banding that is already visible in meiotic chromosomes. The banding pattern is caused by an arrangement of **chromomeres,** dark regions of the chromosomes. Under careful observation the major G bands can be seen to be made of many smaller chromomeres. This banding appearance has led Comings to suggest the mechanism of chromosome folding shown in Figure 13–17.

C bands are Giemsa-stained bands after treatment of the chromosomes with NaOH. The "C" is for "centromere" and these bands represent constitutive heterochromatin surrounding the centromeres. The DNA is also usually satellite rich. (**Satellite DNA** differs in buoyant density from the major portion of cellular DNA; see later.)

R bands are visible with a staining technique that stains the regions between G bands. Since the dark-light pattern is the opposite of the G-band pattern, these bands are called *reverse bands*. From the information supplied by this staining technique (Table 13–3), Comings defines three basic chromatin types: euchromatin, constitutive heterochromatin, and intercalary heterochromatin. Presumably, the only chromatin involved in transcription is **euchromatin. Constitutive heterochromatin** surrounds the centromere and is satellite rich. **Intercalary heterochromatin** is dispersed. Thus, it becomes apparent that the eukaryotic chromosome is a relatively complex structure.

TABLE 13 –3. The Three Major Types of Chromatin in Eukaryotic Chromosomes

	Centromeric Constitutive Heterochromatin	*Intercalary "Heterochromatin"*	*Euchromatin*
Relation to bands	In C bands	In G bands	In R bands
Location	Usually centromeric	Chromosome arms	Chromosome arms
Condition during interphase	Condensed	Condensed	Usually dispersed
Genetic activity	Inactive	Probably inactive	Usually active
Relation to meiotic chromosome chromomeres	Centromeric chromomere	Intercalary chromomeres	Interchromomeric

Quantity of DNA in Eukaryotic Chromosomes

If mutation rates, degree of genetic similarity between species, and time of divergence (evolutionary separation) are known, it is possible to calculate the

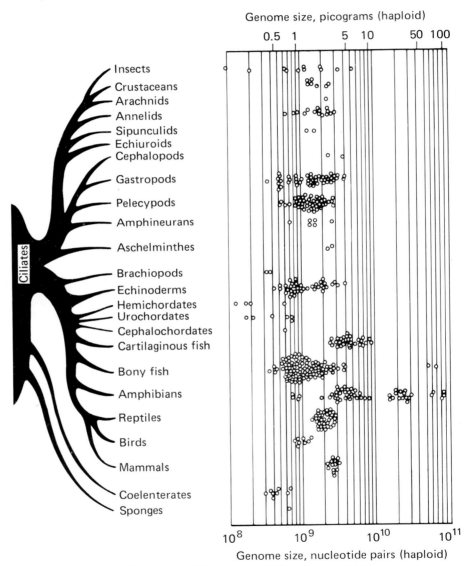

Figure 13–18.
Haploid Genome
Sizes in Eukaryotes

Source: R. Britten and E. Davidson, "Repetitive and Non-repetitive DNA Sequences and a Speculation on the Origins of Evolutionary Novelty," *Quarterly Review of Biology* 46 (1971):111–133. Reproduced by permission.

size of the genome (see Chapter 19). The quantity of DNA within the eukaryotic genome is found to be about 10 times greater than the estimated amount of DNA needed for structural genes. (**Structural genes** are nonregulatory genes.) For example, the human haploid genome has enough DNA to specify almost two million average-sized proteins. Yet, estimates of the true number of genes in the human haploid genome range between 40,000 and 100,000. These estimates are based primarily on mutation rates—that is, if a cell actually had two million loci, deleterious mutations occurring at the usual rate of 10^{-5} to 10^{-6} per generation per locus would destroy the organism. Calculations based on divergence rates of species in known evolutionary time also support the lower number of structural genes.

An interesting, and unexplained, aspect of the quantity of eukaryotic DNA is the fact that very similar species can differ markedly in their genome size (Figure 13–18). Here again, the extra DNA of some of the species must be explained.

A number of hypotheses have been offered to explain all this extra DNA in the genome. Some of it undoubtedly functions as transcriptional control elements or is involved in the structure of the chromosome. Some of the extra DNA occurs within structural genes but is never translated (intervening sequences—see below). These hypotheses do not seem to account for all the extra DNA observed. Alternatively, extra copies of each structural gene could account for the extra DNA. This hypothesis has proved unsatisfactory for several reasons, the most convincing of which is that if individuals contained multiple copies of genes, mutations would soon make most individuals appear as polyploids. We are, therefore, still left with the problem of explaining all the extra DNA.

DNA Repetition in Eukaryotic Chromosomes

DNA–DNA Hybridization. We can further investigate the nature of the extra DNA in eukaryotic chromosomes by determining whether or not DNA within these chromosomes has any amount of repetitiveness. The technique of **DNA–DNA hybridization** is used to measure the amount of repetitiveness of the DNA within the genome. When DNA is heated, it denatures or unwinds at a particular temperature, the melting temperature, or T_m. Melting is noted by a sudden rise in UV absorbance (Figure 13-19). If the DNA solution is then cooled slowly, complementary regions of the DNA can "find" each other and rejoin to form double helixes. Noncomplementary DNA strands will not renature. The speed of renaturation is dependent on at least four parameters: the degree of complementarity, the concentration of the DNA, the amount of time that the DNA is allowed to renature, and the temperature of renaturation. If the temperature is held constant, then the amount of reassociation can be calculated on the basis of the concentration of DNA and time.

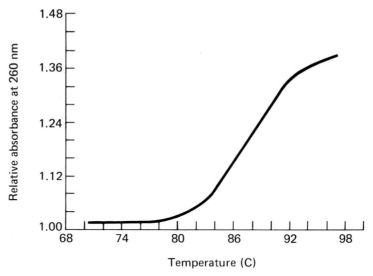

Figure 13–19.   Melting Curves of *E. coli* DNA
The T_m is the midpoint of the rise section (90° C for *E. coli*).

Cot Curves. If C_0 is the original concentration of single-stranded (denatured) DNA in moles per liter and t is time in seconds, then their product provides a useful measure of renaturation called C_0t (or cot). When cot values are plotted against the quantity of remaining single-stranded DNA, the curve (cot curve or cot plot) is very informative. The midpoint, referred to as the $cot_{1/2}$, estimates the amount of the homology of the DNA, or more precisely, the length of unique DNA in the sample. (By **unique DNA,** we mean the length of DNA in a chromosome that has no repeated sequences. We assume that the *E. coli* chromosome is entirely unique.)

In Figure 13–20 fractions of renatured (reassociated) DNA are plotted against cot values (C_0t). The arrows above the complexity (nucleotide pairs) axis indicate the genome size of each organism; under ideal circumstances, the genome sizes coincide with the $cot_{1/2}$ values. As indicated by this figure, all samples of DNA produce cot curves of the same shape, although DNAs of different complexities (different lengths of unique DNA) are located at different points along the abscissa—the further to the right, the more slowly the single-stranded nucleic acids reassociate. If the length of unique (non-repeating) DNA in a particular sample is known, it is possible to estimate the length of unique segments of DNA in new samples by their $cot_{1/2}$ values. For example, *E. coli*, with a chromosome of 4.2×10^6 nucleotides, has a $cot_{1/2}$ of about 10. Poly U and poly A strands have a unique length of only one base pair and this nucleic acid has a $cot_{1/2}$ value of about 2×10^{-6}. From these values and values for the nucleic acids of MS2, T4, and others, the "nucleotide pairs" line above the curve of Figure 13–20 was obtained.

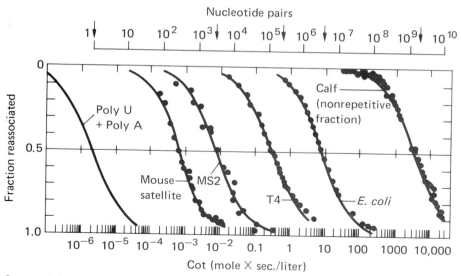

Figure 13-20.
Cot Curves for
Renaturation of DNA

Source: R. Britten and D. Kohne, "Repeated Sequences in DNA," *Science*
161 (9 August 1968):529–540. Copyright 1968 by the American Association for the Advancement of Science. Reproduced by permission.

Highly Repetitive, Intermediately Repetitive, and Unique DNA.
Eukaryotic satellite DNA has a low $cot_{1/2}$ value. Since the basic length of
unique satellite DNA is about 200 nucleotide pairs (Figure 13–20) and since
the quantity of this DNA per cell is much more than this, satellite DNA
must be a highly **repetitive DNA** (containing many copies of a particular
nucleotide sequence). Given the quantity of satellite DNA per cell, there
must be more than 1,000,000 repetitions of the basic sequence of 200 nucleotide pairs.

The cot curve for the whole genome of a eukaryote is different from the
curves of Figure 13–20. The cot curve of mouse DNA is shown in Figure 13–
21. Note that this curve can be divided into three segments on the basis of
the amount of repetitiveness and the concentration of each segment. This
curve indicates three separate types of DNA: a highly repetitive segment
(satellite), a segment that is of intermediate repetitiveness, and a segment of
unique DNA. These segments make up about 10%, 15%, and 70%, respectively, of the total mouse DNA (Figure 13–22).

The highly repetitive satellite DNA is probably involved in the structure
of the chromosome in the centromeric regions. It is not known to be transcribed. The unique DNA makes up the structural genes; it is transcribed.
There is some evidence that DNA of intermediate repetitiveness consists of
sequences that are related but are not identical. If ribosomal RNA genes,
which make up less than 1% of the total DNA, are removed from this inter-

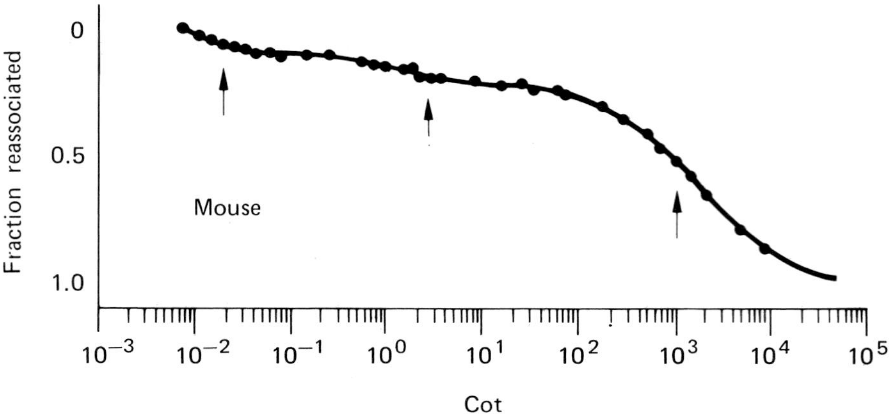

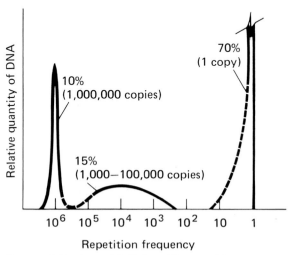

**Figure 13 –21.
Cot Curve for Mouse
DNA**
Arrows indicate
segments of highly
repetitive (left),
moderately repetitive
(middle), and
nonrepetitive
(unique) DNA.

Source: Betty L. McConaughy and Brian J. McCarthy, ''Related Base
Sequences in the DNA of Simple and Complex Organisms. VI. The Extent
of Base Sequence Divergence among the DNA's of Various Rodents,'' *Biochemical Genetics* 4 (1970):425–446. Reproduced by permission of
Plenum Publishing Corporation and the authors.

mediately repetitive DNA, then its transcriptional role is not clear. The
knowledge that, on the basis of nucleotide repetition, there are three types
of eukaryotic DNA and that only the unique (nonrepeating sequences) DNA
is transcribed still does not explain why there is so much excess DNA.

**Figure 13 –22.
Frequency of
Repetition of Mouse
DNA Segments**
Dotted regions are
less certain.

Source: R. Britten and D. Kohne, ''Repeated Sequences in DNA,'' *Science*
161 (9 August 1968):529–540. Copyright 1968 by the American Association for the Advancement of Science. Reproduced by permission.

Intervening Sequences. A possible explanation for the excess of DNA in eukaryotes may be found in the fact that most of the eukaryotic genes studied so far contain **intervening sequences** between translatable units. That is, within the length of the gene, from the codon for the first amino acid to the codon for the last, there are long stretches of DNA that are never translated. The DNA is transcribed into very long mRNA segments, sometimes 5 to 10 times longer than the segment that will be translated. The transcripts have been referred to as **heterogeneous nuclear mRNA,** or hnRNA. The intervening sequences are then removed from the transcript and the remaining mRNA is transported out of the nucleus into the cytoplasm, where it is translated (Figure 13–23). The terms **intron** and **exon** have been suggested to describe an intervening sequence of DNA and a sequence represented in the final mRNA, respectively. An example of a gene with intervening sequences is shown in Figure 13–24.

If it is found that the occurrence of these large intervening sequences is the rule in eukaryotes, we will have an explanation for the large amount of excess DNA in the eukaryotic genome. It then becomes necessary to explain why the intervening sequences exist. Currently, there are at least two major hypotheses. Walter Gilbert of Harvard has suggested that intervening sequences allow for a rapid shuffling of parts of genes during recombination processes and are thus very important from an evolutionary point of view. He has suggested that the parts of genes between the intervening sequences code for functional segments of a protein. Several genes that have been examined so far have supported his hypothesis. For example, the second of three segments of the globin gene is the part of the protein that binds heme.

Other molecular geneticists feel that there is more to intervening sequences than just an evolutionary advantage in recombination. Some have

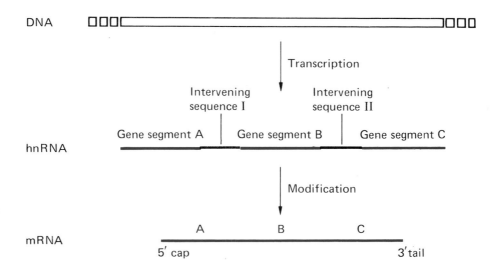

Figure 13 –23. Intervening Sequences in DNA Transcribed, Then Removed from mRNA before Translation

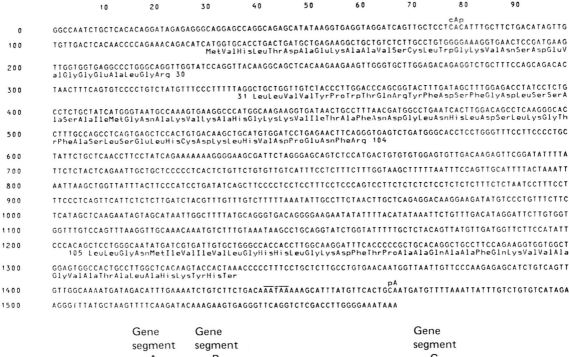

```
                10        20        30        40        50        60        70        80        90
                                                                                  cAp
   0  GGCCAATCTGCTCACACAGGATAGAGAGGGCAGGAGCCAGGCAGAGCATATAAGGTGAGGTAGGATCAGTTGCTCCTCACATTTGCTTCTGACATAGTTG

 100  TGTTGACTCACAACCCCAGAAACAGACATCATGGTGCACCTGACTGATGCTGAGAAGGCTGCTGTCTCTTGCCTGTGGGGAAAGGTGAACTCCGATGAAG
                                   MetValHisLeuThrAspAlaGluLysAlaAlaValSerCysLeuTrpGlyLysValAsnSerAspGluV

 200  TTGGTGGTGAGGCCCTGGGCAGGTTGGTATCCAGGTTACAAGGCAGCTCACAAGAAGAAGTTGGGTGCTTGGAGACAGAGGTCTGCTTTCAGCAGACAC
      alGlyGlyGluAlaLeuGlyArg 30

 300  TAACTTTCAGTGTCCCCTGTCTATGTTTCCCTTTTTAGGCTGCTGGTTGTCTACCCTTGGACCCAGCGGTACTTTGATAGCTTTGGAGACCTATCCTCTG
                                  31 LeuLeuValValTyrProTrpThrGlnArgTyrPheAspSerPheGlyAspLeuSerSerA

 400  CCTCTGCTATCATGGGTAATGCCAAAGTGAAGGCCCATGGCAAGAAGGTGATAACTGCCTTTAACGATGGCCTGAATCACTTGGACAGCCTCAAGGGCAC
      laSerAlaIleMetGlyAsnAlaLysValLysAlaHisGlyLysLysValIleThrAlaPheAsnAspGlyLeuAsnHisLeuAspSerLeuLysGlyTh

 500  CTTTGCCAGCCTCAGTGAGCTCCACTGTGACAAGCTGCATGTGGATCCTGAGAACTTCAGGGTGAGTCTGATGGGCACCTCCTGGGTTTCCTTCCCCTGC
      rPheAlaSerLeuSerGluLeuHisCysAspLysLeuHisValAspProGluAsnPheArg 104

 600  TATTCTGCTCAACCTTCCTATCAGAAAAAAAGGGGAAGCGATTCTAGGGAGCAGTCTCCATGACTGTGTGTGGAGTGTTGACAAGAGTTCGGATATTTTA

 700  TTCTCTACTCAGAATTGCTGCTCCCCCTCACTCTGTTCTGTGTTGTCATTTCCTCTTTCTTTGGTAAGCTTTTTAATTTCCAGTTGCATTTTACTAAATT

 800  AATTAAGCTGGTTATTTTACTTCCCATCCTGATATCAGCTTCCCCTCCTCCTTTCCTCCCAGTCCTTCTCTCTCTCCTCTCTCTTTCTCTAATCCTTTCCT

 900  TTCCCTCAGTTCATTCTCTCTTGATCTACGTTTGTTTGTCTTTTTAAATATTGCCTTCTAACTTGCTCAGAGGACAAGGAAGATATGTCCCTGTTTCTTC

1000  TCATAGCTCAAGAATAGTAGCATAATTGGCTTTTATGCAGGGTGACAGGGGAAGAATATATTTTACATATAAATTCTGTTTGACATAGGATTCTTGTGGT

1100  GGTTTGTCCAGTTTAAGGTTGCAAACAAATGTCTTTGTAAATAAGCCTGCAGGTATCTGGTATTTTTGCTCTACAGTTATGTTGATGGTTCTTCCATATT

1200  CCCACAGCTCCTGGGCAATATGATCGTGATTGTGCTGGGCCACCACCTTGGCAAGGATTTCACCCCCGCTGCACAGGCTGCCTTCCAGAAGGTGGTGGCT
      105 LeuLeuGlyAsnMetIleValIleValLeuGlyHisHisLeuGlyLysAspPheThrProAlaAlaGlnAlaAlaPheGlnLysValValAla

1300  GGAGTGGCCACTGCCTTGGCTCACAAGTACCACTAAACCCCCTTTCCTGCTCTTGCCTGTGAACAATGGTTAATTGTTCCCAAGAGAGCATCTGTCAGTT
      GlyValAlaThrAlaLeuAlaHisLysTyrHisTer
                                                                               pA
1400  GTTGGCAAAATGATAGACATTTGAAAATCTGTCTTCTGACAAATAAAAAGCATTTATGTTCACTGCAATGATGTTTTAAATTATTTGTCTGTGTCATAGA

1500  AGGGTTTATGCTAAGTTTTCAAGATACAAAGAAGTGAGGGTTCAGGTCTCGACCTTGGGGAAATAAA
```

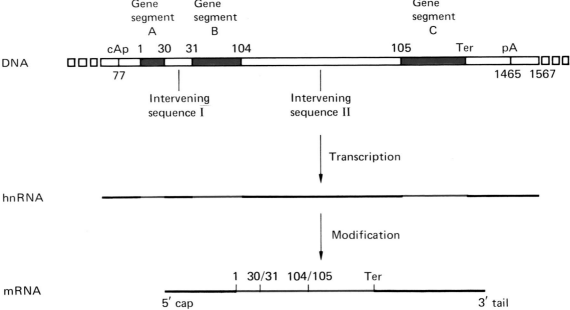

Figure 13–24. Nucleotide Sequence of Mouse β-Globin Major Gene

The DNA strand corresponding to the mRNA strand is shown. cAp indicates the start of the capped mRNA; pA indicates the start of the poly A tail; numbers inside the sequence are adjacent amino acid positions; Ter is the termination codon.

Source: David A. Konkel, Shirley M. Tilghman, and Philip Leder, "The Sequence of the Chromosomal Mouse β-Globin Major Gene: Homologies in Capping, Splicing and Poly (A) Sites," *Cell* 15 (December 1978): 1125–1132. Reproduced by permission.

suggested that intervening sequences serve a regulatory role in gene expression. For example, in several cases where intervening sequences have been removed from genes, the genes were never translated. Perhaps, it is reasoned, one of the functions of intervening sequences is to prevent mRNA from being degraded by the cell.

In summary then, there are no certain answers as to why there is an apparent excess of DNA in the eukaryotic genome. It has been established that the eukaryotic chromosome is made of three kinds of chromatin and three kinds of DNA. The centromeric DNA is the highly repetitive satellite DNA, is probably of structural importance, and is probably not transcribed. Both unique and intermediately repetitive DNAs are scattered on all the chromosomes. Intermediately repetitive DNAs occur in from 10^2 to 10^5 copies per sequence, with unknown function. Unique DNA, composed of nonrepeated nucleotide sequences, includes sequences that are translated as structural genes and also intervening sequences that are never translated. Perhaps these intervening sequences, the universal formation of which is uncertain, account for the excess amount of DNA.

We will now turn our attention briefly to three topics of developmental interest that are major fields of study in themselves: the genetics of cancer in man, endocrinological genetics, and the genetics of immunology.

CANCER

The following discussion is restricted to cancer in humans. **Cancer** is an informal term for a diverse class of diseases marked by abnormal cell proliferation. Control of normal cell development is lost and cells proliferate at an inappropriate rate to form growths known as *tumors* (or *neoplasms*). Benign tumors grow in one place and do not invade other tissues. Malignant tumors continue to proliferate and invade nearby tissues or, by a process called *metastasis,* invade distant parts of the body and start new sites of uncontrolled cell growth wherever they go. There are various theories to explain carcinogenesis. Since cancer is such a heterogeneous group of diseases, each explanation may eventually prove correct for some form of the disease.

Viruses and Human Cancer

While virally caused cancers are known in many organisms, only one human cancer has been positively related to a virus. Burkitt's lymphoma, a malignant tumor of lymphatic tissue in the jaw of African children, has been related to the Epstein-Barr virus, a herpesvirus that among other things causes infectious mononucleosis. Other virus/cancer relationships are being pursued—for example, the relationship between hepatitis-B virus and hepatic cancer. Since leukemia and breast cancer are of viral origin in mice, it is supposed, although not yet demonstrated, that the diseases are virally caused in humans.

Mutation Theory of Cancer

Mutation, both point and chromosomal, has been implicated in carcinogenesis. Many malignancies are associated with alterations in the structure and number of chromosomes. For example, the translocation of the long arm of chromosome 22 (the Philadelphia chromosome) to another chromosome, usually 9, is associated with chronic myelogenous leukemia; renal-cell cancer is associated with a translocation between chromosomes 3 and 8; Wilm's Tumor, a childhood kidney cancer, is associated with a small deletion on chromosome 11. Trisomy of chromosomes 7, 8, and 9 is commonly associated with acute leukemia. Chromosomes 14 and 8 are often translocated in Burkitt's lymphoma. While at present no cause-and-effect relationship has been established, it is known that certain tumor-associated viruses preferentially attach to certain chromosomes. Future work will elucidate the relationship between viruses, aneuploids, chromosome rearrangements, and cancers.

Point mutations are known to cause human cancers. For example, xeroderma pigmentosum (Chapter 14) is caused by mutations that inactivate the UV mutation repair system so that exposure to the sun results in skin lesions that often become malignant. A related disease, ataxia-telangiectasia, is a defect in x-ray induced excision repair mechanisms. Persons with this defect are at risk for acute and chronic leukemia, lymphomas, and (in women) ovarian cancers.

Certain other diseases carry a higher than normal risk for cancer. Fanconi's anemia is a syndrome of malformations of the heart, kidney, and extremities; pigmentary changes of the skin; and changes in the bone marrow. It is associated with acute leukemia and with cancers of the skin, liver, and esophagus. Relatives of patients with Fanconi's anemia are also prone to malignancies; the risk of cancer mortality before age 45 for heterozygotes is increased 3–6 fold. On the basis of risk and gene frequency, it has been calculated that as many as 1% of persons dying before age 45 from any malignancy may be heterozygous for Fanconi's anemia and that as many as 5% may be heterozygous for ataxia-telangiectasia.

There is general disagreement among clinicians as to whether or not malignancies are inherited, except for malignancies, such as xeroderma pigmentosum, that are inherited in a simple pattern. One view holds that there is no genetic element to cancer risk, whereas a second view holds that the human population is heterogeneous: There is a low-risk group and a high-risk group.

TABLE 13–4. Tumors Found in the Two Known Cancer-Family Syndromes

Type I	Type II
Glandular tumors of:	Connective tissue tumors of:
Endometrium	Breast
Colon	Bone
Breast	Brain
Ovary	Leukemia
Prostate	Embryonal neoplasms

Cancer-Family Syndromes

Studies of the incidence of cancer in families show heterogeneity and thus give some support to the second view. Recently, when 4000 clinic registrants were interviewed, almost half reported virtually no family history of cancer, whereas about 7% reported that many family members had cancer. This 7% was considered cancer prone on the basis of 3 or more close relatives of the interviewed person having cancer. The interpretation of the study is that some families are predisposed toward cancer while most are not, rather than everyone in the population having a uniform and low probability of having cancer. Lending support to this interpretation are the *cancer-family syndromes,* where family members seem to inherit a nonspecific predisposition toward tumors of various types. At least two cancer-family syndromes are known (Table 13–4; Figure 13–25).

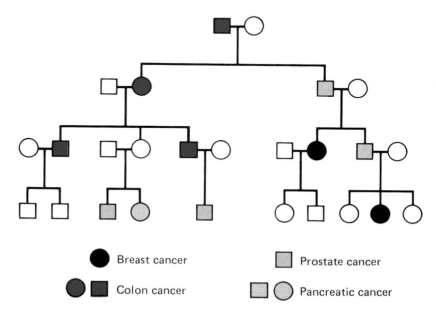

Figure 13–25.
Pedigree of a Type I
Cancer-Family
Syndrome

● Breast cancer

● ■ Colon cancer

□ Prostate cancer

□ ○ Pancreatic cancer

Of current interest is the hypothesis that most cancers are not inherited but result from mutation where multiple mutational events are needed to cause malignancy. This mutation theory explains the fact that cancer incidence increases with age and that most cancers are clonal—arise out of one original defective cell. It is hypothesized that cancer-family syndromes could be the result of the inheritance of one of the mutants; the next mutation would produce the lesion. The hypothesis predicts what is generally found: an earlier age at onset for persons with cancer-family syndrome.

Environmental Causes of Cancer

Environment plays a major role in carcinogenesis, and many environmental carcinogenic agents are known (Table 13–5). Many of the agents in Table 13–5 are also mutagens and thus create support for the mutation theory of cancer. Bruce Ames at Berkeley developed a routine screening test for carcinogens by making use of the relationship between mutagenicity and carcinogenicity. Since there is an apparent relationship between DNA damage, mutation, and carcinogenesis, Ames's system tests various chemicals for cancer-causing ability by seeing if they cause mutations in the bacterium *Salmonella*. A *Salmonella* strain that requires the amino acid histidine is used. Bacteria are exposed to a particular suspect chemical and the number of revertant bacterial colonies is recorded. Only revertants (those that no longer require histidine) can grow on minimal-medium petri plates—therefore, revertants are mutants.

TABLE 13 –5. Carcinogenic Substances in the Environment

Carcinogen	Cancer Site(s)
Aromatic amines	Bladder
Arsenic	Liver, lung, skin
Asbestos	Lung
Benzine	Bone marrow
Chromium	Lung, nose, nasopharynx sinuses
Cigarettes	Lung
Coal products	Bladder, lung
Dusts	Lung
Ionizing radiation	Bone, bone marrow, lung
Iron oxide	Lung
Isopropyl oil	Nasopharynx sinuses, nose
Mustard gas	Lung
Nickel	Lung, nasopharynx sinuses, nose
Petroleum	Lung
Ultraviolet irradiation	Skin
Vinyl chloride	Liver
Wood and leather dust	Nasopharynx sinuses, nose

It has been suggested that the most effective cancer prevention program would be to simply remove carcinogens from the environment. Only about 1% of all known cancers are due to simple Mendelian inheritance; about 25% of all cases are not associated with either a carcinogen or a genetic element; about 75% of all cancers are either entirely environmental or else partly environmental and partly genetic.

GENETICS AND ENDOCRINOLOGY

In vertebrates many developmental and homeostatic controls are mediated through the endocrine system. Hormones act to integrate and control various processes: Their control is spoken of as a **hormone net** to emphasize interaction and coordination. By definition, **hormones** are chemicals that are secreted by one type of cell and act on a second type of cell. The older, stricter definition, requiring action through the blood system, has given way to a looser definition that allows for diffusion of certain chemicals from one type of cell to another without going through the blood.

Currently, about fifty mammalian hormones are known. These hormones fall into three biochemical categories: *polypeptides, steroids,* and *amines* (Figure 13–26). Geneticists are interested in two aspects of endocrinology: (1) the mechanisms of action of hormones (when they have genetic effects such as transcriptional control) and (2) the inheritance and mode of action of mutations in hormone or receptor-site pathways.

The general scheme for hormonal action is shown in Figure 13–27. For polypeptide and amine hormones, an adenyl cyclase model is suggested where the second messenger within the cell is cyclic AMP. (The role of cyclic AMP in catabolite repression of *E. coli* was discussed in Chapter 12.) This cyclic-AMP model basically works through post-transcriptional control of enzyme activity (Figure 13–28). As an example, Sutherland and Rall described the role of cyclic AMP in the action of epinephrine in the liver. The hormone binds to a receptor site on the cell membrane. Once the hormone is bound, adenyl cyclase is activated and converts ATP to cyclic AMP. In the cell phosphorylase *b* kinase exists as two subunits, an active one and a regulatory one. The cyclic AMP reacts with the regulatory subunit and frees the kinase itself to phosphorylate proteins (Figure 13–28). Thus, the effect of the hormone is to increase the phosphorylation of proteins within the cell through the cyclic-AMP second-messenger system.

Steroid hormones seem to have more of a direct effect on transcription (Figure 13–29). For example, estrogen has a general growth effect on the mammalian uterus. The active form, estradiol, passes through the membrane into the cell cytoplasm where it combines with and causes a change in a receptor protein, which then is transported into the nucleus. There it interacts with the transcription apparatus in such a way as to initiate the transcription of specific genes. In the hen oviduct transcription is begun on the

Category	Hormone	Structure

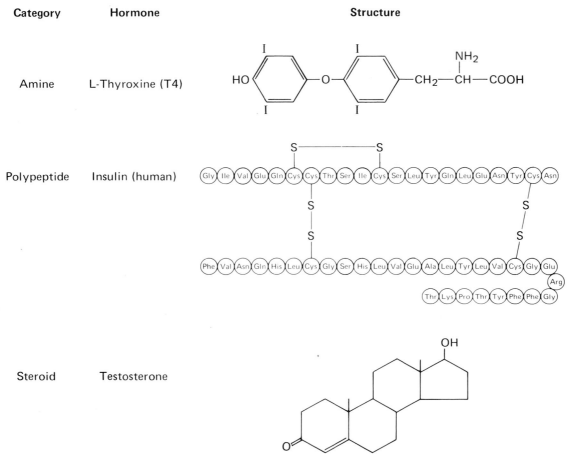

Amine L-Thyroxine (T4)

Polypeptide Insulin (human)

Steroid Testosterone

Figure 13–26. The Three Types of Hormones

gene for ovalbumin. In addition, transcription of ribosomal RNA is increased (as indicated by an increase in the nucleolus size) and there is an increase in production of transfer RNA. Thus, a general protein-synthesizing response occurs as well as the response of specific genes. The general conclusion is that the hormone net is vital for development and homeostasis and that some hormones, notably the steroids, work directly on transcription.

As to the second aspect of endocrine genetics—the inheritance of mutations that affect endocrine function—at least fifty mutants are known that have major endocrine effects in man. Table 13–6 lists endocrine disorders known to be influenced by genetic factors. Many other disorders are known to have major endocrinological effects but are not included in the table because they are not inherited in a simple manner. For example, many aneu-

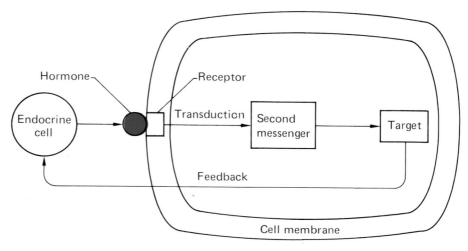

Figure 13–27.   General Scheme of Hormonal Action
The hormone is secreted by the endocrine cell and travels, by diffusion or by blood, to the target cell, where it is recognized by a protein or lipoprotein receptor. The information of the interaction of receptor and hormone is transduced into a second, intracellular messenger, which interacts with the intracellular receptor or target. There is then a feedback to the original endocrine cell.

ploids produce disorders of sexual development that are endocrine disorders. These disorders would include Turner's syndrome, Klinefelter's syndrome, and XYY syndrome. In addition, there are tumors that produce hormones in inappropriate places (ectopic hormone-producing tumors). In the various endocrine disorders known to be hereditary, all the normal rules of inheritance are followed: All inherited forms are either autosomal dominants, autosomal recessives, or X linked.

IMMUNOGENETICS

Immunity is the ability of an organism to resist infection. If a foreign agent **(antigen),** such as a bacterium or virus, invades an organism, an immune response is triggered whereby very specific proteins **(immunoglobulins)** are produced by derivatives of the B lymphocytes (a type of white blood cell) to protect the organism from the antigen by one of three mechanisms. Immunoglobulins can coat foreign agents (antigens) so that they can be taken up by phagocytes. Or, immunoglobulins can combine with the antigens—for example, by covering the membrane-recognition sites of a virus—and thereby directly destroy their functioning. Or, in combination with comple-

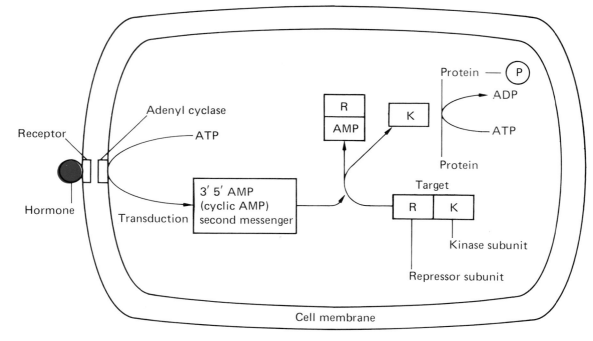

Figure 13 –28.   Cyclic-AMP Model of Polypeptide Hormonal Action
Cyclic AMP is produced when the hormone interacts with its membrane
receptor. Cyclic AMP combines with an enzyme repressor and frees the
enzyme to catalyze its specific reactions, which in this figure is a kinase
that phosphorylates proteins.

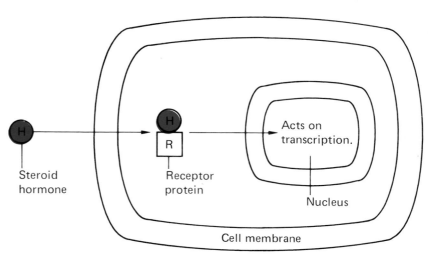

Figure 13 –29.   Model for the Action of a Steroid Hormone

TABLE 13 –6. Endocrine Disorders Influenced by Genetic Factors

Target	Disorder
Pituitary	Diabetes insipidus Pituitary dwarfism Pituitary hypogonadism
Parathyroid	Hypoparathyroidism Albright's hereditary osteodystrophy Hyperparathyroidism Vitamin D-resistant rickets
Thyroid	Goiter Cretinism Myxedema Chronic lymphocytic thyroiditis Thyrotoxicosis
Pancreas	Diabetes mellitus Hypoglycemia
Adrenals	Adrenogenital syndromes Adrenal insufficiency Bartter's syndrome

Sexual development	XX gonadal dysgenesis:
	    True hermaphroditism
	    Reifenstein's syndrome
	    Primary male hypogonadism
	    Sertoli-cell-only syndrome
	    Stein-Leventhal syndrome

Endocrine system	Multiple endocrine adenomatosis Medullary thyroid carcinoma-pheochromocytoma syndromes

Other	Baldness
	Hirsutism
	Myotonic dystrophy
	Werner's syndrome
	Laurence-Moon-Bardet-Biedl syndrome
	Congenital lipodystrophy
	XY gonadal dysgenesis:
	    Testicular feminization
	    Pseudovaginal
	      perineoscrotal hypospadias
	    Precocious puberty

ment, a blood component, immunoglobulins can lead directly to the destruction of antigens. By the nature of their function, immunoglobulins must be very specific: They must recognize foreign substances but must not recognize the normal components of an organism's "self."

Immunoglobulins are large protein molecules composed of two identical *light* (L) polypeptide chains (214 amino acids) and two identical *heavy* (H)

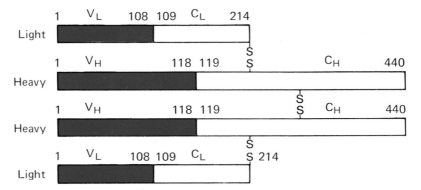

Figure 13–30.
Schematic View of
an Immunoglobulin
Protein
V = variable region;
C = constant region.

chains (440 amino acids) held together by 3 sulfhydryl bonds (Figure 13–30). Each polypeptide chain has a variable (V) and a constant (C) region of amino acid sequences. The variable regions give specificity to the immunoglobulin. There are 5 major types of heavy chains, giving rise to 5 types of immuno-globulin—IgG, IgA, IgM, IgD, and IgE. Mutants of the nonvariable parts of the chains are known; they are called **allotypes** and follow the rules of sim-ple Mendelian inheritance. In the variable region, however, diversity is much greater than two alleles per individual. The variation in this region is referred to as **idiotypic variation** and the average individual has the poten-tial to express about 1,000,000 different immunoglobulins. This number is arrived at through the study of persons with multiple myeloma, a malig-nancy where one lymphocyte divides over and over until it makes up a sub-stantial portion of the lymphocyte pool. From these persons it is possible to isolate a relatively purified immunoglobulin that is the product of a single clone of cells. It is found that a very low proportion of a normal person's lymphocytes produces any one specific immunoglobulin.

There are two ways to account for the high degree of antibody variabil-ity. One way **(germ-line theory)** is that every B lymphocyte cell has all the genes for every type of immunoglobulin but only transcribes one. The other way **(somatic-mutation theory)** is that there is only one basic immuno-globulin gene, but it undergoes such a large amount of somatic mutation that there are finally about 1,000,000 varieties of immunoglobulin-producing lym-phocytes. While the final judgment has not been made, general opinion runs in favor of a modified germ-line theory. There is evidence, however, to refute both theories.

For example, the germ-line theory, which hypothesizes so many genes, runs counter to the concept of the allotype—that is, that genetic variability of the homozygous-heterozygous type exists for the constant portions of the light chains. Perhaps, then, we might think that the immunoglobulins are synthesized in parts—that is, there is a single gene for the constant parts but multiple genes for the variable parts of the immunoglobulin molecule. The proteins are later spliced together to form the complete immunoglobu-lin. This thinking is probably not correct. When pulse labeling is done, such that radioactive amino acids are added for a very short time, it is found that labeled amino acids are found in only one place in the immunoglobulin chain rather than in two places, which would be expected if the splice theory were correct.

A currently popular model to explain immunoglobulin variability suggests numerous loci for the variable region and few loci for the constant region. A translocation to form a whole gene prior to transcription takes place, and any gap regions (intervening sequences) in the mRNA are excised prior to translation (Figure 13–31). Translation would then occur in a normal

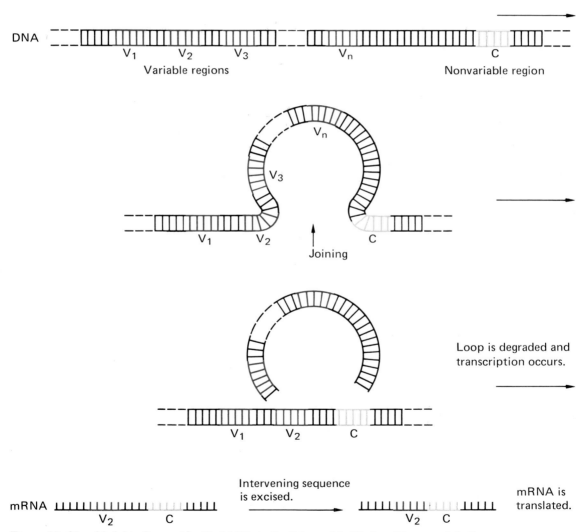

Figure 13 –31. Model to Account for Variability in the Polypeptide Chains of Immunoglobulins
A single constant region is adjacent to multiple variable regions. A looping process will give rise to a cistron with the constant region plus a variable region.

fashion for a whole single polypeptide. Thus, it is possible that various combinations of variable and constant regions would be translated from the same chromosome, which would account for the observed variety in immunoglobulins.

CHAPTER SUMMARY

By karyotyping and by nuclear transplanting, it is possible to demonstrate that, in most cases, development does not proceed simply by the loss of chromosomes or chromosome parts or by the permanent modification of chromosomes. To further study developmental control in eukaryotes, we must understand the eukaryotic chromosome.

The eukaryotic chromosome is apparently uninemic. It consists of one DNA double helix per chromosome arm. Nucleoprotein is composed of DNA, histones, and nonhistone proteins. Histones are involved in a nonspecific packaging of the DNA into nucleosomes. The majority of the nonhistone protein is involved in the structure of the chromosome as a scaffold and is not involved in regulation. Presumably, undetectably small quantities of the nonhistone protein take part in regulation of transcription. Substructuring in the eukaryotic chromosome is demonstrated by G-, C-, and R-banding techniques. It appears that there are C bands (constitutive heterochromatin) around the centromeres. These bands consist primarily of satellite DNA, which seems to have a nontranscriptional role in the structure of the chromosome. G bands (Giemsa stained), presumably, represent intercalary heterochromatin and, also presumably, do not have an active transcriptional role. R bands (reverse bands) appear between the G bands and represent intercalary euchromatin, the site of transcribed, structural genes.

The cot measure of DNA–DNA hybridization reveals that about 10% of the eukaryotic chromosome consists of highly repetitive nucleotide sequences (presumably, satellite DNA), about 15% consists of DNA that is intermediately repetitive, and about 70% consists of unique DNA in which there is no repetition of nucleotide sequences. The number of structural genes in a eukaryote account for about one-tenth of the total DNA. Recent research suggests that the large excess of DNA consists of intervening sequences, but the role played by the intervening sequences is not clear.

The genetics of cancer, endocrinology, and immunology were considered briefly. Predisposition to cancer may be inherited. One commonly held theory suggests that multiple mutational events are needed to produce cancer and that certain families inherit the outcome of one of these events.

While many endocrine disorders are known to be under simple genetic control, many hormones, most notably the steroid hormones, have a role in transcription control.

The great variety of immunoglobulins in any individual may be explained by the same type of intervening sequences that appear to explain the excess amount of DNA. That is, perhaps parts of the genes for the immunoglobulins exist as sections separated by intervening nucleotide sequences. There may then be recombination of these genes, followed by transcription and then excision of the intervening sequences before translation.

EXERCISES AND PROBLEMS

1. Draw a mitotic chromosome during metaphase. Diagram the various kinds of bands that can be brought out by various staining techniques. What information about the DNA contents of these bands is known?

2. Give a 250 Å fiber model of the chromosome to account for G bands. Give a similar model to account for polytene chromosome puffs.

3. Diagram the relation of DNA to the 250 Å fiber of the eukaryotic chromosome.

4. Diagram Hämmerling's experiments to show the role of nuclear cytoplasmic interactions in development.

5. Why do you suppose that so much research on developmental genetics has been done with amphibians?

6. An investigator removes the nucleus from a frog's egg and replaces it with an early blastula nucleus from *Xenopus*. What alternative predictions can you make about the future course of development? Will a frog or a *Xenopus* result, if anything? Will it have frog or *Xenopus* germ cells?

7. Give the formula for cot values. Draw some cot curves of various types of DNA. What do the $cot_{1/2}$ values mean? How can you determine the degree of repetitiveness of mouse DNA from cot curves?

8. Assume that a certain peptide hormone has the effect on a target cell of isomerizing a monosaccharide. Draw a model of this effect and include a second messenger.

9. What enzymes would be required for the steps of Figure 13-31 to take place? Describe these enzymes.

10. Figure 13-29 presented a model of steroid hormone action. Give some possibilities for the details of the reactions that occur within the nucleus.

11. From the pedigree of Figure 13-25, what modes of inheritance would be consistent with the various types of cancers?

SUGGESTIONS FOR FURTHER READING

Abelson, J. 1979. RNA processing and the intervening sequence problem. *Ann. Rev. Biochem.* 48:1035–1069.

Ames, B. 1979. Identifying environmental chemicals causing mutations and cancer. *Science* 204:587–593.

Briggs, R., and T. King. 1952. Transplantation of living nuclei from blastula cells into enucleated frog's eggs. *Proc. Nat. Acad. Sci., USA* 38:455–463.

Britten, R., and D. Kohne. 1968. Repeated sequences in DNA. *Science* 161:529–540.

Chambon, P. 1981. Split genes. *Sci. Amer.* 244:60–71.

Comings, D. 1978. Mechanisms of chromosome banding and implications for chromosome structure. *Ann. Rev. Genet.* 12:25–46.

DuPraw, E. 1970. *DNA and Chromosomes.* New York: Holt, Rinehart and Winston.

Ebert, J., and I. Sussex. 1970. *Interacting Systems in Development.* New York: Holt, Rinehart and Winston.

Evans, H. 1977. Some facts and fancies relating to chromosome structure in man. In *Advances in Human Genetics,* vol. 8. New York: Plenum Publishing, pp. 347–438.

Hämmerling, J. 1953. Nucelo-cytoplasmic relationships in the development of *Acetabularia. Intern. Rev. Cytology* 4:87–127.

Knudson, A. 1977. Genetics and etiology of human cancer. In *Advances in Human Genetics,* vol. 8. New York: Plenum Publishing, pp. 1–66.

Konkel, D., S. Tilghman, and P. Leder. 1978. The sequence of the chromosomal mouse β-globin major gene: Homologies in capping, splicing and poly(A) sites. *Cell* 15:1125–1132.

Kornberg, R., and A. Klug. 1981. The nucleosome. *Sci. Amer.* 244:52–64.

Lewin, B. 1974. *Gene Expression, 2: Eucaryotic Chromosomes.* New York: Wiley.

Roitt, I. 1977. *Essential Immunology,* 3rd ed. Oxford: Blackwell.

Schimke, R. 1978. *Genetics and Cancer in Man.* Edinburgh: Churchill Livingstone.

Steward, F., M. Mapes, and K. Mears. 1958. Growth and organized development of cultured cells. II. Organization in cultures grown from freely suspended cells. *Amer. J. Bot.* 45:705–708.

Sutherland, E., and T. Rall. 1960. The relation of adenosine-3′,5′-phosphate and phosphorylase to the actions of catecholamines and other hormones. *Pharmacol. Rev.* 12:265–299.

Williams, R., ed. 1974. *Textbook of Endocrinology.* Philadelphia: Saunders.

14

Mutation, DNA Repair, and Recombination

Mutation, DNA repair, and recombination are treated together in this chapter because the three processes have much in common. The physical alteration of DNA is involved in all three processes, and they all make use of some of the same enzymes. We progress from mutation—the change in DNA—to repair of damaged DNA and finally to recombination—the new arrangement of independent pieces of DNA.

MUTATION

The concept of mutation (a term coined by DeVries, a rediscoverer of Mendel) is pervasive in genetics. **Mutation** is both the process by which a gene (or chromosome) changes structurally and the end result of that process. Without alternate forms of genes, the biological diversity that exists today would not have evolved. Without alternate forms of genes, it would be difficult for geneticists to determine which of an organism's characteristics are genetically controlled. Only the most modern techniques of molecular biology and biochemical analysis can circumvent reliance on mutation for understanding genetic systems. Even then, the background for current knowledge in molecular genetics was provided by studies of mutation. This chapter will examine the mechanisms of mutation, the ways in which cells repair mutational changes of various sorts, and the process of recombination, which shares some of the same enzymes used by the repair mechanisms.

Fluctuation Test

In 1943 Salvadore Luria and Max Delbrück published a paper entitled "Mutations of bacteria from virus sensitivity to virus resistance." This paper

Salvador E. Luria
(1912–)
Courtesy of Dr.
Salvador E. Luria

ushered in the era of bacterial genetics by demonstrating that the phenotypic variants found in bacteria are actually due to mutations rather than to induced physiological changes. Very little work had previously been done in bacterial genetics because of the feeling that bacteria did not have "normal" genetic systems like the systems of fruit flies and corn. Rather, it was believed that bacteria responded to environmental change by physiological adaptation—a non-Darwinian view. As Luria said, bacteriology remained "the last stronghold of Lamarckism"—the view that acquired characteristics are inherited.

What Causes Genetic Variation?   Luria and Delbrück studied the Ton^r (phage T1 resistant) mutants of a normal Ton^s (phage T1 sensitive) *E. coli* strain. They used an enrichment experiment—as described in Chapter 6— wherein a petri plate is spread with *E. coli* bacteria and T1 phage. If there are enough phages to attack all of the *E. coli* cells, there will be no colony growth on the plate: All the bacteria will be lysed. However, if one of the bacterial cells is resistant to T1 phage, it will produce a bacterial colony, and all descendants of the cells from this colony will be resistant. If descendants of the resistant cells are grown in the absence of T1 and then replated on a medium with T1, they too will be resistant. Two possible explanations exist for the phenomenon:

1. Every *E. coli* cell is capable of being induced to be resistant to T1 but only a very small number actually are induced. That is, all cells are genetically identical, with a very low probability of acting in a resistant manner in the presence of T1 phage. Cells that are then resistant to T1 show a low-probability physiological or metabolic response. When resistance is induced, the cell and its progeny remain resistant.

2. There exists in the culture a small number of *E. coli* cells that are already resistant to T1. When the whole culture is plated with T1, only these cells survive.

If the presumed rates of induction and mutation are the same, there is difficulty in determining which of the two mechanisms is operating. Luria and Delbrück, however, developed a means of distinguishing between the results of these mechanisms. They reasoned as follows: If the T1 resistance were physiologically induced, the relative frequency of resistant *E. coli* cells in a culture of the normal (Ton^s) strain should be a constant, independent of the number of cells in the culture or the length of time that the culture has been growing. If resistance were due to random mutation, the frequency of mutant (Ton^r) cells should depend on the density and age of the culture (Figure 14–1)—that is, the appearance of a mutant cell would be a random event. If mutation occurs early in the growth of the culture, then many cells will descend from the mutant cell and there will thus be many resistant colonies. If the mutation does not occur until late in the growth of the culture,

Max Delbrück
(1906–1981)
Courtesy of Dr. Max
Delbrück

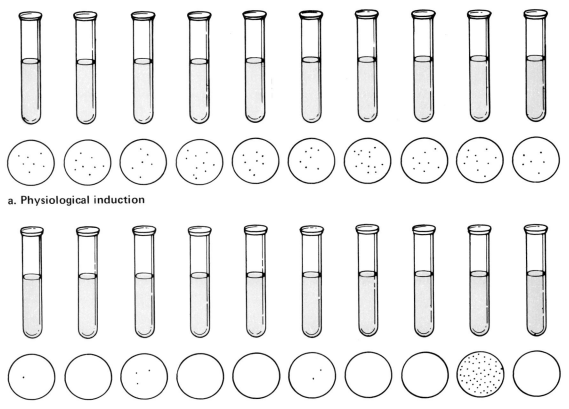

a. Physiological induction

b. Random mutation

Figure 14 – 1.   Occurrence of *E. coli* Ton^r Colonies in Ton^s Cultures
The number of Ton^r cells may be due to physiological induction or
random mutation.

then the subsequent number of mutant colonies will be few. Thus, if the mutation hypothesis is correct, when a significant number of cultures is tested, considerable fluctuation from culture to culture in the number of resistant colonies would be noted.

Results of the Fluctuation Test.   Luria and Delbrück also developed what is known as the **fluctuation test.** They counted the mutants in (1) a large number of small (they referred to them as "individual") cultures and (2) a number of subsamples from a single large or "bulk" culture. All subsamples from a bulk culture should have a similar number of resistant cells— any differences would be due to random sampling error. But, among the individual cultures, if mutation is the case, the number of resistant cells should vary from culture to culture; the number would be related to the time that the mutation occurred during the growth of each culture. If mutation arose early, there would be many resistant cells. If it arose late, there would be relatively few resistant cells.

Luria and Delbrück began with twenty 0.2 ml individual cultures and one 10 ml bulk culture. Each was inoculated with 10^3 *E. coli* cells per ml and incubated until there were about 10^8 cells per ml. Each individual culture was then spread out on a plate containing a very high concentration of T1; ten 0.2 ml subsamples from the bulk culture were plated in the same way. The results, in Table 14–1, show that there was minimal variation in the number of resistant cells among the bulk-culture subsamples but a very large amount of variation, as predicted by random mutation, among the individual cultures.

TABLE 14 –1. Results from the Luria and Delbrück Fluctuation Test

Individual Cultures		Samples from Bulk Culture	
Culture Number	Ton^r Colonies Found	Sample Number	Ton^r Colonies Found
1	1	1	14
2	0	2	15
3	3	3	13
4	0	4	21
5	0	5	15
6	5	6	14
7	0	7	26
8	5	8	16
9	0	9	20
10	6	10	13
11	107		
12	0		
13	0		
14	0		
15	1		
16	0		
17	0		
18	64		
19	0		
20	35		
Mean $(\overline{n})$	11.4		16.7

Source: From S. E. Luria and M. Delbrück, *Genetics* 28, 491 (1943).

Hermann J. Muller
(1890–1967)
Courtesy of National
Academy of
Sciences

Spontaneous versus Induced Mutation

H. J. Muller won the Nobel Prize for his demonstration that x rays can cause mutations. His work was published in 1927 in a paper entitled "Artificial transmutation of the gene." At about the same time, Stadler induced mutation in barley. The basic impetus for Muller's work was the fact that mutation occurs so infrequently that genetic research was hampered by the inability to obtain mutants. Muller exposed flies to varying doses of x rays and then observed their progeny. He came to several conclusions. First, x rays greatly increased the occurrence of mutations. Second, the observable types of mutants (autosomal or sex-linked recessives and dominants or lethals and semilethals) were similar to the types that result from natural or "spontaneous" mutations. Before continuing with Muller's work, it is necessary to define mutation rate.

Mutation Rates. The **mutation rate** is defined as the proportion of mutants per cell division in bacteria or single-celled organisms or as the proportion of mutants per gamete in higher organisms. Mutation rates vary tremendously depending upon the length of genetic material, the kind of mutation, and other factors to be discussed later. For example, Luria and Delbrück found that in *E. coli* the mutation rate per cell division of Ton^s to Ton^r was 3×10^{-8}, while the mutation rate of the wild type to histidine-requiring phenotype (His$^+$ to His$^-$) was 2×10^{-6}. The rate of **reversion** (return of the mutant to the wild type) was 7.5×10^{-9}. Here, the mutation and reversion rates differ because the His$^-$ phenotype can be caused by many different mutations. But, reversion requires specific, and hence more unlikely, changes to correct the His$^-$ phenotype back to the wild type. The lethal mutation rate in *Drosophila* is about 1×10^{-2} per gamete for the total genome. This number is relatively large because, as with His$^-$, many different mutations produce the same phenotype (lethality in this case). We now return to Muller's work with *Drosophila*.

ClB Technique. Muller concentrated on recessive lethal X-linked mutants because they were the most common and easiest mutants to work with. One of the methods he devised for detecting these mutants is known as the **ClB method** (Figure 14–2). Muller developed a strain of flies in which the female had one wild (normal) X chromosome and one X chromosome that carried a mutant recessive lethal gene (*l*), a mutant dominant Bar-eye gene (*B*), and an inverted area (*C*) that prevented crossing over between *B* and *l*. To study the mutagenic effect of x rays, male flies in the P$_1$ generation are x rayed and are then crossed to ClB females, which are heterozygous for the *ClB* X chromosome. The daughters of this cross will be carriers of the irradiated male's X chromosome. Half of the F$_1$ will be bar eyed (ClB) and the other half, wild type. The bar-eyed daughters are then crossed to wild-type males (the F$_1$ cross of Figure 14–2). Normally, half the sons in the F$_2$

Lewis J. Stadler
(1896–1954)
Genetics 41
(1956):frontispiece.

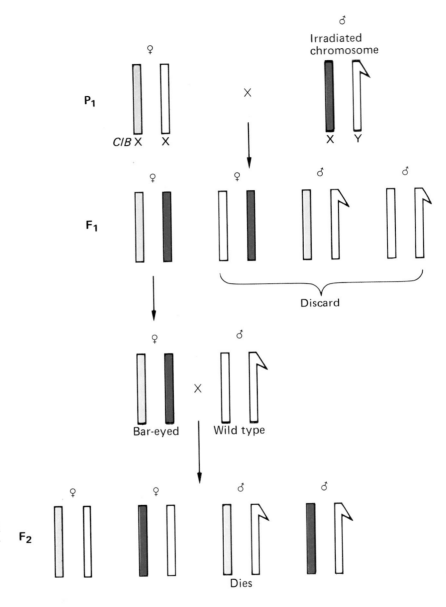

Figure 14 – 2.
Muller's ClB Method
for Detecting X-Ray–
Induced Recessive
Lethal Mutations on
the X Chromosome
of *Drosophila*

would die because they receive the *ClB* chromosome, which is lethal in the hemizygous state. If the irradiated X chromosome also has a lethal, then the other half of the F_2 males will also die. Thus, if a lethal is induced in the P_1 generation in the male's X chromosome in a gamete, this lethal will result in the absence of males in the F_2 progeny. The absence of males is easily discerned at a glance by persons experienced in handling *Drosophila*.

Point Mutations

The mutations of primary concern in this chapter are **point mutations,** which consist of single changes in the nucleotide sequence. Chromosomal mutations, changes in the number and visible structure of chromosomes, have been treated in Chapter 7. The basic mechanism for mutation is a change of a nucleotide (base pair) in DNA. If the change is a replacement of some kind, then a new codon will be created. In many cases, this new codon, upon translation, will result in a new amino acid. As discussed in Chapter 11, the genetic code has the property of "protecting" the cell from mutation: Common amino acids have the most codons; similar amino acids have similar codons; and the wobble position of the codon is the least important position in translation. However, when base changes result in new amino acids, new proteins appear; these new proteins can alter the morphology or physiology of the organism and result in observable or lethal mutants.

Frameshift Mutation.   A point mutation may consist of replacement, addition, or deletion of a base (pair). Point mutations that add or subtract a base are, potentially, the most devastating in their effects on the cell or organism because they change the reading frame of a gene—from the site of mutation on, the whole gene is read differently (Figure 14–3). A **frameshift mutation** causes two problems. First, all the amino acids from the frameshift on will be different and thus yield (most probably) a useless protein. Second, stop-signal information will be misread. One of the new codons may be a nonsense codon, which will cause a premature stoppage of translation. Or, if the translation apparatus reaches the original nonsense codon, it will no longer be recognized as such because it is in a different reading frame, and therefore the translation process will continue beyond the end of the gene (Figure 14–4).

Back Mutation and Suppression.   A second point mutation in the same gene can have one of three possible effects (Figure 14–3). First, the mutation can result in either another mutant codon or one codon that has experienced two changes. The second effect is **back mutation:** if the change is at the same site, the original sequence can be returned—the gene then becomes a revertant—the original function is restored. Third, **intragenic suppression** can take place. Intragenic suppression occurs when a second mutation masks the occurrence of the original mutation without actually restoring the original sequence. The new sequence is a double mutation that appears to have the original (unmutated) phenotype. (Intragenic suppression also produces revertants to the original phenotype.) In the example of Figure 14–3, a T addition is followed by a C deletion that substitutes the CCT sequence for the original CCC. Either of the sequences, when transcribed (GGG, GGA), are codons for glycine. Thus suppression would occur even if the new codon were for a different amino acid, as long as the phenotype of the organ-

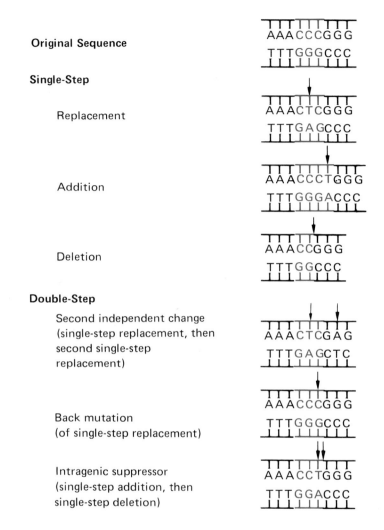

Figure 14 – 3.
Types of Point
Mutations of DNA

ism is reverted approximately to the original. Suppressed mutations can be distinguished from true back mutations either by subtle differences in phenotype or, more exactly, by crossing studies (Chapter 15).

Spontaneous Mutagenesis

Watson and Crick originally suggested that mutation could occur spontaneously during the process of DNA replication if pairing errors occurred. If a base of the DNA shifted into one of its rare tautomeric forms (a **tautomeric shift**) during the replication process, an inappropriate pairing of

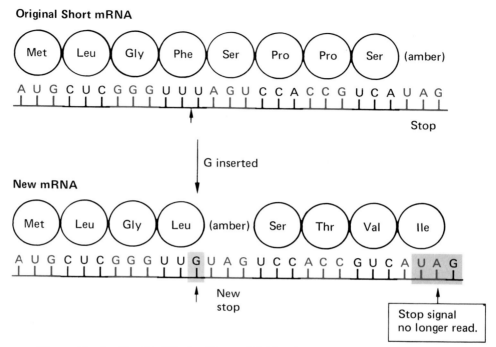

Figure 14 – 4.   Possible Effects of Frameshift Mutation
If the new sequencing had not created a stop signal, reading of the new codons would continue beyond the original end of the gene.

bases would occur. Normally, adenine and cytosine are in the amino (NH_2) form—their tautomeric shifts are to the imino (NH) form. Similarly, guanine and thymine go from a keto (C=O) form to an enol (COH) form (Figure 14–5). Table 14–2 shows the new base pairings that would occur following tautomeric shifts of the DNA bases. An example of the molecular structure of one of these tautomeric pairings is shown in Figure 14–6.

During DNA replication a tautomeric shift in either the incoming base *(substrate transition)* or the base already in the strand *(template transi-*

TABLE 14 –2. Pairing Relationships of DNA Bases in the Normal and Tautomeric Forms

Base	In Normal State, Pairs with:	In Tautomeric State, Pairs with:
A	T	C
T	A	G
G	C	T
C	G	A

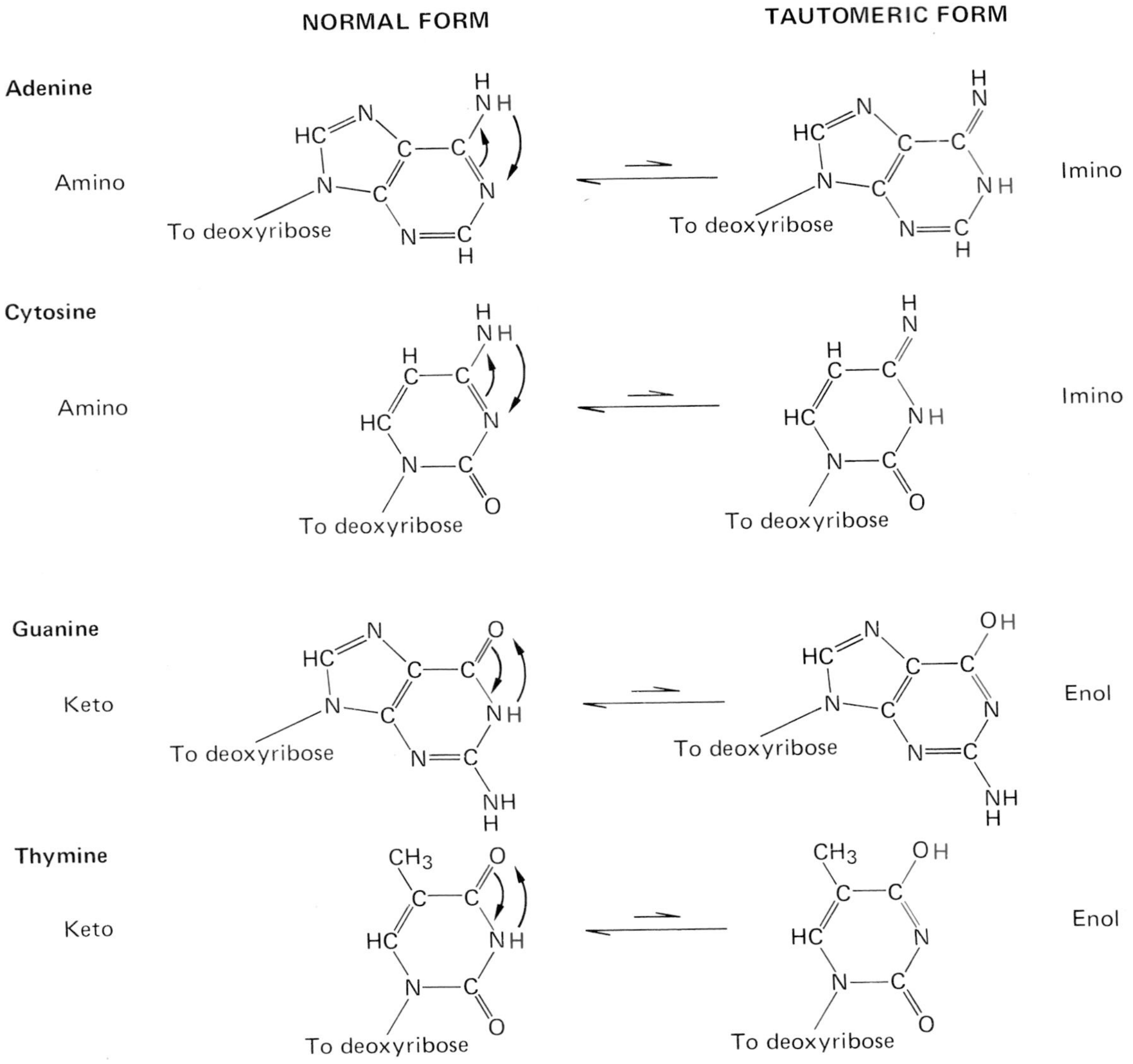

NORMAL FORM

TAUTOMERIC FORM

Figure 14 – 5.   Normal and Tautomeric Forms of DNA Bases

tion) will result in mispairing. The mispairing will be permanent and will result in a new base pair after an additional round of DNA replication. The original strand is unchanged (Figure 14–7).

As can be seen from Figure 14–7, the replacement of one base pair by another maintains the same purine/pyrimidine relationship—that is, AT is replaced by GC and GC by AT. In both examples, a purine-pyrimidine combination is replaced by a purine-pyrimidine combination. (Or, more specifi-

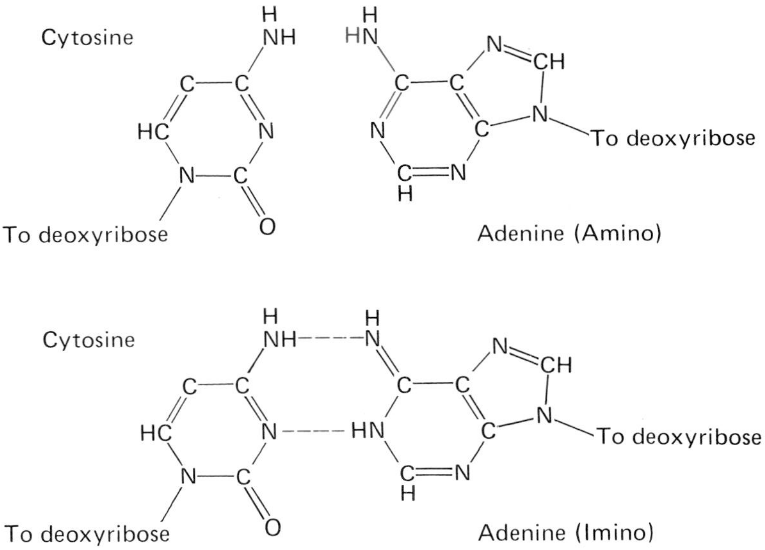

**Figure 14 – 6.
Tautomeric Forms of
Adenine**
In the (normal) amino
form, adenine does
not base pair with
cytosine; in the
tautomeric imino
form, it can.

cally, a purine replaces another purine: Adenine is replaced by guanine in the first example and guanine by adenine in the second.) The mutation is referred to as a **transition mutation:** A purine (or pyrimidine) is replaced by another purine (or pyrimidine) through a transitional state involving a tautomeric shift. The form of replacement, where a purine replaces a pyrimidine or vice versa, is referred to as a **transversion.** It is less clear how transversion mutation occurs.

Chemical Mutagenesis

Muller's original work demonstrated that x rays can cause mutation. Mutation can also be induced by certain chemical and temperature treatments. The following chemicals are commonly used in experimental genetics to induce mutation:

- Base analogues (5-bromouracil, 2-aminopurine)
- Acridine dyes (proflavin, acridine orange)
- Alkylating agents (ethyl methane sulfonate, ethyl ethane sufonate)
- Nitrous acid
- Hydroxylamine

Discussion here of chemical mutagenesis will be restricted to these chemicals.

Template Transition—Tautomerization of adenine in the template

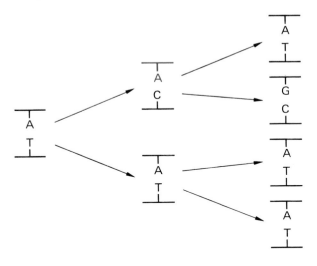

Substrate Transition—Tautomerization of incoming adenine

Figure 14 – 7.
Tautomeric Shifts
Resulting in Copy
Error and
Replacement of One
Base Pair by
Another in DNA
Tautomeric shifts are
shown in blue.

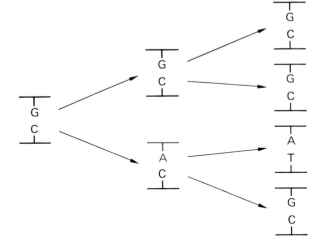

Base Analogues.   The two most widely used base analogues (Figure 14–8) are the pyrimidine analogue 5-bromouracil (5BU) and the purine analogue 2-aminopurine (2AP). The mutagenic mechanism of the two is slightly different. 5BU is incorporated into DNA in place of thymine; it acts just like thymine in DNA replication and, since hydrogen bonding is not changed, it should induce no mutation. However, it seems that the bromine atom causes 5BU to tautomerize more readily than thymine does. Thus, 5BU goes from

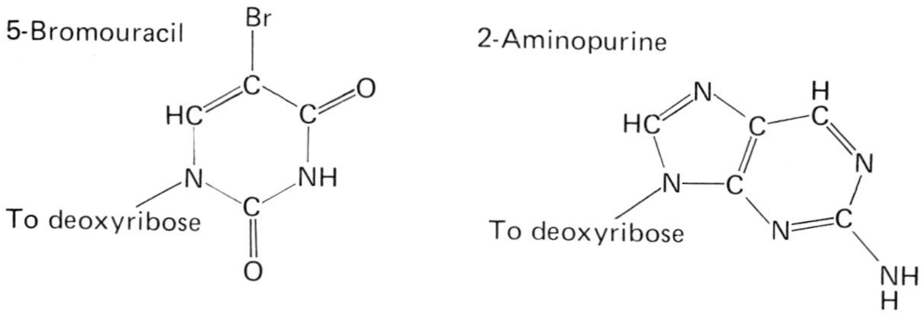

Figure 14 – 8.
Structure of the Base
Analogues
5-Bromouracil (5BU)
and 2-Aminopurine
(2AP)

the keto form (Figure 14–8) to the enol form (as does thymine in Figure 14–5) more readily than thymine. Frequent substrate and template transitions result when the enol form of 5BU pairs with guanine.

2AP is mutagenic by virtue of the fact that it can, like adenine, form two hydrogen bonds with thymine. It can, however, also form a single hydrogen bond with cytosine (Figure 14–9). Thus, at times it replaces adenine and at times, guanine. It promotes both substrate and template transition mutations.

Acridine Dyes.   The molecules of the acridine dyes, such as proflavin and acridine orange (Figure 14–10), are flat. Although the exact mode of action of these molecules is not known, they presumably cause mutation by inserting into the DNA double helix and causing a buckling of the helix in the region of intercalation, which could lead to additions and deletions of bases

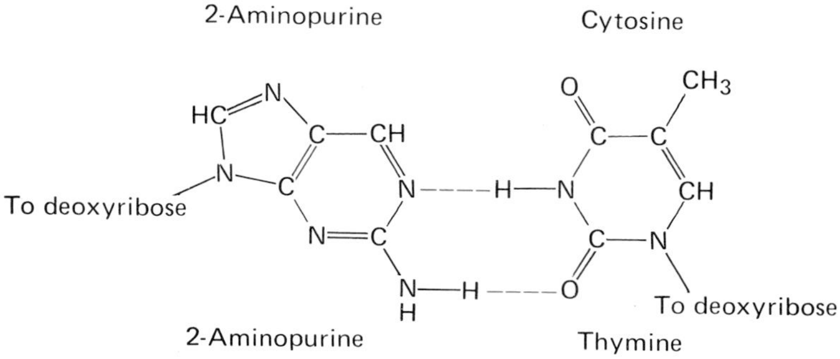

Figure 14 – 9.
Two Possible Pair
Bondings with 2AP

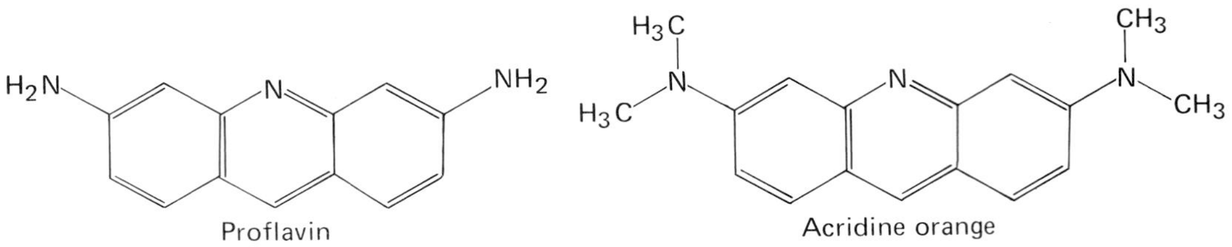

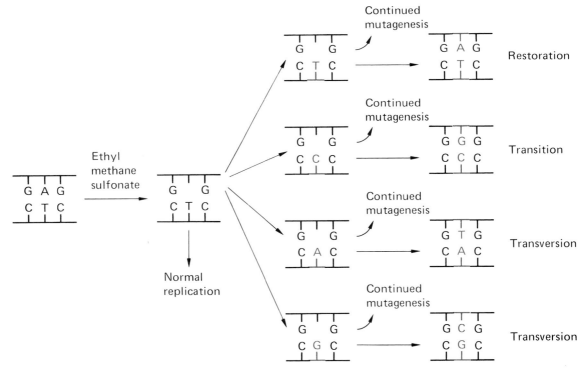

Proflavin   Acridine orange

Figure 14 – 10.   Structure of Two Acridine Dyes

during DNA replication. The use of addition and deletion mutants was mentioned in discussion of the genetic code in Chapter 11. Crick and Brenner used acridine-induced mutations to demonstrate both that the genetic code was read from a fixed point and that it was triplet.

Alkylating Agents. Ethyl methane sulfonate ($CH_3SO_3CH_2CH_3$) and ethyl ethane sulfonate ($CH_3CH_2SO_3CH_2CH_3$) are agents that cause the removal of purine rings from DNA by a multistep process that begins with the ethylating of the purine ring and ends with the hydrolysis of the purine-

Figure 14 – 11.   Four Possible Outcomes after Treatment of DNA with an Alkylating Agent

deoxyribose bond. It is assumed that, when DNA replication and repair (see later discussion) take place, the repair enzyme is free to insert any of the four possible bases into the new strand as a complement to the gap created when these alkylating agents remove a purine (Figure 14–11). If thymine is placed in the newly formed strand, then the original base pair will be restored. Insertion of cytosine will result in a transition mutation; insertion of either adenine or guanine will result in a transversion mutation. Of course, the gap is still there to continue to generate new mutations each generation.

Nitrous Acid.   Nitrous acid (HNO_2) replaces amino groups on nucleotides with keto groups ($-NH_2$ to $=O$), with the result that cytosine is transformed to uracil, adenine to hypoxanthine, and guanine to xanthine. As can be seen from Figure 14–12, transition mutation results from two of the

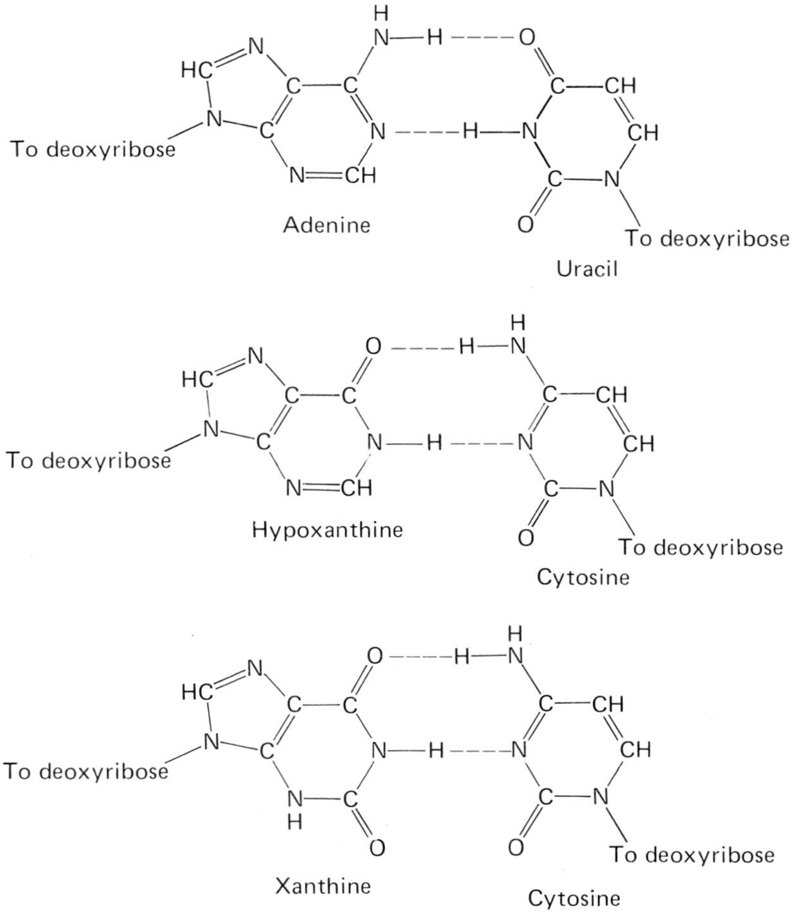

Figure 14 – 12.
Nitrous Acid
Converts Cytosine to
Uracil, Adenine to
Hypoxanthine, and
Guanine to Xanthine
These then base pair
as shown.

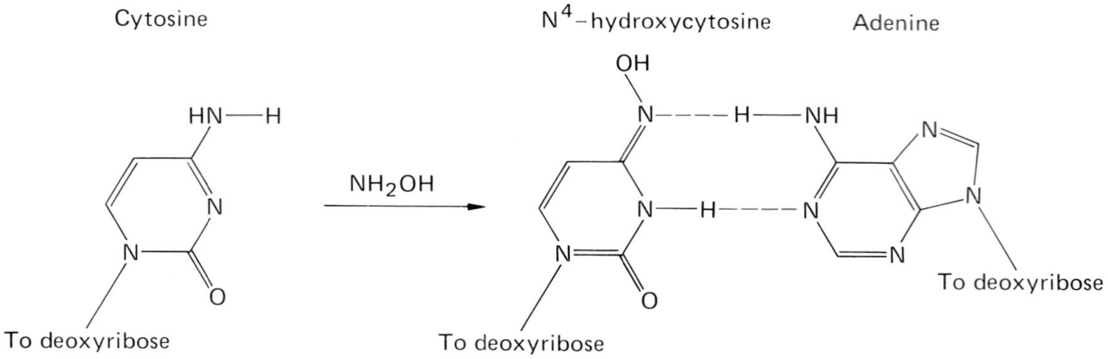

Figure 14 - 13.   Reaction of Hydroxylamine (NH₂OH) with Cytosine
Base pairing of the hydroxylimine derivative with adenine results.

changes. Uracil base pairs with adenine instead of guanine, leading to a UA base pair replacing a CG base pair; hypoxanthine pairs with cytosine instead of thymine, with which the original adenine paired. Thus, an HC base pair replaces an AT base pair. Both of these base pairs (UA and HC) are transition mutations. Xanthine, however, pairs with cytosine just as guanine does. Thus, the change of guanine to xanthine does not result in changes in base pairing. The only change is the actual conversion of guanine to xanthine. Since proper base pairing occurs during DNA replication, the DNA strand containing the xanthine will represent a lower proportion of the total DNA in each generation and will thus be "diluted out" after the nitrous acid is removed.

Like nitrous acid, heat can deaminate cytosine to form uracil and thus bring about transitions (CG to TA). Heat apparently can also bring about transversions. The mechanism is not known.

Hydroxylamine.   Hydroxylamine (NH₂OH) reacts primarily with cytosine and converts it into a hydroxylimine ($=N-OH$), N⁴-hydroxycytosine (Figure 14–13). This derivative pairs with adenine and results in a transition, a CG base pair being replaced by a TA base pair.

Reverse Mutation

Reversion rates can be used to study both the effects of mutagenic agents and the nature of spontaneous mutation. If a mutagen causes a transition mutation, it should also cause—by a transition—reversion of this mutation. For example, Figure 14–14 shows a mutation followed by a reversion, with both caused by nitrous acid. Although any of the mutagens that cause tran-

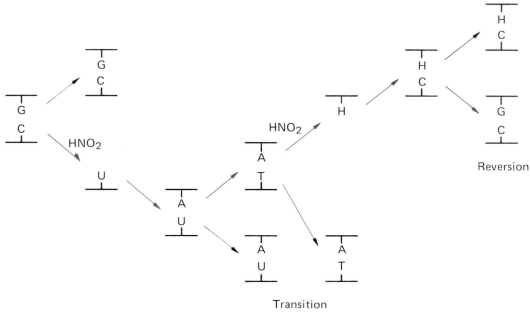

Transition

Figure 14 – 14. Mutation and Reversion of the Mutant by Nitrous Acid

sitions should likewise cause reversions, mutations that involve the addition or deletion of bases are virtually impossible to reverse. This fact can be used as an aid in identifying the process by which both spontaneous and induced mutation occurs.

Freese (1971) studied reversion rates in the rII mutants of phage T4. (See also the work of Benzer in Chapter 15.) Wild type (r⁺) phage produced mutants (rII) either spontaneously or when a mutagen was added. These mutants were tested for their ability to revert to the wild type in the presence or absence of base analogues. The results, presented in Table 14–3, suggest that most mutations caused by transition-inducing mutagens are reversed in the presence of base analogues. Among the transition-inducing mutagens, ethyl ethane sulfonate, which is known to also induce occasional deletions and additions, induced mutations with the lowest reversion rate.

Of the 55 proflavin-induced mutations tested by Freese, only one, or 2%, reverted in the presence of base analogues. This result is consistent with the concept of proflavin acting by causing additions or deletions. The surprising thing is that only 14% of the spontaneous mutations reverted in the presence of base analogues. The implication is that only the low percentage represents simple base-pair transition: 86% must involve additions or deletions of nucleotides.

TABLE 14 –3.   Reversion Percentages of rII Mutants Obtained Either Spontaneously or with Mutagens

Mutagen Used to Induce r^+ → rII (Forward Mutation)	No. of rII Mutants Tested	Percentage of rII Mutants Found Inducible to Revert to r^+ by Base-Analogue Mutagens
2-Aminopurine	98	98
5-Bromouracil	64	95
Hydroxylamine	36	94
Nitrous acid	47	87
Ethyl ethane sulfonate	47	70
Proflavin	55	2
Spontaneous	110	14

Source: From E. Freese, 5th International Congress of Biochemistry, Moscow, 1961.
Note: Reversion was induced by base analogues.

Intergenic Suppression

When a critical mutation occurs in a codon, there are several routes that can lead to survival of the individual. Simple reversion and intragenic suppression have already been considered. A third route is that of **intergenic suppression**—restoration of the function with which a mutated gene is associated by changes at a different gene, called a **suppressor gene.** Suppressor genes can restore proper reading to nonsense, missense, and frameshift mutations. (**Nonsense mutations** convert a codon that originally specified an amino acid into one of the three nonsense codons that do not. **Missense mutations** change a codon so that it specifies a different amino acid. Frameshift mutations were defined earlier.) Suppressor genes are usually tRNA genes. When mutated, intergenic suppressors change the way in which an altered codon is read.

Nonsense Suppression.   Remember that a nonsense mutation results in interruption of translation at the mutation point. The tRNA produced by a nonsense suppressor reads the nonsense codon as if it were a codon for an amino acid. Thus, an amino acid is placed into the protein and reading of the mRNA continues.

Suppression of all three nonsense codons in *E. coli* is known: All nonsense suppressors are altered tRNA genes. For example, at least three suppressors of the mutant amber codon (UAG) are known. One of these suppressors puts tyrosine, one puts glutamine, and one puts serine into the protein chain at the point of an amber codon. Normally, tyrosine tRNA has the anticodon 3′AUG5′. The suppressor that reads amber as a tyrosine codon is a tRNA gene that transcribes a tRNA with the anticodon 3′AUC5′, which is complementary to amber. Hence a mutated tyrosine tRNA reads amber as a tyrosine codon. The question surfaces: If the amber nonsense codon is no longer read as a stop signal, then won't all the genes (or at least all terminating in amber) continue to be translated beyond their ends, thus resulting in the death of the cell? In the tyrosine case, it was found that there are two genes for tyrosine tRNA. One contributes the major fraction of the tRNAs and one, the minor fraction. It is the mutant minor-fraction gene that acts as the suppressor. Thus, most mRNAs are translated normally; therefore, most amber nonsense mutations are terminated prematurely, although a sufficient number are translated to ensure the viability of the mutant cell. In general, the intergenic suppressor mutants would be quickly eliminated in nature because they are inefficient; in the lab they can be cultivated and studied.

Frameshift Suppression. Intergenic frameshift suppressors are also tRNAs. There is, for example, a mutation of one of the glycine codons (GGG) where an additional guanine has been inserted. Normally, this insertion would throw off the whole reading frame; but a suppressor is known—a glycine tRNA that has an anticodon region of four cytosines (Figure 14–15). This suppressor reads the GGGG "codon" as glycine and, in addition, moves the mRNA across the ribosome by four nucleotides, thereby restoring the reading frame. The suppressor itself arose by an insertion of a guanine into the DNA. The mutated tRNA is in all other respects identical to the original tRNA.

Missense Suppression. Missense suppressors also tend to be tRNA mutations. For example, there is a mutation of the *E. coli* tryptophan synthetase gene that changes a GGG codon to an AGG codon and thus substitutes an arginine for a glycine. A suppressor of this mutation is the glycine tRNA that recognizes the AGG sequence (anticodon mutation of 3'CCC5' to 3'UCC5'). GGG is still recognized as glycine because, due to wobble, a second tRNA for glycine recognizes both GGG and GGA.

ram Mutation. Ribosomal mutants that can misread an mRNA are another class of intergenic suppressors. All three nonsense codons are suppressed by **ram** (*r*ibosomal *am*biguity) mutants. Apparently, structural changes in the ribosome make it possible for the wrong tRNA to appear complementary to a particular codon.

Mutator and Antimutator Mutations. While suppressors are mutations that "restore" the normal phenotype, mostly through mutation of tRNA loci, there are also mutations known as **mutator** and **antimutator mutations.** These mutations, as their names imply, cause an increase or decrease in the overall mutation rate of the cell. They seem to be mutations of DNA polymerase, which not only polymerizes DNA nucleotides comple-

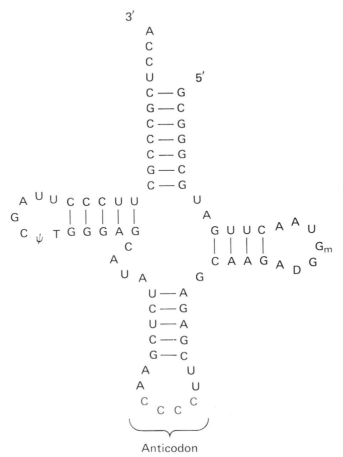

Figure 14 – 15.   Structure of the Glycine Frameshift Suppressor tRNA with a Four-Nucleotide Anticodon

mentary to the template strand but also checks (proofreads) to be sure that the correct base was put in. If, in the proofreading process, the polymerase discovers an error, it can correct this error with its exonuclease activity. Presumably, mutator and antimutator mutations involve changes in the proofreading ability (exonuclease efficiency) of the polymerase.

Phage T4 has its own DNA polymerase, specified by gene 43, with known mutator and antimutator mutants. It has been found that mutator mutants are very poor proofreaders (low exonuclease to polymerase ratios) and thus introduce mutations throughout the phage genome. Antimutator mutants, however, have exceptionally efficient proofreading ability (high exonuclease to polymerase ratios) in their DNA polymerase and, therefore, there is a very low mutation rate for the whole genome.

Radiation Mutagenesis

X Rays. X rays have short, high-energy wavelengths in the range of 0.1 to 10 angstroms. (The wavelength of visible light is about 10,000 times longer.) X rays penetrate and upon contact with atoms cause them to release electrons and become ions; x rays are, therefore, called **ionizing radiation.** Alpha, beta, and gamma rays from radioactive sources also act as ionizing radiation, as do fast-moving particles such as neutrons and protons. Radiation mutagens, then, produce mutation by the direct ionization of DNA or by producing ionized particles in proximity to the DNA and then causing mutation of the DNA.

Target Theory. The **target theory** proposes that the shape of the mutation response curve depends on whether the mutation is due to a single "hit" or to multiple "hits." There is a linear relationship between radiation dose and mutation rate that indicates a single-hit phenomenon. But, with multiple-hit mutations, such as chromosome rearrangements, the response to increased dose is geometric (Figure 14–16). (Chromosome rearrangements such as inversions, deletions, or translocations require DNA to be broken at two points and then put together in a new arrangement. Hence two events—hits or breaks—are needed for this kind of mutation. Two-hit mutation is thus dependent on the square of the dose rate, and hence the response curve is geometric.)

Ionizing radiation is measured in *roentgen units* (r) after their discoverer or, alternatively, in *rads*. A roentgen unit produces about two ionizations per cubic micron of water. Rads are measures of absorbed radiation and are usually more appropriate to biological tissue. At low doses a rad is only slightly more energy than a roentgen unit. The relationship between

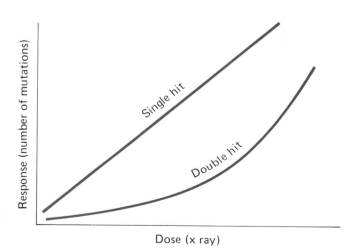

Figure 14 – 16. Single-Hit and Double-Hit Response Curves According to Target Theory

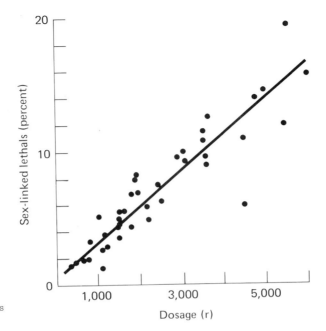

Figure 14 – 17.
Relationship of
X-Ray Dosage, in
Roentgens, to Sex-
Linked Lethals in
Drosophila
Data are from various
sources.

dosage and mutation is shown in Figure 14–17. This linear relationship supports the target theory that x rays act to cause mutations by a single "hit." Also supportive of the target theory is the fact that the mutagenic effect of radiation is the same regardless of whether it is given all at once or spread out over a period of time.

However, straightforward support of target theory breaks down because of repair mechanisms. In higher organisms the amount of mutational damage repaired by these mechanisms depends in part on the length of the time interval between exposure to radiation and fertilization. Thus, the mutation rate is a function of both dosage (direct relationship) and this time interval (inverse relationship). That is, for example, mice given a low radiation dose can repair almost all mutational damage if a sufficient time interval is allowed before mating. Higher doses or shorter intervals will result in less repair. Repair will be discussed in more detail later. Apparently, in eukaryotes, higher organisms—for example, mice—have better repair mechanisms than lower organisms—for example, flies.

Ultraviolet Light.   Ultraviolet light, with wavelengths 100–1000 times longer than x rays, is not an ionizing radiation—it affects only molecules that absorb it directly. Primarily, ultraviolet (UV) light causes linkage, or **dimerization,** of adjacent pyrimidines on DNA (Figure 14–18). While cytosine–cytosine and cytosine–thymine dimers are produced, the principal product of UV irradiation is thymine-thymine dimerization—that is, the dimerization of adjacent thymines on the same strand of DNA. Dimerization creates

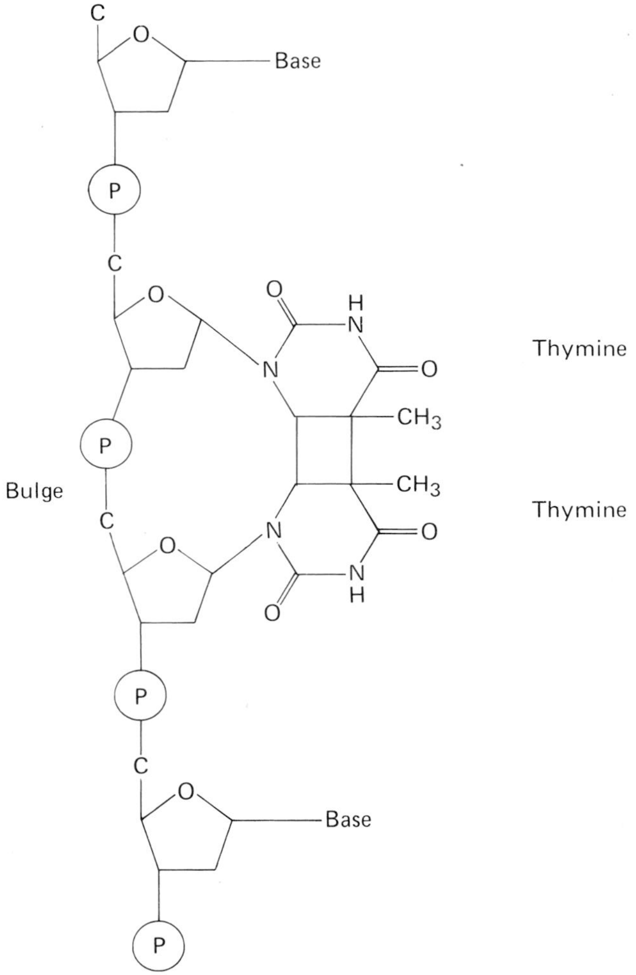

Figure 14 – 18.
UV-Induced
Dimerization of
Adjacent Thymines
in DNA

a distortion in the double helix and this distortion interferes with both DNA replication and RNA transcription. The presence of a dimer will result in breaks and mutations in the newly synthesized DNA. Presumably, the mutations result from misrepair of the DNA.

DNA REPAIR

Several systems are involved in repair of UV damage in *E. coli*. These systems primarily repair forms of damage that produce distortions in the double helix, of which the best known example is the dimerization of thymine. The distortion created in the double helix by dimerization of adjacent thymines

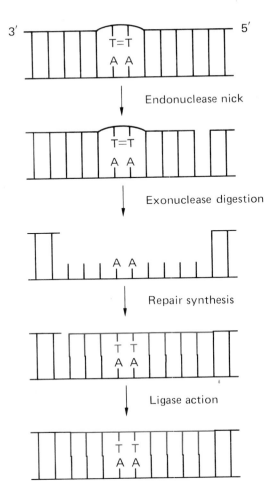

Figure 14 – 19.
Mechanism of
Excision (Dark)
Repair of Thymine
Dimerization

can be detected by repair enzymes. The major forms of repair of thymine dimerization are *photoreactivation, excision repair,* and *postreplicative repair.*

Photoreactivation (Light Repair)

In *E. coli* there is an enzyme, deoxyribodipyrimidine photolyase, the product of the *phr* gene (for **photoreactivation**) that will bind in the dark to dimerized thymine. When light shines on the cells, the enzyme breaks the bonds of the thymine dimer and then falls free of the DNA. This enzyme thus reverses the UV-induced dimerization. Mutants of this gene, *phr⁻*, produce a phenotype that cannot photoreactivate the DNA.

Excision Repair (Dark Repair)

The mechanism of **excision repair** is not dependent on light. If DNA is synthesized in the presence of radioactive thymine and is then exposed to UV light, thymine dimerization will occur. In the absence of photoreactivation, radioactive thymine dimers are found in the cytoplasm after about half an hour. Thus, a mechanism exists whereby there is excision of the dimers rather than an undoing of the dimerization. This excision is accomplished in the following manner. An endonuclease that can detect thymine dimers makes a nick in the DNA strand on the 5′ end of the dimer (Figure 14–19). (This endonuclease is most likely composed of subunits coded by the *uvrA, uvrB,* and *uvrC* genes.) Then, an exonuclease (usually DNA polymerase I in its exonuclease capacity) begins digestion (removal of nucelotides) at the nick in the 5′ to 3′ direction. The digestion continues past the dimer until, usually, about 20 nucleotides are excised. A gap is left in the DNA—the dimer is gone. The DNA polymerase I then repairs the gap by using the remaining strand for complementarity. The gap is filled in the usual 5′ to 3′ direction. When the final nucleotide is added, the remaining ester bond is formed by DNA ligase. Thus, a relatively simple system of locating, excising, and then repairing the damage caused by UV light occurs.

Excision repair systems are found in virtually all organisms, from the largest viruses to eukaryotes. In humans there is an autosomal recessive trait termed *xeroderma pigmentosum.* Persons with this trait freckle heavily when exposed to the UV rays of the sun, and they have a high incidence of skin cancer. The disease is due to an inability to repair thymine dimerization induced by UV light.

Postreplicative Repair

A third mechanism for repair of thymine dimers and other DNA lesions is initiated at replication, just after the polymerase bypasses the lesioned area. Termed **postreplicative repair,** this mechanism allows replication to continue even in regions where damage has been missed by photoreactivation or excision repair. If a lesion, such as a thymine dimer, is not repaired, the normal process of DNA replication will be disrupted because DNA polymerase (Pol III) cannot make a complementary strand opposite some lesions (such as thymine dimers). When the polymerase comes to this dimer, it will skip past it after several seconds; it moves as far as about 800 nucleotides down the strand and leaves a large gap in the new double helix. If allowed to remain, this gap results in broken pieces of DNA after another round of replication. The gap is repaired by enzymes belonging to the *rec* system— mutants of genes at the *rec* loci are *rec*ombination-deficient genes. Following repair by the *rec* system, the original lesion remains, but an intact double helix can be synthesized.

The *rec* system ties together mutation repair and recombination because *E. coli* cells with the *rec⁻* gene do not have the ability for postreplicative repair and are recombinationless—that is, they do not undergo recombination after a new piece of DNA is brought into the cell (as from conjugation or transduction). The *rec* system consists of three genes controlling two known enzymes and of several other genes whose function is unknown. The *recA* gene controls the translation of an enzyme that is involved in aiding homologous single strands of DNA to reform double helixes. The *recB* and *recC* genes produce subunits of an enzyme that has exonuclease activity. This enzyme is called *recBC exonuclease,* or *exonuclease V.* Presumably, the *rec* system repairs the dimer-created gap in DNA as follows. The recBC exonuclease causes the newly synthesized daughter double helix without the lesion to open up in the region of the lesion. The strands are stabilized as single strands by unwinding proteins (Chapter 9).

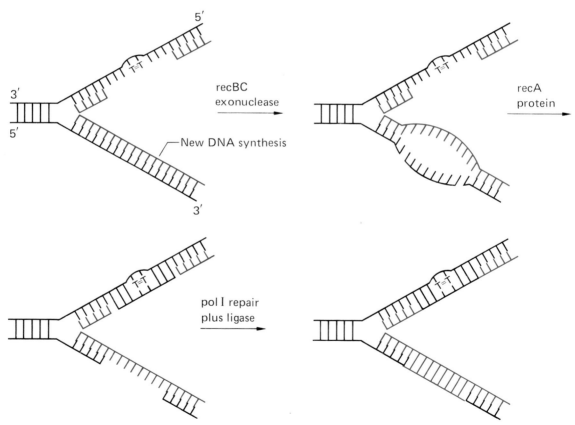

Figure 14 – 20.   A Plausible Scheme for the Roles of recA and recBC Enzymes in Postreplication Repair

The recA enzyme probably joins the proper single strand of the new double helix with the region of the lesion (Figure 14–20). Then, with the aid of repair enzymes and ligase, both double helixes are restored. The scheme is tentative. It has emerged out of much new information and explains how recombination and postreplicative repair are linked. Steps similar to the steps of Figure 14–20 are involved in the recombination process.

There are several other possible postreplicative repair mechanisms and many enzymes involved in the processes of repair and recombination. In 1979 Hanawalt and associates listed 66 known genes involved in the process of DNA repair alone. Another postreplicative repair mechanism that may or may not involve *recA* is termed *SOS repair,* a "last-ditch" effort on the cell's part to complete the gapped double helix, even in the presence of extensive mutation. In this proposed mechanism, after DNA polymerase III skips past a lesion such as a thymine dimer, the cell fills in the gap without using template information for complementarity—the site will contain mutations, but at least the integrity of the double helix will be maintained (Figure 14–21).

The postreplicative repair mechanisms, especially in the region of the lesion itself, seem less efficient than excision repair or normal DNA replication: Repair of the lesions can result in transitions, transversions, additions, and deletions. Thus, error in postreplicative repair can account for some of the mutational effects of some mutagens, such as proflavin and alkylating agents, and makes the occurrence of transversions, deletions, and additions understandable.

RECOMBINATION

Since Rec⁻ *E. coli* cells lack the ability for both recombination and postreplicative repair of mutation damage, they provide some insight into the type of mechanism involved in the recombination process. While **recombina-**

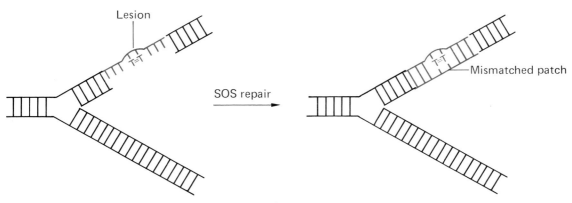

Figure 14 – 21.   SOS Repair of DNA

Figure 14 – 22.
Copy-Choice Model
for DNA Replication

tion, the nonparental arrangement of alleles in progeny, can come about both by independent assortment and crossing over, we will be concerned here with crossing over as the mechanism of recombination.

Copy Choice

From the 1930s to the 1950s, a **copy-choice hypothesis,** originally suggested by Belling for eukaryotes, was considered a possible model for recombination. In essence, this model proposed that as DNA replication took place, the replicating enzyme could switch from one chromosome to its homologue (Figure 14–22). Evidence such as the occurrence of recombination in the virtual absence of new DNA and the relationship of recombination to mutation repair suggested that the copy-choice hypothesis was almost assuredly incorrect.

Breakage and Reunion Phenomena

A more workable model for recombination is offered by the **breakage-and-reunion hypothesis,** namely: Homologous parts of chromosomes come into apposition; both strands are broken and then rejoined (Figure 14–23). This

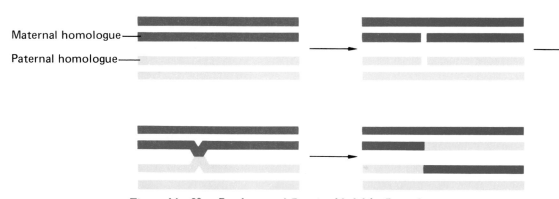

Figure 14 – 23.   Breakage-and-Reunion Model for Recombination in a Tetrad at Meiosis

general model fits what is known about the concordance of recombination and repair—that is, that both involve breakage of the DNA followed by a small amount of repair synthesis. Several breakage-and-reunion models have been sugested.

Apparent Reciprocity. These models must explain several facts, one of which is reciprocity. In Chapter 5 recombination in eukaryotes was described as a breakage and reunion (crossing over) of homologous chromatids. No genetic material was lost, gained, or changed. The sum of the two recombinant chromatids contained the same total genetic material as the original chromatids. Hence **reciprocity** means the conservation of the genetic material in the sense that a recombinant chromatid contains a new arrangement of genetic material but not a change in the DNA. All the remaining alleles are still present in the reciprocal recombinant chromatid. (In a nonreciprocal event the sum of the DNA of the two recombinant chromatids would not exactly equal the sum of the original DNA.)

Nonreciprocity in Recombination. Reciprocity holds also for bacterial and viral "populations." For example, if T2 phage with h and r genotypes are crossed by multiple infection of bacteria, equal numbers of $h\,r$ and h^+ r^+ phage are obtained in the daughter phage. On careful inspection, however, reciprocity does *not* hold in prokaryotes. Hershey found in the phage-cross example just described that, in the progeny phage from any *one* bacterium, the cross had produced only one of the recombinant classes. In bacteria, recombination is a process of integrating foreign DNA into the host chromosome. Thus, by its very nature, the process is nonreciprocal. Nonreciprocity is also found in eukaryotes.

Gene Conversion. In yeast, an ascomycete fungus, the products of meiosis are kept together in the ascus. It is thus possible to determine if there has been reciprocity. If an a mutant has been crossed with the wild type, $+$, we expect a $2a:2+$ ratio in the spores after meiosis. It turns out that there is a $3a:1+$ or a $1a:3+$ ratio among the spores from many of the asci of this cross. The result is called **gene conversion** and is associated with the reciprocal exchange of marker loci on either side of the converted locus. It becomes clear that any model of recombination must explain nonreciprocal recombination in both prokaryotes and eukaryotes.

Negative Interference. A model of recombination should also explain negative interference. In eukaryotes, crossover in one area of a chromosome has the effect of inhibiting a second crossover in the same general area. This effect is called *interference,* or *positive interference.* However, *within a locus,* a single crossover does not prevent several other crossovers in the same vicinity; it actually enhances the probability of other crossovers in the same region. The intralocus effect of a crossover has been termed **negative**

interference. It occurs in prokaryotes and eukaryotes. (Crossing over within a locus has the same mechanism as crossing over between loci and will be analyzed in the next chapter.)

Holliday Mechanism of Breakage and Reunion

General Reciprocity. All currently accepted models of recombination include the formation of hybrid DNA molecules between the two strands about to undergo recombination. The occurrence of complementary base pairings is consistent with the fact that recombination is an exact process: Rarely does it result in additions or deletions of nucleotides. A modified model by Holliday (1964) is described here. This model yields reciprocal recombinants for genes on either side of a crossover.

The first step in recombination is the breaking of two homologous double helixes, each at the same place and each on only one strand of the duplex (double helix). This step is shown in Figure 14–24. The broken strands then pair with their complement on the other duplex and covalent bonds are formed. The crossover point can slide down the duplexes. In order to release the cross-linked duplexes, a second cut in each double helix is required. If the cut occurs on the same strands as were originally cut, there will be no recombination of loci on both sides of the hybrid piece (patch formation). If, however, the cuts are in the other two strands, there will be a reciprocal recombination of loci at the ends (splice formation).

It should be evident that many of the same enzymes involved in mutation repair, primarily postreplicative, could control the recombination process. Two duplexes have to be cut by an endonuclease. The duplexes then have to unfold and the cut single strands rejoined with sister duplexes. Ligation is necessary. Then, further cuts are needed to free the cross-linked DNA. There might even be the necessity for digestion of short segments and refilling these segments with polymerase. The sequence is almost identical to postreplicative repair, which uses the *rec* system.

Nonreciprocity. The Holliday model is easily adapted to a nonreciprocal incorporation of genetic material in bacteria (Figure 14–25). Here, no original cuts are made by an endonuclease. Instead, part of the *E. coli* chromosome separates and complementary pairing between it and the exogenous DNA (exogenote) occurs (possibly under the control of the *rec* system). After this pairing, the unpaired segments of the double helix of the bacteria and the exogenote are both degraded. Finally, the remaining double helix is sealed by polymerase and ligase action.

The result of the recombination is a length of **hybrid DNA.** This hybrid DNA, also called **heterozygous DNA** or **heteroduplex DNA,** has one of two fates, if we assume a difference in base sequences in the two

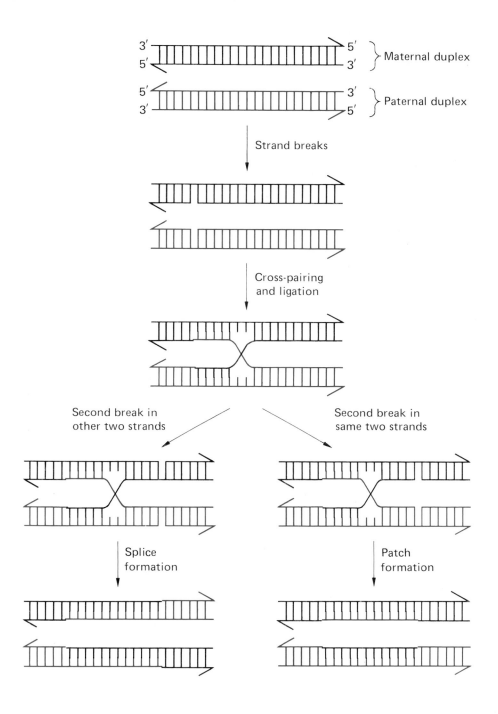

Figure 14 – 24. Basic Holliday Model of Recombination
Two homologous double helixes are shown of the foru present in a
meiotic tetrad. The arrows represent directionality of the strands (3' to 5').

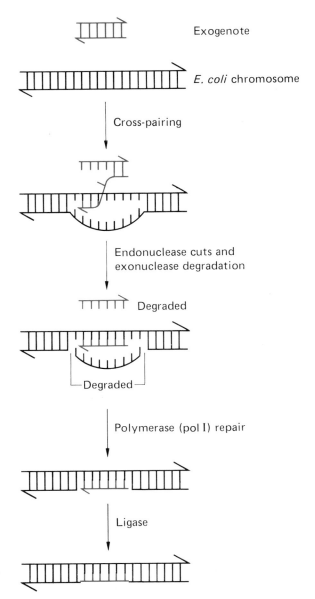

Figure 14 – 25.
Holliday Model
Adapted to
Nonreciprocal
Incorporation of
DNA into the *E. coli*
Chromosome

strands. The heteroduplex can either separate unchanged at the next cell division, or the cell's excision repair mechanisms can attack it (Figure 14–26). The lack of proper base pairing, such as CA in Figure 14–26, may be detectable by an as yet uncharacterized, mismatch repair system or, alternatively, by the excision repair system. In fact, the excision repair system would account for gene conversion in yeast.

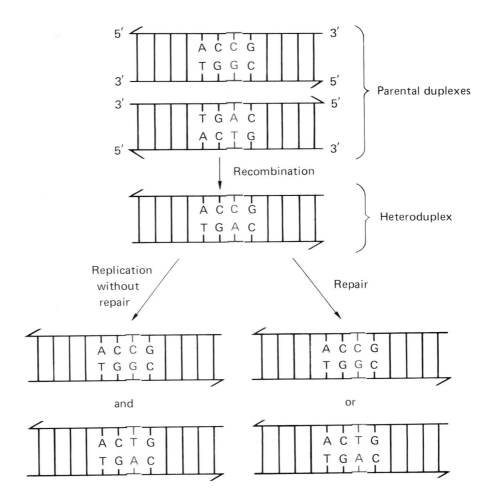

Figure 14 – 26.
Fate of a
Heteroduplex DNA

Gene Conversion. As shown in Figure 14–24, a recombination, even in yeast or any other eukaryote, will generate two heteroduplexes regardless of whether or not there is recombination for outside markers. The repair process can cause gene conversion (Figure 14–27)—the mismatched AC will be changed to an AT or a GC base pair; the TG base pair will be changed to TA or CG. The end result of the repair, as shown in the bottom of Figure 14–27, can be gene conversion from a a + + to a a a + or a + + +. If the heteroduplexes are not repaired, then a single cell will generate both kinds of offspring after one round of DNA replication. Thus the colony from this cell will be half wild type (+) and half a.

Negative Interference. Negative interference can be explained one of two ways. One explanation is that once two double helixes come into close

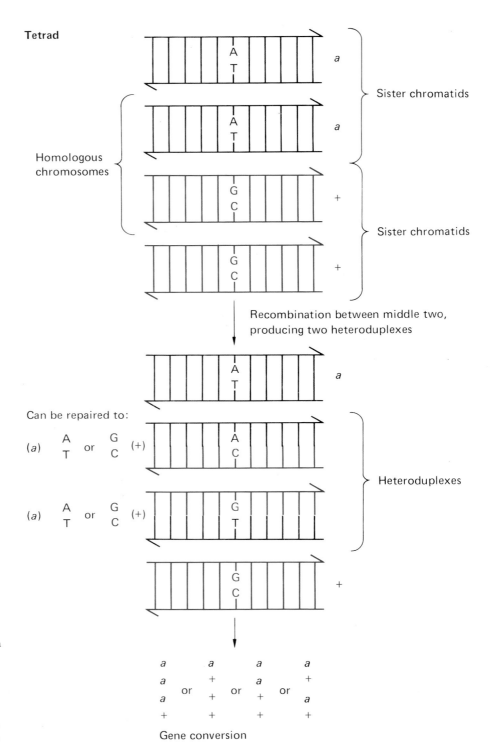

Figure 14 – 27.
Gene Conversion in Yeast Caused by Repair after Recombination
In this figure the difference between two alleles is one base pair (a allele is AT; + allele is CG).

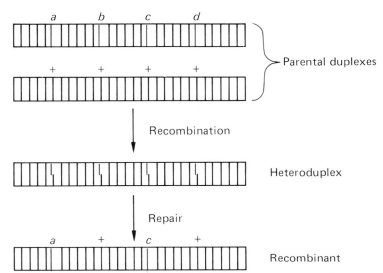

Figure 14 – 28.
High Negative
Interference as a
Consequence of a
Single
Recombination Event
Coupled with Repair

enough proximity to undergo recombination, the enzymes can cause several exchanges to occur in a short segment of DNA. This explanation is consistent with the fact that high negative interference occurs only within short segments. The other explanation is that high negative interference is simply the apparent effect caused by the randomness of the repair process. That is, if several mutations are near one another on the DNA, one recombination can cause the formation of a single heteroduplex that contains the mutant mismatched base pairs. Then, depending on the randomness of the repair process, multiple crossovers will appear to have taken place (Figure 14–28).

Phage Recombination. The final point that a recombination model must explain is the apparent lack of reciprocity in phage crosses. Several models, variants of the cut-heteroduplex-repair models, have been suggested. One of the models, by Broker and Lehman, is shown in Figure 14–29. Homologous regions of the two duplexes line up, and cuts are made by an endonuclease in both duplexes. There is pairing of one strand each from the two double helixes, after some digestion by exonucleases. The resulting figure is H shaped. The region of pairing can progress down the chromosome. Then, endonucleases attack two of the "tails" of the figure. Polymerase and ligase complete the task of forming *one* final double helix. It can be seen from Figure 14–29 that the recombination is not a reciprocal event. Electron microscope evidence for this model is shown in Figure 14–30.

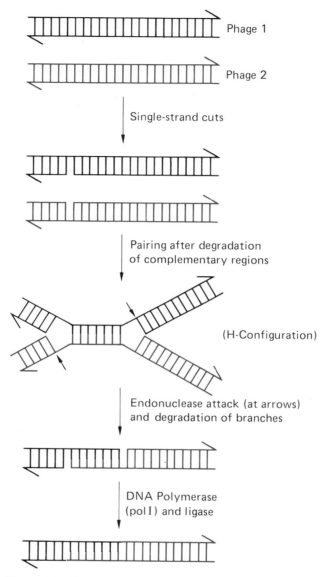

Figure 14 – 29. Model to Explain Nonreciprocal Recombination in Phage

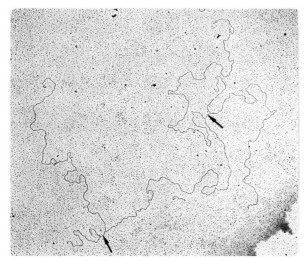

Figure 14 – 30.
Electron Micrograph
of a DNA Molecule
from an *E. coli* Cell
Infected with an
Average of 18 Phage
Two H-joints are
shown at the arrows.
magnification
23,600X

Source: T. R. Broker and I. R. Lehman, "Branched DNA Molecules: Intermediates in T4 Recombination," *Journal of Molecular Biology* 60 (1971):131–149. Reproduced by permission.

CHAPTER SUMMARY

In 1943 Luria and Delbrück demonstrated that bacterial changes are true mutations similar to mutations in higher organisms by showing that there was a high variability in the number of mutants in small cultures as compared to the number of mutants in repeated subsamples of a large culture. Mutations (1) occur spontaneously and (2) are caused by mutagens, which include chemicals and ionizing and nonionizing radiation. This chapter has been primarily concerned with mutations that are nucleotide or point mutations rather than changes in whole chromosomes or chromosome parts.

After a mutation, the normal phenotype, or an approximation to it, can be restored either by simple back mutation or, alternatively, by suppression. Intragenic suppression occurs when a second mutation within the same gene causes a return of normal or nearly normal function. Intergenic suppression occurs when a second mutation happens in a tRNA gene or occasionally in a gene for one of the ribosomal proteins (*ram* mutations). Nonsense, missense, and frameshift mutations can all be suppressed.

Spontaneous mutation primarily occurs because of tautomerization of the bases of DNA. If a base is in the rare form during DNA replication, it can form unusual base pairings and result in mutation. The mechanisms by which the most commonly used mutagens work have been outlined.

The cell has many mechanisms for repairing DNA damage. The most commonly studied lesion of DNA is dimerization of adjacent thymines, which causes a bulge in the double helix that is detectable by several types of enzymes. The simplest repair mechanism is photoreactivation, where the adjacent thymines are undimerized. A second mechanism is excision repair, which involves removal of the region of the DNA strand that has the dimer, followed by polymerization to reconstitute the original double helix. If these two mechanisms do not repair the DNA, then the dimer will still be present at the time of DNA replication. Several other mechanisms then can repair the DNA after the polymerase has bypassed the region of the dimer.

Postreplicative repair is primarily brought about by two enzymes in the *rec* system coded by genes *recA, recB,* and *recC.* The recA protein is involved in rejoining single-stranded DNA and the recBC protein is involved in creating single strands. With several other enzymes, a portion of DNA from the sister helix is presumably inserted into the gap left by the polymerase opposite the lesion. Final repair and ligation then follow. The lesion remains, but the DNA is intact. An SOS repair system may function to simply fill in the gap and thus disregards the necessity for complementarity.

The process of recombination uses many, if not all, of the steps required for postreplicative repair. The Holliday model of recombination explains most of the phenomena that accompany recombination, such as gene conversion and high negative interference. Other models and modifications of the Holliday model contribute to an understanding of nonreciprocity during recombination. Most recombination models agree on a heterozygous intermediate stage. Thus, briefly, there is a battery of enzymes within the cell (the bacterial cell has been most studied) that can modify DNA. These enzymes serve in DNA replication, mutation repair, and recombination.

EXERCISES AND PROBLEMS

1. Construct a data set that Luria and Delbrück might have obtained that would prove the mutation theory wrong.

2. Design a system, similar to *ClB,* that would detect a lethal on a particular autosome in *Drosophila.*

3. Using the chemical mutagens mentioned in the chapter (5BU, 2AP, proflavin, ethyl methane sulfonate, nitrous acid, and hydroxylamine), create a flowchart to show which mutagens can revert mutations caused by the others. Which can revert their own mutations?

4. A point mutation occurs in a particular gene. Describe the types of mutation events that can restore a functional protein. Include intergenic events. Consider missense, nonsense, and frameshift mutations.

5. UV light causes thymine dimerization. Describe the mechanisms, in order of efficiency, that can repair the damage.

6. Diagram, in careful detail, a recombination by way of the Holliday model. Note where specific enzymes are required, and name them.

7. Diagram the tautomeric base pairings in DNA. What base pair replacements occur because of the shifts?

8. What enzymes do photoreactivation, excision repair, and postreplicative repair all use?

9. Describe at least two experiments that could distinguish copy choice from breakage and reunion as the mechanism of recombination.

SUGGESTIONS FOR FURTHER READING

Belling, J. 1933. Crossing over and gene rearrangement in flowering plants. *Genetics* 18:388–413.

Brenner, S., et al. 1961. The theory of mutagenesis. *J. Mol. Biol.* 3:121–124.

Broker, T. R., and I. R. Lehman. 1971. Branched DNA molecules: intermediates in T4 recombination. *J. Mol. Biol.* 60:131–150.

Freese, E. 1971. Molecular mechanisms of mutations. In A. Hollaender, ed., *Chemical Mutagens*, vol. 1. New York: Plenum Publishing, pp. 1–56.

Hanawalt, P., et al. 1979. DNA repair in bacteria and mammalian cells. *Ann. Rev. Biochem.* 48:783–836.

Holliday, R. 1964. A mechanism for gene conversion in fungi. *Genet. Res.* 5:282–304.

Kornberg, A. 1980. *DNA Replication.* San Francisco: Freeman.

Luria, S., and M. Delbrück. 1943. Mutations of bacteria from virus sensitivity to virus resistance. *Genetics* 28:491–511.

Muller, H. J. 1927. Artificial transmutation of the gene. *Science* 66:84–87.

Stadler, L. J. 1928. Mutations in barley induced by X-rays and radium. *Science* 68:186–187.

Stahl, F. 1979. *Genetic Recombination.* San Francisco: Freeman.

Stent, G., and R. Calendar. 1978. *Molecular Genetics: An Introductory Narrative*, 2nd ed. San Francisco: Freeman.

Watson, J. D. 1976. *Molecular Biology of the Gene*, 3rd ed. Menlo Park, Calif.: Benjamin.

15

Gene Function and Structure

Throughout this book, the concept of mutation has been pervasive. In this chapter we pause to examine in depth the ways in which mutation changes proteins and the effects of these changes on the phenotype. Detailed genetic analysis gives us new insights into the meaning of the word *allele*.

EFFECT OF A MUTATION

Mutations of β-Galactosidase

Here, the consequences of mutation will be examined in a bit more detail than in the previous chapter. For example, when a β-galactosidase mutation occurs in the *lac* operon system (z^+ mutating to z^-), the z^- allele does not code for a protein with galactosidase activity. This failure could originate at several different points in the system. First, if no z-gene protein at all is produced (there are immunological techniques that can identify the presence of a protein), the gene or a part of the system that initiates either transcription or translation of the gene may have been deleted. The failure could not have been in the initiation of transcription, however, if the other gene products of the *lac* operon are present. Second, the lack of z-gene protein could be accounted for by a nonsense z-gene mutation such that translation is terminated almost as soon as it begins.

Third, if there is a z-gene product, a frameshift mutation may have occurred such that the protein produced bears little resemblance to the z-gene enzyme in structure. (This protein might not even make its presence known through immunological technique.) Fourth, and most simply, z-gene failure may be the result of a single point mutation whereby a codon is changed in such a way that a new amino acid is placed in the z-gene protein at that point (missense mutation). If this mutation occurs at the active site—the place on the protein where catalysis takes place (Figure 15–1)—it can destroy the enzymatic activity of the protein.

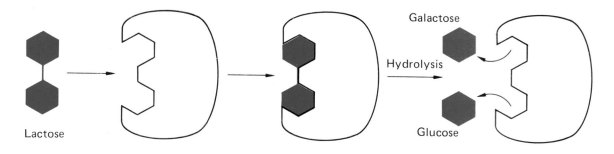

β-galactosidase

Figure 15 – 1.   Active Site of β-Galactosidase
An enzyme-substrate complex is formed; then hydrolysis of lactose into galactose and glucose follows.

Figure 15–1 shows the catalytic role of the z-gene enzyme β-galactosidase in the hydrolysis of lactose into galactose and glucose. If mutation replaced a key amino acid in the active site of the z-gene product, the catalytic process might not take place because the replacement prevented fit of the substrate or because the altered protein failed to induce hydrolysis or did not permit release of the end products. Mutations that change amino acids other than in the active site can also destroy the properties of an enzyme. For example, mutation of amino acids responsible for the final structure of the protein, such as amino acids needed for sulfhydral bonds within peptides or between subunits of the final enzyme, could destroy enzyme activity. Enzyme function could also be lost by mutation of amino acids that determine the final configuration of the active site—without actually being part of it—or by mutation of nonactive-site amino acids required for attachment to membranes—as in enzymes involved in serial transfer of electrons within the mitochondria of eukaryotes. Failure to attach to membranes in the correct position might make the enzyme unavailable when it is needed.

ABO Blood Groups

Other systems have mutations that result in changes in function rather than in the complete absence of function. For example, in the ABO blood-group antigens of humans, a basic mucopolysaccharide found on the surface of red blood cells is acted on in different ways by products of the I^A and I^B alleles. The recessive I^O allele does not control a change in the basic mucopolysaccharide (Figure 15–2). The A enzyme adds a terminal N-acetylgalactosamine while the B enzyme adds a terminal galactose. The *ABO* locus has a non-functioning I^O allele and two alleles, I^A and I^B, with different—although very similar—functions.

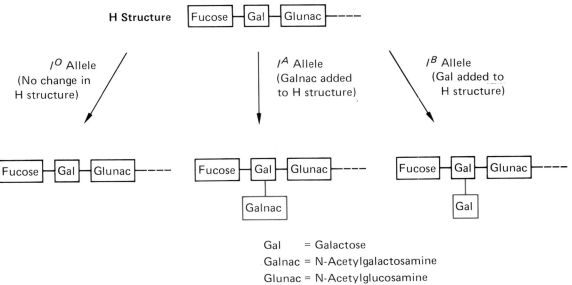

Gal = Galactose
Galnac = N-Acetylgalactosamine
Glunac = N-Acetylglucosamine

Figure 15 – 2. Function of the I^A, I^B, and I^O Alleles of the ABO Locus
The gene products of the I^A and I^B alleles of the ABO locus affect the terminal sugars of a mucopolysaccharide (H structure) found in red blood cells.

Lethal Mutations in *Drosophila*

Chapter 14 described how Muller used a lethal class of mutants to demonstrate that x rays cause mutations. He was thus observing the mutation of genes associated with essential functions. Any change in these particular genes presumably resulted in a critical alteration of essential function and, therefore, in the death of the organism. In the ClB method lethal mutants were isolated on the X chromosome. Since the X chromosome occurs in only one copy in males, all lethals, either recessive or dominant, will show themselves. The majority of lethals are recessives and usually involve loss of function of an enzyme. In order to actually be lethal, the mutants must either be homozygous or, as in the case of the X chromosome, hemizygous.

Conditional Lethality

An interesting class of mutants that has been very useful to geneticists is the **conditional-lethal mutant,** a mutant that is lethal under one condition but not lethal under another condition. **Nutritional-requirement mutants** are a good example and will be discussed later. **Temperature-**

sensitive mutants are conditional lethals whose occurrence has made it possible for geneticists to isolate numerous vital-function mutants that could not otherwise be observed. For example, many temperature-sensitive mutants are completely normal at 25°C but cannot synthesize DNA at 42°C. Presumably, temperature-sensitive mutations produce enzymes with amino acid substitutions, in places other than the active site, such that higher temperature leads to premature denaturing of the mutant enzyme. Thus, the enzyme is perfectly normal at 25°C, the **permissive temperature,** but nonfunctional at 42°C, the **restrictive temperature.**

The interesting thing about most conditional-lethal mutants that cannot synthesize DNA at the restrictive temperature is that they have a completely normal polymerase I (Kornberg's enzyme). Apparently polymerase I is not normally the enzyme used by the cell for DNA replication. When a conditional lethal with a mutant polymerase I was isolated, it was found to replicate its DNA at all normal temperatures, but it was unable to repair damage to the DNA. The conclusion is that polymerase I is primarily involved in repair rather than replication of DNA.

Furthermore, virtually any mutant that results in the loss of enzyme function may be a conditional lethal. Consider, for example, the z^- mutants of *E. coli* discussed earlier. If the cell's sole source of energy was lactose, then z^- would be lethal. But mutant z^- cells can be kept alive and growing simply by providing them with an alternate energy source. Therefore, almost all nutritional mutants can be considered conditional. When supplied with an alternate energy source or a nutrient, such as an amino acid, that the cell cannot synthesize itself, the cell can grow. Without the missing nutrient, the cell dies. This line of thought leads to the next topic, the use of mutants in the analysis and elucidation of biochemical pathways.

BIOCHEMICAL PATHWAYS

Neurospora

George W. Beadle
(1903–)
Courtesy of Dr.
George W. Beadle

The pioneering work in the development of the concept that most genes control enzymes that, in turn, control steps in biochemical pathways was done by George Beadle and Edward Tatum, who eventually shared the Nobel Prize. In their initial work, with *Neurospora,* they demonstrated the existence of conditional lethals—mutants that were viable when a particular nutrient was added to the medium.

Beadle and Tatum presumed that when wild-type *Neurospora*—normally a haploid organism—is grown on a minimal medium, it is manufacturing all of its growth requirements (vitamins, amino acids, nucleotides, and so on) and that if a mutation were to occur that inactivated an enzyme, the specific end product of the enzyme's pathway would not be produced. It followed, therefore, that a spore of their mutant mold would not be able to grow

Edward L. Tatum
(1909–1975)
Courtesy of National
Academy of
Sciences

on minimal medium but that if the required nutrient were provided, the mold should grow.

Nutritional Requirements.   Their presumptions lead to the following technique for determining the type of mutant caused by UV irradiation. A stock of *Neurospora* is irradiated and then crossed to the wild type. After meiosis, sexual spores are isolated and grown on complete medium that contains all of the vitamins, amino acids, and other nutrients normally synthesized by the wild-type *Neurospora*. Both the wild type and a nutritional mutant will grow on this medium. The growth on complete medium is then subsampled. One aliquot is placed on minimal medium. If growth occurs on minimal medium there has been no nutritional mutation. If there is no growth, subsamples are placed on minimal media that contain various additives. In the example in Figure 5–19, growth occurred when amino acids were added to the minimal medium. Thus, assuming that only a single mutational event has taken place, the event must involve a gene controlling an enzyme in a pathway for synthesizing one of the amino acids. Which amino acid is involved is discovered by subsampling the mutant mold onto a set of slants each containing minimal medium plus only one amino acid.

Beadle, Tatum, and their colleagues isolated mutant strains that required many of the amino acids (arginine, leucine, lysine, methionine, and so on), many vitamins (thiamin, pantothenic acid, riboflavin, and so on), and other compounds such as pyrimidines and adenine. Virtually all of these nutritional mutant strains were caused by single mutations. Their results substantiated the original concept that genes control enzymes that control steps in biochemical synthetic pathways.

This technique has been carried one step further by Beadle and Tatum and by others. Not only is it possible to determine in what pathway a particular mutation occurred, but also with mutant analysis it is possible to work out the whole pathway! For example, for isolated niacin-requiring *Neurospora* mutants, other compounds can satisfy the niacin requirement. If the pathway is as shown in Figure 15–3, mutation at B will produce a mutant that will grow if niacin is provided but that will not grow on minimal medium. This mutant will also grow if provided with 3-hydroxy-anthranilic acid rather than niacin. Since the pathway from 3-hydroxy-anthranilic acid to niacin is not blocked by mutation, this compound can be converted to niacin. However, the addition of kynurenine will not allow growth of this mutant because kynurenine cannot be synthesized to niacin—the block is just before 3-hydroxy-anthranilic acid (Table 15–1).

Similarly, if mutation occurs at A of Figure 15–3, the mutant will require niacin for growth. However, growth will occur if either 3-hydroxy-anthranilic acid or kynurenine is added to the minimal medium (Table 15–1). It has thus been demonstrated, by mutant A, that both 3-hydroxy-anthranilic acid and kynurenine are in the pathway of niacin synthesis and, by mutant B, that kynurenine occurs before 3-hydroxy-anthranilic acid. Any

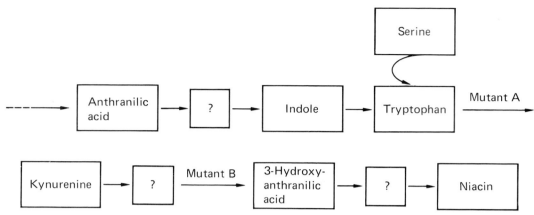

Figure 15 – 3. Pathway of Niacin Synthesis in *Neurospora*
Each arrow represents an enzyme-mediated step. Each question mark
represents a presumed, but unknown, compound.

component of the synthetic pathway that will support growth must of necessity come after the point of mutation. Thus, by analyzing several mutants, all requiring the same nutritional end product (niacin in this case), and by having sufficient understanding of biochemical structure to guess at which substances to try, pathways can be constructed.

Accumulated Substances.   Biochemical pathways can also be discovered by observing what substances accumulate in the cell or in the culture medium. For example, in Figure 15–3 a mutation at B should result in a buildup of the unknown substance between kynurenine and 3-hydroxy-anthranilic acid. A mutation at point A should result in a buildup of tryptophan. However, the accumulation of a substance is not an entirely reliable indicator of biochemical pathways for several reasons. First, since many reactions can go in both directions, a buildup of one product will force the pathway backward in the "wrong" direction without a clear-cut buildup of

TABLE 15 –1.   Growth Performance of *Neurospora* Mutants (Plus sign indicates growth; minus sign indicates no growth.)

	Minimal Medium Additive			
	Tryptophan	Kynurenine	3-Hydroxy-Anthranilic Acid	Niacin
Wild type	+	+	+	+
Mutant A	−	+	+	+
Mutant B	−	−	+	+

any substance. Second, many pathways have other branches. For example, tryptophan is both used in protein synthesis and involved in the synthesis of other amino acids—any excess would be funneled off in different pathways. And third, if the substance being built up is toxic, as virtually any substance in large quantities in a cell is, the cell might take steps to detoxify or break down the substance. Thus, other by-products not directly in the pathway under study would be produced.

Beadle and Tatum concluded from their studies that one gene controls the production of one enzyme. The *one-gene–one-enzyme hypothesis* is an oversimplification that will be modified before the chapter is over. As a rule of thumb, the hypothesis is valid and has served to direct attention to the functional relationship between genes and enzymes in biochemical pathways.

Humans

Archibald E. Garrod
(1858–1936)
Genetics 56
(1967):frontispiece.

Just after the turn of the century, A. E. Garrod, a British physician, had made discoveries similar to the discoveries of Beadle and Tatum regarding the role of genes in controlling metabolic pathways in humans. He termed genetic diseases of metabolic pathways *inborn errors of metabolism.* For example, normal people degrade homogentisic acid (alkapton) into maleylacetoacetic acid. Persons with the disease alkaptonuria are homozygous for a nonfunctional enzyme essential to the process. Absence of the appropriate enzyme blocks the degradation reaction so that there is a buildup of homogentisic acid. This acid darkens upon oxidation. Thus, affected persons can be identified by the black color of their urine after its exposure to air. The eventual effects of alkaptonuria are diseases of the joints and a darkening of cartilage that is visible in the ears and eye sclera.

GENETIC FINE STRUCTURE

Complementation

If two recessive mutations arise independently and both have the same phenotype, how do we know whether they are both mutations of the same gene? That is, how do we know that they are alleles? To answer this question, we must construct a heterozygote and determine whether or not there is **complementation** between the two. Two mutants of the same gene will both produce mutant mRNAs that will result in mutant enzymes. If, however, the mutations are not allelic, each mutant should have a normal (wild-type) copy of the other's mutant gene. As is diagrammed in Figure 15–4, if the two mutant genes are truly alleles, then the phenotype should be mutant. If, however, the two mutant genes are nonallelic, then the a_1 mutant will have

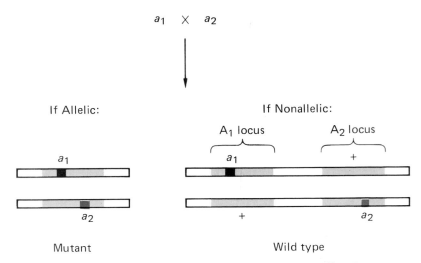

If Allelic:

If Nonallelic:

Mutant

Wild type

Figure 15 – 4.　Determination of Whether Two Mutants That Affect the Same Gene Trait Are or Are Not Allelic
A heterozygote is formed (a_1/a_2) and the phenotype noted. (If two loci are involved, they do not have to be on the same chromosome.)

contributed the wild-type allele at the A_2 locus and the a_2 mutant will have contributed the wild-type allele at the A_1 locus. Thus, the two mutants complement each other, and the result is the wild type. Mutants that fail to complement each other are termed **functional alleles.** The test for defining alleles strictly on the basis of function is termed the **cis-trans complementation test.**

There are two different configurations in which a heterozygous double mutant can be formed (Figure 15–5). In the test above, only the *trans* configuration was used to determine whether or not the two mutations were allelic. In reality, the *cis* configuration is almost never tested; it is the control, where wild-type activity with recessive mutations is always expected. The test is thus sometimes simply called a *trans* test. In any event, from the terms *cis* and *trans*, Seymour Benzer coined the term **cistron,** which is the

Figure 15 – 5.
Recessive Mutations in Heterozygous Double Mutant
The mutant lesions can be in either the *cis* or *trans* configuration.

Trans

Cis

If wild type, nonallelic; if mutant, allelic

Always wild type

smallest unit of function that exhibits a *cis-trans* position effect. We thus have a new word for the gene—one where function is more explicit. We have in essence brought down the concept of one-gene–one-enzyme to one-cistron–one-polypeptide. The cistron is thus the smallest unit that codes for messenger RNA that is then translated into a single polypeptide.

From functional alleles we can go one step further in recombination analysis by determining whether two allelic mutations occur at exactly the same place in the cistron. That is, when two mutations are found to be functional alleles, are they also **structural alleles?** The methods used to analyze complementation at the beginning of this section can be used here. Crosses are carried out to form a heterozygote, which is then tested for recombination between the two mutants. If no recombination occurs, then the two alleles probably contain the same structural change and are thus structural alleles. If a small amount of recombination occurs, then it is possible to generate the wild type, which means that the two alleles are not mutations at the same point (Figure 15–6). Alleles that were functional but not structural were first termed **pseudoalleles** because it was believed that loci were complex—made of subloci. Fine-structure analysis led to understanding that a locus is a length of genetic material, divisible by recombination, rather than a bead on a string.

As a nonbacterial example of complementation analysis, eye-color mutants of *Drosophila* can be studied. The white-eye locus has a series of alleles for varying shades of red. This locus is sex linked and located at about map unit 3.0 on the X chromosome. There are several other eye-color loci on the X chromosome—for example, prune and ruby. If an apricot-eyed female

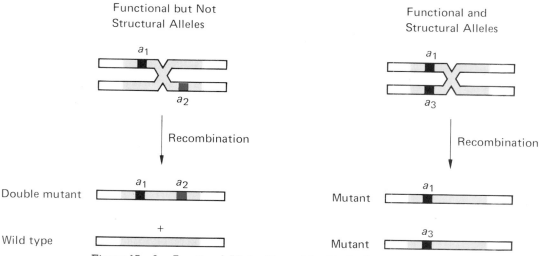

Figure 15 – 6.   Functional Alleles May or May Not Be Structurally Allelic

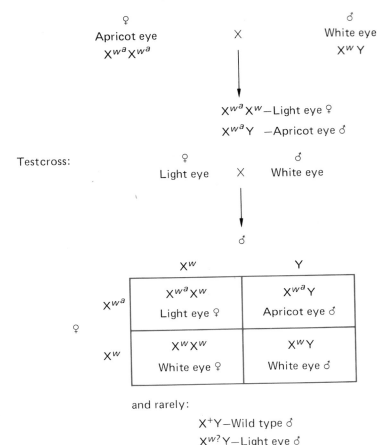

Figure 15 – 7.
Crosses Demonstrating That Apricot and White Eyes Are Functional but Not Structural Alleles in *Drosophila*

is crossed to a white-eyed male, the female offspring are all heterozygous and have light-colored eyes (Figure 15–7). Thus, apricot and white are functional alleles: They do *not* complement (Table 15–2). To determine if apricot and white are structural alleles, light-eyed females are crossed with white-eyed males and the offspring are observed for the presence of wild-type or light-eyed males. Rarely, these males do appear. Their rate of appearance is less than 0.001%, which is, however, significantly above the background mutation rate. The conclusion is that apricot and white are functional but not structural alleles.

Fine-Structure Mapping

After Beadle and Tatum established that the gene controls the production of an enzyme that controls a step in a biochemical pathway, Seymour Benzer

TABLE 15 –2. Complementation Matrix of X-Linked *Drosophila* Eye-Color Mutants (Plus sign indicates that female offspring are wild type; minus sign indicates that they are mutant.)

	White	Prune	Apricot	Buff	Cherry	Eosin	Ruby
White (*w*)	−	+	−	−	−	−	+
Prune (*pn*)		−	+	+	+	+	+
Apricot (*wᵃ*)			−	−	−	−	+
Buff (*wᵇᶠ*)				−	−	−	+
Cherry (*wᶜʰ*)					−	−	+
Eosin (*wᵉ*)						−	+
Ruby (*rb*)							−

used analytical techniques to dissect the fine structure of the gene. Fine-structure mapping means an examination of the size and number of sites within a gene that are capable of mutation and recombination. At a time when biochemical techniques did not permit DNA sequencing, Benzer used classical recombination and mutation techniques on bacterial viruses to provide reasonable answers to the questions of fine structure and to give insight into the nature of the gene. He coined the terms **muton** for the smallest mutable site and **recon** for the smallest unit of recombination. It is now known that a single nucleotide is both the muton and the recon.

Prior to Benzer's work, genes were thought of as beads on a string. Analysis of mutation sites within a gene by recombination was hampered by the very low rate of recombination between sites within a gene: If two alleles involve different mutation sites on the same gene, there is a distinct probability of getting both mutations on the same chromosome by recombination, but, in view of the very short distances within a gene, this probability is very low (see Figure 15–6). (Note that for every double mutant formed, there is the same probability of reconstituting the wild type.) Although it certainly seemed desirable to map sites within the gene as well as between genes (Chapters 4–6), there remained the problem of finding an organism that would allow fine-structure analysis.

*r*II Screening Techniques. Benzer used the bacteriophage T4 because of the growth potential of phage, where a generation takes about an hour and the increase is about a hundredfold. Actually, any prokaryote or virus should suffice, but Benzer made use of other unique screening properties of the phage that made it possible to recognize one particular mutant in about a billion. Benzer used the *r*II mutants of T4. These mutants produce large, smooth-edged plaques on *E. coli,* whereas the wild type produces smaller, less smooth-edged plaques (see Figure 6–9). The *r* of *r*II stands for rapid lysis; the II implies that there is more than one gene for this characteristic.

The screening system employed by Benzer made use of the fact that *r*II mutants do not grow on *E. coli* strain K12. The normal strain, *E. coli* B,

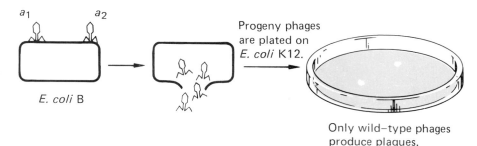

Figure 15 – 8.
Using *E. coli* K12 and B to Screen for Recombination at the *r*II Locus of Phage T4
In this case we are looking for wild-type progeny from a cross of two mutants·

Progeny phages are plated on *E. coli* K12.

Only wild–type phages produce plaques.

allows growth of both the wild-type and *r*II mutants. The wild type can grow on K12. Thus, various mutants can be crossed by multiple infection of *E. coli* B cells, and Benzer could screen for wild-type recombinants by plating the resultant phage on *E. coli* K12 (Figure 15–8). The count of the number of phage that are put on the plate can be obtained by taking a small subsample and plating it on *E. coli* B. Only the wild-type recombinant will produce a colony on the K12 plate. Thus, it is possible to detect one recombinant in a billion, all in an afternoon's work. Think how much labor and materials would be needed to get the same amount of information by using fruit flies or corn.

Benzer sought to map the number of recombinable and mutable sites within the *r*II region of T4. The term *region* will be used here until the number of loci it contains is established. Benzer began to isolate independently derived *r*II mutants and crossed them among themselves. The first thing he found was that the *r*II region was composed of two cistrons. That is, almost all of the mutations belonged to one of two *complementation groups*. The exceptions were mutations that seemed to belong to both cistrons and were soon found to be deletions—that is, part of each cistron was missing (Table 15–3). The *A* cistron mutations would not complement each other but would complement all the mutations in the *B* cistron.

Deletion Mapping. As the number of independently isolated mutants increased, it became obvious that to cross every possible combination would entail millions of crosses. To overcome the problem, Benzer isolated mutants that were partial or complete deletion mutations of each cistron. Deletion mutations were easily discovered because they acted like structural alleles of many alleles that were not themselves structurally allelic. Once a sequence of deletion mutations covering the *A* and *B* cistrons was isolated, a minimum of crosses was required to localize a new mutation to a portion of one of the cistrons. A second series of smaller deletions within the region was then isolated, and further localization was accomplished (Figure 15–9).

Then, each new mutant was crossed with each of the mutants so far isolated in a region and the mutation was localized within this same region. If the new mutant were structurally allelic to a previously isolated mutant,

TABLE 15 –3. Complementation Matrix of 10 *r*II Mutations (Plus sign indicates complementation; minus sign indicates no complementation. The two cistrons are arbitrarily designated *A* and *B*. Mutants 4 and 9 must be deletions that cover parts of both cistrons.)

	1	2	3	4	5	6	7	8	9	10
1	−	−	+	−	−	−	+	−	−	−
2		−	+	−	−	−	+	−	−	−
3			−	−	+	+	−	+	−	+
4				−	−	−	−	−	−	−
5					−	−	+	−	−	−
6						−	+	−	−	−
7							−	+	−	+
8								−	−	−
9									−	−
10										−

Alleles: *A* cistron: 1, 2, 4, 5, 6, 8, 9, 10; *B* cistron: 3, 4, 7, 9.

it was scored as an independent isolation of the same mutation. If it were not a structural allele of any of the known mutants of the subregion, it was added to this region as a new mutation point. The exact position of each new mutation within the region was determined by the relative frequency of recombination between it and the known mutants of this region (Chapter 4). Deletion mapping techniques allow the calculation of relative position, which also gives relative gene order. Benzer eventually isolated about 350 mutants from 80 different subregions defined by deletion mutations. An abbreviated map is shown in Figure 15–10.

What conclusions did Benzer draw from his work? First, he concluded that since all of the mutants in both *r*II cistrons can be ordered in a linear fashion, the original Watson-Crick model of DNA as a linear molecule was correct. Second, he concluded that reasonable inroads had been made toward saturation of the map; saturation would occur when at least one mutation had been located at every mutable site. Benzer reasoned that since many sites were represented by only one mutant, there must be sites represented by zero mutants. Given that he had mapped about 350 sites, he calculated that there were at least another 100 sites still undetected by mutation. We now know that a total of about 450 sites is an underestimate. Since no protein was isolated, there was no independent estimate of the number of nucleotides in these cistrons (number of amino acids times 3 nucleotides per codon). Thus, although Benzer has not saturated the map with mutations, he certainly has made respectable progress in dissecting the gene.

Hot Spots. Benzer also looked into the lack of uniformity in the occurrence of mutations (note the "hot spot" at B4 of Figure 15–10). Presuming that all base pairs are either AT or GC, this lack of uniformity is unexpected.

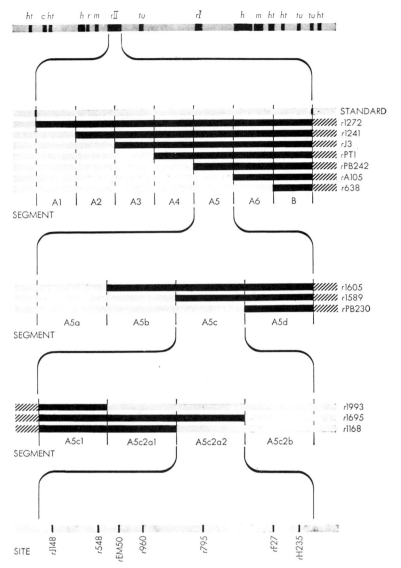

Figure 15 – 9.   Localization of an rII Mutant by Deletion Mapping
If the new mutant (for example, r960) is located in the A5c2a2
region, it first would not complement r1272, r1241, rJ3, rPT1, or
rPB242. It would complement rA105 or r638 and thus would be
localized to the A5 region. Then the mutant would be crossed with
r1605, r1589, and rPB230. It would only complement with rPB230
and thus would be localized to the A5c region. Then crossed with
r1993, r1695, and r1168, the mutant would complement r1993 and
r1168 and would be localized to the A5c2a2 region. Finally the
mutant would be crossed pairwise with all the known mutants of this
region to determine relative arrangement and distance.

Source: From Seymour Benzer, "The Fine Structure of the Gene."
Copyright © January 1962 by Scientific American, Inc. All rights
reserved. Reprinted by permission.

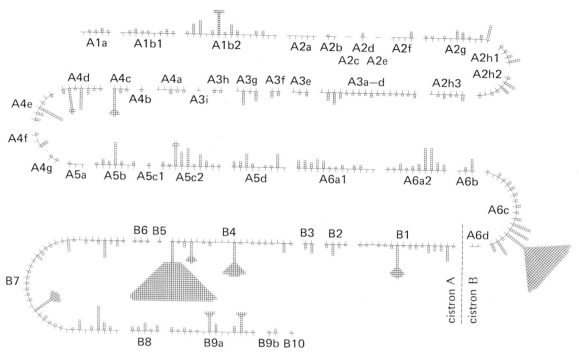

Figure 15 – 10.   Abbreviated Map of Spontaneous Mutations of the *A* and *B* Cistrons of the *rII* Region of T4

Each square represents one independently isolated mutation.

Source: From S. Benzer, ''On the Topography of the Genetic Fine Structure,'' *Proceedings. The National Academy of Sciences, USA* 47(1961): 403–415. Reprinted by permission of the author.

Benzer suggested that spontaneous mutation is not just a function of the base pair itself, but of the surrounding bases as well. To further investigate the issue and to look at the way that mutagens act at the fine-structure level, Benzer repeated his original experiments using various mutagens to generate mutations rather than depending on their spontaneous occurrence. He used many mutagens, including UV light, proflavin, nitrous acid, 5-bromouracil, and 2-aminopurine.

Benzer discovered that the mutagens showed a high degree of specificity within the cistron. The mutation rate of some sites was increased by as much as a factor of 10,000; at other sites the rate was not increased at all. Although this specificity is to be expected of some mutagens because of the way in which they work, the degree of specificity shown by the mutagens was surprising. Thus, the concept of many mutagens acting nonspecifically, or at least being able to act at many points on the DNA (Chapter 14), may have to be modified to take into account "hot spots" of greatly increased mutation rate, which may be accounted for by the specific arrangement of neighboring base pairs in an area.

Thus, to recapitulate, Benzer's work supports the model of a gene as a linear arrangement of DNA whose nucleotides are the smallest units of mutation, where the link between any adjacent nucleotides is capable of being broken in the recombination process. The smallest unit of function, determined by a complementation test, is the cistron. Mutagenesis is not uniform over the cistron but may depend on the particular arrangement of bases in a given region.

Polypeptide Complementation. Benzer warned that certainty is elusive in the complementation test because sometimes two mutations of the same functional unit can result in partial activity of the same order of magnitude as is expected of the *cis* control. The problem can be traced to the interactions of subunits at the polypeptide level. That is, some proteins are made up of subunits, and it is possible that certain mutant combinations produce subunits that interact to restore the enzymatic function of the protein (Figure 15–11). With this problem in mind, geneticists use the complementation test to determine functional relationships between mutants.

COLINEARITY

*r*II Mutants of T4

The last topic to be covered in this chapter is the colinearity of the gene and the polypeptide. Benzer's work established that the gene was a linear entity

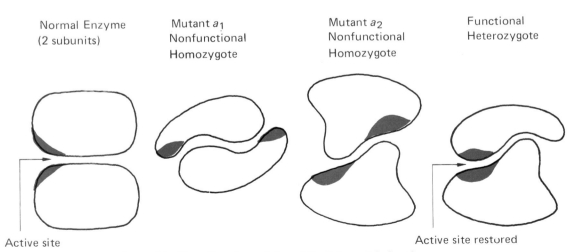

| Normal Enzyme (2 subunits) | Mutant a_1 Nonfunctional Homozygote | Mutant a_2 Nonfunctional Homozygote | Functional Heterozygote |

Active site   Active site restored

Figure 15 – 11. Complementation at the Polypeptide Level

as had been predicted by Watson and Crick. However, Benzer's work could not bridge the gap between the gene and its protein product. To do this, it is necessary to show that for every mutational change of the DNA, a corresponding change takes place in the protein product of the gene. Colinearity would be established by showing that nucleotide and amino acid changes occurred linearly and in the same order in the protein and in the cistron.

Conceptually, the simplest way to analyze colinearity is to gather several mutants, sequence both the DNA and the protein product, and compare them point for point. If the changed nucleotides corresponded to changed amino acids in the same place, then colinearity could be established. Nevertheless, the fact of colinearity was established before our technical sophistication reached the point of DNA sequencing.

Ideally, Benzer himself might have solved the colinearity issue. He was halfway there, with his 350 or so isolated mutants of T4. But Benzer did not have a protein product to analyze. No particular mutant protein had been isolated from *r*II mutants. In the midst of competition to find just the right system, Charles Yanofsky of Stanford University and his colleagues emerged in the early to mid 1960s with the required proof that the ordering of a polypeptide's amino acids corresponded to the nucleotide sequence in the gene that specified it. Yanofsky's success rested with his choice of an amenable system.

Tryptophan Synthetase

Charles Yanofsky (1925–   )
Courtesy of Dr. Charles Yanofsky and News and Publications Service, Stanford University

Yanofsky did his research on the tryptophan biosynthesis pathway in *E. coli*. The last enzyme in the pathway, tryptophan synthetase, catalyzes the reaction of indole-3-glycerol-phosphate plus serine to tryptophan plus 3-phosphoglyceraldehyde. The enzyme itself is made of four subunits specified by two separate cistrons. Each polypeptide is present twice.

Yanofsky and his colleagues concentrated on the *A* subunit. They mapped *A*-cistron mutations with techniques similar to Benzer's techniques, except that the mutations were of a bacterial gene instead of a phage. Instead of mating the bacteria to get recombination distances, Yanofsky used the more precise technique of transduction (Chapter 6) with the transducing phage P1. Each new mutant was first tested against a series of deletion mutations to establish in what region it was located. Then, mutants within a region were crossed among themselves to establish relative positions and distances.

The protein products of the bacterial genes were isolated through biochemical techniques (see Chapter 11). Electrophoresis and chromatography were then used to establish the fingerprint patterns of the mutant and wild-type proteins. Assuming a single missense mutation, a comparison of the mutant and the wild-type fingerprints would show a difference of just one amino acid spot (Figure 15–12). The process bypassed an entire sequencing

Ala-Arg-Trp-Ser-Ser spot

Ala-Lys-Trp-Ser-Ser spot

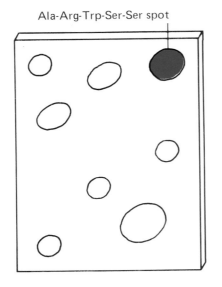

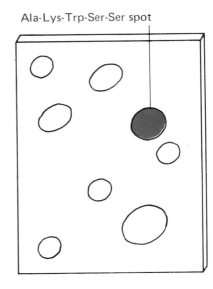

Wild type

Mutant

Figure 15 – 12.
Difference in
"Fingerprints"
between Mutant and
Wild-Type
Polypeptide Digest

operation. The mutation was immediately located in a region of the protein, and the new amino acid could be identified by analysis of this one spot.

Yanofsky and his colleagues found that the *A* protein of tryptophan synthetase consists of 267 amino acids. The detail of nucleotide and amino acid change are shown for 9 of the mutants in Figure 15–13.

This figure demonstrates that a sequence of 9 mutations in the linear *A* cistron are colinear with 9 amino acid changes in the protein itself. In two cases, two mutations mapped so close as to be almost indistinguishable. In both cases, the two mutations proved to be in one codon: The same amino acid position was altered in each (A23–A46; A58–A78). Thus, exactly as predicted and expected, there is colinearity between the gene and protein. This work was independently confirmed at the same time by Brenner and his colleagues, who used head-protein mutants of T4 phage.

CHAPTER SUMMARY

A mutation in a gene can have many effects on the function of the protein coded for by this gene and, therefore, many possible effects on the phenotype. Function of the protein can be eliminated, attenuated, enhanced, or changed completely. Of special interest to genetic analysis are lethal mutations and conditional-lethal mutations. Nutritional-requirement mutants and temperature-sensitive mutants have made possible detailed study of cell

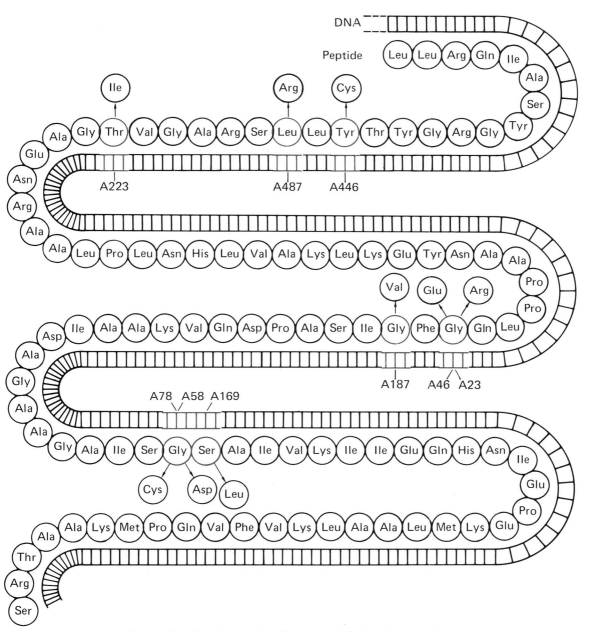

Figure 15 – 13. Amino Acid Sequence of Carboxy Terminal End of the Tryptophan Synthetase *A* Protein and the DNA of the Cistron
Mutant amino acids are noted with the mutations (for example, A446) shown in the positions to which they map on the DNA.

function that would not have otherwise been possible. Genes of critical functions, such as DNA replication, can be studied through temperature-sensitive mutations; metabolic pathways have been worked out with nutritional mutants.

Allelism is defined by the *cis-trans* complementation test. Complementation implies independent loci, or nonallelism. Functional alleles that involve the same nucleotide are called *structural alleles*. Fine-structure mapping was done by Benzer on T4 phage. Colinearity of the gene and protein was demonstrated by Yanofsky.

EXERCISES AND PROBLEMS

1. A *Neurospora* mutant is isolated that requires arginine for growth. Give a comprehensive list of the kinds of mutations that could have taken place to make the mold *arg⁻*. Include events that do or do not yield protein products.

2. A mutant of T4 phage is isolated that does not lyse *E. coli* cells at 42°C. Within the bacterium, however, fully formed phage are present. At 25°C the life cycle of this phage is completely normal. Explain the phenomenon.

3. What types of enzymatic functions are best studied using temperature-sensitive mutants?

4. The following table shows the growth (+) of 4 mutant strains of *Neurospora* on minimal medium with various additives. The additives are in the pathway of niacin synthesis (vitamin B₃). Diagram the pathway and show which steps the various mutants block. What compound would each mutant accumulate?

Mutant	Minimal Medium Additive					
	Nothing	Niacin	Tryptophan	Kynurenine	3-Hydroxy-Anthranilic Acid	Indole
10575	—	+	+	+	+	—
4008	—	+	+	+	+	+
4540	—	+	—	—	+	—
3416	—	+	—	—	—	—

When you complete this problem, compare it with Figure 15–3. What effect on growth would be observed following a mutation in the pathway of serine biosynthesis?

5. The following table shows the growth (+) or the lack of growth (−) of various mutants in another pathway of biosynthesis. Determine this pathway, the point of blockage of each mutant, and the substrate accumulated by each mutant.

Mutant	Minimal Medium Additive					
	Nothing	A	B	C	D	E
1	−	−	−	−	−	+
2	−	+	+	+	−	+
3	−	+	−	−	−	+
4	−	+	+	+	+	+
5	−	+	−	+	−	+

6. Seven arginine-requiring mutants of *E. coli* were independently isolated. All pairwise matings were done (by transduction) to determine the number of loci (complementation groups) involved. If a + indicates growth and a − no growth on minimal medium, how many complementation groups are involved here?

	1	2	3	4	5	6	7
1	−	+	+	+	+	−	−
2		−	+	+	−	+	+
3			−	−	+	+	+
4				−	+	+	+
5					−	+	+
6						−	−
7							−

Why is only "half a table given? Must the upper left to lower right diagonal be −?

7. A group of *r*II mutants (M to T) have been localized to the *A* cistron because of their failure to complement with a known deletion of the *A* cistron. These mutants are then mated pairwise with the following series of subregion deletions. The mating is done on *E. coli* B and plated out on *E. coli* K12. A plus shows the presence of plaques on K12 while a minus shows an absence of growth. Localize each of the *r*II *A* mutations.

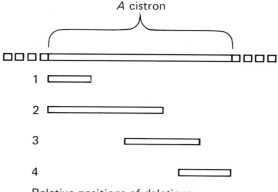

Relative positions of deletions

Mutant	Deletion			
	1	2	3	4
M	+	+	−	+
N	+	−	−	+
O	+	+	+	−
P	+	+	−	−
Q	+	−	+	+
R	−	−	+	+
S	−	−	−	+
T	+	−	−	−

8. A *Drosophila* worker isolates four eye-color forms of the fly: wild type, white, carmine, and ruby. (The worker does not know that white, carmine, and ruby are three separate loci on the X chromosome.) What crosses should be made to determine allelic relations of the loci? What results should be expected? A new mutant, eosin, is isolated. What crosses with what results should be carried out to determine that eosin is an allele of white?

9. What is the relevance to both Benzer's and Yanofsky's work of the occurrence of intervening sequences? Did Benzer and Yanofsky work with genes that do not have intervening sequences?

SUGGESTIONS FOR FURTHER READING

Beadle, G. W., and E. L. Tatum. 1941. Genetic control of biochemical reactions in *Neurospora*. *Proc. Nat. Acad. Sci., USA* 27:499–506.

Benzer, S. 1959. On the topology of the genetic fine structure. *Proc. Nat. Acad. Sci., USA* 45:1607–1620.

Benzer, S. 1961. On the topography of the genetic fine structure. *Proc. Nat. Acad. Sci., USA* 47:403–415.

Epstein, R. H., et al. 1963. Physiological studies of conditional lethal mutations of bacteriophage T4D. *Cold Spr. Harb. Symp. Quant. Biol.* 28:375–392.

Garrod, A. E. 1909. *Inborn Errors of Metabolism*. London: Henry Frowde, Hodder and Stoughton.

Sarabhai, A., et al. 1964. Colinearity of the gene with the polypeptide chain. *Nature* 201:13–17.

Srb, A., R. Owen, and R. Edgar, eds. 1970. *Facets of Genetics*. San Francisco: Freeman. (A collection of *Scientific American* articles. Note especially articles by Beadle, Benzer, Yanofsky, and Edgar and Epstein.)

Yanofsky, C., et al. 1967. The complete amino-acid sequence of the tryptophan synthetase A protein (α subunit) and its colinear relationship with the genetic map of the A gene. *Proc. Nat. Acad. Sci., USA* 57:296–298.

16 Extrachromosomal Inheritance

We have already examined two modes of phenotype control—chromosomal genes and environment. This chapter deals with a third mode of phenotype control—namely, extrachromosomal inheritance (also called non-Mendelian, cytoplasmic, nonchromosomal, or maternal inheritance). The mechanisms of extrachromosomal inheritance fall into two broad groups: **Maternal inheritance** is controlled by non-DNA cytoplasmic substances (examples of which are snail coiling and moth pigmentation); **cytoplasmic inheritance** is controlled by nonnuclear genomes (found in chloroplasts and mitochondria, infective agents, and plasmids).

Maternal effects result from the assymetric contribution of the female parent to the development of zygotes. That is, although both male and female parents contribute equally to the zygote in terms of chromosomal genes (with the exception of sex chromosomes), the sperm rarely contributes anything to development other than chromosomes. The female parent usually contributes the initial cytoplasm and organelles of the zygote. Zygotic development, therefore, usually begins within a maternal milieu, and the maternal cytoplasm directly affects zygotic development. Maternal and cytoplasmic inheritance are often one and the same. For example, nongreen plants may be due to mutation of the chloroplast itself, while the inheritance of this chloroplast is through the female gamete.

Cytoplasmic inheritance refers to the inheritance pattern of organelles and parasitic or symbiotic particles that have their own genetic material. Chloroplasts, mitochondria, bacteria, viruses, and, of course, plasmids all have their own genetic material. These genomes are open to mutation. As we shall see, their inheritance pattern will not follow Mendel's rules for chromosomal genes.

DETERMINING NON-MENDELIAN INHERITANCE

It is not unreasonable to ask how one determines that a trait is inherited. The question does not have as obvious an answer as might be expected. Environmentally induced traits can mimic inherited phenotypes—as with the phenocopies discussed in Chapter 4. For example, the inheritance of vitamin-D–resistant rickets is mimicked by lack of vitamin D in the diet. It is possible to determine that the dietary rickets was not inherited by simply administering adequate quantities of vitamin D. Rickets that is inherited does not respond until about 150 times the normally adequate amount of vitamin D is administered.

Some environmentally induced traits persist for several generations. For example, a particular *Drosophila* strain that normally grows at 21°C was exposed to 36°C for 22 hours. Dwarf progeny were produced. When they were mated among themselves, fewer and fewer dwarfs appeared in each generation, but smaller-than-normal flies were seen into the fifth generation. The phenomenon of an environmentally induced trait that persists for several generations has been termed **dauermodification.**

Extrachromosomal inheritance is usually identified by the odd results of reciprocal crosses. If the progeny of simple reciprocal crosses are not fol-

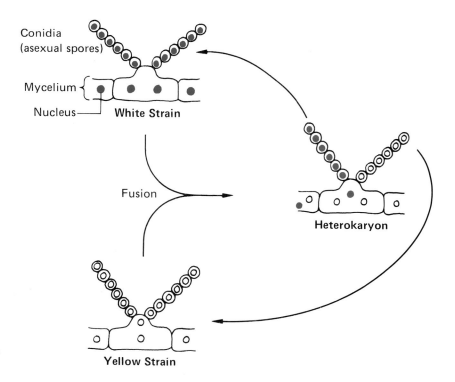

Figure 16 – 1.
Heterokaryon Test in
Aspergillus

lowed for several generations, the results can be misleading where extrachromosomal inheritance is concerned. The technique of **nuclear transplantation,** where feasible, has proved useful in identifying extrachromosomal inheritance. In this technique the nucleus of a cell, such as an amoeba or frog egg, is removed by microsurgery and another nucleus is substituted. Thus, not only can a nucleus be isolated from its cytoplasm, but various nuclei can be implanted in the same cytoplasm. A similar experiment, called a *heterokaryon test,* can be done with various fungi such as *Neurospora* and *Aspergillus,* in which mycelia of the same mating type can fuse, forming a **heterokaryon,** which is a cell containing nuclei from different strains (Figure 16–1). Thus, nuclei of both strains exist in the mixed cytoplasm. Subsequently, spores (conidia) that have one or the other nucleus in the mixed cytoplasm can be isolated. The phenotype of the colonies produced from these isolated conidia will show whether the trait under observation is controlled by the nucleus or the cytoplasm. Chromosomal genes in a particular cytoplasm can also be isolated by repeated backcrossing of offspring with the male-parent type. In each cross the content of the female chromosomal genes will be halved, but, presumably, the cytoplasm will remain similar to the female line. Thus, after several generations, male genes can be isolated in female cytoplasm. The phenotypic results of the final cross will show whether inheritance was chromosomal or extrachromosomal.

MATERNAL INHERITANCE

Snail Coiling

Snails, of which a well-known example is the pond snail, *Limnaea peregra,* are coiled either to the right (dextrally) or to the left (sinistrally). (If the snail is held so that one looks directly at the opening, the snail is dextrally coiled when the opening comes from the right-hand side and sinistrally coiled when it comes from the left-hand side: See Figure 16–2.) The inheritance pattern of the coiling is at first perplexing.

In the left half of Figure 16–2, a dextral snail provides the eggs while a sinistral snail provides the sperm. The offspring are all dextral; presumably, therefore, dextral is dominant. When the F_1 are self-fertilized (snails are hermaphroditic), all the offspring are dextrally coiled. The result is unexpected. Nevertheless, when the F_2 are self-fertilized, 1/4 will produce only sinistral offspring and 3/4 will produce only dextral offspring. If self-fertilization is continued through ensuing generations, this 3:1 phenotype ratio will be revealed as a Mendelian 1:2:1 genotype ratio, which reaffirms the notion of a single locus with two alleles—with dextral dominant. Something interfered with the expected inheritance pattern.

When the reciprocal cross is made (Figure 16–2, right), the F_1 have the same genotype as just described but are coiled sinistrally as the egg parent

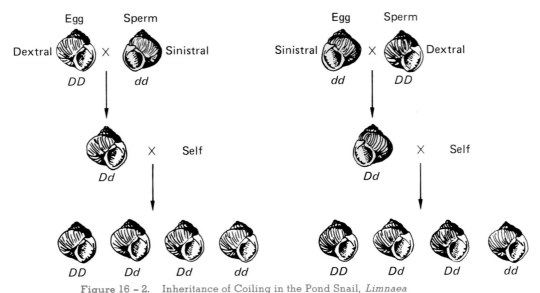

Figure 16 – 2.   Inheritance of Coiling in the Pond Snail, *Limnaea peregra*
Reciprocal crosses (D = dominant dextral and d = recessive sinistral).

is. From here on, the results are exactly the same for both crosses. A pattern emerges where the F_1 are phenotypically similar to the female parent regardless of their genotype. The explanation is that the phenotype of a snail is due to the genotype of its mother, for whom dextral is dominant. Thus, the DD mother in Figure 16–2 produces F_1 progeny that are dextral with a Dd genotype, and the dd mother produces progeny with the same Dd genotype but a sinistral phenotype because the mother was dd. Why does this pattern occur?

A process of **spiral cleavage** has been found to occur in the zygote of mollusks and some other invertebrates. The spindle at mitosis is tipped in relation to the axis of the egg. If the spindle is tipped one way, a snail will be coiled sinistrally; if it is tipped the other way, the snail will be coiled dextrally. The direction of tipping is determined by the maternal cytoplasm, which is under the control of the maternal genotype. Obviously, maternal control affects only one generation—in each generation the coiling is dependent on the maternal genotype.

Moth Pigmentation

There are other examples of maternal effects, where the cytoplasm of the mother, under the control of chromosomal genes, controls the phenotype of her offspring. In the flour moth, *Ephestia kühniella*, kynurenin, which is a precursor for pigment, is accumulated in the eggs. The recessive allele, a,

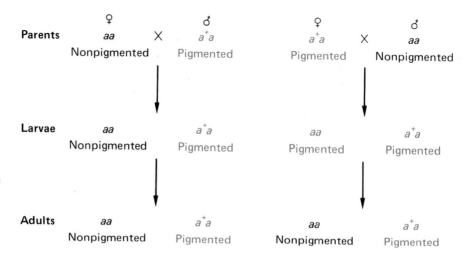

Figure 16 – 3.
Inheritance Pattern
of Larval and Adult
Pigmentation in the
Flour Moth, *Ephestia*
kühniella
The *a* locus controls
the presence (a^+) or
absence (*a*) of
kynurenin.

when homozygous, results in a lack of kynurenin. Reciprocal crosses give different results for larvae and adults: When a nonpigmented female is crossed with a pigmented male, the results are strictly Mendelian; but when the mother is pigmented (a^+/a), all the larvae are pigmented regardless of their genotype (Figure 16–3). The initial larval pigmentation comes from residual kynurenin in the eggs, which is then diluted out so that the adults' pigmentation phenotype follows the genotype.

CYTOPLASMIC INHERITANCE

Mitochondria

The **mitochondrion** is a cellular organelle of eukaryotes in which the Krebs cycle and electron transport reactions take place. The actual number of mitochondria per cell is difficult to determine because serial sectioning of whole cells under the electron microscope is required. A mitochondrion (Figure 16–4) is an organelle with an inner and outer membrane. The inner membrane shows a large amount of folding, and there is some controversy as to whether or not the outer membrane is part of the endoplasmic reticulum of the cell rather than an integral part of the mitochondrion. There is also controversy over the origins of this organelle, with some individuals claiming that it originated from an endosymbiotic bacterium, while others have suggested that it evolved from protoplasmic components of the cell. We will examine this issue later. As far as we are concerned, however, the most interesting aspect of the mitochondrion is that it has its own DNA, which occurs as one or more circles about 5 μm—or about 15,000 base pairs—in length (Figure 16–5).

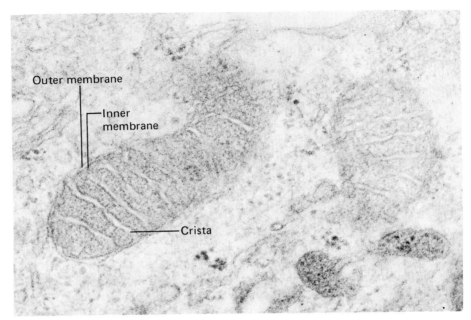

Figure 16 – 4.
Electron Micrograph
of Hamster Lung
Macrophage Cell
with Its
Mitochondria
magnification
20,000X

Source: Courtesy of Wayne Rosenkrans.

The mitochondrion contains unique components for protein synthesis and DNA replication. There are ribosomes, several tRNA species, DNA polymerase, and RNA polymerase that are mitochondria specific—they differ from the cell's cytoplasmic components. The mitochondrial components are similar to the cytoplasmic components of bacteria (they are prokaryotic). This affinity (close resemblance) has sparked the mentioned controversy over the origin of the mitochondrion. Proponents of the symbiosis school point to bacterial affinity to support the contention that mitochondria were originally free-living bacteria. Proponents of the intracellular school suggest that the mitochondrion evolved within the cell prior to the evolution of eukaryotes and would, therefore, have the original prokaryote affinities. (The different codon usage by mitochondria was discussed in Chapter 11.)

In all eukaryotic cells the mitochondria house the enzymes and processes of the Krebs cycle and the electron transport chain—reactions collectively referred to as *respiration*. The Krebs cycle enzymes are not membrane bound within the mitochondria; the electron transport enzymes are regularly ordered on the inner membrane. The Krebs cycle consists of about eight enzymes and the electron transport chain consists of at least nine proteins (Figure 16–6).

The amount of DNA within the mitochondria is very limited: 15,000 base pairs cannot even control the synthesis of the proteins for the translation apparatus. The ribosomal RNA and 12 to 20 transfer RNAs are known

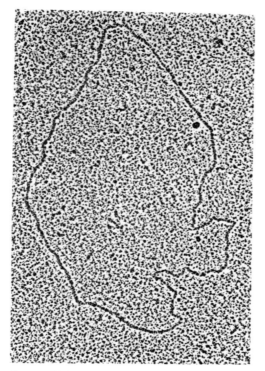

Figure 16 – 5.
Electron Micrograph
of Circle DNA from
within a
Mitchondrion
magnification
48,000X

Source: M. M. K. Nass, "The Circularity of Mitochondrial DNA," *Proceedings. The National Academy of Sciences USA* 56 (1966):1215–1222.
Reproduced by permission of the author.

(from hybridization studies) to be encoded by the mitochondrial DNA. The Krebs cycle enzymes and many of the electron transport proteins, such as cytochrome c, are coded for by the cell's nuclear chromosomes. Thus, the genetic system here is interactive. Although the mitochondria have their own DNA coding for their own rRNA and tRNA, which have prokaryotic properties, the majority of their respiration enzymes are coded for by the nuclear genes. Once the interaction within this genetic system is clearly understood, we could conceive of several inheritance patterns that would follow either cytoplasmic or nuclear lines and be involved phenotypically with respiration. Among the best-studied phenotypes with these patterns are the *petite* mutations of yeast and the *poky* mutations of *Neurospora*.

Petites.   Under aerobic conditions yeast grow with a distinctive colony morphology. Under anaerobic conditions the colonies are smaller and the structure of the mitochondria becomes more reduced. Occasionally, when growing aerobically, small, anaerobic-like colonies will appear, but in these

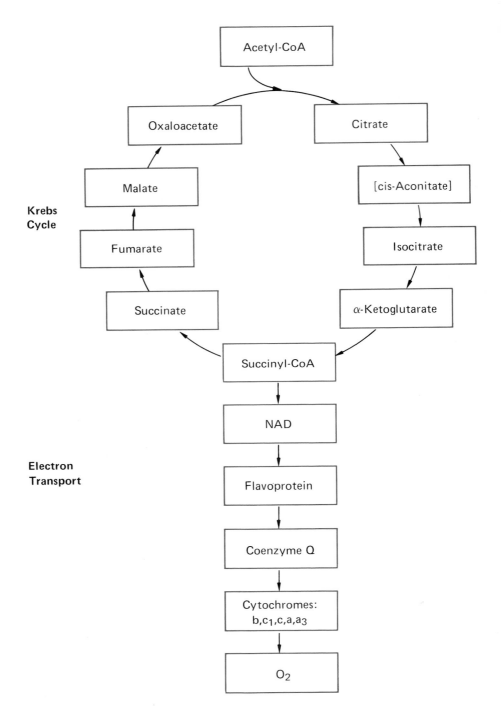

Figure 16 – 6. Krebs Cycle and Electron Transport Chain within the Mitochondria

colonies the mitochondria appear perfectly normal. These colonies have been termed **petites,** French for small. When petites are crossed with the wild type, three modes of inheritance are observed (Figure 16–7). The *segregational petite* is due to mutation of a chromosomal gene and exhibits Mendelian inheritance. The *neutral petite* is lost immediately upon crossing to the wild type. The *suppressive petite* shows variability in expression, depending upon how suppressive the trait is, but it is able to convert the wild-type mitochondria to the petite form. The petites are usually found to lack one or another cytochrome. All petites are failures of mitochondrial function, whether the function is controlled by the mitochondria themselves or by the cell's nucleus.

While the mechanisms of neutral and suppressive petites are not known with certainty, observation of their DNA has had some interesting results. In some neutral and suppressive petites, there is no change in the buoyant density of the DNA. (**Buoyant density** is merely another term for the position at which the DNA equilibrates during density gradient centrifugation and is a measure of the size and molecular weight of the molecule.) In other petites, changes in buoyant density range from very small to the complete absence of DNA.

Petites, therefore, can be the result of an approximation to a point mutation (where there is no measurable change in the DNA), or of marked changes in the DNA, or of the total absence of DNA. In most petites, protein synthesis within the mitochondrion is lacking.

Neutral petites seem to have mitochondria that entirely lack DNA. When neutral petites are crossed with the wild type to form a diploid cell,

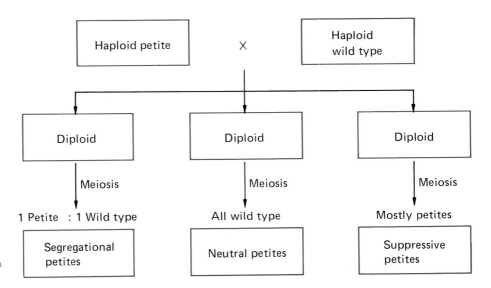

Figure 16 – 7.
Petite Yeasts
Categorized on the
Basis of Segregation
Patterns

the normal mitochondria take over. During meiosis, virtually every spore receives enough normal mitochondria to dominate this cell; the progeny are, therefore, all normal.

Suppressive petites could exert their influence over normal mitochondria in one of two ways. First, the suppressive mitochondria might simply out compete the normal mitochondria and take over—that is, they might simply grow faster. Alternatively, crossing over between the DNA of the suppressive petite and the wild type might affect the normal DNA if the suppressive petite's DNA were severely damaged (and if crossing over could occur among mitochondrial DNAs). If large portions of the DNA from the suppressive mitochondria were missing or altered, then crossing with the normal mitochondria's DNA might exchange some of this damaged DNA. Several experiments have been done where a suppressive petite and a normal, each with mitochondrial DNA of known buoyant density, were crossed. The DNAs of the offspring colonies, which were petites, were of intermediate buoyant densities. For example, when a normal strain with mitochondrial DNA of 1.684 g/cm^3 buoyant density was crossed with a suppressive petite of buoyant density 1.677 g/cm^3, there were offspring colonies whose mitochondrial DNA had buoyant densities of 1.671, 1.674, and 1.683 g/cm^3. Such information supports the notions that mitochondrial DNA is open to recombination and that the suppressive character takes over a colony by way of recombination.

Antibiotic Influences.   Since the machinery of mitochondrial protein synthesis is prokaryotic in nature, mitochondrial protein synthesis can be inhibited by antibiotics such as chloramphenicol and erythromycin. These antibiotics induce a petite-type growth response and antibiotic-resistant strains can be obtained through mutation by growing yeast on the antibiotic. The resistance appears to be inherited on the mitochrondrial DNA, not the cellular DNA. Such an inheritance pattern is shown with crosses involving a resistant and a sensitive (wild) type yeast. The resulting diploid colonies segregate both resistant and sensitive cells (Figure 16–8). Although not expected on the basis of a chromosomal gene, the random sorting of diploid mitochondria could result in a wild-type cell containing only sensitive mitochondria. Since yeast have only one to ten mitochondria per cell, this sorting out of sensitive mitochondria can be expected to occur at a relatively high rate.

After the buoyant-density data just cited showed that recombination does occur among mitochondrial DNAs, a mitochondrial **polarity gene** was discovered—alleles associated with this gene were usually found in daughter mitochondria. By using the polarity gene and several antibiotic resistance loci, some preliminary mapping work has been done in yeast mitochondria. A tentative map of some loci is shown in Figure 16–9.

Mitochondrial mutations are known in many other organisms. In *Neurospora*, for example, there is a petite phenotype known as **poky.**

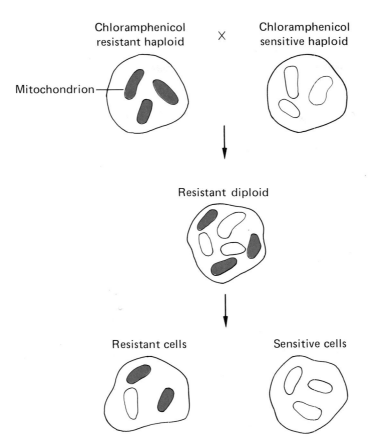

Figure 16 – 8.
Inheritance of
Antibiotic
(Chloramphenicol)
Resistance in Yeast

Chloroplasts

The **chloroplast** is the organelle that carries out photosynthesis and starch-grain formation in plants. Prior to the development of chlorophyll, chloroplasts are referred to as **plastids.** There are many similarities, both morphologically and genetically, between plastids and mitochondria. Normally, the chloroplast (Figure 16–10) contains chlorophyll, the photosynthetic pigment. However, when grown in the dark (and under some other circumstances), plastids do not develop and remain reduced in size and complexity. These mutant plastids, referred to as **proplastids,** are each about the size and shape of a mitochondrion.

Like mitochondria, chloroplasts contain DNA and ribosomes, both of the prokaryotic type. The DNA is most likely a circle $\approx 40 \ \mu$m in length—about 10 times the size of the mitochondrial DNA. Also found within the chloroplast are ribosomal RNA and transfer RNA, both made from the genome of the chloroplast and both of the prokaryotic type. Presumably, the

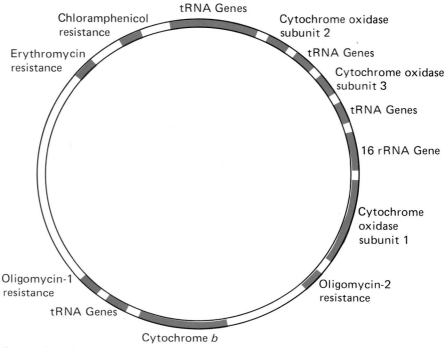

tRNA Genes

Chloramphenicol resistance

Cytochrome oxidase subunit 2

tRNA Genes

Erythromycin resistance

Cytochrome oxidase subunit 3

tRNA Genes

16 rRNA Gene

Cytochrome oxidase subunit 1

Oligomycin-1 resistance

Oligomycin-2 resistance

tRNA Genes

Cytochrome *b*

Figure 16 – 9.
Map of Several Loci on the Mitochondrial DNA of Yeast
The polarity gene is near chloramphenicol resistance.

Source: Data from Tzagoloff et al., (1979).

chloroplast evolved from symbiotic blue-green algae, which have many affinities with the chloroplast. The ribosomal RNA of blue-green algae will hybridize with the DNA of chloroplasts.

The similarities between mitochondria and chloroplasts make it possible to predict the inheritance patterns of chloroplast mutations on the basis of existing knowledge of mitochondrial genetics: We should find both chromosomal and plastid mutants of the chloroplast function. There should be simple segregation in the chromosomal mutants and cytoplasmic patterns in the chloroplast DNA mutants. Investigation of these inheritance patterns is complicated by the fact that plant cells have both mitochondria and chloroplasts and, since both have prokaryotic affinities, it is sometimes difficult to determine whether a genetic trait is due to the genetic systems of the chloroplasts or the mitochondria.

Lesions in the photosystems (I and II) of the chloroplast result in proplastid formation, with a loss of green color. When proplastid formation occurs in a particular tissue of a plant, variegation will result—that is, there will be both green and white parts, often as stripes. Some interesting genetic studies have been done on the inheritance of variegation, especially in the area of the interaction of chloroplast and chromosomal genes.

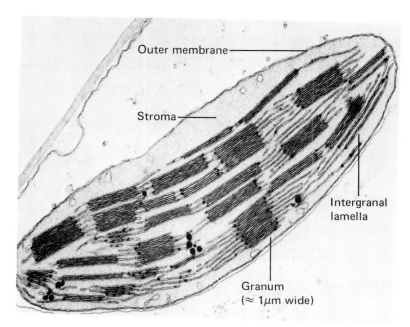

Figure 16 – 10. Electron Micrograph of a Lettuce Chloroplast

Source: C. Arntzen, figure 22–1, p. 590 in A. L. Lehninger, *Biochemistry* (New York: Worth Publishing, 1975). Reproduced by permission.

Marcus M. Rhoades (1903–) Courtesy of Dr. Marcus M. Rhoades.

Zea Mays. Rhoades worked on the variegation in corn *(Zea mays)* controlled by the *iojap* chromosomal locus, a recessive trait that, when homozygous, prevents the proplastids from developing as chloroplasts, which results in variegation. The interaction of chromosomal and extrachromosomal inheritance is shown in the reciprocal crosses of Figure 16–11, where one cross produces results exactly as would be predicted on the basis of simple Mendelian inheritance, with the homozygous recessive *(ij/ij)* inducing variegation. When the reciprocal cross is carried out, blotch variegation is seen in both the F_1 and F_2 that carry the dominant *Ij* allele.

This inheritance pattern is caused by the fact that, in corn, the pollen grain does not carry any chloroplasts whereas the ovule does. Thus, the first cross in Figure 16–11 (left) deals with the passage into the F_2 of normal chloroplasts only. In the F_2 the *ij ij* genotype then induces variegation. The chloroplasts of the pollen parent are unimportant because they do not enter into the F_1. In the reciprocal cross, however, because the stigma parent is variegated, the F_1 has the genotype of the heterozygous green plant but carries proplastids from the ovule that will remain proplastids even under the *Ij* allele. Therefore, in cells where segregation of chloroplasts leads to all colorless cells, these proplastids will produce a white spot, which causes blotchy variegation. Once the *ij* allele induces chloroplasts to become proplastids, they do not revert to the normal type even under the *Ij* allele. Thus,

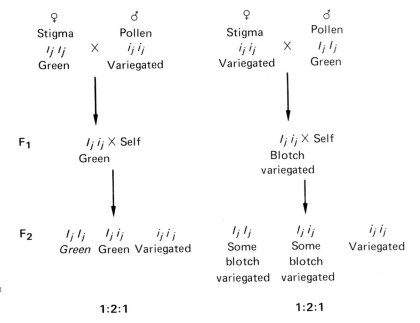

Figure 16 – 11.
Reciprocal Crosses
Involving the
Chromosomal Gene
iojap
The homozygous
recessive condition
(*ij/ij*) induces
variegation. Blotch
variegation produces
white spots rather
than striping.

there is the interaction of a chromosomal gene and the chloroplast itself, which "inherits" a transformed condition.

There is some indication that *iojap* may suppress the chloroplast rather than cause a mutation of some function. There are loci in corn and in other species that can induce back mutation in the chloroplasts. It is more likely that removal of suppression occurs rather than an actual reversion. It is difficult to envision a locus that would correct one small DNA lesion at a very high rate.

The situation in corn, where the pollen does not carry chloroplasts, has been termed *status albomaculatus*. This situation is more common in plants. In some species, however, plastids are carried by the pollen; the situation is referred to as *status paralbomaculatus*. In this latter situation we would expect different patterns in the inheritance of an *iojap*-type locus, and in fact that is precisely the case.

Four-O'clocks.   The first work with corn variegation was done by Correns, one of Mendel's rediscoverers. He found maternal inheritance of variegation in the four-o'clock plant, *Mirabilis jalapa*. Correns could predict color and variegation of offspring solely on the basis of the region of the plant on which the stigma parent was located. A flower from a white sector, when pollinated by any pollen, would produce white plants; a flower on a green sector or a variegated sector produced green or variegated plants, respectively, when pollinated by pollen from any region of a variegated plant. We thus see the simple maternal nature of the inheritance of the variegation. A chromosomal gene, *iojap*, induces variegation to occur. Inheritance of this induced variegation follows the "maternal" pattern of chloroplast inheritance in *status albomaculatus* plants.

Chlamydomonas.   The single-celled green alga *Chlamydomonas reinhardi* has been used extensively in the study of extrachromosomal inheritance, for several reasons. It has a single, large chloroplast; it can survive by culture technique even when the chloroplast is not functioning; and it shows some interesting non-Mendelian inheritance patterns related to the mating type. Sager has done extensive work on inheritance of streptomycin resistance in *Chlamydomonas*.

Streptomycin resistance can be selected for in *Chlamydomonas* in several ways. Normal cells are sensitive to the antibiotic and are killed in its presence. If cells are grown in low levels of the antibiotic (100 μg/ml), some cells will show resistance to it. When these cells are crossed to the wild type, the resistance will segregate in a 1 : 1 ratio; the implication is that it is controlled by a chromosomal locus. The same experiment can be repeated using high levels of the antibiotic in the medium (500–1600 μg/ml). Again, resistant colonies are found. If they are crossed to the wild type, a 1 : 1 ratio does not ensue.

Chlamydomonas does not have sexes but does have mating types, mt^+ and mt^-. Only individuals of opposite type can mate. Mating type is inherited as a single locus with two alleles. When two haploid cells of opposite mating type fuse, they form a diploid zygote, which then undergoes meiosis to produce four haploid cells, two of mt^+ and two of mt^-. The high-level resistance always segregates with the mt^+ parent (Figure 16–12). It is as if the mt^+ parent were contributing the cytoplasm to the zygote in a manner similar to that of the *status albomaculatus* plants. The mt^- parent is acting like a pollen parent by making a chromosomal contribution but not a cytoplasmic one.

The mechanism of the extrachromosomal inheritance of *Chlamydomonas* is not known. It is especially mystifying in that both parent cells fuse and appear to contribute equally to the zygote. The current belief is that the target of the streptomycin resistance is the chloroplast.

More recent work has shown that the mt^+ inheritance is only 99.98% effective. That is, 0.02% of the offspring in crosses of the type shown in Figure 16–12 have the streptomycin phenotype of the mt^- parent. Thus the possibility is left open of studying recombination in chloroplast genes. While most of the evidence is only indirect and plagued by the previously mentioned problems of separating chloroplast and mitochondrial effects, some initial mapping studies have been done.

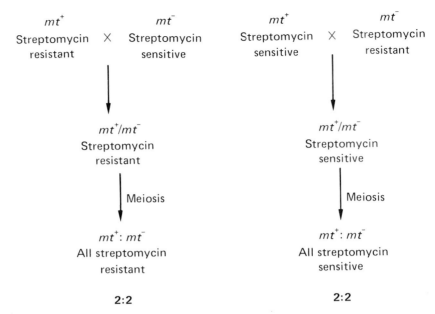

Figure 16 – 12.
Inheritance Pattern
of Streptomycin
Resistance
Dependent on the
Genotype of the *mt⁺*
Parent

Figure 16 – 12. Inheritance Pattern of Streptomycin Resistance Dependent on the Genotype of the mt^+ Parent

Infective Particles

Paramecium. Sonneborn discovered the killer trait in *Paramecium.* Before analyzing this trait, we must digress a moment to look at the life cycle of *Paramecium,* a ciliated protozoan familiar to most biologists. Ciliates have two types of nuclei: macronuclei and micronuclei. In *Paramecium* there are two micronuclei, which are primarily reproductive nuclei, and one macronucleus, which is a polyploid nucleus concerned with the vegetative functions of the cell. During simple cell division, termed **binary fission,** the micronuclei divide by mitosis and the macronucleus simply constricts and is pulled in half.

Paramecia undergo two types of sexual processes, **conjugation** and **autogamy.** In conjugation two mating types come together and form a bridge between themselves. The nuclear events are shown in Figure 16–13. Briefly, the macronuclei of each cell disintegrate while the micronuclei undergo meiosis. Of the resulting eight micronuclei per cell, seven disintegrate and one remains; this one undergoes mitosis to form two haploid nuclei per cell. A reciprocal exchange of nuclei across the bridge then occurs. Next, with each cell having two haploid nuclei, one original and one migrant, they fuse to form a diploid nucleus. The diploid nuclei in the two conjugating cells are genetically identical because of the reciprocity of the process. These nuclei then undergo two mitoses to form four diploid nuclei: Two will become macronuclei, which will separate at the next cell division; two will

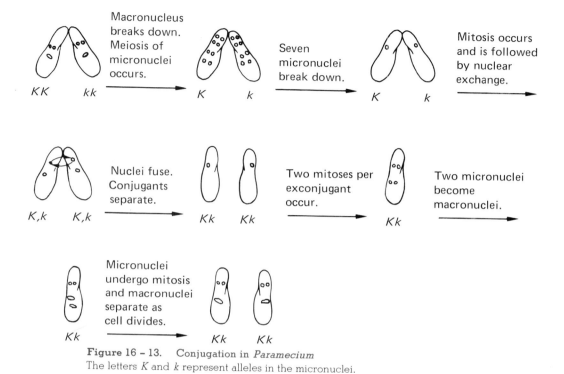

KK   *kk*

Macronucleus
breaks down.
Meiosis of
micronuclei
occurs.

K   *k*

Seven
micronuclei
break down.

K   *k*

Mitosis occurs
and is followed
by nuclear
exchange.

K,k   *K,k*

Nuclei fuse.
Conjugants
separate.

Kk   *Kk*

Two mitoses per
exconjugant
occur.

Kk

Two micronuclei
become
macronuclei.

Kk

Micronuclei
undergo mitosis
and macronuclei
separate as
cell divides.

Kk   *Kk*

Figure 16 – 13.   Conjugation in *Paramecium*
The letters *K* and *k* represent alleles in the micronuclei.

remain as micronuclei, which will divide by mitosis at the next cell division. The two cells will separate and are known as **exconjugants.** Primarily depending on the amount of time conjugating paramecia remain united, there may or may not be an exchange of cytoplasm along with the exchange of nuclei.

In the second type of sexual process, autogamy, and as the name implies, only one *Paramecium* is involved (Figure 16–14). The nuclear events are the same as in conjugation except that, at the point where a reciprocal exchange of nuclei would take place, the two haploid nuclei simply fuse. All cells after autogamy are homozygous.

Sonneborn and his colleagues found that when specific stocks of *Paramecium* are mixed together, one stock has the ability to cause the death of individuals of another stock. Individuals causing death were called "killers" and individuals dying were referred to as "sensitives." During conjugation, the sensitives are temporarily resistant to the killers. If cytoplasm is not exchanged during the conjugation, the exconjugants retain their original phenotype: Killers stay killers and sensitives stay sensitives. When there is an exchange of cytoplasm between sensitive and killer cells, both exconjugants are killers. The transfer of some cytoplasmic particle seems to be

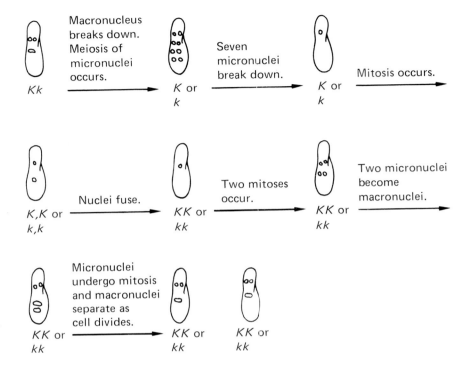

Figure 16 – 14.
Autogamy in
Paramecium
The letters *K* and *k*
represent alleles in
the micronuclei.

implied. Indeed, Sonneborn observed such particles in the cytoplasm of killers and called them **kappa particles** (Figure 16–15).

Although the occurrence of killer *Paramecium* does not appear to involve chromosomal genes, Sonneborn reported one case in which exconjugant killer *Paramecia* of hybrid origin underwent autogamy—half of the resulting cells had no kappa particles and had become sensitives. The implication is that a gene is required for the presence of kappa particles—a conclusion that has subsequently been verified by numerous crosses. Figure 16–16 illustrates the sequence of genetic events that would produce a heterozygous killer *Paramecium* that upon autogamy would have a 50% chance of becoming sensitive.

Recently, Preer and his colleagues have studied kappa itself. Although not yet cultured outside of a *Paramecium*, it is presumably a bacterium and has many attributes of a bacterium including size, cell wall, presence of DNA, and presence of certain prokaryotic reactions. Preer and his colleagues have named it *Caedobacter taeniospiralis* (Figure 16–17). Kappa occurs in at least two forms. The *N* form is the infective form that is passed from one *Paramecium* to another. It does not confer killer specificity on the host cell. This *N* form is attacked by bacteriophages that induce formation of inclusions, called *R* bodies, inside the kappa particle and thus convert it to the *B* form. (These *R* bodies, see Figure 16–17, are visible under the light micro-

Tracy M. Sonneborn
(1905–1981)
Photograph by
Dellenback.

Figure 16–15.
(a) Normal
(Sensitive)
Paramecium
(b) Kappa-
Containing (Killer)
Paramecium
A *Paramecium* is
about 200 µm long.

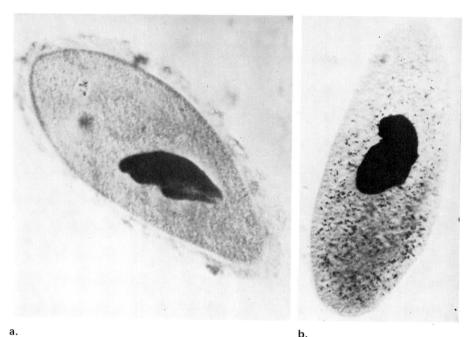

a.

b.

Source: T. M. Sonneborn, figure 29–3, p. 373 in I. H. Herskowitz,
Genetics, 2nd ed. (Boston: Little, Brown, 1965). Reproduced by
permission.

scope as refractile bodies within the kappa particles.) In the *B* form, kappa
can no longer replicate; it is often lysed within the cell, and it confers killer
specificity on the host cell. The sensitives are killed by a toxin (**paramecin**)
that is liberated into the environment. Precisely what steps are involved in
its formation are not known, although it is plain that the virus plays an
integral role. Whether the virus DNA or the kappa DNA codes the toxin is
also not at present known.

Kappa is not the only bacterial infective agent known in *Paramecium.*
Another interesting agent is seen in the **mate-killer** infection. Here again,
killer paramecia have visible bacteria-like particles in the cytoplasm. They
are called **mu particles;** Preer and his colleagues have named them *Caedobacter conjugatus.* Mate-killer paramecia do not liberate a toxin into the
environment but instead kill their mates during conjugation. One of two
unlinked dominant genes is required for the presence of mu particles. The
two dominants are M_1 and M_2. An interesting phenomenon happens when
a mate-killer becomes homozygous $m_1m_1m_2m_2$ by autogamy: While the offspring will eventually lose their mu particles, there is virtually no loss of particles until about the eighth generation, when some offspring lose all their
mu. Up to this generation all the cells had maintained a full complement of
mu. In the fifteenth generation only about 7% of the cells still have mu
particles.

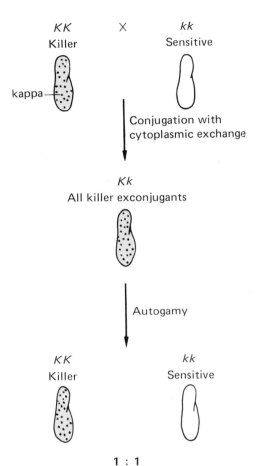

Figure 16 – 16.
Autogamy in a
Heterozygous *(Kk)*
Killer *Paramecium*
Upon autogamy, the
heterozygote has a
50% chance of
becoming a
homozygous *(KK)*
killer or a
homozygous *(kk)*
sensitive that loses its
kappa particles.

This phenomenon is explained as the dilution out, not of the mu themselves, but of a factor called **metagon,** which is necessary for the maintenance of mu in the cell. Once the cell becomes homozygous recessive, no further metagon production occurs. That metagon is subsequently diluted out is verified by examining fifteenth generation cells that still have their mu. We would expect that after fission one daughter cell would have a metagon and the other cell would not. What we expect in fact happens. The rate of dilution is consistent with an original number of about 1000 metagons per cell. The metagon appears to be mRNA because it is destroyed by RNase. Its protein product is presently unknown.

We thus see several instances of infective particles that interact with the genome of *Paramecium* with interesting phenotypic results. Similar interactions are known in other organisms—for example, the killer trait in yeast.

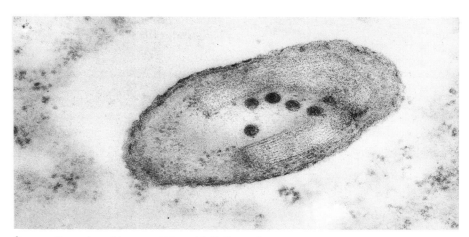

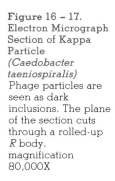

Figure 16 – 17. Electron Micrograph Section of Kappa Particle *(Caedobacter taeniospiralis)* Phage particles are seen as dark inclusions. The plane of the section cuts through a rolled-up *R* body. magnification 80,000X

Source: Reproduced by permission of J. R. Preer, Jr.

Drosophila. There are several instances in insects where infective particles mimic patterns of inheritance. Two infection/genome relationships are the CO_2 sensitivity of *Drosophila* that is mediated by a virus known as sigma and the sex-ratio trait, also in *Drosophila.*

In *Drosophila* there are several forms of the **sex-ratio phenotype,** where females produce mostly, if not only, daughters. One form is inherited as a chromosomal gene. Another form, however, is not chromosomal. In this form females usually produce a few sons. These sons do not pass on the sex-ratio trait. The daughters of sex-ratio females do pass on the trait, which was shown to be extra-chromosomal by the fact that it persisted even after all the chromosomes had been substituted out of the stock by appropriate crosses.

About half the eggs of a sex-ratio female do not develop. Cytoplasm can be withdrawn from the undeveloped eggs and used to infect other females. The trait, then, is caused by some cytoplasmic factor that could infect other females and is not passed on by sperm. Upon detailed cytological examination, a spiroplasma has been found in the sex-ratio females (Figure 16–18). When the bacterium is isolated, it can be used to infect other female *Drosophila* with the sex-ratio trait—it is, therefore, the causal agent.

PLASMIDS

Earlier, in Chapters 6 and 12, plasmids were discussed in regard to their role in the study of prokaryotic genetics and their use in recombinant DNA work. They are mentioned again here because they represent extrachromosomal genetic systems in prokaryotes. The autonomous segments of DNA known

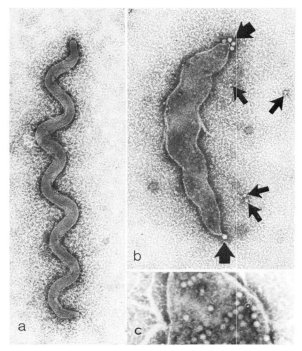

Figure 16 – 18.
Electron Micrograph
of Spiroplasma
Associated with
Extrachromosomal
Sex-Ratio Trait in
Drosophila
magnification
32,200X

Source: K. Oishi and D. F. Poulson, ''A Virus Associated with SR-spiro-
chetes of *Drosophila nebulosa*,'' *Proceedings. The National Academy of
Sciences USA* 67 (1970):1565–1572. Reproduced by permission of the
authors.

as plasmids are primarily known from bacteria, where they occur as circles
of DNA within the host cell (noncircular DNA is soon degraded by the bacterial cell). When plasmids become integrated into the chromosomes, they
become indistinguishable from chromosomal material. Plasmids that can
associate with the host's chromosome are called episomes (Chapter 6).

F Factor

The F factor, discussed in detail in Chapter 6, is a plasmid that contains
genes for the construction of the F-pilus (probably the bridge through which
the F factor is passed into an F^- cell during conjugation), genes for the proteins needed for the transfer process, and several loci that involve resistance
to certain F^--specific phages as well as resistance to other F factors entering
the same cell. The F factor can attach to the host chromosome at various
points to convert the host to Hfr, which then has the ability to pass the entire
host chromosome into the F^- cell. The F factor can dissociate from the Hfr
chromosome. If it is not correctly removed from the host at the time when

it dissociates, it can undergo sexduction—that is, carry certain host loci with it—a process that has been useful in mapping and physiological studies.

Insertion Sequences

The F factor can attach at many points on the host chromosome because both the host and the F factor have certain regions in common. These regions are now called **insertion sequences** (IS). These insertion sequences are also found on other plasmids and phages. Many are known, such as IS1, IS2, and IS3. IS1 is about 800 bases long, whereas IS2 and IS3 are each about 1300 bases long. The F factor contains IS2 and IS3. *Escherichia coli* has as many as 15 copies of IS1 and IS2. There is some evidence that these insertion sequences are relatively mobile. That is, there are many places they can occur on the *E. coli* genome and there is evidence that their number and position are not constant from one bacterial strain to the next. They are believed to be important in the recombination process because they provide sites for attachment of plasmids to the bacterial chromosome and to each other (Figure 16–19).

R and Col Plasmids

Two other kinds of plasmids found frequently in bacteria are the R and Col plasmids. The R plasmids carry resistance to various antibiotics and the Col plasmids have genes that are responsible for producing proteins called *colicins,* which are toxic to strains of *E. coli* (Figure 16–20). Plasmids containing Col-like toxins specific for other bacterial species are also known. Col and R plasmids can exist in two states. In one state the plasmid has a sequence of genes called the **transfer operon** *(tra),* which makes the plasmids similar to F factors: They can transfer their genes from one bacterium to the next. In the other state the plasmids lack this operon and cannot transfer their

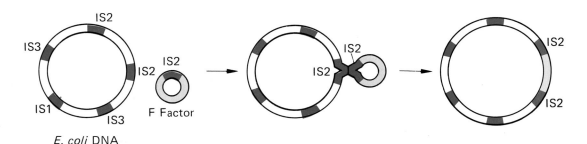

Figure 16 – 19.   Function of Insertion Sequences in the Attachment of the F Factor to the *E. coli* Chromosome

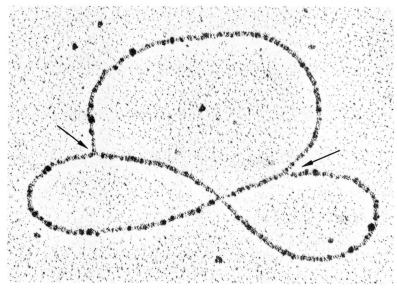

Figure 16 – 20.
Electron Micrograph
of Replication of Col
E1 Circular Plasmid
The arrows mark the
branch points.
magnification
120,000X

Source: J. I. Tomizawa, Y. Sakakibara, and T. Kakefuda, "Replication of Colicin E1 Plasmid DNA in Cell Extracts: Origin and Direction of Replication," *Proceedings. The National Academy of Sciences USA* 71 (1974):2260–2264. Reproduced by permission.

loci to another cell. Thus, Col and R plasmids are actually made of two parts: the loci for antibiotic resistance or colicin production and the part responsible for infectious transfer. In R plasmids the infectious transfer part is abbreviated as RTF (resistance transfer factor).

The occurrence of resistance plasmids was observed in Japan in the late 1950s when it was discovered that bacteria were simultaneously acquiring resistance to several antibacterial agents. When *Shigella* cultures were exposed to streptomycin, sulfonamide, chloramphenicol, or tetracycline, the bacteria developed resistance not only to that agent but to one or more of the others as well. The plasmid responsible for this multiple resistance was named R222.

The Col plasmids contain loci that produce proteins that are toxic, for various reasons, to strains of bacteria not carrying the plasmids. Colicins attack sensitive bacterial cells by attack at bacterial surface receptors. On the basis of the types of receptors they attack, colicins have been classified into 20 or more categories. Some colicins may enter the cell directly; some do not. For example, colicin K appears to kill sensitive cells by inhibiting DNA, RNA, and protein synthesis while not directly entering the cell. Colicin E3, however, acts as an intracellular ribonuclease that cleaves off about 50 nucleotides from the 3′ end of the 16S rRNA within the ribosome. The cleavage inactivates the sensitive cell's ribosomes, and is, of course, lethal.

Since many R plasmids, Col plasmids, and F factors, as well as host chromosomes have insertion sequences, there is a good deal of exchange among the plasmids, and many are able to integrate into the host chromosome. While their mobility makes it easier to map and study plasmids, it also poses a human health problem for two reasons. First, resistance to various antibacterial agents is easily transferred among enterobacteria worldwide. Transfer of resistance can even occur outside of host organisms (people) where there is pollution or sewerage. Second, resistance found in harmless enterobacteria, such as *E. coli,* can easily be passed to more pathogenic bacteria, such as *Shigella* and *Salmonella.* Since every time we use antibacterial drugs we are selecting for resistance, we should not use them indiscriminantly. There has been recent concern over excessive medical use of antibacterial drugs as well as the use of large quantities of antibiotics in animal feed.

Uncovering Plasmids

How do we know when we are dealing with a plasmid rather than with the chromosomal genes of a bacterium? Most directly, plasmids may be looked for with an electron microscope or by density gradient centrifugation. But several, less direct lines of evidence will also supply the answer. To begin with, multiple aspects of the phenotype (resistance to several antibacterial agents) change simultaneously, as with plasmid R222. Then, the phenotype change is infectious—Lewin (1977) has stated: "Resistance is infectious . . ."; the Japanese workers found that with R222, resistant cells converted nonresistant cells.

There are several other clues to the presence of a plasmid. In linkage studies, such as transduction studies, plasmid loci show no linkage to host loci. Plasmids themselves can be mapped because their loci will be linked to each other, but they will not be linked to any of the normal host loci. In a population of cells with a plasmid, there are many spontaneous losses of the plasmid and, therefore, of its loci. Since the plasmid DNA replicates at its own speed, it can miss being incorporated into a daughter cell. And finally, certain treatments—with acridine dyes, for example—have little effect on the host chromosome replication but selectively prevent the plasmid from replicating; thus the plasmid will disappear from the cell population. Thus, the existence of plasmids in a bacterial population can be verified with morphological, physiological, and analytical evidence.

CHAPTER SUMMARY

Patterns of non-Mendelian inheritance fall into two categories: maternal inheritance and cytoplasmic inheritance. Maternal inheritance is illustrated by snail-shell coiling. The direction of coiling is determined by the genotype of the maternal parent, where dextral coiling is dominant to sinistral coiling.

Cytoplasmic inheritance is usually demonstrated by organelles, symbionts, or parasites that have their own genetic material. Chloroplasts and mitochrondria have small circular chromosomes with prokaryotic affinities. There is an interaction between organelles and nuclei partly because the organelles do not encode all their enzymes. Mitochondrial defects (manifest as petite or poky phenotypes) can be inherited through nuclear genes or through the mitochondrion itself. A similar pattern is seen for chloroplasts. The processes of cytoplasmic inheritance are exemplified by symbiotic bacteria (kappa, mu) in *Paramecium*.

Plasmids are autonomous segments of DNA. In prokaryotes the sex factor, F, and R and Col plasmids have been well studied. They usually carry an operon for transfer, as well as insertion sequences for attachment to cell chromosomes and to each other. They thus represent highly mobile segments of genetic material.

EXERCISES AND PROBLEMS

1. Two strains of *Drosophila melanogaster* differ in their ability to tolerate CO_2. One strain is resistant while the other is sensitive: They are killed by CO_2 anesthetization. What genetic experiments would you do to determine that the trait is caused by a virus? How would you rule out chromosomal genes?

2. How would you rule out a viral origin for snail-shell coiling?

3. How would you determine that a segregative petite in yeast is controlled by a chromosomal gene?

4. What genetic tests could you do to show that mate-killer in *Paramecium* requires a dominant allele at any one to *two* loci?

5. Christian and Lemunyan (1958) have shown that mice raised under crowded conditions produce two generations with reduced growth rates. What sort of genetic control might exist, and how could this control be demonstrated?

6. Give the genotypes such that a sinistral female snail can produce dextral offspring. What genotypes could the male parent of the sinistral female have?

7. What results would be obtained by making all possible pairwise crosses of the three types of yeast petites?

8. A killer and a nonkiller strain of *Paramecium* are mixed. Cytoplasmic exchange occurs during conjugation. Approximately 25% of the exconjugants are sensitive and the remaining 75% are killers. What are the genotypes of the two strains and what ratios of sensitives and killers would result if the various exconjugants underwent autogamy?

9. In *Chlamydomonas* 0.02% of the meiotic products of a streptomycin sensitive (str^s) crossed with a streptomycin resistant (str^r) cell are of the mt$^-$ type. How can you use this information in mapping?

10. An ornamental spider plant has green and white striped leaves. How can you determine whether maternal inheritance is responsible and whether there is interaction with an *iojap*-type chromosomal gene?

11. Describe the types of evidence that could be gathered to determine whether a trait in *E. coli* were controlled by chromosomal or plasmid genes.

12. Snail coiling is called a maternal trait. Is it possible that it is simply sex linked?

SUGGESTIONS FOR FURTHER READING

Beale, G., and J. Knowles. 1978. *Extranuclear Genetics*. Baltimore, Md.: University Park Press.

Broda, P. 1979. *Plasmids*. San Francisco: Freeman.

Christian, J., and C. Lemunyan. 1958. Adverse effects of crowding on lactation and reproduction of mice and two generations of their progeny. *Endocrinology* 63:517–529.

Gillham, N. 1978. *Organelle Heredity*. New York: Raven Press.

Grun, P. 1976. *Cytoplasmic Genetics and Evolution*. New York: Columbia University Press.

Lewin, B. 1977. *Gene Expression, 3: Plasmids and Phages*. New York: Wiley.

Margulis, L. 1970. *Origin of Eukaryotic Cells*. New Haven, Conn.: Yale University Press.

Preer, J., L. Preer, and A. Jurand. 1974. *Kappa* and other endosymbionts in *Paramecium aurelia*. Bact. Rev. 38:113–163.

Rhoades, M. 1946. Plastid mutations. *Cold Spr. Harb. Symp. Quant. Biol.* 11:202–207.

Sager, R. 1972. *Cytoplasmic Genes and Organelles*. New York: Academic Press.

Sonneborn, T. 1950. Methods in the general biology and genetics of *Paramecium aurelia*. *J. Exp. Zool.* 113:87–148.

Stuttard, C., and K. Rozee, eds. 1980. *Plasmids and Transposons*. New York: Academic Press.

Tzagoloff, A., G. Macino, and W. Sebald. 1979. Mitochondrial genes and translation products. *Ann. Rev. Biochem.* 48:419–441.

Williamson, D., and D. Poulson. 1979. Sex ratio organisms (spiroplasmas) of *Drosophila*. *The Mycoplasmas* 3:175–208.

PART III

EVOLUTIONARY GENETICS

17

Population Genetics: Equilibrium, Mutation, Migration, and Genetic Drift

J. B. S. Haldane (1892–1964) *Genetics* 52 (1965):frontispiece (drawing by Margaret Alice Egerton).

Sewall Wright (1889–   ) Courtesy of Dr. Sewall Wright

To many people the word evolution evokes a picture of dinosaurs or the lineage of fossil horses from the small *Hyracotherium* of 55 million years ago to the modern horses of today (Figure 17–1). But there is another side to evolutionary theory—population genetics, which is the algebraic description of the way in which allelic frequencies change in populations over time. Population genetics gives insights into the mechanisms of evolution. This chapter will look at what population genetics can tell us about the way in which evolution acts.

Almost all of the theory concerned with the mathematical description of genetic changes in populations was developed in a short period of time during the 1920s and 1930s by three men: R. A. Fisher, J. B. S. Haldane, and Sewall Wright. Some measure of disagreement emerged among these men, but they disagreed on which evolutionary processes were more important than others, rather than on how processes worked. Recently, new excitement has arisen in the field of population genetics, primarily on two fronts. First, the advent of the high-speed computer made it possible to process a large amount of arithmetic in a very short period of time so that complex simulations of real populations could be added to the repertoire of the experimental geneticist. Second, the advent of the technique of electrophoresis provided a means of gathering the large amount of empirical data necessary for checking some of the assumptions used for many of the mathematical models. The information derived from the electrophoretic data was not what the theorists had expected, and new controversy arose as to the role of "neutral" changes in natural populations. We will examine these changes in more detail later.

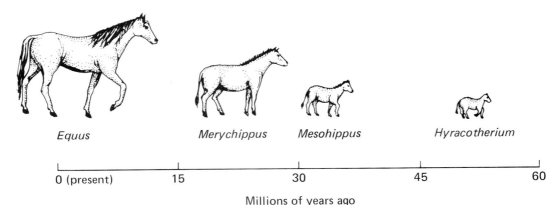

Figure 17 – 1.   Evolutionary Lineage of the Horse from the Eocene
Epoch 55 Million Years Ago

HARDY-WEINBERG EQUILIBRIUM

Wilhelm Weinberg
(1862–1937)
Genetics 47
(1962):frontispiece.

In 1908 G. H. Hardy, a British mathematician, and W. Weinberg, a German physician, independently discovered that an equilibrium in allelic and genotypic frequencies will arise in a diploid population that adheres to several suppositions (random mating, large size, no mutation or migration, and no selection). The equilibrium has three facets:

1. The allelic frequencies at an autosomal locus will not change from one generation to the next (allelic frequency equilibrium).
2. The genotypic frequencies of the population are determined on the basis of the allelic frequencies (genotypic frequency equilibrium).
3. If the equilibrium is disturbed, it will recover after just one generation of random mating.

Calculating Allelic Frequency

If we consider an autosomal locus in diploids, allelic frequencies then can be measured in either of two ways. The first way is simply by counting genes:

$$\text{frequency of the } a \text{ allele } (q) = \frac{\text{number of } a \text{ alleles}}{\text{total number of alleles}}$$

The expression "frequency of" can be shortened to $f(\ )$. For example, the frequency of the a allele will be written as $f(a)$. Since the homozygotes have two of a given allele and heterozygotes have only one and since the total number of alleles is twice the number of individuals (each person carries two alleles), we can calculate allelic frequencies in the following manner. Con-

DE FINETTI DIAGRAM

The triangle devised by B. De Finetti in 1926 and known as the De Finetti diagram is an interesting way to look at the frequencies of three genotypes in a population. Any point within the triangle represents a population, and the lengths of perpendiculars extended to the three sides from such a point represent the relative proportion of the genotypes in this population. For example, in population 1 the lines D, R, and H represent the relative frequencies of AA, aa, and Aa, respectively. If the triangle is drawn with an altitude of 1, the three perpendiculars can be measured directly because their sum will also be 1. There are two other interesting properties of this triangle. First, the populations in Hardy-Weinberg equilibrium trace a parabola within the triangle: Population 2 is in equilibrium; population 1 is not. Second, the heterozygote perpendicular (H) divides the base of the triangle into the proportions of the allelic frequencies, p and q.

The shape of the parabola within the triangle shows several things. Its peak is directly over the center of the base. Since the perpendicular to the base, H, is the proportion of heterozygotes, it is apparent that heterozygotes are most numerous

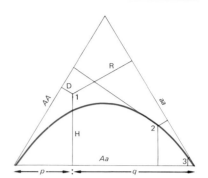

**Figure 1.
The De Finetti
Diagram**

when the base is bisected when $p = q = 0.5$. As allelic frequency increases or decreases, the parabola approaches the sides of the triangle. That is, as allelic frequency increases or decreases, the relative proportion of heterozygotes to rarer homozygotes rapidly increases. Thus in populations in Hardy-Weinberg equilibrium with allelic frequencies far from 0.5, the less-common allele is more often found in heterozygotes than in homozygotes. (See the relative lengths of R and H in populations 2 and 3.) Chapter 18 will discuss the importance of this allelic distribution for selection and eugenics.

sider, for example, the phenotypic distribution of M/N blood types among 200 persons chosen randomly in Columbus, Ohio:

$$MM = 114$$
$$MN = 76$$
$$NN = \underline{10}$$
$$200$$

Then, $$p = f(M) = \frac{2(114) + 76}{2(200)} = \frac{304}{400} = 0.76$$

Similarly, $$q = f(N) = \frac{2(10) + 76}{2(200)} = \frac{96}{400} = 0.24$$

Alternatively, knowing p allows calculation of q as

$$q = 1 - p \quad \text{since} \quad p + q = 1$$

An alternative way of calculating allelic frequency is based on knowledge of the genotypic frequencies, which in this example are

$$f(MM) = \frac{114}{200} = 0.57$$

$$f(MN) = \frac{76}{200} = 0.38$$

$$f(NN) = \frac{10}{200} = 0.05$$

Then allelic frequencies can be calculated as follows:

$$p = f(M) = f(MM)\text{homozygotes} + \tfrac{1}{2}f(MN)\text{heterozygotes}$$

$$= 0.57 + \tfrac{1}{2}(0.38) = 0.76$$

$$q = f(N) = f(NN)\text{homozygotes} + \tfrac{1}{2}f(MN)\text{heterozygotes}$$

$$= .05 + \tfrac{1}{2}(0.38) = 0.24$$

Assumptions of Hardy-Weinberg Equilibrium

We will consider a population of diploid organisms with a single locus segregating two alleles (every individual is one of three genotypes—*MM, MN,* or *NN*). Later on, the discussion will generalize to many loci, multiple alleles, sex linkage, haploids, and polyploids. For the moment, the focus is on a genetic system such as the *MN* locus in humans. The following four major assumptions are necessary for the Hardy-Weinberg equilibrium to hold

Random Mating.   The first of these assumptions is random mating. **Random mating** assumes that individuals mate according to the product rule of probability (Chapter 3). That is, the mating of individuals of any two genotypes is the combining of independent events: the product of the frequencies (or probabilities) of the mating genotypes in the population. If the *MM* genotype makes up 90% of a population, then any individual has a 90% chance (probability = 0.9) of mating with a person with an *MM* genotype. The probability of an *MM* by *MM* mating is (0.9)(0.9), or 0.81.

   Deviations from random mating come about for two reasons—choice or circumstance. If members of a population choose individuals of a particular phenotype as mates more often or less often than at random, the population is engaged in **assortative mating.** If mates with similar phenotypes are chosen disproportionately, *positive assortative mating* is in force; if mates with dissimilar phenotypes are chosen, *negative assortative mating,* or **disassortative mating,** is at work. To the extent that the phenotype reflects the genotype, genetic changes result from assortative mating.

Deviations from random mating also arise when mating individuals are genetically related to each other either more closely or less closely than to any other individual that could be chosen at random from the population. **Inbreeding** is the mating of individuals who are related and **outbreeding** is the mating of genetically unrelated individuals. Often, inbreeding is a consequence of small population size. The consequences of inbreeding are similar to those of assortative mating; the consequences of outbreeding resemble those of disassortative mating. (The next chapter deals at length with inbreeding.)

Deviations from random mating alter the genotypic frequency but *not* the allelic frequency. Envision a population in which every individual is the parent of two children. Each child of each individual will have one of this individual's alleles. On the average, each individual will pass on one copy of each of his or her alleles. Assortative mating and inbreeding will change the zygotic (genotypic) combinations from one generation to the next but will not change which alleles are transported into the next generation.

No Selection. The second assumption necessary to the Hardy-Weinberg equilibrium is that no genotype will result in an individual with a better survival or reproduction than individuals of any other genotype—that is, that there is no selection.

Large Population Size. A sample may not be an accurate representation of a population, especially if the sample is small. Thus, the third assumption is that the population is infinitely large. The larger the population, the greater the probability that a random sample of a generation's gametes will accurately represent the allelic frequency in the population. When populations are small or when alleles are rare, changes in allelic frequency take place due to chance alone. These changes are referred to as **random genetic drift,** or just *genetic drift*.

No Mutation or Migration. Allelic and genotypic frequencies may be changed by the loss or addition of alleles through genetic mutation or through migration (immigration and emigration) of individuals from or into a population. The fourth Hardy-Weinberg assumption is that there is no such allelic loss or addition in the population.

In summary, the Hardy-Weinberg equilibrium holds (is exactly true) in an infinitely large, randomly mating population where no selection, mutation, or migration occur. (Mutation and migration equilibrium conditions as described later can allow the Hardy-Weinberg equilibrium to hold even in the face of mutation and migration.) In view of the assumptions, it seems that such equilibrium is not a characteristic of natural populations. However, this is not the case. Hardy-Weinberg equilibrium is routine in natural populations for two major reasons. First, the consequences of violating some of the assumptions (for example, no mutation, infinitely large population size) are small. For example, mutation rates are on the order of one change per locus per generation per 10^5 gametes. Thus, the measurable effect of mutation in a single generation is negligible. In addition, populations do not have to be infinitely large to overcome the major problems of sampling error. That is, a smaller-than-infinite population will still achieve the Hardy-Weinberg equilibrium. Minor deviations from the other assumptions will still give a good fit to the equilibrium. Second, the Hardy-Weinberg equilibrium is extremely resilient to change because, regardless of the perturbation, the equilibrium is reestablished after only one generation of random mating.

Proof of Hardy-Weinberg Equilibrium

The three facets of the Hardy-Weinberg equilibrium are that (1) allelic frequencies do not change from generation to generation; (2) the equilibrium is achieved in one generation of random mating; and (3) allelic frequencies determine genotypic frequencies. We will concentrate for a moment on the third facet. In a population segregating the A and a alleles at the A locus, each individual will be one of three genotypes: AA, Aa, or aa. If $p = f(A)$ and $q = f(a)$, then the genotypic frequencies in the next generation can be predicted. If all the assumptions of the Hardy-Weinberg equilibrium are met, then the three genotypes should occur in the population in the frequency at which gametes would be randomly drawn in pairs from a **"gene pool."** (Gene pool is defined as all of the alleles available among the reproductive members of the population from which gametes can be drawn.) Thus,

$$f(AA) = (p \times p) = p^2$$
$$f(Aa) = (p \times q) + (q \times p) = 2pq$$
$$f(aa) = (q \times q) = q^2$$

which is a restatement of the third facet of the Hardy-Weinberg equilibrium.

All three facets of the Hardy-Weinberg equilibrium can be proved for the one-locus, two-allele case in sexually reproducing diploids by simply observing the offspring of a randomly mating, infinitely large population. If the initial frequencies of the three genotypes are any arbitrary values—for example, X, Y, and Z for the AA, Aa, and aa genotypes—then the proportions of offspring after one generation of random mating are as shown in Table 17–1. For example, the probability of an AA individual mating with

TABLE 17 – 1. Proportions of Offspring in a Randomly Mating Population Segregating the A and a Alleles at the A locus: $X = f(AA)$, $Y = f(Aa$, and $Z = f(aa)$

Mating	Proportion	Offspring		
		AA	Aa	aa
$AA \times AA$	X^2	X^2		
$AA \times Aa$	XY	$\frac{1}{2}XY$	$\frac{1}{2}XY$	
$AA \times aa$	XZ		XZ	
$Aa \times AA$	XY	$\frac{1}{2}XY$	$\frac{1}{2}XY$	
$Aa \times Aa$	Y^2	$\frac{1}{4}Y^2$	$\frac{1}{2}Y^2$	$\frac{1}{4}Y^2$
$Aa \times aa$	YZ		$\frac{1}{2}YZ$	$\frac{1}{2}YZ$
$aa \times AA$	XZ		XZ	
$aa \times Aa$	YZ		$\frac{1}{2}YZ$	$\frac{1}{2}YZ$
$aa \times aa$	Z^2			Z^2
Sum	$(X + Y + Z)^2$	$(X + \frac{1}{2}Y)^2$	$2(X + \frac{1}{2}Y)(Z + \frac{1}{2}Y)$	$(Z + \frac{1}{2}Y)^2$

an AA individual is $X \cdot X$, or X^2. Since all the offspring of this mating are AA, they are counted only under the AA column of offspring in Table 17–1. When all possible matings are counted, the offspring with each genotype are summed. Then the proportion of AA offspring is $X^2 + \frac{1}{2}XY + \frac{1}{4}Y^2$, which factors to $(X + \frac{1}{2}Y)^2$. Recall that the frequency of the homozygotes of an allele plus half the frequency of the heterozygotes is this allele's frequency. Hence, $(X + \frac{1}{2}Y)$ is the frequency of A, or p; $(X + \frac{1}{2}Y)^2$ is then p^2. Thus, after one generation of random mating, the proportion of AA homozygotes is p^2. Similarly, the frequency of aa homozygotes after one generation of random mating is $\frac{1}{4}Y^2 + \frac{1}{2}YZ + \frac{1}{2}YZ + Z^2$, which factors to $(Z + \frac{1}{2}Y)^2$, or q^2. The frequency of heterozygotes when summed and factored (Table 17–1) is $2(X + \frac{1}{2}Y)(Z + \frac{1}{2}Y)$, or $2pq$. Hence, after one generation of random mating, the three genotypes (AA, Aa, and aa) occur as p^2, $2pq$, and q^2.

Has the allelic frequency changed from one generation to the next (parents to offspring)? Before random mating,

$$f(A) = f(AA) + \frac{1}{2}f(Aa) = X + \frac{1}{2}Y = p$$

After random mating,

$$f(A) = f(AA) + \frac{1}{2}f(Aa) = p^2 + \frac{1}{2}(2pq) = p^2 + pq = p(p + q) = p$$

Thus, the allelic frequency, p, does not change from generation to generation. Here, by observing the offspring of a randomly mating population, all three facets of the Hardy-Weinberg equilibrium are proved.

Generation Time

The concept of "generations" is in reality a very difficult concept. Demographers have complex formulas relating generation time to age of reproduction, amount of reproduction in each age group, and probabilities of survival in each age group. Here, the concept will be simplified for purposes of discussion by the use of **discrete generations,** unless otherwise noted. That is, it is assumed that all the individuals drawn from a sample population for purposes of determining allelic and genotypic frequencies are drawn from the same generation and that, in resampling the population, the second sample represents the next generation—offspring of the first sample. The discrete-generation type of model holds for organisms, such as annual plants and fruit flies in bottles, where generations can be controlled by simply discarding the parents—generations usually overlap in populations of humans and most other organisms. More precise mathematical models, which involve integral calculus, are required to properly describe overlapping generations. Therefore, in this and the next chapter, discrete generations are assumed.

Testing for Fit to Hardy-Weinberg Equilibrium

There are several ways to determine whether a given population conforms to the Hardy-Weinberg equilibrium at a particular locus. One way would be to sample the next generation to see whether the allelic frequencies have changed. However, the question usually arises when there is just a single sample from a population representing only one generation. Can the Hardy-Weinberg equilibrium be determined with just one sample? The answer is yes, since to ask whether a population is in Hardy-Weinberg equilibrium is, in essence, to ask whether the three genotypes (*AA, Aa,* and *aa*) occur with the frequencies, $p^2, 2pq,$ and q^2. If they do, then the population is considered to be in Hardy-Weinberg equilibrium; if they do not, then the population is considered not in Hardy-Weinberg equilibrium.

M/N Blood Types.   In order to determine whether observed and expected (Hardy-Weinberg) frequencies are the same, the chi-square statistical test can be used. Chi-square compares an observed number with an expected number. In this case the observed values are the actual numbers of the three genotypes in the sample population, and the expected values come from the prediction of the genotypes occurring as $p^2, 2pq,$ and q^2. An analysis for the Ohio M/N blood-type data is presented in Table 17–2, which shows that the agreement between observed and expected is very good even before calculation of the chi-square value. Since there is one degree of freedom for the chi-square calculation in Table 17–2 and the critical chi-square for one degree of freedom at the 0.05 level is 3.841 (Table 3–4), it is found that the

TABLE 17 – 2. Chi-Square Test of Goodness-of-Fit to the Hardy-Weinberg Equilibrium of a Sample of 200 Persons for M/N Blood Types Where $p = 0.76$ and $q = 0.24$

	MM	MN	NN	Total
Observed numbers	114	76	10	200
Expected proportions	p^2	$2pq$	q^2	1.0
	(0.57)	(0.37)	(0.06)	1.0
Expected numbers	115.5	73.0	11.5	200.0
$\chi^2 = (O - E)^2/E$	0.02	0.12	0.20	0.34

Ohio population does not deviate from Hardy-Weinberg equilibrium at the *MN* locus. (A simple rule of thumb when chi-square tests are done with allelic frequencies is that the number of degrees of freedom equals the number of phenotypes minus the number of alleles. This rule holds because an extra degree of freedom is lost every time an independent parameter is estimated. In this analysis one extra degree of freedom is lost by estimating p or q.)

The chi-square analysis in Table 17–2 may seem paradoxical. Because the observed allelic frequencies calculated from the original genotypic data are used to calculate the expected genotypic frequencies, it may appear to some individuals that the analysis must, perforce, show that the population is in Hardy-Weinberg equilibrium. To demonstrate that this is not necessarily the case, a counterexample is presented in Table 17–3, which uses data similar to the Ohio data except that the number in the heterozygote class has been distributed among the two homozygote classes so that the same allelic frequencies are maintained while quite a different genotypic distribution is created. The chi-square value of 200.38 for this data demonstrates that the population represented in Table 17–3 is not in Hardy-Weinberg equilibrium. Thus, a chi-square analysis of Hardy-Weinberg equilibrium is by no means circular.

TABLE 17 – 3. Chi-Square Test of Goodness-of-Fit to the Hardy-Weinberg Equilibrium of a Second Sample of 200 Persons for M/N Blood Types Where $p = 0.76$ and $q = 0.24$ and Heterozygotes are Absent

	MM	MN	NN	Total
Observed numbers	152	0	48	200
Expected proportions	p^2	$2pq$	q^2	1.0
	(0.57)	(0.37)	(0.06)	1.0
Expected numbers	115.5	73.0	11.5	200.0
$\chi^2 = (O - E)^2/E$	11.53	73.00	115.85	200.38

PKU. Circumstances do exist in which the Hardy-Weinberg equilibrium cannot be proved or disproved. In the case of a dominant trait, for example, without breeding data, allelic frequencies cannot be calculated from the genotypic classes because, since both have the same phenotype, the dominant homozygote cannot be distinguished from the heterozygote. However, allelic frequencies can be estimated by assuming that Hardy-Weinberg equilibrium exists and, thereby, assuming that the frequency of the recessive homozygote is q^2, from which q and then p can be determined.

If, for example, Hardy-Weinberg equilibrium is assumed for a disease such as phenylketonuria (PKU), which occurs only as a homozygous recessive genotype, it is possible to determine what proportion of the population consists of heterozygous carriers of one recessive PKU allele. But, is it fair to assume Hardy-Weinberg equilibrium here? There was, until recent medical practices intervened, a good deal of selection against individuals with PKU: It produced mental retardation. Thus, the "no selection" assumption required for equilibrium is violated. However, 1 child in 10,000 live births has a PKU genotype, and, when a homozygote is as rare as 1 in 10,000, selection is having a negligible effect on allelic frequency. Therefore, because of the rarity of the trait, it is fair to assume Hardy-Weinberg equilibrium here and to calculate as follows:

$$\text{homozygous recessive} = q^2 = 1/10{,}000 = 0.0001$$

so

$$q = 0.01 \quad \text{and} \quad p = 1 - q = 0.99$$

Therefore,

$$\text{homozygous normal} = p^2 = (0.99)^2 \approx 0.98 \quad \text{or} \quad 98 \text{ in } 100$$

$$\text{heterozygote} = 2pq = 2(0.01)(0.99) \approx 0.02 \quad \text{or} \quad 2 \text{ in } 100$$

By assuming the Hardy-Weinberg equilibrium, we have discovered something not intuitively obvious: A recessive trait as rare as 1 in 10,000 is carried in the heterozygous state by 1 in 50 individuals. Obviously, then, chi-square cannot be used here to verify the assumption of Hardy-Weinberg equilibrium since the allelic frequencies were derived by assuming the frequencies of the Hardy-Weinberg equilibrium to begin with. Furthermore, number of phenotypes ($= 2$) − number of alleles ($= 2$) $= 0$ degrees of freedom, and a chi-square test cannot be done with zero degrees of freedom.

EXTENSIONS OF HARDY-WEINBERG EQUILIBRIUM

The Hardy-Weinberg equilibrium can be extended to cover more than the case of one autosomal locus with two alleles in a diploid population. We can begin the extensions with multiple alleles.

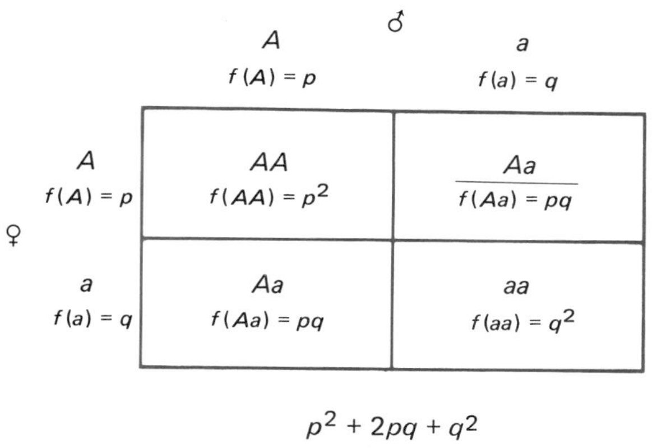

Figure 17 – 2.
Gene Pool Concept
of Zygote Formation

Multiple Alleles

Multinomial Expansion.   The expected genotypic array under Hardy-Weinberg equilibrium is p^2, $2pq$, and q^2 ($p^2 + 2pq + q^2 = 1$)—these are the terms of the binomial expansion, $(p + q)^2$. If males and females are viewed as each having two alternate alleles in the proportions of p and q, then genotypes will be distributed as a binomial expansion and will generate homozygotes and heterozygotes in the next generation in the frequencies p^2, $2pq$, and q^2 (see Figure 17–2). To generalize to more than two alleles, all that is needed is to add terms to the binomial expansion and thus create a multinomial expansion. For example, with alleles a, b, c, ... with frequencies p, q, r, ..., the genotypic distribution should be $(p + q + r + \dots)^2$ or

$$p^2 + 2pq + 2pr + q^2 + 2qr + r^2 \dots$$

Homozygotes will occur as the terms p^2, q^2, r^2, ... and heterozygotes will occur with frequencies such as $2pq$, $2pr$, $2qr$,.... The ABO blood-type locus in humans has multiple alleles and dominance.

ABO Blood Groups.   The example of the ABO system will be considered as consisting of three alleles, I^A, I^B, and I^O, where the I^A and I^B alleles are codominant and both are dominant to the I^O allele. These alleles control the production of a red blood cell surface antigen. Table 17–4 contains data from a sample of 500 persons from Massachusetts. Is the population in Hardy-Weinberg equilibrium? The answer is not apparent from the data in Table 17–4 alone since there are two possible genotypes for both the A and B phenotypes. No estimate of the allelic frequencies is possible without making assumptions about the genotype numbers within the two phenotypic classes. Is it possible to estimate the allelic frequencies? The answer is yes, if we assume that Hardy-Weinberg equilibrium exists.

TABLE 17 – 4. ABO Blood-Type Distribution in 500 Persons from Massachusetts

Blood Type	Genotype	Number
A	$I^A I^A$	
	$I^A I^O$	199
B	$I^B I^B$	
	$I^B I^O$	53
AB	$I^A I^B$	17
O	$I^O I^O$	231

One procedure is as follows. Blood type O has the genotype $I^O I^O$ and, if the population is in Hardy-Weinberg equilibrium, this genotype should occur in a frequency of r^2, where $r = f(I^O)$. Thus,

$$f(I^O I^O) = 231/500 = 0.462 = r^2$$

and

$$r = f(I^O) = \sqrt{0.462} = 0.680$$

From Table 17–4 phenotype A and phenotype O represent the genotypes $I^A I^A$, $I^A I^O$, and $I^O I^O$. These together should be $(p + r)^2$, where $p = f(I^A)$, $p^2 = f(I^A I^A)$, $2pr = f(I^A I^O)$, and $r^2 = f(I^O I^O)$:

$$(p + r)^2 = (199 + 231)/500 = 0.860$$
$$p + r = \sqrt{0.860}$$

and

$$p = \sqrt{0.860} - r = 0.927 - 0.680 = 0.247$$

The frequency of allele I^B, q, can be obtained by similar logic and by using blood types B and O, or simply by subtraction:

$$q = 1 - (p + r) = 1 - (0.927) = 0.073$$

Thus, the Hardy-Weinberg equilibrium can be extended to include multiple alleles and used to obtain estimates of the allelic frequencies in the ABO blood groups. With ABO, it is statistically feasible to do a chi-square test because there is 1 degree of freedom (number of phenotypes − number of alleles = 4 − 3 = 1). We are really only testing the AB category: None of the other categories will contribute to the chi-square (the observed and expected values of phenotypes A, B, and O will be equal).

		♂ A $f(A) = p^{♂}$	♂ a $f(a) = q^{♂}$
♀	A $f(A) = p^{♀}$	AA $f(AA) = p^{♂}p^{♀}$	Aa $f(Aa) = p^{♀}q^{♂}$
♀	a $f(a) = q^{♀}$	Aa $f(Aa) = p^{♂}q^{♀}$	aa $f(aa) = q^{♂}q^{♀}$

$$AA \quad + \quad 2Aa \quad + \quad aa$$

$$p^{♂}p^{♀} + (p^{♂}q^{♀} + p^{♀}q^{♂}) + q^{♂}q^{♀}$$

Figure 17 – 3. Random Mating When Allelic Frequencies Differ between the Sexes

Four other extensions of the Hardy-Weinberg equilibrium are worth considering: (1) an autosomal locus with different frequencies in the two sexes, (2) a sex-linked locus, (3) multiple loci, and (4) polyploids other than diploids.

Autosomal Locus with Different Frequencies in the Two Sexes

Consider the A locus with the A and a alleles with the frequencies $p^{♂}$ and $q^{♂}$ in males and $p^{♀}$ and $q^{♀}$ in females. Hardy-Weinberg equilibrium occurs, but it requires two generations. In the first generation after random mating, the allelic frequency in each sex becomes the average of the frequency in the previous population:

$$p = \tfrac{1}{2}(p^{♂} + p^{♀})$$

In the second generation Hardy-Weinberg equilibrium is achieved. Figure 17–3 shows that, with allelic frequencies different in the two sexes, the genotypes after one generation of random mating will be

$$f(AA) = p^{♂}p^{♀}$$
$$f(Aa) = p^{♂}q^{♀} + p^{♀}q^{♂}$$
$$f(aa) = q^{♂}q^{♀}$$

The frequency of the A allele in both sexes will be

$$p = p^{♂}p^{♀} + \tfrac{1}{2}(p^{♂}q^{♀} + p^{♀}q^{♂})$$

Substituting $(1 - p)$ for all q's gives

$$p = p^\delta p^\female + \tfrac{1}{2}p^\delta(1 - p^\female) + \tfrac{1}{2}p^\female(1 - p^\delta)$$

which simplifies to

$$p = \tfrac{1}{2}(p^\delta + p^\female)$$

Hence, in the first generation of random mating, allelic frequencies p and q are averaged between the sexes. The population is not, however, in Hardy-Weinberg equilibrium. Although the allelic frequencies are now identical in the two sexes, the genotypes are not distributed binomially. For example, the frequency of AA homozygotes in each sex is $p^\delta p^\female$, which is not equal to p^2, given $p^\delta \neq p^\female$. However, with the allelic frequencies now identical in the two sexes, a second generation of random mating will establish Hardy-Weinberg equilibrium.

Sex-Linked Locus

If a trait is sex linked and if the two sexes have different frequencies of the two alleles, then the approach to equilibrium is gradual and oscillatory: After one generation of random mating, the allelic frequencies in each sex are not the average frequencies of the two sexes. Rather, since males receive their X chromosomes from their mothers, they lag one generation behind females. Since females get their X chromosomes from both their parents, they average the allelic frequencies each generation. Thus, the females bring allelic distribution halfway to equilibrium each generation while the males lag a generation. Algebra may be used here for clarification. Let t = the current generation and $t - 1$ = the previous generation. Then,

$$p_t^\delta = p_{t-1}^\female \tag{17.1}$$

$$p_t^\female = \tfrac{1}{2}p_{t-1}^\delta + \tfrac{1}{2}p_{t-1}^\female = \tfrac{1}{2}(p_{t-1}^\delta + p_{t-1}^\female) \tag{17.2}$$

The average allelic frequency, $\overline{p}_t$, will be

$$\overline{p}_t = \tfrac{1}{3}p_t^\delta + \tfrac{2}{3}p_t^\female \tag{17.3}$$

(Remember that females have two X chromosomes while males have only one.) Substitute equations 17.1 and 17.2 into equation 17.3:

$$\overline{p}_t = \tfrac{1}{3}p_{t-1}^\female + \tfrac{2}{3}[\tfrac{1}{2}(p_{t-1}^\delta + p_{t-1}^\female)] \tag{17.4}$$

$$= \tfrac{1}{3}p_{t-1}^\delta + \tfrac{2}{3}p_{t-1}^\female$$

Thus,

$$\overline{p}_t = \overline{p}_{t-1}$$

and, by extension,

$$\overline{p}_t = \overline{p}_{t-1} = \overline{p}_{t-2} = \ldots = \overline{p}$$

where $\overline{p}$ is thus a constant mean allelic frequency that is unchanged each generation.

Since allelic frequency in males simply follows allelic frequency in females by a generation, only the change in female allelic frequency will be examined here, noting in particular how far female allelic frequency, $p_t^{\female}$, is from equilibrium, $\overline{p}$, each generation. The procedure is as follows: Subtracting $\overline{p}$ from both sides of equation 17.2 gives

$$p_t^{\female} - \overline{p} = \tfrac{1}{2}(p_{t-1}^{\male} + p_{t-1}^{\female}) - \overline{p} \qquad \textbf{(17.5)}$$

For $p_{t-1}^{\male}$ substitute as follows from equation 17.4:

$$p_{t-1}^{\male} = 3\overline{p} - 2p_{t-1}^{\female}$$

Equation 17.5 then becomes

$$p_t^{\female} - \overline{p} = \tfrac{1}{2}[(3\overline{p} - 2p_{t-1}^{\female}) + p_{t-1}^{\female}] - \overline{p}$$

$$= \tfrac{1}{2}\overline{p} - \tfrac{1}{2}p_{t-1}^{\female}$$

$$= (-\tfrac{1}{2})(p_{t-1}^{\female} - \overline{p})$$

Since this relationship is a general one between any two generations, the equation can be extended to two or more generations as follows:

$$p_t^{\female} - \overline{p} = (-\tfrac{1}{2})^2(p_{t-2}^{\female} - \overline{p})$$

or,

$$p_t^{\female} - \overline{p} = (-\tfrac{1}{2})^n(p_0^{\female} - \overline{p}) \qquad \textbf{(17.6)}$$

where $p_0^{\female}$ is the initial allelic frequency in females and n is the number of generations from time 0 to time t.

In words, equation 17.6 says that the allelic frequency in females gets halfway to equilibrium each generation, but in each generation it approaches equilibrium from a different direction (first from above, then from below). If the male allelic frequency, which lags a generation, is also calculated, Figure 17–4 results.

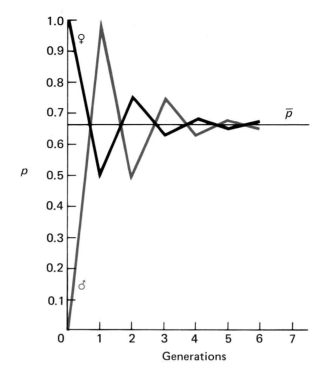

Figure 17 – 4.
Approach to
Equilibrium of
Alleles at a Sex-
Linked Locus
We start with $p_0^\delta =$
0.0 and $p_0^\varphi = 1.0$.
Then $\bar{p}$ will be 2/3.

It is worth emphasizing here that, at equilibrium, the two phenotypes resulting from a sex-linked locus occur in males as p and q (hemizygosity) while the females have the normal array of genotypes: p^2 of *AA*, $2pq$ of *Aa*, and q^2 of *aa*. If q is the frequency of a deleterious recessive, q of the males will be affected but only q^2 of the females will be. For example, if q is 0.1, 10% of males but only 1% of females will have the trait, which helps to explain why so many more males than females are color-blind or have hemophilia—both traits are sex-linked recessive traits.

Multiple Loci

The Hardy-Weinberg equilibrium can also be extended for consideration of several loci at the same time in the same population, a situation deserving mention because the whole genome is likely involved in evolutionary processes and we must, eventually, consider simultaneous allelic changes in all loci in an organism. (Even with the high-speed computer, simultaneous consideration of all loci is a bit far off in the future.) When two loci, *A* and *B*, on the same chromosome are in equilibrium with each other, the combinations of alleles on a chromosome (or in a gamete) follow the product rule of probability. For example, the chromosome with the A_i and B_j alleles will occur in the frequency $p_i q_j$. The condition is referred to as **linkage equilib-**

rium. When alleles of different loci are out of equilibrium (that is, not randomly distributed in gametes), the condition is referred to as **linkage disequilibrium.** The approach to linkage equilibrium is gradual and is a function of the recombination distance between the two loci. Unlinked genes appearing 50 map (recombination) units apart also approach linkage equilibrium gradually.

Polyploids Other Than Diploids

Extension of the Hardy-Weinberg equilibrium to predict allelic frequencies in polyploids is considerably more complex than the procedures just discussed. One of the problems with polyploids is the phenomenon known as **double reduction,** where a crossover between a gene and its centromere can give rise to a homozygous gamete in a completely heterozygous individual (an $a_i a_j a_k a_l$ tetraploid can result in an $a_i a_i$ gamete). In polyploids the approach to equilibrium is gradual and dependent on the rate of double reduction. Discussion of allelic frequencies in polyploids will be left to more advanced texts in population genetics.

MODELS FOR POPULATION GENETICS

The balance of this chapter will be devoted to discussing some of the effects of violating, or relaxing, the assumptions of the Hardy-Weinberg equilibrium. Although the results of relaxing each of the assumptions will be discussed, the effect on selection is probably the most important to evolution because selection gives rise to organisms that are better adapted to their environments. This chapter will consider the effects of mutation, migration, and small population size on the Hardy-Weinberg equilibrium; the effects of nonrandom mating and selection will be discussed in Chapter 18. In almost all cases the algebra used will follow the same general pattern: Derive an expression of allelic frequency next generation by knowing allelic frequency of the current generation and by knowing how the particular phenomenon (for example, mutation, selection) affects allelic frequency; then, derive an expression for change in allelic frequency between the two generations; finally, express the equilibrium allelic frequency—that is, the allelic frequency when the change in allelic frequency between generations is zero.

The steps to be taken in solving for equilibrium in models for population genetics can be outlined as follows:

1. Set up an algebraic model.
2. Calculate next generation's allelic frequency, q_{n+1}.
3. Calculate change in allelic frequencies between generations, Δq.
4. Calculate the equilibrium condition, $\hat{q}$(q-hat), where $\Delta q = 0$.

A COMPUTER PROGRAM TO SIMULATE THE APPROACH TO EQUILIBRIUM OF ALLELES AT A SEX-LINKED LOCUS

It is surprising how much insight into the processes of population genetics can be gained by modeling them on the computer. The simple computer program and the programmable calculator program presented here calculate changing allelic frequencies due to random mating when alleles at a sex-linked locus are in different frequencies in males and females. The computer program is written in FORTRAN; similar logic is used in the programmable calculator program. The interested student can simulate any of the processes described in these chapters by using either of these programs as a model. Output should be graphed. The program should be rerun several times with various sets of values for the variables (p^{δ}, p°, number of generations).

COMPUTER

In the computer program p^{δ} is 0.0, p° is 1.0, and the number of generations is 25. The program calculates $\overline{p}$ and prints it along with the initial p^{δ} and p° and then calculates the new allelic frequencies for females and males after one generation and prints them along with the generation number.

Computer Program

```
/LOAD FORTG
      PM = 0.
      PF = 1.
      PBAR = PM/3 + 2*PF/3
      I = 0
      PRINT 10,PM,PF,PBAR
 10 FORMAT(///'P-MALES = 'F7.5/'P-FEMALES = 'F7.5/'P-BAR = 'F7.5///)
      PRINT 20
 20 FORMAT('GENERATION P-FEMALES P-MALES'/30(1H-))
      PRINT 30,I,PF,PM
 30 FORMAT (I6,7X,F7.5,2X,F7.5)
      DO 25 I = 1,25
      TEMP = PM
```

```
      PM = PF
      PF = .5*PF + .5*TEMP
      PRINT 30,I,PF,PM
 25 CONTINUE
      END
```

After the command to execute, the following output results.

Computer Output

```
P-MALES = 0.0
P-FEMALES = 1.00000
P-BAR = 0.66667
```

GENERATION	P-FEMALES	P-MALES
0	1.00000	0.0
1	0.50000	1.00000
2	0.75000	0.50000
3	0.62500	0.75000
4	0.68750	0.62500
5	0.65625	0.68750
6	0.67188	0.65625
7	0.66406	0.67188
8	0.66797	0.66406
9	0.66602	0.66797
10	0.66699	0.66602
11	0.66650	0.66699
12	0.66675	0.66650
13	0.66663	0.66675
14	0.66669	0.66663
15	0.66666	0.66669
16	0.66667	0.66666
17	0.66666	0.66667
18	0.66667	0.66666
19	0.66667	0.66667
20	0.66667	0.66667
21	0.66667	0.66667
22	0.66667	0.66667
23	0.66667	0.66667
24	0.66667	0.66667
25	0.66667	0.66667

It thus takes 19 generations to reach equilibrium to a 5-place accuracy.

The programmable calculator program uses similar logic. (In this case, the calculator is the Texas Instruments' TI59 with the PC100A printer.) The program is entered, the reset button is pressed (RST), and the data are entered. First, p^δ is entered and R/S is pressed. Then, p° is entered and R/S is pressed again. The program then prints $\overline{p}$, followed by the indefinite sequence of: generation number, p°, p^δ.

Program

000	42	STO		040	04	04
001	04	04		041	65	×
002	91	R/S		042	93	.
003	42	STO		043	05	5
004	05	05		044	95	=
005	65	×		045	85	+
006	02	2		046	43	RCL
007	95	=		047	05	05
008	55	÷		048	65	×
009	03	3		049	93	.
010	95	=		050	05	5
011	85	+		051	95	=
012	43	RCL		052	99	PRT
013	04	04		053	42	STO
014	55	÷		054	07	07
015	03	3		055	43	RCL
016	95	=		056	05	05
017	99	PRT		057	99	PRT
018	00	0		058	42	STO
019	99	PRT		059	04	04
020	85	+		060	43	RCL
021	01	1		061	07	07
022	95	=		062	42	STO
023	42	STO		063	05	05
024	06	06		064	61	GTO
025	43	RCL		065	00	00
026	05	05		066	31	31
027	99	PRT		067	91	R/S
028	43	RCL				
029	04	04				
030	99	PRT				
031	43	RCL				
032	06	06				
033	99	PRT				
034	85	+				
035	01	1				
036	95	=				
037	42	STO				
038	06	06				
039	43	RCL				

Output

```
.6666666667                    13.
0.                             .6666259766
1.                             .6667480469
0.                             14.
1.                             .6666870117
0.5                            .6666259766
1.                             15.
2.                             .6666564941
0.75                           .6666870117
0.5                            16.
3.                             .6666717529
0.625                          .6666564941
0.75                           17.
4.                             .6666641235
0.6875                         .6666717529
0.625                          18.
5.                             .6666679382
0.65625                        .6666641235
0.6875                         19.
6.                             .6666660309
0.671875                       .6666679382
0.65625                        20.
7.                             .6666669846
0.6640625                      .6666660309
0.671875                       21.
8.                             .6666665077
0.66796875                     .6666669846
0.6640625                      22.
9.                             .6666667461
0.666015625                    .6666665077
0.66796875                     23.
10.                            .6666666269
.6669921875                    .6666667461
0.666015625                    24.
11.                            .6666666865
.6665039063                    .6666666269
.6669921875                    25.
12.                            .6666666567
.6667480469                    .6666666865
.6665039063
```

Mutation

Mutational Equilibrium. Mutation affects the Hardy-Weinberg equilibrium by changing one allele to another and thus changing allelic and genotypic frequencies. Consider a simple, realistic model where two alleles A and a exist. A mutates to a at a rate u, and a back mutates to A at a rate v:

$$A \underset{v}{\overset{u}{\rightleftharpoons}} a$$

If p_n is the frequency of A in generation n and q_n is the frequency of a in generation n, then the new frequency of a, q_{n+1}, is the old frequency of a plus the addition of a alleles from forward mutation and the loss of a alleles by back mutation—that is,

$$q_{n+1} = q_n + up_n - vq_n \tag{17.7}$$

Thus, up_n is the increment of a alleles added by forward mutation. Equation 17.7 takes into account not only the rate of forward mutation, u, but also p_n, the frequency of A alleles available to mutate. Similarly, the loss of a alleles to A alleles is the product of both the rate of back mutation, v, and the frequency of the a allele, q_n. Equation 17.7 completes the second modeling step, derivation of an expression for q_{n+1}, allelic frequency after one generation of mutation pressure. Step 3 is to derive an expression for change in allelic frequency between two generations. This change (Δ) is simply the difference between the allelic frequency at generation ($n + 1$) and the allelic frequency at generation n: Thus, for a,

$$\Delta q = q_{n+1} - q_n = (q_n + up_n - vq_n) - q_n \tag{17.8}$$

Equation 17.8 simplifies to

$$\Delta q = up_n - vq_n \tag{17.9}$$

The final step in the mutation model is to calculate the equilibrium condition $\hat{q}$, which is the allelic frequency that occurs when there is no change in allelic frequency from one generation to the next—that is, when Δq (equation 17.9) is equal to zero:

$$\Delta q = up_n - vq_n = 0 \tag{17.10}$$

Thus,

$$up_n = vq_n \tag{17.11}$$

Then, substituting $(1 - q_n)$ for p_n, (since $p = 1 - q$) gives

$$u(1 - q_n) = vq_n$$

or, by rearranging,

$$\hat{q} = \frac{u}{u + v} \qquad \textbf{(17.12a)}$$

$$\hat{p} = \frac{v}{u + v} \qquad \textbf{(17.12b)}$$

Equations 17.12 say (1) that an equilibrium of allelic frequencies will indeed be reached and that, as long as u and v are both greater than zero, the equilibrium will maintain both alleles in the population; and (2) that the equilibrium value $\hat{q}$ (of allele a) is directly proportional to the relative size of u, the rate of forward mutation toward a. Then, if $u = v$, the equilibrium frequency of a ($\hat{q}$) will be 0.5. As u gets larger, the equilibrium value shifts toward higher frequencies of the a allele.

A shortcoming of this model is that there is no obvious information revealing the time frame for reaching allelic equilibrium. It is beyond the scope of this book to derive equations to determine this parameter. In a large population it takes an extremely long time to effect any great change in allelic frequency by mutation pressure alone. Most mutation rates are on the order of 10^{-5}, and equation 17.9 shows that change will be very slow with values of this magnitude.

Stability of Mutational Equilibrium. Having demonstrated that allelic frequencies reach equilibrium following mutation, it is of interest to know whether or not the mutational equilibrium is stable. A stable equilibrium is one that when perturbed returns to the original equilibrium point. An unstable equilibrium is one that will not return when perturbed but, rather, will continue away from the equilibrium point. The Hardy-Weinberg equilibrium is a neutral equilibrium; it remains at the allelic frequency to which it is moved when perturbed.

Stable and unstable equilibrium points can be visualized as a marble either in the bottom of a bowl (stable) or on the top of an inverted bowl (unstable). Stable, unstable, and neutral equilibriums are illustrated in Figure 17–5. Although elegant mathematical ways exist for determining whether an equilibrium is stable, unstable, or neutral, mutational equilibrium can be investigated by moving the allelic frequency slightly away from equilibrium and then determining the direction of the change in allelic frequency by using equation 17.9. A graphical analysis of mutational equilibrium is illustrated in Figure 17–6.

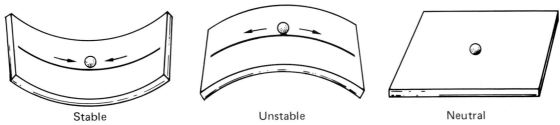

Stable   Unstable   Neutral

Figure 17 – 5.   Types of Equilibriums

Figure 17–6 introduces the process of graphical analysis, whereby an understanding of the dynamics of an event can be obtained by the proper representation of the event in graphical form. In Figure 17–6 the ordinate, or y axis is Δq. Above Δq of zero, q increases, and below, q decreases. The abscissa, or x axis is q, or allelic frequency. The solid diagonal line graphs the relationship of Δq and q—it crosses the $\Delta q = 0$ point at the equilibrium

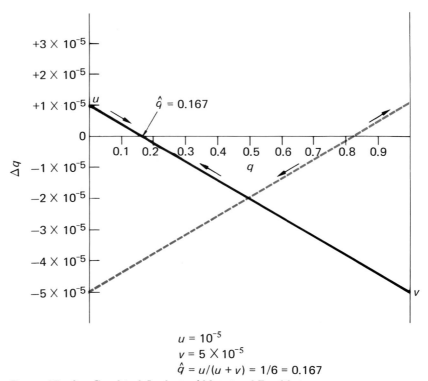

$$u = 10^{-5}$$
$$v = 5 \times 10^{-5}$$
$$\hat{q} = u/(u + v) = 1/6 = 0.167$$

Figure 17 – 6.   Graphical Analysis of Mutational Equilibrium

value ($\hat{q}$) of 0.167. If q is increased above $\hat{q}$ without changing u or v, we move to the right along the diagonal line that enters the negative Δq area. Thus, q will decrease, moving the allelic frequency back toward the equilibrium point. If decreased, q is below the equilibrium, Δq is positive and q will increase toward the equilibrium point. Thus the mutation equilibrium is a stable one. We should be able to determine that the dotted line in Figure 17–6 represents an unstable equilibrium condition. When a population in this condition is moved away from equilibrium, it continues away in the same direction. The solid line, the mutational dynamics of this model, was obtained simply by graphing the results derived by substituting various values for q in equation 17.9. Note that the maximal decrease in q occurs when q is 1.0 and is simply the back mutation rate (v). Similarly, the maximal increase of q occurs at $q = 0.0$ and is the forward mutation rate (u).

General Conclusions. In general, mutation rate by itself has an insignificant effect on the Hardy-Weinberg equilibrium in any one generation because of the very low values that mutation rates have. Mutation rate can, however, determine the eventual allelic frequencies at equilibrium if no other factors act to perturb the gradual changes that mutation rates cause. In addition, mutation provides the alternate alleles upon which natural selection acts.

Migration

Migration is similar to mutation in the sense of changing allelic frequency by adding or subtracting alleles. Many human populations are the product of migrant conglomerates. Let us look at the algebra of migration.

Assume two populations, both containing alleles A and a, but at different frequencies (p_1 and q_1 versus p_2 and q_2), as shown in Figure 17–7. Assume a group of individuals moves from population 2 and joins population 1 and further that this group of migrants makes up a fraction m of the new conglomerate population. Thus, the old residents will make up a fraction ($1 - m$) of the combined population. The conglomerate a-allele frequency, q_c, will be the weighted average:

$$q_c = mq_2 + (1 - m)q_1 \tag{17.13}$$

$$q_c = q_1 + m(q_2 - q_1) \tag{17.14}$$

The change in allelic frequency from before to after the migration event is (for a),

$$\Delta q = q_c - q_1 = [q_1 + m(q_2 - q_1)] - q_1 \tag{17.15}$$

$$\Delta q = m(q_2 - q_1) \tag{17.16}$$

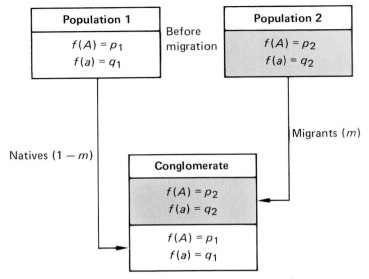

Figure 17 – 7.   Diagrammatic View of Migration

Source: Data from Wallace (1968). Reprinted courtesy of Dr. Bruce Wallace and BSCS from *Population Genetics,* BSCS pamphlet 12.

The migration model is completed by finding the equilibrium value $\hat{q}(\Delta q = 0)$. Δq will be zero when either

$$m = 0 \quad \text{or} \quad q_2 - q_1 = 0; \; q_2 = q_1$$

The conclusions to be drawn from this model are intuitive. Migration can upset the Hardy-Weinberg equilibrium. Allelic frequency in a population under the influence of migration will not change if either the size of the migrant group drops to zero or the allelic frequencies in the migrant and resident groups are identical.

This migration model can be used to determine many changes, such as the amount of admixture of alleles from white populations that has occurred in American blacks. For example, blood-type and other loci can be used to determine allelic frequencies in African black, American black, and American white populations. These frequencies can be used in equation 17.14 to solve for m, the migration proportion. Table 17–5 shows that the migration rates are from 25 to 32%. That is, from 25 to 32% of the alleles in the American black population have been brought in by genetic mixture with American whites. (The same type of analysis could be performed to determine the degree to which various gene pools have contributed to the American white population.)

TABLE 17 – 5. Migration of White Alleles into the American Black Population

Locus	African Black	American White	American Black	m[a]
Rh (R' allele)	0.06	0.42	0.15	0.25
PTC (t allele)	0.18	0.55	0.30	0.32

Source: Data from Glass and Li (1953) and Wallace (1968).

[a] Weyl (1970) has suggested that m is overestimated because it is based on the urban northern black population where gene flow was high.

Small Population Size

Besides migration, another variable that can upset the Hardy-Weinberg equilibrium is small population size. The influence of small population size on the equilibrium of allelic frequencies is important for the simple reason that every population on earth is finite. The Hardy-Weinberg equilibrium assumes an infinite population because, as we have defined it, the Hardy-Weinberg equilibrium is deterministic, not stochastic. That is, it predicts exactly what the allelic and genotypic frequencies should be after one generation; it ignores variation due to sampling error. To some extent every population of organisms on earth violates the Hardy-Weinberg assumption of infinite population size.

Sampling Error.   The term *sampling error* is used frequently in genetics, as elsewhere. The zygotes of every generation are a sample of gametes from the parent generation; the changes in allelic frequency from one generation to the next that are due to inexact sampling of the parent generation's alleles are sampling errors. Toss a coin 100 times and chances are it will not land heads exactly 50 times. However, as the number of coin tosses increases, the percentage of heads will approach 50%, a percentage that is reached with certainty only after an infinite number of tosses. The same applies to any sampling problem, from drawing cards from a deck to drawing gametes from a gene pool.

  If small population size is the only factor causing deviation from Hardy-Weinberg equilibrium, errors in accurately sampling the small gene pool cause the allelic frequencies of a given population to fluctuate from generation to generation in a process known as random genetic drift. The end result will be either fixation or loss of any given allele ($q = 1$ or $q = 0$). The amount of time it will take to reach this fixation–loss endpoint depends on the size of the population. The following discussion briefly investigates the relationship of population size to the speed of random genetic drift.

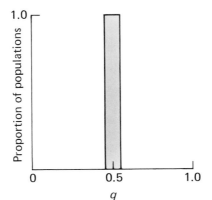

Figure 17 – 8.
Initial Conditions of
Random Drift Model
1000 populations,
each of size 100 and
each with an allelic
frequency (q) of 0.5.

Fokker-Planck Equation. The chance fluctuation (drift) of allelic frequency in a finite population is represented by an equation used by physicists to describe diffusion processes such as Brownian motion and is known as the **Fokker-Planck equation.**

$$\frac{\partial \varphi(q,t)}{\partial t} = \left(\frac{1}{2}\right) \frac{\partial^2}{\partial q^2} [\sigma^2_{\Delta q} \varphi(q,t)] - \frac{\partial}{\partial q} [\Delta q \varphi(q, t)]$$

where $\varphi(q, t)$ is the probability the allelic frequency will be q in generation t

$\sigma^2_{\Delta q}\varphi(q, t)$ is the variance in the change in allelic frequency by chance forces per generation

$\Delta q\varphi(q, t)$ is the mean change in allelic frequency per generation from deterministic forces such as selection, migration, or mutation

While the details of this equation will not be discussed here because of its complexity, some of the conclusions drawn from it are worth noting.

The process of genetic drift can be investigated by starting with a large number of populations of the same finite size and observing how the distribution of allelic frequencies among the populations changes in time due to random genetic drift. For example, start with 1000 hypothetical populations with each containing 100 individuals and in each population have the frequency of allele a be $q = 0.5$ (see Figure 17–8). Measure time in generations, t, as a function of the population size, N ($= 100$ in this example). For instance, $t = N$ is generation 100, $t = N/5$ is generation 20, and $t = 3N$ is generation 300. Then, by using the Fokker-Planck equation or a computer simulation (the results are the same) generate a series of graphs as shown in Figure 17–9. These graphs show that as the number of generations increases, the populations begin to diverge from $q = 0.5$. Some populations go to q values above 0.5 and some go to q values below. Approximately a one to one relation exists: The distribution spreads symmetrically. When a distribution hits the sides of the graph, some populations become fixed for the a allele

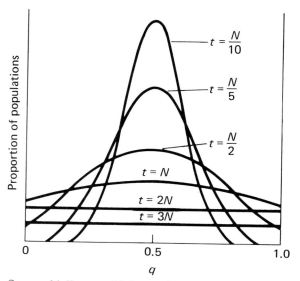

Figure 17 – 9.
Genetic Drift in
Small Populations
$N = 100$; at $t = 1$,
q = 0.5.

Source: M. Kumura. "Solution of a Process of Random Genetic Drift with a Continuous Model," *Proceeding The National Academy of Sciences USA* 41 (1955):144–150. Reproduced by permission of the author.

and some lose the a allele. In a sense, the sides act as sinks: Any population that has lost or "fixed" the a allele will be permanently removed from the process of random genetic drift. Without mutation to bring one or the other allele back into the gene pool, these populations will maintain a constant allelic frequency of 0 or 1.0.

After a point between N (100) and $2N$ (200) generations, the distribution of allelic frequency flattens out and begins to lose populations to the edges (of the graph) at a constant rate as shown in Figure 17–10. The rate

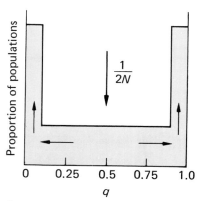

Figure 17 – 10.
Continued Genetic
Drift in 1000
Populations, Each of
Size 100

Source: S. Wright, "Evolution in Mendelian Population," *Genetics* 97 (1931):114. Reprinted by permission.

of loss of populations to the edges is about $1/2N$ $(1/200)$, or 0.5% of the populations per generation. If the initial $(t = 1)$ frequency is not 0.5, everything is shifted in the distribution (Figure 17–11), but the basic process is the same—in all populations sampling error will cause allelic frequencies to drift toward fixation or elimination. If no other factor counteracts this drift, every population is destined eventually to be either fixed for or deficient in any given allele. The amount of time the process takes is dependent on the size of the population. The example used here was based on small populations of 100. If 1,000,000 is substituted in Figure 17–9 for the population

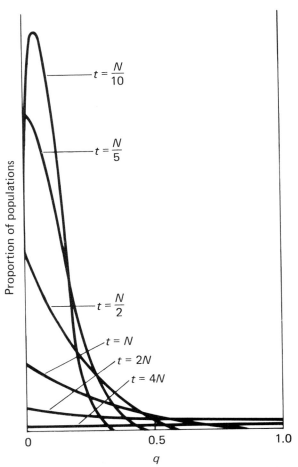

Figure 17 – 11.
Random Genetic
Drift in 1000
Populations
$N = 100$;
at $t = 1$, $q = 0.1$.

Source: M. Kimura, "Solution of a Process of Random Genetic Drift with a Continuous Model," *Proceedings. The National Academy of Sciences USA* 41 (1955):144–150. Reproduced by permission of the author.

sizes of 100, a flat distribution of populations would not be reached until 2,000,000 generations, rather than 200 generations, and so on. Thus, a population experiences the effect of random genetic drift in inverse proportion to its size—small populations rapidly reach fixation or loss of a given allele, whereas large populations take longer to show the same effects. Genetic drift also shows itself several other ways.

Founder Effects and Bottlenecks. Several well-known genetic phenomena are caused by populations starting at or proceeding through small numbers. When a subpopulation is initiated by a small and, therefore, genetically unrepresentative sample of the main population, the genetic drift observed in the subpopulation is referred to as a **founder effect.** A classic human example is the population founded on Pitcairn Island by several of the *Bounty* mutineers and some Polynesians. The particular blend of Caucasian and Polynesian traits that characterizes today's Pitcairn Island population results from there being only a small number of founders of the population.

Sometimes populations go through **bottlenecks** of small population size, with predictable genetic results. After the bottleneck the parents of the next generation have been reduced to a small number and may not be genetically representative of the original population. The field mice on Muskeget Island, Massachusetts have a white forehead blaze of hair not commonly found in nearby mainland populations. Presumably the island population went through a bottleneck at the turn of the century when cats existed on the island and reduced the number of mice to near zero. The population was reestablished by a small group of mice that happened by chance to contain several animals with this forehead blaze.

CHAPTER SUMMARY

In a large randomly mating population of diploids, without the influence of mutation, migration, or selection, an equilibrium will be achieved for an autosomal locus with two alleles. This equilibrium, the Hardy-Weinberg equilibrium, predicts that (1) allelic frequency will not change from generation to generation; (2) genotypes will occur according to the binomial distribution $p^2 = f(AA)$, $2pq = f(Aa)$, and $q^2 = f(aa)$; and (3) if perturbed, equilibrium will be reestablished in just one generation of random mating.

To determine whether a population is at equilibrium, the observed and expected distribution of genotypes can be compared by the chi-square test. In some circumstances, where it is reasonable to assume equilibrium, allelic and genotypic frequencies can be estimated even when dominance occurs.

The Hardy-Weinberg equilibrium is easily extended to a prediction of the frequencies of multiple alleles. It also predicts that two generations are required for equilibrium to be reached when the allelic frequency is different between the sexes. In the case of an X-linked trait with different allelic frequencies in the sexes, the approach to equilibrium is gradual. Females get halfway to equilibrium each generation while males reflect the female allelic frequency one generation back. The Hardy-Weinberg equilibrium also makes predictions where multiple loci and polyploids are concerned.

The effects of relaxing some of the assumptions of the equilibrium are studied. Both mutation and migration transport alleles in and out of a population. Mutation provides the variability on which natural selection acts, but it usually does not directly affect the equilibrium because of the miniscule effect of mutation per generation.

Small population size is a source of sampling error, which results in a changing allelic frequency known as random genetic drift. The smaller the population, the more rapidly it approaches allelic fixation. The dynamics of genetic drift are studied graphically. The next chapter considers the effects of nonrandom mating, as well as selection, on the Hardy-Weinberg equilibrium.

EXERCISES AND PROBLEMS

1. In the following two sets of data, calculate gene and genotypic frequencies and determine whether the populations are in Hardy-Weinberg equilibrium. Do a statistical test if one is required.
   a. Allele A is dominant to a: A- 91
       aa 9
   b. Alleles F and S are codominant in the alcohol dehydrogenase system in *Drosophila*: FF 137
       FS 196
       SS 87

2. The ability to taste PTC is due to the dominant allele T. Among a group of 215 students, 150 could detect the taste of PTC and 65 could not. Calculate the allele frequencies of T and t. Is the population in Hardy-Weinberg equilibrium?

3. The frequency of children homozygous for the recessive allele for the lethal Tay-Sachs disease is about 1 in 215,000. What is the percentage of heterozygotes in the population?

4. The following data are of ABO phenotypes from a population sample of 100 persons. Determine the frequency of the three alleles:

Type A = 7
Type B = 72
Type AB = 12
Type O = 9
What did you have to assume? Is the population in Hardy-Weinberg equilibrium?

5. How quickly and in what manner is the Hardy-Weinberg equilibrium achieved under the following initial conditions?
   a. 1 locus, 5 alleles
   b. 1 locus, 2 alleles, different initial frequencies in the two sexes
   c. 1 sex-linked locus, 2 alleles, different initial frequencies in the two sexes
   d. 2 loci, 2 alleles each, not linked

6. The following data of MN blood types were obtained in an Alaskan town recently established by pipeline crews.

	Men	Women
MM	35	17
MN	50	46
NN	15	37

   a. Is this population in Hardy-Weinberg equilibrium?
   b. What will the frequencies of the three genotypes be after one generation of random mating? Is the population now in equilibrium?
   c. Answer part b of the problem for a second generation of random mating.

7. From the same group of people, the following data were obtained for the glucose-6-phosphate dehydrogenase locus (a and b alleles), which is X linked.

	Men		Women
a	80	aa	9
b	20	ab	42
		bb	49

   a. Is this population in Hardy-Weinberg equilibrium?
   b. Graph the dynamics of the change in allelic frequency in the two sexes.

8. Give a qualitative assessment of the linkage equilibrium between the two loci in problems 6 and 7.

9. Consider a locus with alleles A and a in a large, randomly mating population under the influence of mutation. If the mutation rate to a is 19 times the back-mutation rate, what is the equilibrium frequency of a?

10. The following data refer to the R^0 allele in the Rh blood system:

frequency in African blacks   = 0.62
frequency in American blacks = 0.45
frequency in American whites = 0.03

What is the total proportion of alleles that have entered the American black population? What data would you need to calculate the proportion of black alleles that have entered the American white population?

11. In a population of 500 individuals with a frequency of allele A of 0.7, what is the ultimate fate of the A allele? What is the probability that the population will eventually lose the A allele? How many are N/S generations? 4N generations?

12. PTC tasting is dominant in humans.
    a. Should most human populations be heading toward a $3:1$ ratio of tasters to nontasters? Explain.
    b. Confronted with a population sample of humans of unknown origin, would you expect more or less than half the sample to be tasters?

13. One way of measuring the frequency of allele A is by the expression:

$$p = f(A) = f(AA) + \tfrac{1}{2}f(Aa)$$

Given that

$$p = f(A) = \frac{A \text{ alleles}}{\text{total alleles}}$$

derive the preceding expression.

14. Derive an expression for mutation equilibrium when there is no back mutation.

15. a. Prove that the perpendicular to the base divides the base of the De Finetti diagram in the ratio of $p:q$.
    b. Prove that populations in Hardy-Weinberg equilibrium form a parabola on the De Finetti diagram.

SUGGESTIONS FOR FURTHER READING

Cavalli-Sforza, L. L., and W. F. Bodmer. 1971. *The Genetics of Human Populations*. San Francisco: Freeman.

Crow, J. F., and M. Kimura. 1970. *An Introduction to Population Genetics Theory*. New York: Harper & Row.

De Finetti, B. 1926. Considerazioni matematiche sul l'ereditarieta mendeliana. *Metron* 6:1–41.

Fisher, R. A. 1930. *The Genetical Theory of Natural Selection*. Oxford: Clarendon Press.

Glass, B., and C. C. Li. 1953. The dynamics of racial intermixture—An analysis based on the American Negro. *Amer. J. Hum. Genet.* 5:1–20.

Haldane, J. B. S. 1932. *The Causes of Evolution*. New York: Harper & Row.

Hardy, G. H. 1908. Mendelian proportions in a mixed population. *Science* 28:49–50.

Hartl, D. 1981. *A Primer of Population Genetics*. Sunderland, Mass.: Sinauer.

Jaffe, R. 1979. *A Clear Introduction to FORTRAN IV*. N. Scituate, Mass.: Duxbury Press.

Kimura, M. 1955. Solution of a process of random genetic drift with a continuous model. *Proc. Nat. Acad. Sci. USA* 41:144–150.

Kohn, P., and R. Tamarin. 1978. Selection at electrophoretic loci for reproductive parameters in island and mainland voles. *Evolution* 32:15–28.

Li, C. C. 1955. *Population Genetics*. Chicago: University of Chicago Press.

Mettler, L. E., and T. G. Gregg. 1969. *Population Genetics and Evolution*. Englewood Cliffs, N.J.: Prentice-Hall.

Spiess, E. 1977. *Genes in Populations*. New York: Wiley.

Wallace, B. 1968. *Topics in Population Genetics*. New York: Norton.

Weinberg, W. 1908. Über den Nachweis der Vererbung beim Menschen. *Jahresh. Ver. Vater. Naturk. Wuerttemb.* 64:368–382.

Weyl, N. 1970. Some genetic aspects of plantation slavery. *Perspect. Biol. Med.* 13:618–625.

Wright, S. 1931. Evolution in Mendelian populations. *Genetics* 16:97–159.

18

Population Genetics: Nonrandom Mating and Natural Selection

In Chapter 17 the Hardy-Weinberg equilibrium and its underlying assumptions were introduced, and the effects of mutation, migration, and finite population size on this equilibrium were developed. This chapter will conclude our analysis of the Hardy-Weinberg equilibrium by looking at the ways it is affected by nonrandom mating and natural selection.

NONRANDOM MATING

Random mating is a requirement of the Hardy-Weinberg equilibrium. Deviations from random mating come about when phenotypic resemblance or relatedness influences mate choice. When phenotypic resemblance influences mate choice, either *assortative* or *disassortative mating* occurs, depending on whether individuals choose mates on the basis of similarity or dissimilarity, respectively. (For example, in humans there is assortative mating for height: Short men tend to marry short women and tall men tend to marry tall women.) When relatedness influences mate choice, either *inbreeding* or *outbreeding* occurs, depending on whether mates are more or less related than two randomly chosen individuals from a population. (An example of inbreeding in humans is the marriage of first cousins.) Both general types of nonrandom mating (assortative/disassortative mating and inbreeding/outbreeding) have the same qualitative effects on the Hardy-Weinberg equilibrium: Assortative mating and inbreeding increase homozygosity without changing allelic frequencies, while disassortative mating and outbreeding increase heterozygosity without changing allelic frequencies. However, two differences are apparent between the two types. First, assortative or disassortative mating will disturb the Hardy-Weinberg equilibrium only when the phenotype and genotype are closely related. That is, if assortative mating occurs for a nongenetic trait, then there will be no dis-

629

tortions of the Hardy-Weinberg equilibrium. Inbreeding and outbreeding affect the genome directly. A second difference between the two types of mating is that the effects of inbreeding or outbreeding are felt across the whole genome whereas the disturbances to the Hardy-Weinberg equilibrium by assortative and disassortative mating occur only for the particular trait being assorted.

Inbreeding

Here, only the deviation from random mating that results from inbreeding will be considered in detail. Inbreeding comes about in two ways: (1) the systematic choice of relatives as mates and (2) the subdivision of a population whereby small subunits present individuals with less of a mating choice and more of a likelihood that a mate will be related. Inbreeding will be discussed as if it occurred in the first way, the systematic choice of relatives as mates—the consequences of both ways are similar. (Note that inbreeding can be either the cause or the result of subdivision of a population.)

Common Ancestry. An inbred individual is one whose parents are related—there is **common ancestry** in the family tree. The extent of inbreeding is thus a function of how much common ancestry is shared by the parents of an inbred individual. When mates share parental genes, each parent may pass on copies of the same allele. An inbred individual who is homozygous carries identical copies of a single ancestral allele. Or, expressed in other words, an individual of *aa* genotype is homozygous and, if the *a* allele from each parent is a length of DNA originally copied from the same DNA of a common ancestor, the *aa* individual is inbred. The first observable effect of inbreeding is the expression of hidden recessives.

Each human carries, on the average, about 4 lethal-equivalent alleles, alleles that kill the homozygous individual. In most societies children are protected from these lethals by a cultural pattern of outbreeding (mating with nonrelatives). Rarely does an outbred zygote receive the same recessive lethal from each parent and dominance serves to protect children from deleterious recessive alleles. But, in the process of inbreeding, where the zygote may receive identical ancestral alleles from each parent, there is a substantial increase in the probability that a deleterious allele will become homozygous (Figure 18–1). Inbreeding often results in spontaneous abortions (miscarriages), fetal deaths, and congenital deformities.

Autozygosity. There are, therefore, two types of homozygosity—**allozygosity,** where two alleles are alike but unrelated and **autozygosity,** where two alleles have **identity by descent** (that is, are identical copies of the same ancestral allele). Thus, an **inbreeding coefficient, *F*,** can be defined as the probability of autozygosity—that is, the probability that the two alleles in an individual are identical by descent. This coefficient can range

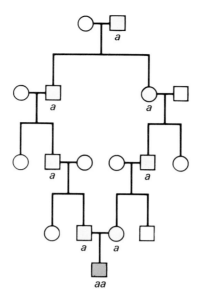

Figure 18 – 1.
Homozygosity by
Descent of Copies of
the Same Ancestral
Allele, *a*

from 0, where there is no inbreeding, to 1, where an individual is autozygous with certainty.

Increased Homozygosity from Inbreeding. What are the effects of inbreeding on the Hardy-Weinberg equilibrium? In a population with a particular locus having two alleles, genotypes can be produced in two ways. (1) Random mating yields the three genotypes (*AA, Aa, aa*) in the proportions of the binomial expansion. This production of genotypes occurs with the probability of $(1 - F)$. (2) Inbreeding produces homozygotes in proportion to their allelic frequencies with a probability of F. Table 18–1 summarizes genotypic proportions in a population with inbreeding.

Several points emerge from Table 18–1. First, when the inbreeding coefficient is zero (= complete random mating), the table reduces to the Hardy-Weinberg equilibrium. Second, there are additions to the population's homozygote categories by inbreeding, but not to the heterozygote category:

TABLE 18 – 1.   Genotypic Proportions in a Population with Inbreeding

Genotype	Random Mating Segment $(1 - F)$		Inbred Segment (F)
AA	$p^2(1 - F)$		pF
Aa	$2pq(1 - F)$		
aa	$q^2(1 - F)$		qF
Sum	$(p^2 + 2pq + q^2)(1 - F) =$ $(1)(1 - F)$	$+$	$(p + q)(F) =$ $(1)(F) = 1$

THE DETERMINATION OF LETHAL EQUIVALENTS

The average person carries about 4 lethal equivalents that are hidden as recessives. Four lethal equivalents means 4 alleles that are lethal when homozygous or 8 alleles conferring a 50% chance of mortality when homozygous or any similar combination of lethal and semilethal alleles. (The exact arrangement cannot be determined with current analytical methods.) The estimate of hidden defective and lethal alleles is arrived at by using inbreeding data.

Crow and Kimura, in 1970, analyzed data showing that in Swedish families where marriages occurred between first cousins, between 16% and 28% of the offspring had genetic diseases. For unrelated parents the comparable figure is between 4% and 6%. Thus, it is estimated that the offspring of first cousins have an added risk of 12% to 22% of having a genetic defect. The children of first cousins have an inbreeding coefficient of 1/16. Thus, for an individual who is completely autozygous, the risk of genetic defect is increased sixteenfold over an individual whose parents are unrelated. If 100% risk is considered one lethal equivalent, then an autozygous individual would carry 2 to 3.5 lethal equivalents (16 × 12% or 16 × 22%). However, a completely autozygous individual is, in essence, a doubled gamete. Since our interest is in the number of deleterious alleles carried by a normal heterozygote, it is necessary to further multiply the risk by a factor of 2 to determine the number of lethal-equivalent alleles carried by a normal heterozygous individual. The conclusion is that the average person carries the equivalent of 4 to 7 alleles that would, in the homozygous state, cause a genetic defect.

A similar calculation can be made using viability data to determine the occurrence of lethal equivalents rather than genetic defects. A study from rural France (see Crow and Kimura, 1970) showed that the mortality rate of offspring of first cousins was 25% whereas the analogous figure for the offspring of unrelated parents was about 12%, an increased risk of 13% for the offspring of cousins. Multiplying this risk figure of 0.13 by 32 (16 × 2) presents a figure of 4 lethal equivalents per average person in the population. Cavalli-Sforza and Bodmer, in 1971, by using a preponderance of data from Japanese populations, reported this number to be about 2 lethal equivalents per heterozygote. Despite some interpopulation differences in these estimates, they are about the same order of magnitude—2–7 lethal equivalents per heterozygote.

Motoo Kimura
(1924–   )
Courtesy of Dr.
Motoo Kimura

Inbreeding does not produce heterozygotes (identity by descent implies homozygosity). Thus, inbreeding increases the proportion of homozygotes in the population. With complete inbreeding ($F = 1$), there are only homozygotes in the population. This condition (lack of heterozygotes) follows from Table 18–1 and from the definition of the inbreeding coefficient, F, as the probability of autozygosity. When $F = 1$, all individuals are autozygous—that is, homozygous by descent.

How does inbreeding affect allelic frequency? Recall from the previous chapter that a new allelic frequency, p_{n+1}, is calculated as the frequency of homozygotes of this allele plus half the frequency of the heterozygotes. With inbreeding,

$$p_{n+1} = p^2(1 - F) + pF + pq(1 - F)$$
$$= p^2 + pq + F(p - p^2 - pq)$$
$$= p(p + q) + pF(1 - p - q)$$
$$= p(1) + pF(0)$$
$$= p$$

Thus, inbreeding does not change allelic frequency. That inbreeding affects zygotic combinations (genotypes) but not allelic frequencies is also apparent intuitively since, although inbreeding may determine the genotype of the offspring, it does not change the number of each allele that an individual transmits into the next generation.

In summary, inbreeding causes an increase in homozygosity, has no effect on allelic frequency, and affects all loci in a population equally. The results of inbreeding can be seen by the appearance of recessive traits that are often deleterious: Inbreeding increases the rate of fetal deaths and congenital malformations in humans. In agricultural crops and farm animals, decreases in size, fertility, vigor, and yield often result from inbreeding.

Assortative Mating

Assortative mating also causes an increase in homozygosity with no effect on allelic frequency. It influences only the genetic traits that are being considered by the mates: It is not effective across the whole genome. Assortative mating does not tend to expose hidden recessives because the mates are not relatives. Disassortative mating and outbreeding increase heterozygosity in a manner analogous to the ways that assortative mating and inbreeding increase homozygosity.

Pedigree Analysis

Path Diagram Construction.   The inbreeding coefficient of an individual can be determined by pedigree analysis. The F coefficient for an individual has the same definition as the one for the population: the probability of autozygosity. It is determined by converting a pedigree to a **path diagram** by eliminating all extraneous individuals—that is, individuals who cannot contribute to the inbreeding coefficient of the individual in question. A path diagram shows the direct line of descent from common ancestors. An example of the conversion of a pedigree to a path diagram is shown in Figure 18–2, where C and F are omitted from the path of descent because they are not related to anyone on the other side of the family tree and, therefore, do not contribute to the "common ancestry" of I. The pedigree in Figure 18–2 shows an offspring who is the daughter of first cousins. Since first cousins are the

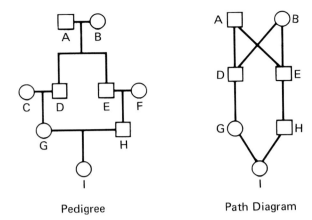

Figure 18 – 2.
Conversion of a
Pedigree, Which
Involves the Mating
of First Cousins, to a
Path Diagram

Pedigree   Path Diagram

offspring of siblings, they share a set of common grandparents. Thus, individual I can be autozygous for alleles from either A or B, her great-grandparents. The path diagram shows the only routes by which autozygosity can occur. Figure 18–3 is another example and shows the inbred offspring of the mating of a brother and sister.

For the mating of first cousins, the inbreeding coefficient of the offspring can be calculated as follows. The path diagram of Figure 18–2 is shown again in Figure 18–4, where the lowercase letters designate gametes. There are two paths of autozygosity in this diagram, one path for each grandparent as a common ancestor:

$$I\ G\ D\ \underset{\uparrow}{A}\ E\ H\ I \quad \text{and} \quad I\ G\ D\ \underset{\uparrow}{B}\ E\ H\ I$$

In the path with A as the common ancestor, A contributes a gamete to D and a gamete to E. The probability is ½ that D and E each carry the same allele. That is, there are four possible allelic combinations for the two gametes, a_1 and a_2: *AA, Aa, aA,* and *aa.* Of these combinations, the first and last (*AA*

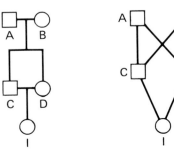

Figure 18 – 3.
Conversion of a Sib-
Mating Pedigree to a
Path Diagram

Pedigree   Path Diagram

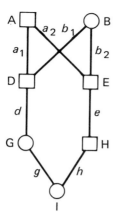

Figure 18 – 4.
Path Diagram of
Mating of First
Cousins

and aa) will give the same allele to the two offspring, D and E, and can thus contribute to autozygosity. The probability that gametes a_1 and d carry the same allele is ½ and the probability that d and g carry the same allele is also ½. Similarly, on the other side of the pedigree, the probability is ½ that a_2 and e carry the same allele and is ½ that e and h carry the same allele. Thus, the overall probability that the alleles carried by g and h are identical (autozygous) is $(½)^5$. In general, it would be $(½)^n$, where n is the number of ancestors in the path.

The reader may have spotted an additional factor here. There is a probability of ½ that the two gametes of the common ancestor A, a_1 and a_2, each carried a copy of one of A's two alleles such that both of A's alleles were represented once. Normally, no autozygosity could result from the situation. However, there is a probability that the two alleles carried by A are identical copies of an ancestral allele. This probability is the inbreeding coefficient of A, F_A. The actual probability of A's transmitting a copy of the same ancestral allele to D and to E is thus $½ + ½F_A$ or $½(1 + F_A)$. In other words, there is a ½ probability that the alleles transmitted from A to D and E are identical. This leaves ½ the time when these alleles can be identical only if A is inbred. The probability of identity of A's two alleles is F_A. The expression for F_I can now be changed from $(½)^n$ to

$$F_I = (½)^n(1 + F_A)$$

This equation accounts only for the inbreeding of I by the path involving the common inbred ancestor, A, and does not account for the symmetrical path with B as the common ancestor. To obtain the total probability of inbreeding, the values from each path must be added (rule #1, not rule #2, of Chapter 3). Thus, the complete formula is

$$F_I = \Sigma[(½)^n(1 + F_J)] \qquad (18.1)$$

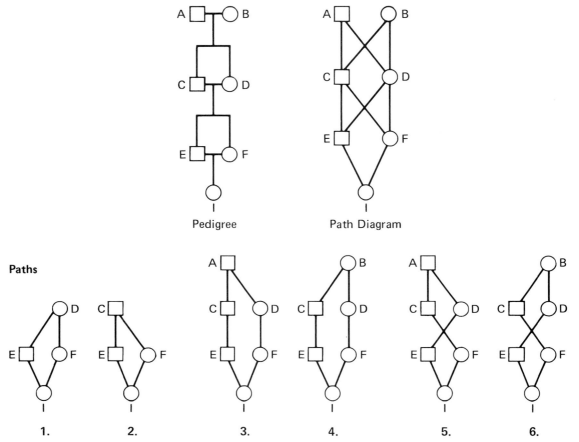

Figure 18 – 5.   Path Diagram of Two Generations of Sib Matings with the Associated Paths to Autozygosity
$F_A = 0.05$

where n is the number of ancestors in a given path, F_J is the inbreeding coefficient of the common ancestor of this path, all paths are summed, and F_I is the probability that the two alleles in I are identical by descent.

In the example of first cousins mating (Figure 18–4),

$$F_I = (\tfrac{1}{2})^5(1 + F_A) + (\tfrac{1}{2})^5(1 + F_B)$$

If we assume that F_A and F_B are zero (which we must assume from a lack of any alternative evidence), then

$$F_I = 2(\tfrac{1}{2})^5 = (\tfrac{1}{2})^4 = 0.0625$$

which can be interpreted as either that about 6.25% of individual I's loci are autozygous or that there is a 6.25% chance of autozygosity at any one of I's loci.

The inbreeding coefficient of the offspring of siblings, as diagrammed in Figure 18–3, can be calculated—again assuming that A and B are not themselves inbred, that is, that F_A and F_B are zero—as

$$F_I = 2(\tfrac{1}{2})^3 = 0.25$$

Thus, about 25% of the alleles in an offspring of a sibling mating are autozygous.

Path Diagram Rules. The following points should be kept in mind when an inbreeding coefficient is calculated:

1. All possible paths must be counted. A path is possible if gametes can actually pass this way. Paths that violate the rules of inheritance cannot be used. For example, in Figure 18–4 the following path is unacceptable: I G D A E B E H I.
2. In any path, an individual can be counted only once.
3. Every path must have one and only one common ancestor. The inbreeding coefficient of any other person in the path is immaterial.

Figure 18–5 presents a complex pedigree produced from repeated sib mating, a pattern found in livestock and laboratory animals. This pedigree has several interesting points. First, there are common ancestors in several different generations. Thus it is necessary to be sure to count *all* paths (paths 5 and 6 might not be immediately obvious). Second, although not shown in Figure 18–5, one of the common ancestors, A, is also inbred—a fact that must be taken into consideration in paths 3 and 5. Thus, F_I is as follows:

$$
\begin{array}{lll}
\text{From path 1:} & (\tfrac{1}{2})^3 & = 0.1250 \\
\qquad\qquad 2: & (\tfrac{1}{2})^3 & = 0.1250 \\
\qquad\qquad 3: & (\tfrac{1}{2})^5(1 + .05) & = 0.0328 \\
\qquad\qquad 4: & (\tfrac{1}{2})^5 & = 0.0313 \\
\qquad\qquad 5: & (\tfrac{1}{2})^5(1 + .05) & = 0.0328 \\
\qquad\qquad 6: & (\tfrac{1}{2})^5 & = 0.0313 \\
\hline
\qquad\quad F_I & & = 0.3781
\end{array}
$$

NATURAL SELECTION

While migration, mutation, and genetic drift all influence allelic frequencies, they do not of necessity produce populations that are better adapted to their environments. Natural selection, however, is a relentless process that weeds out the less suitable organisms in a given environment. The consequences of natural selection, Darwinian evolution, will be discussed in detail in the next chapter. The algebra of the process of natural selection will be discussed in this chapter.

Ways in Which Natural Selection Acts

Fitness.   **Selection,** or **natural selection,** is a process whereby one genotype leaves more offspring than another genotype. Selection is thus a matter of **reproductive success,** which is measured in terms of both fertility and survival. Genotypes that have greater reproductive success will produce more offspring. In the genetic models that follow, different reproductive-success values will be assigned to the different genotypes of a given locus. The genotype that leaves the most offspring is given the highest value for reproductive success. This value is called the **fitness** of the particular genotype. Synonymous terms are: **relative Darwinian fitness** or **adaptive value.** The letter W is routinely used for this value. The fitness varies from 0 to 1 and is always relative to a given population at a given time. For example, in a normal environment fruit flies with long wings may be more fit than fruit flies with short wings. But in a very windy environment a fruit fly with limited flying ability may do better than the long-winged genotype that will be blown around by the wind. Thus, fitness is relative to a given circumstance. In a given environment the genotype that leaves the most offspring has a fitness of 1 and a lethal genotype has a fitness of 0. Any other genotype has a fitness value in between. A number of factors can decrease this fitness value, W, below 1. The sum of forces acting to prevent reproductive success provides a **selection coefficient,** which is usually given the letter s or t and is defined by the equation

$$W = 1 - s \tag{18.2}$$

Components of Fitness.   Natural selection can act at any stage of the life cycle of an organism in four ways. (1) A genotype's reproductive success can be affected by its prenatal survival, juvenile survival, or adult survival—that is, by **zygotic selection.** (2) A heterozygote can have differential success of its gametes such that one allele fertilizes more often than the other. The phenomenon is termed **gametic selection.** A well-studied case is the t-allele system in house mice, where T/t heterozygote males have as many as 95% of their gametes containing the t allele. Selection can also take place in two areas of the reproductive segment of the organism's life cycle. (3) Some genotypes may mate more often than others **(sexual selection).** (4) Or, some genotypes may be more fertile than other genotypes **(fecundity selection).** The particular variable of the life cycle upon which selection acts is termed a **component of fitness.**

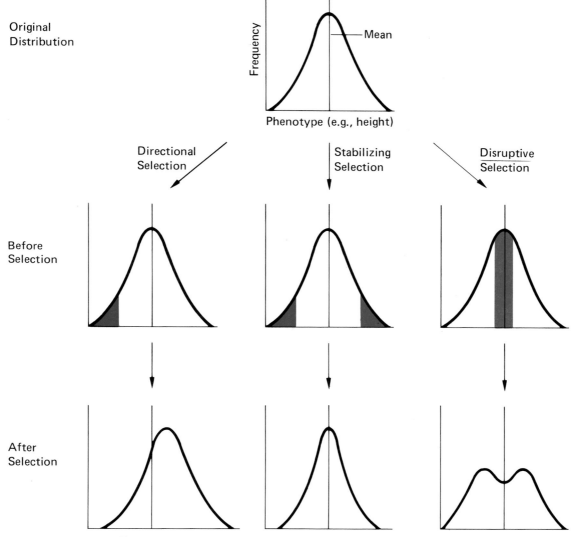

Figure 18 – 6. Directional, Stabilizing, and Disruptive Selection
Colored area shows the group being selected against.

Types of Selection. Figure 18–6 shows the three main ways that the sum total of selection can act. **Directional selection** works by continuously removing individuals from one end of the phenotype distribution (for example, short individuals are removed). Thus the mean is constantly shifted toward the other end of the phenotype distribution (for example, the mean shifts toward the tall end). An example from the geologic record would be the size of horses (Figure 17–1). About 50 million years ago *Hyracotherium,* the oldest known member of the horse family, Equidae, stood less than a foot high at the shoulder. The line of evolution has obviously been toward an increase in size to today's large horses and happened by a gradual favoring of the tall individuals. Tall horses, therefore, had a higher fitness than short ones.

Stabilizing selection works by constantly removing individuals from both ends of a phenotype distribution and thus maintaining the same mean over time. An example of stabilizing selection is height in humans, where the exceptionally tall and the exceptionally short individuals have more difficulty in finding mates than individuals of intermediate height. **Disruptive selection** works by favoring individuals at both ends of a phenotype distribution at the expense of individuals in the middle. It, like stabilizing selection, should maintain the same mean value of the phenotype distribution. Disruptive selection is seen in the occurrence of different discrete forms (polymorphism) within the same species. This selection occurs in mimetic butterflies where a species (for example, the African Swallowtail, *Papilio dardanus*) simultaneously mimics different distasteful species (for example, *Panaus chrysippus, Amauris niavius*). *Papilio dardanus* has discrete mimics of both of these distasteful species. Since intermediate forms would not mimic anything, they would be selected against. Mimicry will be discussed in the next chapter.

Selection against the Homozygous Recessive

Selection can be analyzed by using the same technique used in Chapter 17 to analyze mutation and migration: Define the initial conditions; allow selection to act; calculate the allelic frequency after selection (q_{n+1}); then calculate Δq (change in allelic frequency from one generation to the next) and the equilibrium condition, $\hat{q}$, when Δq becomes zero. The analysis that follows will consider a single locus with two alleles and assumes that selection acts directly on the genotypes in a simple fashion (occurs at a single stage in the life of the organism—for example, larval mortality in *Drosophila*). After selection the population mates at random to form a new generation in Hardy-Weinberg equilibrium.

Selection Model.   Table 18–2 outlines the model for the case of selection against the homozygous recessive genotype. The initial population is in Hardy-Weinberg equilibrium. Even with selection acting during the life cycle of the organism, the Hardy-Weinberg equilibrium will be reestablished anew after each round of random mating (although with a new allelic frequency). Fitnesses are assigned depending on the way in which natural selection is acting. (All selection models will start out the same way. They will diverge at the point of assigning fitnesses.) In the model of Table 18–2, both the dominant homozygote and the heterozygote have the same fitness (W

TABLE 18 – 2.   Selection against the Recessive Homozygote: One Locus with Two Alleles, A and a

	Genotype			
	AA	Aa	aa	Total
Initial genotypic frequencies	p^2	$2pq$	q^2	1
Fitness (W)	1	1	$1 - s$	
Ratio after selection	p^2	$2pq$	$q^2(1 - s)$	$1 - sq^2 = \overline{W}$
Genotypic frequencies after selection	$\dfrac{p^2}{\overline{W}}$	$\dfrac{2pq}{\overline{W}}$	$\dfrac{q^2(1 - s)}{\overline{W}}$	1

$= 1$). Natural selection cannot differentiate between the two genotypes because they both have the same phenotype. The recessive homozygote, however, is being selected against, which means that it has a lower fitness than the two other genotypes.

After selection the ratio of the different genotypes is determined by multiplying their frequencies (Hardy-Weinberg proportions) by their fitnesses. The procedure follows from the definition of fitness as a relative survival value. Thus, only $1 - s$ of the aa genotype survive for every one of the other two genotypes. For example, if s were 0.4, then the fitness of the aa type would be $1 - s$, or 0.6. For every 10 AA or Aa that survived, only 6 of the aa would have survived. The total of the three classes is $1 - sq^2$. That is,

$$p^2 + 2pq + q^2(1 - s) = 1 - sq^2$$

Mean Fitness of a Population.   The value $(1 - sq^2)$ is referred to as the **mean fitness of the population,** $\overline{W}$, because it is the sum of the fitnesses of the genotypes multiplied (weighted) by their proportion. That is, it is a weighted mean of the fitnesses (weighted by their proportions). The new ratios of the three genotypes can be returned to genotypic frequencies by simply dividing by this mean fitness of the population, $\overline{W}$, as in the last line of Table 18–2. (A set of numbers can be converted to proportions of unity by dividing them by their sum.) The new genotypic frequencies are thus the products of their original frequencies times their fitnesses divided by the mean fitness of the population.

After selection the new allelic frequency (q_{n+1}) is the proportion of aa homozygotes plus half the proportion of heterozygotes, or

$$q_{n+1} = \frac{q^2(1 - s)}{\overline{W}} + \frac{pq}{\overline{W}}$$

$$= \frac{q(q - sq + p)}{1 - sq^2}$$

$$q_{n+1} = \frac{q(1 - sq)}{1 - sq^2} \tag{18.3}$$

This model can be simplified somewhat by assuming that the *aa* genotype is lethal. Its fitness would be 0 and *s*, the selection coefficient, would be 1. Thus equation 18.3 would change to

$$q_{n+1} = \frac{q(1 - q)}{1 - q^2} \tag{18.4}$$

Since $(1 - q^2)$ is factorable into $(1 - q)(1 + q)$, equation 18.4 becomes

$$q_{n+1} = \frac{q(1 - q)}{(1 - q)(1 + q)}$$
$$q_{n+1} = \frac{q}{1 + q} \tag{18.5}$$

The change in allelic frequency is then calculated as

$$\Delta q = q_{n+1} - q = \frac{q}{1 + q} - q$$

To solve this equation, *q* is multiplied by $(1 + q)/(1 + q)$ so that both parts of the expression are over the common denominator $(1 + q)$:

$$\Delta q = \frac{q - q(1 + q)}{1 + q}$$
$$\Delta q = \frac{-q^2}{1 + q} \tag{18.6}$$

which is the expression for the change in allelic frequency caused by selection. Since selection will not act again until the same stage in the life cycle next generation, equation 18.6 is also an expression for the change in allelic frequency between generations.

Two facts are immediately apparent from equation 18.6. First, the allelic frequency of the recessive, *q*, is declining, as indicated by the negative sign in front of it. This fact should be intuitive from the way that selection has been defined in the model. Second, the change in allelic frequency is proportional to q^2, which appears in the numerator of the expression. That is, allelic frequency is declining as a relative function of the number of homozygous recessive individuals in the population. This fact is consistent with the method in which the selection model (against the homozygous recessive genotype) was set up. Thus, the final formula confirms the methodology of the model making.

Equilibrium Condition.   The last step in the model is to calculate the equilibrium *q* by setting the Δq equation equal to zero:

$$-\frac{q^2}{1 + q} = 0 \tag{18.7}$$

For a fraction to be zero, the numerator must equal zero. Thus, $-q^2 = 0$, and $q = 0$. Thus, at equilibrium the a allele should be entirely removed from the population. If the aa homozygotes are being removed and if there is no mutation to return a alleles to the population, then eventually the a allele will disappear from the population.

Time Frame for Equilibrium. One shortcoming of this model is that there is no time frame readily apparent in it. That is, from the model it is not immediately apparent how many generations will be required to remove the a allele. The deficiency can be compensated for by using a computer simulation or introducing a calculus differential into the model. Either method would produce the frequency/time graph of Figure 18–7. This figure clearly shows, first, that the a allele is removed more quickly when selection is stronger (when s is larger) and, second, that the curve appears to be asymptotic: the a allele is not immediately eliminated and would not be entirely removed until an infinitely long amount of time has passed. The reason is that as the a allele becomes rarer and rarer, it is more often found as a heterozygote (Table 18–3). Since selection can only remove aa homozygotes, an a allele hidden in an Aa heterozygote will not be selected against. When $q = 0.5$, there are two heterozygotes for every homozygote. When q

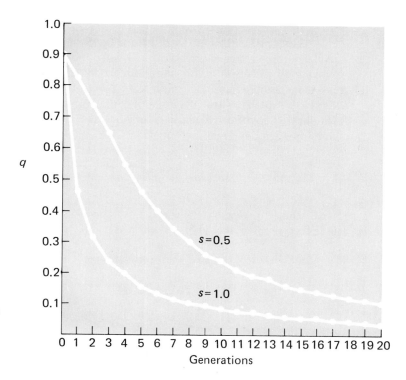

Figure 18 – 7.
Decline in q (the Frequency of the a Allele) with Selection against the aa Homozygote under Different Intensities of Selection

TABLE 18 – 3. Relative Occurrence of Heterozygotes and Homozygotes as Allelic Frequency Declines: $q = f(a)$

q	Aa (%)	aa (%)	Aa/aa
0.5	50	25	2
0.2	32	4	8
0.1	18	1	18
0.01	1.98	.01	198
0.001	.1998	.0001	1998

= 0.001, there are almost 2000 heterozygotes per homozygote. Remember, only the homozygote can be selected against. Natural selection cannot distinguish the dominant homozygote from the heterozygote.

Selection- Mutation Equilibrium

From the foregoing discussion it is apparent that although a deleterious allele may slowly be eliminated from a population, the time element is so great that there is opportunity for mutation to bring the allele back. It is predictable that recessive deleterious alleles may appear in a population at a rate that approaches an equilibrium between selection against the allele and its reappearance by mutation. Given a population in which alleles are removed by selection while being added by mutation, the point at which there is no change in allelic frequency ($\Delta q = 0$)—that is, there is **selection-mutation equilibrium** ($\hat{q}$)—may be determined as follows. The new allelic frequency (q_{n+1}) of the recessive a after nonlethal selection against the recessive homozygote is given by equation 18.3:

$$q_{n+1} = \frac{q(1 - sq)}{1 - sq^2}$$

(Recall that $1 - sq^2 = \overline{W}$, the mean fitness of the population.) Change in allelic frequency will thus be

$$\Delta q = q_{n+1} - q = \frac{q(1 - sq)}{\overline{W}} - \frac{q\overline{W}}{\overline{W}}$$

$$= \frac{q - sq^2 - q + sq^3}{1 - sq^2}$$

$$\Delta q = \frac{-sq^2(1 - q)}{1 - sq^2} \tag{18.8}$$

Equation 18.8 is the general form of equation 18.6 for any value of s. The change in allelic frequency due to mutation can be found by using equation 17.9:

$$\Delta q = up - vq$$

where u and v are the rate of forward mutation and back mutation, respectively. When equilibrium exists, the change from selection will just balance the change from mutation. Thus,

$$up - vq + \left(-\frac{sq^2(1-q)}{1-sq^2} \right) = 0$$

and

$$up - vq = \frac{sq^2(1-q)}{1-sq^2}$$

(18.9)

Now, some judicious simplifying is justified since, in a real situation, q will be very small—the a allele is being selected against so that vq will be close to zero while $1 - sq^2$ will be close to unity. Thus, equation 18.9 becomes

$$up \approx sq^2(1-q)$$

and further simplifying gives

$$u(1-q) \approx sq^2(1-q)$$
$$q^2 \approx u/s$$
$$\hat{q} \approx \sqrt{u/s}$$

(18.10)

In the case of a recessive lethal, s would be unity; so

$$q^2 \approx u \quad \text{and} \quad \hat{q} \approx \sqrt{u}$$

If a homozygous recessive has a fitness of 0.5 ($s = 0.5$) and a mutation rate, u, of 1×10^{-5}, then the allelic frequency at selection–mutation equilibrium will be

$$\hat{q} = \sqrt{\frac{u}{s}} = \sqrt{\frac{1 \times 10^{-5}}{0.5}} = \sqrt{2 \times 10^{-5}}$$
$$= 0.004$$

If the recessive phenotype was lethal, then

$$\hat{q} = \sqrt{\frac{u}{s}} = \sqrt{\frac{1 \times 10^{-5}}{1}}$$
$$= 0.003$$

Types of Selection Models

It can be seen from Table 18–2 that there are only a limited number of selection models possible, based on the limited ways that fitnesses can be assigned. Table 18–4 is an inclusive list of all possible selection models. This list contains only 6 models if we assume that fitnesses are constants and the highest fitness is 1. (The reader might now go through the list of models and determine the equilibrium condition for each.) The outcome of each of these models is described in the following paragraphs.

In both models 1 and 4 (Table 18–4), selection is against genotypes containing the a allele. Model 1, which was just derived in detail, is the case of a deleterious recessive. Almost any enzyme defect in a metabolic pathway fits this model, such as PKU, albinism, alkaptonuria, and so on. In model 4, however, natural selection can detect the heterozygote, which is the case with deleterious alleles that are not complete recessives. An example would be the hemoglobin anomaly called thallasemia, which produces a severe anemia in homozygotes and a milder anemia in heterozygotes. The disorder is common in some European and Oriental populations. It should be clear that selection can more quickly eliminate a partially recessive allele than a completely recessive allele because the allele can no longer "hide" in the heterozygote.

Models 3 and 6 are analogous to models 1 and 4 but involve the dominant allele or, as in model 6, the semidominant allele. Dominants are usually more quickly removed from a population because they are completely open to selection. It takes an infinite number of generations to remove a recessive lethal but only one generation for natural selection to remove a dominant lethal (see model 6, where $s_1 = s_2 = 1$). Examples of human dominant deleterious traits are Huntington's chorea, facioscapular muscular dystrophy, chondrodystrophy, and retinoblastoma.

Model 2 is interesting because selection against the heterozygote leads to an unstable equilibrium at $q = 0.5$. If one heterozygote is removed by selection, one each of the two alleles is eliminated. However, if p and q are

TABLE 18 - 4. All Possible 1-Locus, 2-Allele Selection Models (Assuming All Parameters Are Constants)

		Genotypic Fitness		
	Type of Selection	*AA*	*Aa*	*aa*
1	Against homozygous recessives	1	1	$1 - s$
2	Against heterozygotes	1	$1 - s$	1
3	Against homozygous dominant·	$1 - s$	1	1
4	Against the a allele	1	$1 - s_1$	$1 - s_2$
5	Against homozygotes	$1 - s_1$	1	$1 - s_2$
6	Against the A allele	$1 - s_1$	$1 - s_2$	1

TABLE 18 – 5. Selection Model of Heterozygote Advantage: the A Locus with A and a Alleles

	Genotype			
	AA	Aa	aa	Total
Initial genotypic frequencies	p^2	$2pq$	q^2	1
Fitness (W)	$1 - s_1$	1	$1 - s_2$	
Ratio after selection	$p^2(1 - s_1)$	$2pq$	$q^2(1 - s_2)$	$1 - s_1p^2 - s_2q^2 = \overline{W}$
Genotypic frequencies after selection	$\dfrac{p^2(1 - s_1)}{\overline{W}}$	$\dfrac{2pq}{\overline{W}}$	$\dfrac{q^2(1 - s_2)}{\overline{W}}$	

not equal (and thus not equal to 0.5), then one A allele is not the same proportion of the A alleles as one allele of a is of the a alleles. That is, in a population of 50 individuals with $q = 0.1$ and $p = 0.9$, one a allele is 10% (1/10) of the a alleles while one A allele is only 1.1% (1/90) of the A alleles. Thus, removing one each of the two alleles causes a decrease in q. That is, model 2 is stable at $p = q = 0.5$, but at any other allelic frequency the rarer allele is selected against. An example is the maternal-fetal incompatibility at the Rh locus in humans. Erythroblastosis occurs only in heterozygous offspring (Rh$^+$/Rh$^-$ in Rh$^-$/Rh$^-$ mothers). Thus, heterozygotes are selected against. (See the next chapter.)

In model 5 selection is against homozygotes. This model is called **heterozygote advantage** and the equilibrium condition will be derived in detail because the results are important to evolutionary theory.

Heterozygote Advantage

The selection model for heterozygote advantage is presented in Table 18–5. The first quantity to be solved for is q_{n+1}:

$$q_{n+1} = \frac{q^2(1 - s_2)}{\overline{W}} + \frac{1}{2} \cdot \frac{2pq}{\overline{W}}$$

$$q_{n+1} = \frac{q^2(1 - s_2) + pq}{\overline{W}} \tag{18.11}$$

Next, we solve for Δq:

$$\Delta q = q_{n+1} - q = \frac{q^2(1 - s_2) + pq}{\overline{W}} - q\frac{\overline{W}}{\overline{W}}$$

$$= \frac{q^2(1 - s_2) + pq - q(1 - s_1p^2 - s_2q^2)}{\overline{W}}$$

Now, by clearing all parentheses and grouping terms,

$$\Delta q = \frac{q^2 + pq - q + s_1 p^2 q - s_2 q^2 + s_2 q^3}{\overline{W}} \tag{18.12}$$

This equation can be simplified by noting the following:

a. $q^2 + pq - q = q(q + p - 1) = q(0) = 0$

b. $s_1 p^2 q - s_2 q^2 + s_2 q^3 = q[s_1 p^2 - s_2(q - q^2)]$

And, since $q - q^2 = q(1 - q) = qp$, b becomes

$$q(s_1 p^2 - s_2 pq) = pq(s_1 p - s_2 q)$$

Substituting a and b back into equation 18.12 gives

$$\Delta q = \frac{pq(s_1 p - s_2 q)}{\overline{W}}$$

At equilibrium

$$\Delta q = \frac{pq(s_1 p - s_2 q)}{\overline{W}} = 0 \tag{18.13}$$

For this expression to be zero, either

$$p = 0, \quad q = 0, \quad \text{or} \quad (s_1 p - s_2 q) = 0$$

If $p = 0$ or $q = 0$, the result is trivial—the equilibrium exists only because of the absence of one of the alleles. The more meaningful equilibrium occurs when $(s_1 p - s_2 q) = 0$—that is,

$$s_1 p = s_2 q \quad \text{or} \quad s_1(1 - q) = s_2 q$$

and

$$\hat{q} = \frac{s_1}{s_1 + s_2} \tag{18.14a}$$

$$\hat{p} = \frac{s_2}{s_1 + s_2} \tag{18.14b}$$

Several interesting conclusions follow. First, unlike the other models of selection, this model maintains both alleles in the population. That this equilibrium is stable can be demonstrated by graphing the Δq value against q. Such a graph is shown in Figure 18–8, where q is the frequency of allele a and the fitness of genotypes AA, Aa, and aa are assumed to be 0.8, 1, and 0.7, respectively. Note that when the equilibrium of allelic frequency is per-

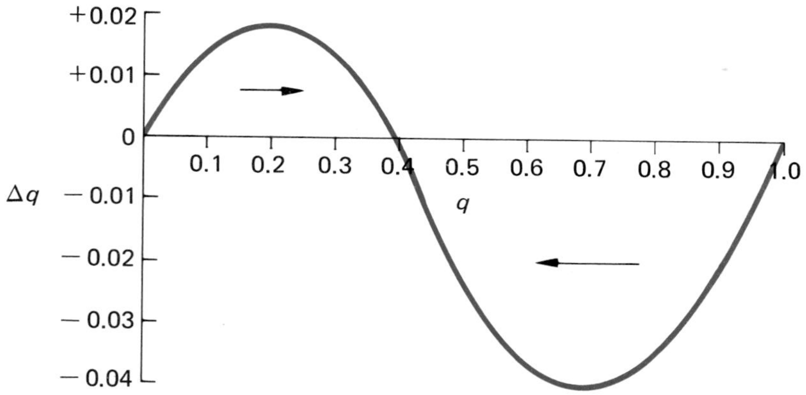

Figure 18 – 8.
Plot of Allelic
Frequency (q) versus
Change in Allelic
Frequency (Δq) for a
Hypothetical Stable
Polymorphism
Maintained by
Heterozygote
Advantage

turbed by an increase or decrease of q, the population will return to the point of equilibrium. Second, the equilibrium is independent of the original allelic frequencies since it involves only the selection coefficients, s_1 and s_2. Lastly, the equilibrium for each allele is dependent on the coefficient of selection against the *other* allele. As the selection against A increases (s_1 increases), the equilibrium shifts toward a higher value of q (more a alleles).

Genetic Load

As long as there is selection in a population, different genotypes will have different reproductive success. One or more genotypes will have poorer survival or reproduction than the most fit genotype. The reduced fitness of a population can be evaluated by comparing this population to one composed entirely of the most fit genotype. The decrease in the mean fitness of a population compared to the fitness of a population composed only of individuals with the best genotypes is termed the **genetic load, L.** Algebraically,

$$L = \frac{W_{max} - \overline{W}}{W_{max}} \qquad (18.15)$$

Since W_{max} is usually conceived of as unity, the formula for calculation of the load is

$$L = 1 - \overline{W} \qquad (18.16)$$

While every source of selection and variability in a population contributes to genetic load, there are basically two main sources—mutational and segregational.

Mutational Load. Deleterious alleles brought into a population that is selecting against them create a **mutational load** that contributes to the genetic load. (Mutational load is defined as the relative loss of fitness of a population caused by the creation of deleterious alleles by mutation.) If a recessive allele is selected against, the mean fitness (from Table 18–2) is

$$\overline{W} = 1 - sq^2$$

If we substitute for q in this equation, the equilibrium value of q—that is, $\hat{q} \approx \sqrt{u/s}$ from equation 18.10—gives

$$\overline{W} = 1 - s(\sqrt{u/s})^2$$
$$= 1 - u$$

Since the genetic load is

$$L = 1 - \overline{W}$$

we can substitute to obtain

$$L = 1 - (1 - u)$$
$$= u$$

Here, at mutation–selection equilibrium, the genetic load is the mutation rate. Mutation is providing alleles to be removed by selection and is thus causing the genetic load.

Segregational Load. **Segregational load** is the relative loss of fitness of a population because segregation of less fit genotypes is occurring. This load is due to loci with heterozygous advantage. Each generation, these loci produce homozygotes that are then selected against. In heterozygous advantage the mean fitness is (Table 18–5)

$$\overline{W} = 1 - s_1 p^2 - s_2 q^2$$

Substituting $\hat{p}$ and $\hat{q}$ from equation 18.14 gives

$$\overline{W} = 1 - s_1 \left(\frac{s_2}{s_1 + s_2} \right)^2 - s_2 \left(\frac{s_1}{s_1 + s_2} \right)^2$$
$$= 1 - \frac{s_1 s_2^2 + s_1^2 s_2}{(s_1 + s_2)^2}$$
$$= 1 - \frac{s_1 s_2 (s_1 + s_2)}{(s_1 + s_2)^2}$$
$$= 1 - \frac{s_1 s_2}{s_1 + s_2}$$

Then, genetic load becomes

$$L = \frac{s_1 s_2}{s_1 + s_2}$$

In a balanced lethal system, $s_1 = s_2 = 1$. The genetic load would thus be 0.5. Hence the fitness of a population with a balanced lethal is only half the fitness of a population without a balanced lethal.

CHAPTER SUMMARY

Random mating is required for the Hardy-Weinberg equilibrium to hold. Deviations from random mating fall into two categories, depending on whether phenotypic resemblance or relatedness are involved in mate choice. Phenotypic resemblance is the basis for assortative and disassortative mating, in which individuals choose similar or dissimilar mates, respectively. Assortative mating causes increased homozygosity only among loci controlling the traits for which selection is made. There is no change in allelic frequency. Disassortative mating causes increased heterozygosity.

Mating among relatives is termed inbreeding and is measured by F, the inbreeding coefficient. F measures the probability of autozygosity (homozygosity by descent). It can be calculated from pedigrees by using the formula

$$F = \Sigma(\tfrac{1}{2})^n(1 + F_A)$$

Inbreeding exposes recessive deleterious traits and also causes homozygosity uniformly across the genome. It too does not, by itself, change allelic frequency.

Natural selection is defined by differential reproductive success and is measured by either survival or reproductive characteristics. Depending upon which genotypes are most fit, natural selection can act in several ways to change allelic and genotypic frequencies. Selection against the homozygous recessive acts to remove the allele. Mutation will bring the allele back into the population. Thus, there exists a selection–mutation equilibrium, which will maintain the unfavorable allele at a relatively low frequency.

Heterozygote advantage will maintain both alleles in a population. The equilibrium condition was defined. Both heterozygote advantage and mutation contribute heavily to the genetic load of a population, the proportional decrease in fitness due to factors that produce variation and selection in a population.

1. What is the inbreeding coefficient of I in the following pedigree? Assume that the inbreeding coefficients of other members of the pedigree are zero unless other information tells you differently.

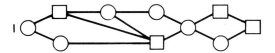

2. What is the inbreeding coefficient of individual I? $F_A = 0.01$; $F_B = 0.02$; $F_C = 0.02$.

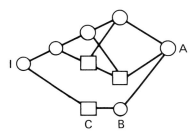

3. The following is the pedigree of the child produced by the mating of full siblings. The individuals in the pedigree have the following inbreeding coefficients: $F_A = 0.2$; all others $= 0.0$. Convert the pedigree to a path diagram and determine the inbreeding coefficient of individual E.

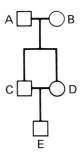

4. Differentiate among stabilizing, directional, and disruptive selection.

5. Derive a model of selection where the fitness of the heterozygote is half the fitness of one of the homozygotes and twice the fitness of the other. Give expressions for the following:
   a. Mean population fitness
   b. Equilibrium allelic frequency (stable?)
   c. Genetic load

6. Derive an expression for the equilibrium allelic frequency under a model where selection is against the heterozygotes. Is the equilibrium stable?

7. Derive the equilibrium allelic frequency for selection that favors heterozygotes. Show that the equilibrium is stable.

8. The following table describes selection at the A locus in a given diploid species where $p = f(A)$ and $q = f(a)$.

	Genotypes			Total
	AA	Aa	aa	
Before selection	p^2	$2pq$	q^2	1
Fitness (W)	$1 - s_1$	1	$1 - s_2$	
After selection	$p^2(1 - s_1)$	$2pq$	$q^2(1 - s_2)$	$\overline{W} = 1 - s_1 p^2 - s_2 q^2$

a. Describe the type of selection that is occurring here. Why does the total equal 1 before selection but $\overline{W}$ after selection?

b. Derive an equation for q after one generation of selection (q_{n+1}).

c. This system will reach equilibrium, with $p = s_2/(s_1 + s_2)$. If selection is twice as strong against aa as against AA, what will the equilibrium allelic frequencies be? If $s_1 = 0.1$ and $s_2 = 0.2$, what will the percentage of heterozygotes be at equilibrium? What will the segregational load be?

9. Given, a locus with alleles A and a in a population in Hardy-Weinberg equilibrium: Set up a model and the initial formula for the allelic frequency of the dominant (p_{n+1}) after one generation if selection acts only against the dominant homozygote.

10. Given, a locus with alleles A and a in a large randomly mating population: Allele A mutates to a at a rate of u and there is no back mutation. However, the aa homozygote is selected against with a fitness of $1 - s$. Give a formula for the equilibrium condition. If $u = 5 \times 10^{-5}$ and $s = 0.15$, what are the equilibrium allelic frequencies? What is the genetic load at equilibrium?

11. If a locus has alleles A_1 and A_2, what will be the equilibrium frequency of A_1 if both homozygotes are lethal? What is this type of selection called?

12. The following data were collected from a population of *Drosophila* segregating sepia (s) and wild-type $(+)$ eye colors. A sample was taken when the eggs were deposited and later among adults. Reconstruct the model of selection. What is the genetic load?

	s/s	s/+	+/+
Egg	25	50	25
Adult	30	60	10

13. The following data are taken from Dobzhansky's work with chromosomal inversions in *Drosophila pseudoobscura* and represent 4 samples from various altitudes in the Sierra Nevada in California. What would you say about and what would you do in the lab to test the fitness of the inversions? What factors could cause the changes in fitness?

Elevation of Sample	Inversion			
	ST	AR	CH	Others
6800 ft	26	44	16	14
4600 ft	32	37	19	12
3000 ft	41	35	14	10
800 ft	46	25	16	13

(ST = Standard; AR = Arrowhead; CH = Chiricahua)

SUGGESTIONS FOR FURTHER READING

Cavalli-Sforza, L., and W. Bodmer. 1971. *The Genetics of Human Populations*. San Francisco: Freeman.

Crow, J., and M. Kimura. 1970. *An Introduction to Population Genetics Theory*. New York: Harper & Row.

Dobzhansky, T. 1958. Genetics of natural populations. XXVII. The genetic changes in populations of *Drosophila pseudoobscura* in the American Southwest. *Evolution* 12:385–401.

Li, C. 1955. *Population Genetics*. Chicago: University of Chicago Press.

Mettler, L., and T. Gregg. 1969. *Population Genetics and Evolution*. Englewood Cliffs, N.J.: Prentice-Hall.

Spiess, E. 1977. *Genes in Populations*. New York: Wiley.

Wallace, B. 1968. *Topics in Population Genetics*. New York: Norton.

Wallace, B. 1970. *Genetic Load*. Englewood Cliffs, N.J.: Prentice-Hall.

Wright, S. 1969. *Evolution and the Genetics of Populations*, vol. 2. *The Theory of Gene Frequencies*. Chicago: University of Chicago Press.

19 Genetics of the Evolutionary Process

The previous two chapters laid the theoretical groundwork for an understanding of the process of evolution in natural populations. Populations change, or evolve, through the process of natural selection and the other forces that perturb the Hardy-Weinberg equilibrium. The merger of population genetics theory with classical Darwinian views of evolution is known as **neo-Darwinism** or the "new synthesis."

DARWINIAN EVOLUTION

Charles Darwin (Figure 19–1) was a British naturalist who published his theory of evolution in 1859 in a book entitled *The Origin of Species*. Darwin had been greatly influenced by the writings of the Reverend Thomas Malthus, who is best known for his statement that populations increase exponentially while their food supplies only increase arithmetically. Malthus (*An Essay on the Principle of Population*, 1798) was referring specifically to human populations and was trying to encourage people to reduce their birthrate rather than have their offspring starve to death. Malthus's writings impressed upon Darwin the realization that under limited resources—the usual circumstance in nature—not all organisms survive: There will be *competition* among organisms to survive.

Darwin was the naturalist aboard the ship *Beagle,* which sailed around the world from 1831 to 1836 with the primary purpose of charting the coast of South America. During the travels of *Beagle,* Darwin amassed great quantities of observations (especially on the Galápagos Islands) that led him to suggest a theory wherein organisms become adapted to their environment by the process of **natural selection.** In outline, the process, as proposed by Darwin, has the following steps:

Figure 19 – 1.   Charles Darwin

Source: Reprinted from Colin Patterson: *Evolution.* Copyright © Trustees of the British Museum (Natural History) 1978. Used by permission of the publisher, Cornell University Press.

1. *Variation is a characteristic of virtually every group of animals and plants.* Darwin saw variation as an inherent property of all populations.

2. *Every group of organisms overproduces offspring.* In most populations, which, over time, maintain a relatively constant density, every parent, on the whole, just replaces itself. Thus, most offspring produced by the individuals of a population will die before they reproduce. Hence, in every group of organisms, there is an overabundance of young.

3. *Competition will occur among the young.* Competition can be for food, space, or any other resource. It can be an active fight for available resources (**interference competition**) or it can be in the form of a superior ability to gather necessary resources (**exploitation competition**).

4. *The most fit will survive.* This step is the cornerstone and the best-known part of Darwin's theory. Among all the organisms competing for a limited array of resources, only the organisms best able to obtain and utilize these resources will survive: the **survival of the fittest.**

5. *To whatever degree the characteristics of the most fit are inherited, the "favored traits" will be passed on to the next generation.*

DARWIN'S MISTAKE

From time to time, attacks on Darwinism are mounted, usually by persons who see evolutionary theory as antireligious or by persons who basically misunderstand Darwin's theory. One recent attack is entitled "Darwin's Mistake," by Tom Bethell, and was published in the February, 1976, issue of *Harper's* magazine. (Stephen Gould countered this attack with an article in *Natural History* magazine the same year.)

Bethell began by pointing out that Darwinian theory is a tautology rather than a predictive theory, where tautology means a statement that is true by definition. That is, evolution is the survival of the fittest. But who are the fittest? Obviously, the individuals who survive. Thus, without an independent criterion for fitness, other than survival, we are left with the statement that evolution is the survival of the survivors. This is indeed a tautology. But, it is possible to assign independent criteria for fitness. Darwin wrote extensively about artificial selection in pigeons, where the breeders' choice was the criterion for fitness. (Many novel breeds of pigeons have been created in this way.) The same sort of selection (artificial) has been practiced by horticulturists, farmers, and animal breeders. Here too, survival is not the criterion for fitness—productivity is.

It is a bit more difficult to establish independent criteria of fitness a priori in nature. Often, uncontrolled or unseen vagaries have major impacts on the course of events in nature. Surely the temperature became colder before mammoths became wooly. Is it then reasonable, or safe, to predict that elephants would get wooly if the climate became colder in Africa today? The answer is no, for several reasons. First, the elephants might adapt to colder weather in any of a large number of different ways: They could get fatter, they could migrate south, they could develop feathers, and so on. To some extent, their adaptation depends not only on the changing environment, but also on the species' reserve variation from which the adaptive traits must come. Second, the elephants could simply go extinct: They might not be able to adapt at all. And third, if the climatic changes were not exceptionally severe, the elephants might not change at all. Predicting the exact course of evolution is nearly impossible! It is, therefore, very difficult in nature to provide independent criteria for fitness; many modern evolutionary biologists, while not doubting Darwinism, do worry to some extent about the difficulties in testing modern evolutionary theory.

Bethell then went on to try to refute Darwinism by the following argument: Survival of the fittest can be redefined to mean that some organisms have more offspring than others. Thus, natural selection cannot be a creative force because the only thing it works on is organisms alive now, some having more offspring than others. How, asks Bethell, can this possibly give us tigers and horses from ancestors that did not look like tigers and horses? The answer is that mutation produces variants in the population. The organisms best able to compete will leave the most offspring (not *have* the most offspring, but *leave* the most offspring). Given an array of different genotypes in a population, natural selection is the process that determines which genotypes will increase in future generations. Traits that give the bearer an advantage increase in the population and evolution takes place. Natural selection was the force behind the gradual evolution of horses from the small Eocene horse to the modern *Equus* (Figure 17–1).

Misinterpretation of mutation is the basis for other attacks on Darwinism. For example, Darwinian evolution has been attacked as not feasible since most mutations are deleterious. How, the argument goes, can evolution proceed by a combination of deleterious events? The answer is that although most mutations are deleterious, some are not. This is especially true in changing environments where yesterday's deleterious mutant may be today's favored mutant.

Thus, over time, the characteristics of a population will change through the process of natural selection—a population will either evolve through directional selection or have its less fit individuals removed through stabilizing or disruptive selection (see Figure 18–6). Nonrandom mating, genetic drift, migration, and mutation may also play a role in the population differentiation.

Evolution and Speciation

The Darwinian definition of **evolution** is a gradual change in phenotypic frequencies, which results in a population of individuals that are better adapted to the environment than their ancestors were. **Speciation** comes in two different forms: (1) It may be the evolution of a population over time until a point is reached where the current population cannot be classified as belonging to the same species as the original population, or (2) it may be the divergence of a population into two distinct forms **(species)** that exist simultaneously. Before we continue, the term *species* should be defined.

Prior to Darwin's time, organisms were grouped in accordance with the **type-species (or morphological-species) concept,** which defined a species as a group of organisms that were morphologically similar. All variants were considered to be imperfections. One of Darwin's greatest contributions to modern biological theory was to treat variation as a normal part of the description of a group of organisms. The modern species concept groups together as members of the same species only organisms that can interbreed. That is, a species is a group of organisms that can, as a group, produce fertile offspring. Taxonomists, however, often use the morphological-species concept as a working definition—two similar organisms belong to different species if they are as different as two organisms belonging to two recognized species. The definition of species on the basis of interbreeding unfortunately falls down in many places—mostly due to technical problems of applying it. For example, since speciation is a dynamic process, there will be subgroups of a population in various stages of becoming new species; the rate of successful interbreeding among these subgroups may range from 0 to 100%. How should the in-betweens be classified? There is no correct answer. It depends on the circumstance.

Other problems make it necessary to turn to the morphological-species concept. Many biologists work with preserved specimens that cannot be tested for interbreeding ability. Also, two organisms that will not interbreed in nature may in a laboratory setting. Thus, the interbreeding test carried out in the lab is not necessarily an adequate criterion of speciation. And lastly, much data on the process of evolution comes from the fossil record, where, of course, it is impossible to apply a test of interbreeding. Other problems arise in classifying groups that are geographically isolated from each other, such as populations on islands. So, although there is a good theoretical definition of a species (interbreeding individuals), more often than not it is necessary to apply morphological criteria to determine whether or not two populations belong to different species; that is, do they look different? Sometimes, there are cases where no decision can be made about the species status of a population. It is clear that a population has evolved but it is not clear whether it has evolved enough to be called a new species.

Mechanisms of Speciation

Reproductive Isolation.   How does one species become two? Two processes must take place. First, there must arise a barrier to free reproduction and gene flow such that a population of the original species can begin to evolve independently. Second, once this barrier to gene flow has arisen, **reproductive isolating mechanisms** must evolve. Reproductive isolating mechanisms are environmental, behavioral, mechanical, and physiological barriers that prevent two individuals of two species from producing viable offspring. Stebbins has listed the barriers that separate similar species as follows:

A. Prezygotic mechanisms: Fertilization and zygote formation are prevented.
   1. Habitat—The populations live in the same region but occupy different habitats.
   2. Seasonal or temporal—The populations exist in the same regions but are sexually mature at different times.
   3. Ethological (in animals only)—The populations are isolated by different and incompatible behavior before mating.
   4. Mechanical—Cross-fertilization is prevented or restricted by incompatible differences in reproductive structures (genitalia in animals, flowers in plants).
B. Postzygotic mechanisms: Fertilization takes place and hybrid zygotes are formed but are inviable or give rise to weak or sterile hybrids.
   1. Hybrid inviability or weakness.
   2. Developmental hybrid sterility—Hybrids are sterile because gonads develop abnormally or meiosis breaks down before it is completed.
   3. Segregational hybrid sterility—Hybrids are sterile because of abnormal distribution to the gametes of whole chromosomes, chromosome segments, or combinations of genes.
   4. F_2 breakdown—F_1 hybrids are normal, vigorous, and fertile, but F_2 contains many weak or sterile individuals.

Allopatric, Parapatric, and Sympatric Speciation.   Barriers to gene flow can arise in three different ways; each way defines a different type of speciation. If a geographic barrier, such as a river or mountain, appears, populations of a species will be physically isolated. Physical isolation can also occur if migrants of a species cross a particular barrier and found a new population. The physically isolated populations of the species can then evolve independently. If reproductive isolating mechanisms evolve, then two distinct species will be formed. If they come together in the future, they will remain as distinct species. Speciation in which the evolution of reproductive isolating mechanisms occurs during physical separation of the populations is called **allopatric speciation** (Figure 19–2). Until recently, evolutionary biologists believed that this mode of speciation was the general rule. It is now believed that two other modes of speciation occur at least as frequently.

Parapatric speciation occurs when a population of a species that occupies a large range enters a new niche, or habitat (Figure 19–2). Although no physical barrier arises, occupancy of the new niche will result in a barrier to gene flow between the population in the new niche and the rest of the species. Here again, the evolution of reproductive isolating mechanisms will produce two species where there was only one before. Parapatric speciation is seen often in relatively nonvagile animals, such as snails, and certain grasshoppers, and in annual plants.

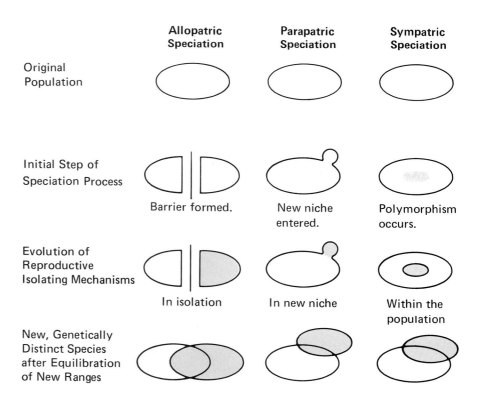

Figure 19 – 2.
The Three General
Modes of Speciation

Sympatric speciation occurs when a polymorphism arises within an interbreeding population before a shift to a new niche. (A polymorphism is the occurrence of alternate alleles in the same population.) This mode of speciation seems to be common in parasites. For example, if a polymorphism arises within a parasite species such that a certain genotype can adapt to a new host, this genotype may be the forerunner of a new species. If the parasite not only feeds on the new host but also mates on the new host, a barrier to gene flow arises although the parasite may be surrounded by other members of the species with the other genotype. Sympatric speciation can thus occur in the middle of a species range rather than at the edges (Figure 19–2).

The end result of the speciation process is the divergence of a homogeneous population into two (or more) species. The classic example of the phenomenon is the case of the ground finches of the Galápagos Islands. The birds are very well studied not only because they present a striking case of speciation but also because they were studied by Darwin and were a strong influence on his views. Figure 19–3 is a map of the Galápagos Islands, and Figure 19–4 is a diagram of the species of "Darwin's Finches."

An original finch somehow came over from South America and with time spread to the various islands of the Galápagos. Given the limited ability of the birds to get from island to island, allopatric speciation took place. On each island each population evolved reproductive isolating mechanisms while changing morphologically to fill certain niches not available to the parent species in South America. For example, in South America no finches have evolved to be like woodpeckers because there are many woodpecker species already there that do quite a good job of being woodpeckers. But the Galápagos, being isolated from South America, exhibit what is called a **depauperate fauna**—that is, a fauna lacking many species found on the mainland. The islands lacked woodpeckers, and a very useful food resource for birds—insects beneath the bark of trees—was going unused. Finches that could make use of this resource would be at an advantage and thus favored by natural selection—they could leave more offspring. On one island a finch did evolve to use this food resource. The woodpecker finch functions as a woodpecker by inserting cactus needles into holes in dead trees to extract insects. Darwin wrote, "Seeing this gradation and diversity of structure in one small, intimately related group of birds, one might really fancy that from an original paucity of birds in this archipelago, one species had been taken and modified for different ends."

MAINTENANCE OF GENETIC VARIATION

Darwinian evolution depends on the occurrence of variation within a population. While selection against a certain genotype usually leads to the removal of an allele and, therefore, to the reduction of variation within a

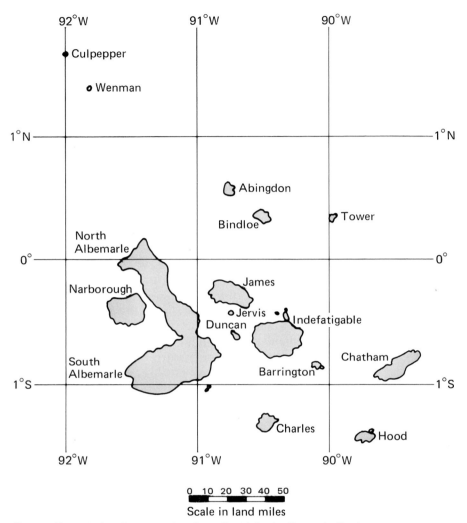

Figure 19 - 3.
The Galápagos
Archipelago

Source: Reprinted with permission from David Lack, *Darwin's Finches* (Cambridge: Cambridge University Press, 1947).

population, selection favoring the heterozygotes allows for the maintenance of more than one allele at a given locus—and permits variation to occur. E. B. Ford, a British evolutionary biologist, applied the term **genetic polymorphism** to the occurrence of more than one allele at a given locus, with the least frequent allele occurring in the population more frequently than can be accounted for by mutation. Mutation of course brings new alleles into a population. If natural processes are continued long enough, an equilibrium allelic frequency will be reached as a function of forward (u) and backward

Scientific Name	Descriptive Designation
1. *Geospiza magnirostris* Gould	Large ground-finch
2. *Geospiza fortis* Gould	Medium ground-finch
3. *Geospiza fuliginosa* Gould	Small ground-finch
4. *Geospiza difficilis* Sharpe	Sharp-beaked ground-finch
5. *Geospiza scandens* (Gould)	Cactus ground-finch
6. *Geospiza conirostris* Ridgway	Large cactus ground-finch
7. *Camarhynchus crassirostris* Gould	Vegetarian tree-finch
8. *Camarhynchus psittacula* Gould	Large insectivorous tree-finch
9. *Camarhynchus pauper* Ridgway	Large insectivorous tree-finch on Charles
10. *Camarhynchus parvulus* (Gould)	Small insectivorous tree-finch
11. *Camarhynchus pallidus* (Sclater and Salvin)	Woodpecker-finch
12. *Camarhynchus heliobates* (Snodgrass and Heller)	Mangrove-finch
13. *Certhidea olivacea* Gould	Warbler-finch
14. *Pinaroloxias inornata* (Gould)	Cocos-finch

Figure 19 – 4.
Species of Darwin's
Finches

Source: Reprinted with permission from David Lack, *Darwin's Finches* (Cambridge: Cambridge University Press, 1947).

Edmund Brisco Ford
(1901–   )
Courtesy of Dr.
Edmund Brisco Ford

(v) mutation rates, such that $\hat{q} = u/(u + v)$. Cases where mutation is counterbalanced by selection, such that one allele is maintained at a frequency very close to the mutation rate, are not considered cases of polymorphism.

Prior to the mid-1960s, the general belief was that only a very few loci were polymorphic in any individual or any population. For example, humans exhibit variation in height; skin, hair, and eye color; hair texture; and other easily visible traits. What percentage of the whole genome these variable traits represent will not be known until the whole genome is sampled. It may be that such traits are obvious phenotypic characteristics that could be expected to be variable but that the whole genome is not very variable. Or it may be that almost all the genome is variable, given the sample of these visible traits.

In 1966 two University of Chicago researchers found a way to randomly sample the genome. R. C. Lewontin (now at Harvard) and J. L. Hubby used acrylamide-gel electrophoresis to investigate variability directly in the gene products of the fruit fly *Drosophila pseudoobscura*. (Independent, similar work in humans was reported by H. Harris.) Electrophoresis (see Chapter 4) is a method of separating proteins on the basis of their different electric charges. When two alleles differ by producing proteins that have different amino acids with different charges, the products of these two alleles can be differentiated by electrophoresis. Lewontin and Hubby reasoned that choosing enzymes and general proteins that are amenable to separation by electrophoresis is, in fact, choosing a random sample of the genome of the fruit fly. If this is the case, then the degree of polymorphism found by electrophoretic sampling would provide an estimate of the amount of variability occurring in the individual organism and in the population. Their results were startling.

Of 18 loci examined, Lewontin and Hubby found that the species was polymorphic at 39% of the loci, the average population was polymorphic at 30% of the loci, and the average individual was heterozygous at 12% of the loci. The high rate of polymorphism sparked two interrelated controversies. The first was whether or not electrophoresis does, in fact, randomly sample the genome; the second was whether or not most electrophoretic alleles are maintained in the population by natural selection. Let us return to the arguments after looking at the ways in which genetic polymorphism can be maintained in natural populations.

Richard C. Lewontin
(1929–   )
Courtesy of Dr.
Richard C. Lewontin

Maintaining Polymorphism

Mutation.   Polymorphism due solely to mutation pressure can occur but is a rare event because of the very slow rate at which mutation pressure changes allelic frequency. A polymorphism maintained by mutation could occur only in alleles that are not affected by natural selection. A mutation–selection balance is ruled out by the definition of polymorphism. The addition of new variants by mutation would provide the raw materials for selection, but the appearance of new variants by mutation is not considered to be polymorphism.

Heterozygote Advantage. When selection acts against both homozygotes, it favors the heterozygote (Chapter 18)—that is, there exists a *heterozygote advantage*. An equilibrium is achieved, dependent solely on the selection coefficients, which will maintain both alleles. The classic example of heterozygote advantage in humans is the case of sickle-cell anemia. Sickle-cell hemoglobin (Hb^S) differs from normal hemoglobin (Hb^A) by having a valine in place of a glutamic acid in position number 6 of the beta chain of the globin molecule. Under reduced oxygen the erythrocytes containing sickle-cell hemoglobin change from round to sickle-shaped cells (Figure 19–5). There are two unfortunate consequences of this change of cell shape: (1) sickle-shaped cells are rapidly broken down, which causes anemia as well as hypertrophy of the bone marrow; (2) the sickle cells clump, which blocks capillaries and produces local losses of blood flow resulting in tissue damage.

Figure 19 – 5.   Sickle-Shaped Red Blood Cells from a Person with Sickle-Cell Anemia
Normal red blood cell is about 7–8 μm in diameter.

Source: Reproduced courtesy of Dr. Patricia N. Farnsworth.

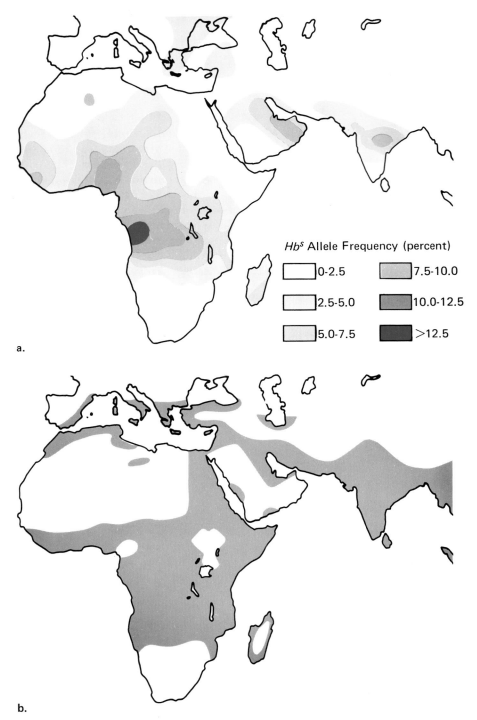

Hb^s Allele Frequency (percent)

0-2.5	7.5-10.0
2.5-5.0	10.0-12.5
5.0-7.5	>12.5

a.

b.

Figure 19 – 6. Distribution of (a) the Sickle-Cell Allele (*Hb^s*), and (b) Falciparum Malaria (Caused by *Plasmodium falciparum*) in Africa, Europe, the Middle East, and India

Source: From L. L. Cavalli-Sforza, "The genetics of human populations." Copyright © 1974 by Scientific American, Inc. All rights reserved.

Such a condition of reduced fitness leads to the assumption that the sickle-cell allele would be selected against in all populations and, therefore, would be rare. But this is not the case. The sickle-cell allele is common in many parts of Africa, India, and southeast Asia. What could possibly maintain this detrimental allele? In the search for an answer to this question, the major discovery was that the distribution of the sickle-cell allele correlated well with the distribution of malaria (Figure 19–6). The following facts have now been uncovered. The sickle-cell homozygote (Hb^S/Hb^S) almost always dies of anemia. The sickle-cell heterozygote (Hb^S/Hb^A) is only slightly anemic and is resistant to malaria. The normal homozygote (Hb^A/Hb^A) is not anemic and has no resistance to malaria. Thus, in areas of malaria, the best genotype of the three appears to be the sickle-cell heterozygote, which results in a minor anemia coupled with a resistance to malaria.

This conclusion is reinforced by the changes in allelic frequency that occur when a population from a malarial area is moved to a nonmalarial area: The normal homozygote is no longer at risk for malaria and selection acts mainly on the sickle-cell homozygote and to a slight extent on the heterozygote. Table 19–1 shows data for African versus American blacks. The African population is, of course, under malarial risk while the American population is not.

Heterozygote advantage is an expensive mechanism for maintaining polymorphism. Losses must occur in both homozygote groups in order for the polymorphism to exist. Thus, part of the reproductive output of a population is lost each generation to maintain each system of heterozygote advantage. In the case of sickle-cell anemia, a tragic loss of human life due to either anemia or malaria results. The necessary loss of individuals to maintain polymorphism at a particular locus is part of the *genetic load* of a population: In the sickle-cell case, it is a segregational load.

Frequency-Dependent Selection.   All the selection models discussed so far have had selection coefficients that were constants. This is not always the case. For example, Lee Ehrman of the State University of New York has shown that when a female fruit fly has a choice of mates of different genotypes, she will choose the rare genotype. **Frequency-dependent selection** is selection whereby a genotype is at an advantage when rare and at a disadvantage when common. A model for frequency-dependent selection can be

TABLE 19 - 1. Sickle-Cell Anemia Frequencies in American and African Blacks

	Percent of Homozygotes $Hb^A Hb^A$	Percent of Heterozygotes $Hb^A Hb^S$	Frequency of Hb^S q
African blacks (Mid-central Africa)	82	18	.09
American blacks	92	8	.04

constructed by assigning fitnesses that are not constants. The easiest way to do this is to assign fitnesses that are a function of allelic frequency. Thus, the assigned fitnesses for one locus with two alleles could be $(1.5 - p)$ and $(1.5 - q)$ for the AA and aa genotypes respectively, and 1 for the Aa genotype. An interesting outcome of this model is that at $p = q = 0.5$, the system is in equilibrium and there is no selection (all the fitnesses = 1). There is evidence that this type of model may be relatively common in nature.

Transient Polymorphism. An *apparent* genetic polymorphism can result when an allele is being eliminated by selection. For example, if we start out with a population homozygous for the a allele and a mutation brings in a more favored A allele, the population will gradually become all A through directional selection. However, during the process of replacement, both alleles will be present.

Balanced Systems. Selection at one stage in the life cycle of an organism can be balanced by a different form of selection at another stage in the life cycle. An allele can be favored in a larva but selected against in an adult. There can also be a balance of selection in different parts of the habitat. For example, an allele can be favored in a wet part of the habitat but selected against in a dry part of the habitat.

Maintaining Many Polymorphisms

In summary, the classically accepted ways to account for the continued existence of allelic polymorphism in a population are mutation, heterozygote advantage, frequency-dependent selection, and balanced selection. Until the work of Lewontin and Hubby, heterozygote advantage was believed to be the most common method of maintaining a polymorphism at a given locus. The maintenance of an allele by heterozygote advantage costs the population a certain number of its offspring due to the mortality of the homozygotes (segregational genetic load). A population can afford the loss if polymorphism is maintained at only a small number of loci. After Lewontin and Hubby reported that polymorphism seemed to exist at a large proportion of the loci, new explanations were needed to account for it. The new explanations addressed three possibilities:

1. Electrophoresis does not randomly sample the genome, and thus a large amount of variability does not really exist.
2. New population genetic models can be derived that explain how this large amount of variability is maintained.
3. Electrophoretic alleles are not under selection pressure. That is, isozymic forms of an enzyme all perform this enzyme's function equally well—the **neutral gene hypothesis.**

Sampling the Genome. Does electrophoresis randomly sample the genome? Since, on the basis of DNA content, the genome of higher organisms has the potential to contain several million genes, there will always be a question as to whether or not it is really being randomly sampled by electrophoresis. Several lines of evidence suggest that the results from electrophoresis are underestimates of the true amount of genetic variability present in a population. For example, the majority of amino acid substitutions do not change the charge of the protein; so what appear as single bands on an electrophoretic gel could actually be heterogeneous mixtures of the products of several alleles. (However, newer techniques are improving the precision of electrophoresis.) Also, it is now known that glycolytic enzymes are less polymorphic than other enzymes. And, since glycolysis is a limited process in which most enzymes are not involved, it follows that the average heterozygosity over all loci should be slightly higher than the original estimate that included glycolytic enzymes. Since the original report of Lewontin and Hubby, numerous studies on many different organisms agree, for the most part, on the high amount of polymorphism in natural populations (Table 19–2). Recent technical advances of multidimensional electrophoresis and DNA sequencing also support the hypothesis that electrophoresis does randomly sample the genome. If anything, electrophoresis underestimates variability.

TABLE 19 – 2.   Survey of Genic Heterozygosity

Species	Number of Populations	Number of Loci	Proportion of Loci Polymorphic per Population	Heterozygosity per Locus	Standard Error of Heterozygosity
Homo sapiens	1	71	.28	.067	.018
Mus musculus musculus	4	41	.29	.091	.023
M. m. brevirostris	1	40	.30	.110	
M. m. domesticus	2	41	.20	.056	.022
Peromyscus polionotus	7 (regions)	32	.23	.057	.014
Drosophila pseudoobscura	10	24	.43	.128	.041
D. persimilis	1	24	.25	.106	.040
D. obscura	3 (regions)	30	.53	.108	.030
D. subobscura	6	31	.47	.076	.024
D. willistoni	2–21	28	.86	.184	.032
	10	20	.81	.175	.039
D. melanogaster	1	19	.42	.119	.037
D. simulans	1	18	.61	.160	.052
Limulus polyphemus	4	25	.25	.061	.024

Note: See Lewontin (1974) for individual references.

Source: R. C. Lewontin, *The Genetic Basis of Evolutionary Change* (New York: Columbia University Press, 1974), p. 17. Reprinted by permission of the publisher and the author.

Multilocus Selection Models.   Can the high degree of variability be accounted for by standard genetic models? If each locus is considered separately, then, for each polymorphic locus different young in a population must be lost in order to maintain that polymorphism by heterozygote advantage. The loss would soon outstrip the reproductive capacity of any species. Models proposed since Lewontin and Hubby's report have suggested that natural selection favors the individuals that are the most heterozygous overall. Individuals selected against because of their homozygosity would be the individuals with many homozygous loci. That is, natural selection acts on the entire genome, not on each locus separately. It can be shown algebraically that the large number of polymorphisms that exist in natural populations could be maintained by these models.

Neutral Alleles.   It may be that the high incidence of polymorphism revealed by electrophoresis is not important from an evolutionary point of view. If all or most electrophoretic alleles are neutral alleles (that is, if no allele is more fit than its alternative), there will be no selection at these loci and the variation observed in the population is merely a chance accumulation of mutations. This model is a clearcut alternative to the natural-selection model. That is, does evolution proceed, for the most part, by way of natural selection or by way of random processes? The resolution of the conflict will reveal a great deal about the process of evolution. Unfortunately, at this time no definitive data exist.

Which Hypothesis Is Correct?

Researchers favoring the concept that most electrophoretic alleles are neutral do not deny that selection exists. They do not hold that evolution is nonadaptive but say merely that most of the variation found in nature is nonadaptive. The distinction is important, and the demonstration that selection actually exists, in electrophoretic systems or otherwise, is not proof against the neutralist view. There are examples of selection for nonelectrophoretic loci—sickle-cell anemia, for one. Selection for electrophoretic systems is also known. For example, Koehn showed that different alleles of an esterase locus in a freshwater fish in Colorado had different enzyme activities at different water temperatures. Koehn then showed that the alleles were distributed as would be predicted on the basis of the temperature of the water. The cold-adapted enzyme (Es-I^b) was prevalent in the fish in colder waters (higher latitudes) and the warm-adapted enzyme (Es-I^a) was prevalent in the fish in warmer waters (lower latitudes; Figure 19–7). But isolated instances of selection are not adequate to prove the case for evolution by means of natural selection or disprove the case for neutral alleles and evolution by means of random mutational processes. Both theories recognize natural selection as the guiding force in producing adapted organisms. What

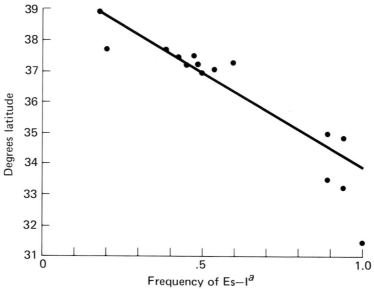

Figure 19 – 7.
Relation of Latitude
and Frequency of
the Warm-Adapted
Esterase Allele *Es-I*[a]
in Populations of the
Fish *Catostomus
clarki*

Source: Richard Koehn, "Functional and Evolutionary Dynamics of Polymorphic Esterases in Catostomid Fishes," *Transactions of the American
Fisheries Society* 99 (1970):223. Reprinted by permission.

is needed is proof that the majority of polymorphic loci are either being
selected or are neutral (and not being selected); for this proof, each locus
must be examined independently—a very difficult undertaking.

Also failing to adequately support either one or the other evolutionary
theory are data on the geographic distribution of alleles. Often, a single allele
predominates over the range of a species (Figure 19–8). Changes in allelic
frequency from one geographic area to another can often be attributed to
clinical selection (selection along a geographic gradient), where allelic frequency changes as altitude, latitude, or some other geographic parameter
changes. But, in line with the neutralist view, it can be shown that a geographic pattern similar to the pattern in Figure 19–8 can also be produced
by a very low level of migration, as little as 1 individual per 1000 per generation, even with neutral alleles.

What, then, is the true nature of the large amount of variability, provided by polymorphism? Is it selectively maintained or is it neutral and randomly distributed? The answer to this question will strongly affect the final
view of the evolutionary process. That the question is presently unanswera-
ble should not be cause for discouragement. Rather, this is an exciting time
to be interested in evolutionary genetics: There are major questions still left
to be answered and a growing armament of tools with which to seek the
answers.

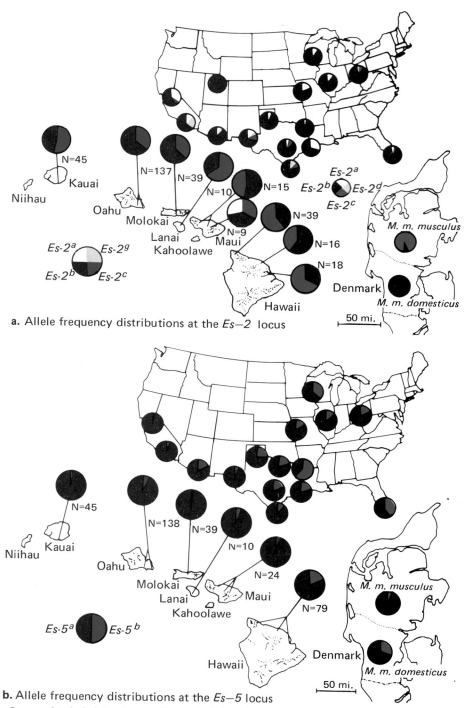

a. Allele frequency distributions at the *Es–2* locus

b. Allele frequency distributions at the *Es–5* locus

Figure 19 – 8.
Esterase Allele
Frequency
Distribution

Source: Linda L. Wheeler and Robert K. Selander, "Genetic Variation in Populations of the House Mouse, *Mus musculus,* in the Hawaiian Islands," *Studies in Genetics* VII (Austin: University of Texas Publications, 1972), pp. 275, 277. Reprinted by permission.

ECOLOGICAL GENETICS

The remainder of this chapter will be devoted to examples of evolution in natural populations. There are two sources for such examples: fossil record and extant populations. Some of the strongest evidence that evolution has in fact occurred comes from the fossil record. The study of fossil horses, for example, illustrates how much information the fossil record can provide about the past course of evolution.

The examples offered here, however, concern evolution as it occurs today. A careful study of extant populations, sometimes combined with a look at historical or fossil evidence, can give strong insight into the way that natural selection actually works. The studies can support, quantify, and enhance the theory developed in the previous two chapters. This chapter will examine how natural selection and other mechanisms have worked to produce organisms that are the most fit in their particular ecological niches.

Molecular Evolution of Cytochrome C

The advancing technology that made it possible to detect the sequence of amino acids in a protein also made it possible to discover by how much the proteins of various species differ. The similarities and differences of proteins from species to species reveal which species are closely related—and the degree of relatedness—so that a phylogenetic tree based on amino acid substitutions can be constructed and compared to the currently accepted phylogenetic tree based on morphological differences. Knowledge of the changes in amino acid sequencing can be used to estimate the rate of evolutionary change. That is, since the amino acid data show how many substitutional changes have occurred between two known groups of organisms and since the geologic record gives relatively firm dates for the divergence of certain groups, we can calculate molecular **evolutionary rates:** amino acid substitutions per million years.

Most studies concerned with the rate of amino acid substitutions have been done on hemoglobin, cytochrome c, and a class of proteins called fibrinoproteins. The way in which an amino acid sequence changes from species to species is shown in Figure 19–9. A phylogenetic tree constructed from

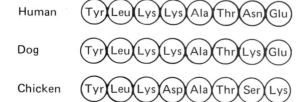

Figure 19 – 9. The Amino Acids Making Up the Terminal Portion of Cytochrome C in Three Species

Human   Tyr Leu Lys Lys Ala Thr Asn Glu

Dog   Tyr Leu Lys Lys Ala Thr Lys Glu

Chicken   Tyr Leu Lys Asp Ala Thr Ser Lys

combined information for the three types of proteins is shown in Figure 19–10. Table 19–3 gives some idea of the magnitude of protein changes between species. Figure 19–10 is in good agreement with the known divergence of the various species (as dete: .ined by radioactive dating of fossils). An interesting outcome of this analysis of amino acid substitutions is the discovery that not all amino acids are substituted at the same rate.

The evidence suggests that there are three classes of amino acids in terms of substitution rate: invariant, moderately variant, and hypervariant. It seems probable that there will be no amino acid substitutions in and around the active site of an enzyme, since any amino acid change in the area might be deleterious and thus lethal. For example, there is a segment of cytochrome c, from amino acids 70 to 80, that is invariant in all organisms tested. This area includes a binding site of the enzyme. Thus, analysis of amino acid sequencing and substitution in proteins such as cytochrome c indicates that the rate of evolution varies from part to part of an enzyme and that amino acid sequence data are useful for determining evolutionary history.

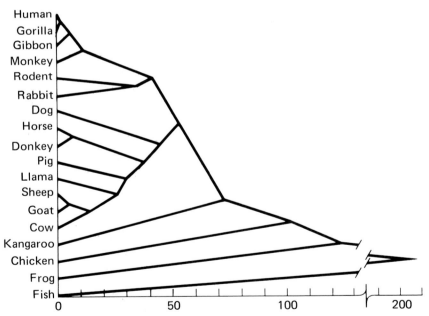

Figure 19 - 10. Composite Evolution of Hemoglobin, Cytochrome C, and Fibrinopeptide A
The total number of amino acid substitutions is given on the horizontal axis.

Source: C. H. Langley and W. M. Fitch, ''An Examination of the Constancy of the Rate of Molecular Evolution,'' *Journal of Molecular Evolution* 3 (1974):168. Reprinted by permission.

TABLE 19 – 3. Amino Acid Differences (%) in Cytochrome C between Different Organisms

	Human	Pig	Horse	Chicken	Turtle	Bullfrog	Tuna	Carp	Lamprey	Fruit fly	Screw-worm	Silkworm	Sesame	Sunflower	Wheat	C. krusei	Yeast	N. crassa	R. rubrum
Human	0	10	12	13	14	17	20	17	19	27	25	29	35	38	38	46	41	44	65
Pig, bovine, sheep		0	3	9	9	11	16	11	13	22	20	25	38	40	40	45	41	43	64
Horse			0	11	11	13	18	13	15	22	20	27	39	41	41	46	42	43	64
Chicken, turkey				0	8	11	16	14	17	23	21	26	40	41	41	45	41	44	64
Snapping turtle					0	10	17	13	18	22	22	26	38	39	41	47	44	45	64
Bullfrog						0	14	13	20	20	20	27	41	42	43	46	43	45	65
Tuna fish							0	8	18	23	22	30	42	43	44	43	43	45	65
Carp								0	12	21	20	25	40	41	42	45	42	43	64
Lamprey									0	27	26	30	44	44	46	50	45	47	66
Fruit fly										0	2	14	42	41	42	43	42	38	65
Screw-worm fly											0	13	41	40	40	43	42	38	64
Silkworm moth												0	39	40	40	40	44	44	65
Sesame													0	10	13	47	44	48	65
Sunflower														0	13	47	43	49	67
Wheat															0	45	42	48	66
Candida krusei																0	25	39	72
Baker's yeast																	0	38	69
Neurospora crassa																		0	69
Rhodospirillum rubrum																			0

Source: M. O. Dayhoff, ed. Taken with permission from the Atlas of Protein Sequence and Structure, vol. 5 (Washington, D.C.: National Biomedical Research Foundation, 1972), p. D–8.

Mimicry

Mimicry is a phenomenon whereby an individual of one species gains an advantage by resembling an individual of a different species. There are two types of mimicry.

In *Batesian mimicry,* named after H. W. Bates, a vulnerable organism (mimic) gains a selective advantage by looking like a dangerous or distasteful organism (model). The classical example of Batesian mimicry involves the Monarch and Viceroy butterflies (Figure 19–11). Although the Viceroy is smaller and, on close examination, looks quite different from the Monarch, the resemblance is striking at first glance. Monarch butterflies feed on milkweed plants and from them obtain noxious chemicals called cardiac glycosides, which the Monarchs store in their bodies. When a bird tries to eat a Monarch, it becomes sick and regurgitates what it has eaten. Thereafter, the bird will not only avoid eating Monarchs but also avoid eating any butterflies that look anything like Monarchs. The mimetic Viceroy butterflies thereby gain a selective advantage over their relatives and prosper. There are many examples of this phenomenon in the animal kingdom, especially in butterflies and other insects.

In *Müllerian mimicry,* named after F. Müller, several groups of organisms gain an advantage by looking like one another. This mimicry occurs

Figure 19 – 11.
Monarch Butterfly
(bottom) and One of
Its Mimics, a Viceroy
Butterfly (top)

Source: Edward S. Ross, figure 43–18, p. 818 in H. Curtis, *Biology,* 3rd
ed. (New York: Worth Publishing, 1979). Reproduced by permission.

among organisms where all the mimetic species are offensive or obnoxious.
The classical example is the general similarity of bees, wasps, and hornets.

Both forms of mimicry depend upon a selective pressure generated by
predation. Certain requirements must be met in order for each system to
work properly. Batesian mimicry has the following requirements:

1. The model species must be conspicuous and inedible or dangerous.
2. Both model and mimic species must occur in the same area, with the
   model being very abundant. If the model is rare, predators do not have
   sufficient opportunity to learn that its pattern is associated with a bad
   taste, and in fact, the reverse can happen: The model can be at a selective disadvantage if it is rare because predators will learn from the
   mimic that the pattern is associated with something good to eat.
3. The mimic should be very similar to the model in morphological characteristics easily perceived by predators, but not necessarily similar in
   other traits. The mimic is not evolving to be the model, only to look
   like it.

Müllerian mimicry requires that all the species be similar in appearance
and distinctively colored. They can, however, be equally numerous. And, as
the British geneticist, P. M. Sheppard pointed out, the resemblance among
Müllerian mimics need not be as good as between the mimic and model of

SOCIOBIOLOGY

In 1975 Edward Wilson published a mammoth tome entitled, *Sociobiology: The New Synthesis.* This book has been the center of a major controversy that is spreading to the fields of sociology, psychology, anthropology, ethology, and political science. The basic premise of the book is that behavior is under genetic control as are most other traits. While Wilson's book is primarily concerned with the animal kingdom, the controversy has arisen because of its extension to humans.

Prior to the mid-1960s, there existed the theory of **group selection,** which explained certain animal behaviors on the basis of the benefit that these behaviors had on the whole population or species. For example, under crowded conditions many birds cease reproducing. The interpretation of this phenomenon was that these birds were being altruistic: Their failure to breed was for the ultimate good of the species. In 1966 George Williams published a book *(Adaptation and Natural Selection: A Critique of Some Current Evolutionary Thought)* refuting the altruistic view with the charge that the individuals that performed altruistic acts would be selected against (natural selection would favor the cheaters): organisms not performing altruistic acts would have a higher fitness. Williams held that altruism had to be interpreted on the basis of benefits accruing to individuals; after his book, the idea of doing something for the good of the species became passé. How, then, can apparent altruism be accounted for?

W. D. Hamilton offered the concept of **inclusive fitness**—that is, that natural selection not only favors alleles that benefit an individual, but also will favor alleles that benefit close relatives of an individual because close relatives share many alleles in common. Thus, an act that might appear on the surface to be altruistic may in fact be benefiting an individual's inclusive fitness.

There are two other explanations for altruism that are consistent with benefits to individual fitness. One is that many altruistic acts are in reality selfish: They just look altruistic. The other involves Robert Triver's concept of **reciprocal altruism.** In species where individuals can recognize each other, and remember, an altruistic act may be done with the full expectation that the recipient will reciprocate in the future. Perhaps human friendship is best explained on the basis of reciprocal altruism. Hence, apparently altruistic acts can be explained as either selfish acts—acts that improve an individual's inclusive fitness—or acts of reciprocal altruism.

This turn-around in thought, from group selection to individual selection, has been an intellectual revolution in modern evolutionary biology. Prior to this revolution, many of the behaviors and phenomena in nature that involved apparent altruism were difficult to explain. Now, sociobiological reasoning provides the explanation.

The reason that so much controversy has sprung up over sociobiology is the implication the theory has for social, political, and legal issues. For example, human conflicts such as husband-wife, parent-child, and child-child conflicts may be built into the genes. Altruism, our highest nobility, has been exposed as mere selfishness. The alternate to the theory of sociobiology is the theory that most human behavioral acts are a result of cultural learning. At present, although much evidence remains to be gathered, the sociobiology concept is very attractive to many evolutionists.

Edward O. Wilson
(1929–   )
Photo by Pat Hill/
OMNI Publications
Int'l Ltd.

a Batesian pair because Müllerian mimics are not trying to deceive a predator, only to remind the predator of the relationship.

Although there have been some critics of mimicry theory, especially critics of the way in which the system could evolve, the general model put forward by the population geneticist and mathematician R. A. Fisher is generally accepted. According to Fisher, any new mutation that gave a mimic any slight advantage would be selected for. As time proceeded, other loci that might favorably modify expression of mimetic genes would also be selected for in order to improve the similarity of mimic and model. This mechanism surmounts the criticism that a single mutation could not produce a mimic that so closely resembled its model.

Industrial Melanism

Industrial melanism is the darkening (becoming melanic) of moths in England and Europe during the period of industrialization and is an interesting case study of natural selection: It illustrates the type of selection that can be caused by human intervention in natural systems; it is a spectacular example of very rapid evolution in nature; it shows the importance to natural populations of having a reserve of genetic variability. This phenomenon was first observed in England and gets its name from the increase in frequency of dark-color morphs of several species of moths concurrent with a change in the environment due to industrialization. Industrial melanism has occurred in more than 50 species of moths, of which the best studied is the peppered moth, *Biston betularia*. Until the middle of the nineteenth century, there was little industrialization in England, and the only known form of the peppered moth was the *typical* form, which is white with a peppering of black (Figure 19–12). A black form, known as *carbonaria,* was discovered in 1848 and, as industrial pollution increased, so did the frequency of carbonaria, until it now often makes up 95% of a population. The black pigmentation is controlled by a single dominant allele.

Prior to industrialization, trees were covered by lichens that had a pattern remarkably similar to the pattern of the typical form of the moths. (Actually, it is the other way around: The moths are remarkably similar in pattern to the lichens. The moths had the advantage that, in the lichen, they were hidden from predators. This advantage is called **cryptic coloration** and will be selected for. The lichens, however, gained no advantage by looking like the moths. Thus, the lichen pattern most probably came first and was later copied by the moths.)

The typical moths were well hidden from predators when on the lichen-covered tree bark. With the advent of industrialization, pollution killed the lichens and blackened the trunks of the trees; the typical moths were then obvious to predators, while the black carbonaria form had become nearly invisible. And so, the carbonaria form increased in frequency as a result of

a.   **b.**

Figure 19 – 12.   Peppered Moths on (a) Lichen-Covered Tree Trunk from Unpolluted Rural Area and (b) Soot-Covered Tree Trunk from Polluted Industrial Area
Each tree trunk has one typical and one melanic moth.

Source: Reprinted from Colin Patterson: *Evolution.* Copyright © Trustees of the British Museum (Natural History) 1978. Used by permission of the publisher, Cornell University Press.

predators selectively eating the typical form. H. B. D. Kettlewell, a British biologist, demonstrated that predation by birds was the selective force here. He released equal numbers of both forms of the moth and observed the feeding of birds on these moths as well as the recapture rates of the two forms. He found that in an industrial area 43 typical moths were taken by birds to 15 carbonaria; in a pollution-free area 164 carbonaria were taken to 26 typicals. In recent years industrial cities have lowered pollution levels and, as expected, the frequency of the carbonaria form has decreased.

CHAPTER SUMMARY

The classical theory of evolution was put forward by Charles Darwin. He saw as natural the variation among individuals within a population of similar organisms. He noted as well that offspring are overproduced in nature, and that this inevitably leads to competition for scarce resources. Darwin assumed that, when competition occurs, the most fit will survive and that, through time, a population would become better adapted to its environment through the process of natural selection—that is, survival of the fittest. Applying population-genetics algebra to this theory leads to the modern concept of evolution, neo-Darwinism.

Speciation occurs when gene flow in a population is blocked. Different populations of a species can then evolve independently. When the point is reached where individuals from the isolates can no longer interbreed, then speciation has taken place. If the isolates then comingle, they will remain as separate species.

A major area of study in modern evolutionary biology is the explanation of the amount of genetic variation that occurs in natural populations. In 1966 Lewontin and Hubby showed that there was a tremendous amount of heterozygosity in natural populations. Attempts to explain this variation have led to two major competing theories. One theory is that the variation is being maintained selectively. The other is that the variation observed by Lewontin and Hubby, most of it electrophoretic, is not under selective pressure but is instead neutral. Current evidence does not favor either hypothesis.

Three studies serve to demonstrate the action of selection in natural populations. The relative rates of amino acid substitutions through geologic time can be estimated by studying the molecular evolution of cytochrome c. Rates of amino acid substitutions aid in determining phylogenetic relationships and show that evolution acts at different speeds on different parts of a gene.

Mimicry occurs when organisms obtain an advantage by looking like an obnoxious or dangerous organism. Batesian mimics are harmless while Müllerian mimics are all noxious. Industrial melanism is the change in frequency of color morphs of various moth species due to industrial pollution. Originally, light moth forms were protected from predators by blending in with the lichen on tree trunks. When pollution killed off the lichens, dark moths were protected by blending in with the soot-covered bark. Complete changes in morph frequencies, from light to dark, have taken place in recent times and show very rapid evolution.

EXERCISES AND PROBLEMS

1. What would you look for in data to determine among frequency-dependent selection, heterozygote advantage, or transient polymorphism? How would you expect the data to agree or disagree in terms of (a) selection coefficients and (b) approaches to equilibrium.

2. What mechanisms permit the maintenance of genetic variability in natural populations? Give examples where possible.

3. Outline the Darwinian mechanism of the process of evolution. What is meant by neo-Darwinism?

4. Describe how the process of sympatric speciation could take place.

5. Discuss the "neutral gene hypothesis." What is its alternative?

6. Koehn showed that different functioning alleles of an esterase system were correlated to water temperature. What sorts of selection can you imagine to affect the same types of alleles in mammals, which are homeothermic (warm-blooded) and hence maintain a relatively constant internal temperature?

7. Outline the salient points of the cytochrome c, mimicry, and industrial melanism systems discussed in the chapter.

SUGGESTIONS FOR FURTHER READING

Bethell, T. 1976. Darwin's Mistake. *Harper's*, February, pp. 70–75.

Bush, G. 1975. Modes of animal speciation. *Ann. Rev. Ecol. Syst.* 6:339–364.

Cavalli-Sforza, L. 1974. The genetics of human populations. *Sci. Amer.* 231: 81–89.

Darwin, C. 1845. *The Voyage of the Beagle,* 1962 ed. Garden City, N.Y.: Doubleday.

Darwin, C. 1859. *The Origin of Species by Means of Natural Selection or the Preservation of Favored Races in the Struggle for Life.* London: John Murray.

Dayhoff, M., ed. 1972. *Atlas of Protein Sequence and Structure,* vol. 5. Washington, D.C.: National Biomedical Research Foundation.

Ford, E. 1964. *Ecological Genetics.* London: Methuen.

Koehn, R. 1970. Functional and evolutionary dynamics of polymorphic esterases in catostomid fishes. *Amer. Fish. Soc., Trans.* 99:219–228.

Lack, D. 1961. *Darwin's Finches.* New York: Harper & Row. (First published in 1947.)

Langley, C., and W. Fitch. 1974. An examination of the constancy of the rate of molecular evolution. *J. Mol. Evol.* 3:161–177.

Lewontin, R. 1974. *The Genetic Basis of Evolutionary Change.* New York: Columbia University Press.

Lewontin, R., and J. Hubby. 1966. A molecular approach to the study of genic heterozygosity in natural populations. II. *Genetics* 54:595–609.

Ramshaw, J., J. Coyne, and R. Lewontin, 1979. The sensitivity of gel electrophoresis as a detector of genetic variation. *Genetics* 93:1019–1037.

Stebbins, G. 1966. *Processes of Organic Evolution.* Englewood Cliffs, N.J.: Prentice-Hall.

Thoday, J., and J. Gibson. 1962. Isolation by disruptive selection. *Nature* 193:1164–1166.

Wheeler, L., and R. Selander. 1972. Genetic variation in populations of the house mouse, *Mus musculus,* in the Hawaiian Islands. *Univ. Texas, Studies in Genetics, No. VII,* pp. 269–296.

Wilson, E. 1975. *Sociobiology: The New Synthesis.* Cambridge, Mass.: Harvard University Press.

Brief Answers to Selected Exercises and Problems

Chapter 1. Mendel's Principles

1-1. Dwarf F_2 when selfed produce all dwarf offspring. Tall F_2 when selfed fall into two classes: one-third (homozygotes) produce only tall offspring, two-thirds (heterozygotes) produce tall and dwarf offspring in a $3:1$ ratio.

1-2. As an example, we can look at the F_2 with round, green seeds. One-third are *RRyy* homozygotes that produce only round, green offspring when testcrossed. Two-thirds are *Rryy* and produce equal numbers of round, green seeds and wrinkled, green seeds.

1-3. $119:32:9$ is very close to a $12:3:1$ ratio. Thus two loci probably control the trait.

1-4. 1/16 *RRtt*; 3/16 *RRT–*; 2/16 *Rrtt*; 6/16 *RrT–*; 1/16 *rrtt*; 3/16 *rrT–*.

1-5. 1/16 *ccTT*; 3/16 *C–TT*; 2/16 *ccTt*; 6/16 *C–Tt*; 1/16 *cctt*; 3/16 *C–tt*.

1-6. The proportion of 6/16 represents *A–bb* and *aaB–* genotypes.

1-7. All.

1-8. When it can be determined that the allele of paternal origin could not have come from the male in question. A type A woman cannot have a type AB child with a type A man.

1-9. The F_1 would produce 2^4, or 16, different gametes. The F_2 would have 2^4, or 16, different phenotypes. The F_2 would have 3^4, or 81, different genotypes. $1/(16)^2$, or 1/256, of the F_2 would be *aabbccdd*.

1-10. Wingless gene (wild-type allele): Wi^+ or $+$. Wingless mutant: Wi. (W is already the allele designation for wrinkled.)

1-12. See Figure 15–2.

1-13. y^+, rv^+, and tuf^+ are wild-type alleles of recessive mutants; Ax^+ and $M(2)e^+$ are wild-type alleles of dominant mutants; *dow* and *bur* are recessive mutant alleles; *Hw, Co,* and *J* are dominant mutant alleles.

1-14. The nature of DNA and the nature of the biochemical pathway that the protein functions in.

1-15. $12:3:1$; $12:4$.

Chapter 2.   Mitosis and Meiosis

2-2. Excluding crossovers, 2^3, or 8, different gametes can arise during meiosis. A crossover between the A locus and its centromere does not alter gametic combinations.

2-3.   **a.** Prophase of mitosis, $2n = 6$ or prophase of meiosis II, $2n = 12$.
   b. Metaphase of mitosis, $2n = 6$ or metaphase of meiosis II, $2n = 12$.
   c. Anaphase of mitosis, $2n = 6$ or anaphase of meiosis II, $2n = 12$.
   d. Prophase of meiosis I, $2n = 6$.
   e. Anaphase of meiosis I, $2n = 6$.
   f. Anaphase of meiosis II, $2n = 6$.

2-4.   **a.** 46; **b.** 0; **c.** 23; **d.** 23; **e.** 23.

2-6. All possible triploid combinations can result (e.g., $AAaBbb$). Since the endosperm results from the combination of two genotypically identical haploid female (polar) nuclei and one haploid male (sperm) nucleus, the endosperm genotype does tell the sporophyte genotype. For example, if the endosperm is $AAaBbb$, the female nuclei must have been Ab (because the A and b alleles appear twice each) and the male nucleus aB. Thus the sporophyte will be $AaBb$.

2-7.   40; 10.

2-8. In *Drosophila* and corn, maternal cytoplasm predominates. In *Neurospora* both parents (mating types) contribute cytoplasm.

2-9. $AaBb \times ab$. Four types of daughters and four types of sons (unfertilized eggs) result.

Chapter 3.   Probability and Statistics

3-1.   **a.** $Rr \times Rr$ produces $R-:rr$ in a 3:1 ratio. Chi-square $= 0.263$, $p > 0.05$.
   b. $Rr \times Rr$ produces $R-:rr$ in a 3:1 ratio. Chi-square $= 0.474$, $p > 0.05$.
   c. $RR:Rr$ are in a 2:1 ratio. Chi-square $= 0.175$, $p > 0.05$.
   d. $VvLl \times vvll$ produces a 1:1:1:1 ratio. Chi-square $= 1.084$, $p > 0.05$.

3-2.   **a.** $\dfrac{5!}{3!2!} (1/2)^3 (1/2)^2 = 0.3125$
   b. $(1/2)^5 = 0.03125$
   c. $2(1/2)^5 = 0.0625$
   d. $(1/2)^5 = 0.03125$
   e. $2(1/2)^5 = 0.0625$
   f. 4 daughters, 1 son plus 5 daughters: $\dfrac{5!}{4!1!} (1/2)^4(1/2) + (1/2)^5 = 0.1875$
   g. $(1/2)^2 = 0.25$

3–3. **a.** $1/4 = 0.25$

b. $1/8 = 0.125$

c. $\dfrac{6!}{2!2!2!} (1/8)^2(1/8)^2(3/8)^2 = 0.0030899$

d. $3/8 = 0.375$

3–4. (Remember that albinos have blue eyes.)

a. $(1/4)^5 = 0.0009765$

b. $(1/8)^5 = 0.0000305$

c. $\dfrac{5!}{4!1!} (7/32)^4(9/32) = 0.0032199$

d. $\dfrac{4!}{2!2!} (1/8)^2(1/8)^2 = 0.0014648$

3–5. **a.** $1/10,000 = 0.0001$

b. $1/10,000 = 0.0001$

c. $(1/10,000)^2 = 0.00000001$

3–6. **a.** $2(1/2)^4 = 0.125$

b. $(1/2)^4(1/2)^4 = 0.0039062$

3–7. Assume that Mendel was observing the phenotype of offspring produced from seeds obtained by testcrossing the plant with a dominant phenotype ($A- \times aa$). As soon as a plant with the recessive phenotype appears, he would be completely certain that the dominant parent was heterozygous. If he tested five seeds and all produced plants with the dominant phenotype, there would be a 3% chance that the plant were still heterozygous. Testing seven seeds would lower the chance of heterozygosity to less than 1%.

3–8. 1/2; one-half the families would stop at two children, one-half of the remainder would stop at three children, and so on, giving

$$\text{average family size} = \sum_{n=1}^{\infty} \left(\frac{1}{2^n} \right)(n + 1)$$

$$= \sum_{n=1}^{\infty} \left(\frac{n + 1}{2^n} \right)$$

$$= 3.0$$

Chapter 4.   Sex Determination, Sex Linkage, and Pedigree Analysis

4–1. In the F_2, all females are wild type while half the males are lozenge and half are wild type. In the F_2 of the reciprocal cross, half of each sex are lozenge and half are wild type.

4–2. If the parental female is pale colored, all the F_2 males are wild type while the females (XY) are pale and wild type in a 1:1 ratio. In the F_2 of the reciprocal cross, half of each sex are pale and half wild type.

4–3. The protein is probably a dimer, which, in the heterozygote, can be of fast–fast, fast–slow, or slow–slow subunits. A female heterozygous for a sex-linked dimeric enzyme should show the pattern of slot 3 in whole blood and slots 1 or 2 in individually cloned cells.

4-4. **a.** 0; human female, male fly.
b. 1; human female, female fly.
c. 0; human male, male fly.
d. 1; human male, female fly.
e. 2; human female, female fly.
f. 4; human female, female fly.
g. 1/0 mosaic; human male/female mosaic, male/female mosaic fly.

4-5. **a.** The phenotype has the propensity to have twin offspring. It could be recessive or dominant, sex-linked or autosomal.
b. Autosomal dominant or possibly autosomal recessive.
c. Autosomal recessive or sex-linked recessive.
d. Autosomal recessive.

4-7. Early.

4-9. Females: 3/4 wild type, 1/4 fuzzy. Males: 3/8 wild type, 3/8 cut, 1/8 fuzzy, 1/8 cut and fuzzy.

Chapter 5.   Linkage and Mapping in Eukaryotes

5-1. **a.** 1.1 map units apart.
b. Given the map units, gametes are produced on the average as follows: $gr +$ 49.45%, $+ ro$ 49.45%, $gr ro$ 0.55%, and $+$ 0.55%. Constructing the Punnett square, the phenotypes of the offspring would be as follows: wild type 49.99%, groucho and rough 0.003%, groucho 24.99%, and rough 24.99%.

5-2. The two loci are 16.0 map units apart on the X chromosome.

5-3. Use a dihybrid female testcrossed to a hemizygous male with both recessives. Five percent of each recombinant class should be found in each sex. The same results will be found for an autosomal locus; however, reciprocal crosses will give the same results for an autosomal locus but not for the sex-linked locus.

5-4. **a.** Yes.
b. Gene order has fl in the center. From ca to fl is 35.0 map units; fl to cu is 6.0 map units. The coefficient of coincidence is 10/21 = 0.48.

5-5. **a.** The albino locus is in the middle, 16.0 map units from either end locus.
b. The trihybrid was $b\ a\ c/+ + +$.
c. Interference is $1 - (20/26) = 0.23$.

5-6. **a.** Brittle and glossy are linked; red is assorting independently. The b to g map distance is 8.0.
b. The b and g alleles are in coupling phase.
c. Not relevant.

5-7. **a.** Work backward from the 0.61% double recombinants.
b. With a coefficient of coincidence of 0.60, only 0.366% double recombinants will occur. For example,

ancon, spiny, arctus oculus	421
wild type	422
ancon, spiny	28
arctus oculus	29
ancon	48
spiny, arctus oculus	48
ancon, arctus oculus	2
spiny	2

5–8. Notchy is known on the X chromosome by independent crosses that give nonreciprocal results. See also the answer to question number 5–3.

5–11. The measured map units are 1.1, 16.0, and 50. Haldane's function gives 1.1, 19.3, and an indeterminate number of map units. For Kosambi's function, the values are approximately 1.1, 17, and 270 map units.

5–12. PD, 1, 2, 4, 6, 8–10; NPD, 3; TT, 5, 7. The loci are 20 map units apart.

5–13. **a.** PD, 1, 2, 4, 6, 8, 10, 11; NPD, 9; TT, 3, 7; unscorable, 5, 12.
b. Yes.
c. 20 map units apart.

5–14. If we use *Neurospora* terminology, PD are first-division segregants, NPD are also first-division segregants, and TT would be scored as second-division segregants. Hence, map distance between *m* and the centromere would be 10 map units. We see that the yeast formula gives more accurate results because of NPD.

5–15. The distance between *a* and its centromere is 25 map units.

5–16. **a.** FDS, 1, 4–6, 8, 9, 11; SDS, 2, 7, 10; unscorable, 3, 12.
b. The distance between *f* and its centromere is 15 map units.

5–17. NPD + $\frac{1}{2}$TT = 12%

5–18. SDS = 24%

5–19. The best first-order estimate is that *a* and *b* are on the same side of the same centromere. *a* is 11 map units down the chromosome and *b* is 15 map units from *a* (Table 5–8).

5–20. For example: 1, 722; 2, 10; 3, 127; 4, 128; 5, 11; 6, 0; 7, 2.

5–21. **a.** chromosome 6; **b.** chromosome 1; **c.** chromosome 2; **d.** chromosome 7; **e.** chromosome 8 or above. These are assignment tests.

5–22. 20 map units apart.

Chapter 6. Linkage and Mapping in Prokaryotes and Viruses

6–1. Far. If near, both the Hfr and F$^-$ members of a conjugation event can be killed by the antibiotic.

6–4. *lac* to *gal*, 9 minutes; *gal* to *his*, 27 minutes; *his* to *argG*, 24 minutes; *argG* to *xyl*, 11 minutes; *xyl* to *ilv*, 4 minutes; *ilv* to *thr*, 17 minutes; *thr* to *lac*, 8 minutes.

6–5. Colony 1, *xyl$^-$ arg$^-$*; colony 2, *his$^-$ arg$^-$*; colony 3, +; colony 4, *arg$^-$*; colony 5, *ilv$^-$* and maybe *thr$^-$*; colony 6, *his$^-$* and maybe *gal$^-$*.

6–6. *trp his tyr*: *trp* to *his*, 0.34; *his* to *tyr*, 0.13.

6–7. *ara leu ilvH: ara leu,* 0.92; *leu ilvH,* 0.97.

6–9. The infecting phage has DNA slightly longer than one whole genome: there is terminal redundancy. Replication of the DNA, followed by recombination, produces very long molecules of phage DNA (numerous genomes per molecule). This DNA is then cut into pieces to fit the phage head. The result is a circular permutation of the linear molecule.

6–10. *m r tu: m r,* 12.8%; *r tu,* 20.8%; coefficient of coincidence, 1.21.

Chapter 7. Variation in Chromosome Structure and Number

7–1. All chromosomes form linear bivalents.

7–2. Yes.

7–3. No.

7–5. Diagram this to show that a crossover between a centromere and the center of the cross can change the consequences of the pattern of centromere separation.

7–6. Problems occur during meiosis.

7–7. 92; 45.

7–8. Usually allopolyploids: amphidiploids should have little or no meiotic problems.

7–9. 28.

7–10. XX, 0, and X (unaffected) eggs will result.

7–11. One normal chromosome; one normal inversion chromosome; one dicentric with duplications and deficiencies; one acentric with duplications and deficiencies.

7–12. Reciprocal translocation (some effects occur only in the heterozygous condition). Look for the cross-shaped figure at meiosis or in salivary gland chromosomes.

7–13. Inversion (some effects occur only in the heterozygous condition). Look for a loop at meiosis or in salivary gland chromosomes.

Chapter 8. Quantitative Inheritance

8–1. The y variable is follicles; the x variable is eggs laid. $\bar{x} = 38.4$, $\bar{y} = 37.5$, $\text{cov}(x,y) = 113.2$, $s_x^2 = 116.6$, $s_y^2 = 163.0$, $r = 0.82$, $b = 0.97$, $a = 0.3$.

8–2. Three loci. Each effective allele contributes about 1/2 lb over the 2 lb base.

8–3. Four loci. Each effective allele contributes about 1/4″ above 2″.

8–4. Independent assortment; regression to the mean.

8–7. Improved nutrition, medicine, and sanitation.

8–10. Yes; yes; if each effective allele adds color, then individuals with the base color (white) who marry each other presumably cannot have children with darker skin.

8–11. Begin by graphing the data. The general form of the data indicates that additive genes predominate. Excessive skew is caused by nonadditive effects. The number of loci involved is difficult to determine because the long-eared parent was so variable. More information is needed. (Probably 4 to 8 loci are involved.)

8-13. Thorax length: $V_D + V_I = 6$; $H_N = 0.43$; $H_B = 0.49$. Eggs laid: $V_D + V_I = 44$; $H_N = 0.18$; $H_B = 0.62$.

8-14. $H = 0.38$; H (high line) $= 0.2$; H (low line) $= 0.56$. Part of the difference may be due to the number of alleles available for selection in each direction and nonadditive factors.

8-15. $r = 0.36$; $H_N = 0.36/0.50 = 0.72$.

8-16. $H = $ final difference in means divided by the summed cumulative selection differentials $= (23 - 18)/(13 + 14) = 5/27 = 0.19$.

Chapter 9.   Chemistry of the Gene

9-2. The newly formed DNA needs to be cleaved at the replication fork by an endonuclease.

9-4. Yes. The code could be the number of tetranucleotide units (e.g., one unit is alanine, 2 units is arginine, and so on).

9-5. None other were ever really taken seriously.

9-10. T to G; T to T; A to C. If similar polarity, G to T; T to T; C to A. Randomness only.

Chapter 10.   Gene Expression: Transcription

10-1. By determining nucleotide sequence similarities among different species of organisms (Chapter 19).

10-2. Higher than in DNA replication.

10-3. Transcription start signal is promoter, recognized with the aid of sigma factor of RNA polymerase. Transcription stop signal is terminator, recognized with the aid of the rho factor. Polycistronic transcripts are routine in prokaryotes.

10-6. Random start of transcription; failure to properly terminate transcription.

Chapter 11.   Gene Expression: Translation

11-1. 5′AUGUUACCGGGAAAAUAG3′; anticodons: 3′UAC5′, 3′AAU5′, 3′GGC5′, 3′CCU5′, 3′UUU5′. Methionine, leucine, proline, glycine, lysine.

11-2. 5; $C = G$.

11-3. 12/27 phenylalanine; 6/27 serine; 6/27 leucine; 3/27 proline.

Chapter 12.   Control in Prokaryotes and Phages

12-1. **a.** Inducible; **b.–d.** constitutive; **e.** neither, superrepressed; **f.** inducible.

12-2. **a. i.** z^+; **ii.** no transcription.
  b. i. z^+, z^-; **ii.** no transcription.

c. i. z^+, z^-, **ii.** no transcription.
d. i. z^+, z^-, **ii.** z^-
e. i. z^+, z^-, **ii.** z^+

12–5. Repression; lytic response.

12–6. Assuming that the mutants produce inactive proteins: *cI,* lytic response; *cII,* lytic response; *N,* neither lytic nor lysogenic responses possible; *cro,* no lytic response possible; *att,* no lysogenic response possible; *Q,* no lytic response possible.

Chapter 13. Gene Expression: Control in Eukaryotes (Developmental Genetics)

13–9. To move V_2 next to C requires the same types of enzymes involved in looping out a prophage or plasmid. These could be an endonuclease and repair enzymes, including ligase. RNA polymerase then transcribes the gene. Removal of the mRNA gap could involve a similar looping out process.

13–11. Assuming that each cancer might be caused by a single locus and assuming that breast cancer appears only in women and prostate cancer appears only in men, colon and prostate cancer can be by any form of inheritance. Pancreatic cancer can only be by an autosomal recessive and breast cancer cannot be by anything but an X-linked dominant.

Chapter 14. Mutation, DNA Repair, and Recombination

14–1. For example, if the 20 individual cultures of Table 14–1 had values of 15, 13, 15, 20, 17, 14, 21, 19, 16, 13, 27, 14, 15, 26, 12, 21, 14, 17, 12, 14, then the mutation theory would not be supported.

14–9. For example, copy choice makes the prediction that recombinant chromatids will be made only of newly synthesized DNA. Phage lambda can be grown on *E. coli* with heavy nitrogen (^{15}N) for several generations and then switched to light nitrogen (^{14}N) for one generation. We predict only lambda with ^{14}N DNA will be recombinant. The lambda can be separated by density gradient centrifugation and tested for recombinant phenotypes. Light, heavy, and heterogeneous phage all show recombination.

Chapter 15. Gene Function and Structure

15–2. Probably a temperature-sensitive mutant of the lysing enzyme.

15–3. Enzymes with critical functions.

15–4. Mutant 3416 is blocked just before niacin. Mutant 4540 is blocked just before 3-hydroxy-anthranilic acid, which precedes niacin. Mutant

10575 is blocked between indole and tryptophan. Mutant 4008 is blocked before indole and will accumulate an unknown precursor. Mutant 10575 should accumulate indole, 4540 should accumulate kynurenine or tryptophan, and 3416 should accumulate 3-hydroxy-anthranilic acid.

Since serine is needed to convert tryptophan to kynurenine, a mutant in the serine pathway would act as a mutant in the niacin pathway at tryptophan.

15–5. The pathway is: precursor–D–B–C–A–E and the mutant order in the steps of the above pathway is 4, 2, 5, 3, 1. Presumably the substrate just before the blocked step is accumulated (mutant 5 accumulates B).

15–6. Three complementation groups are present (1, 6, 7; 2, 5; 3, 4). The half table missing is a mirror image because a cross of 1 and 3 is the same as a cross of 3 and 1 (reciprocity). The diagonal is always (–) because every mutant is a structural allele of itself.

15–7. There are six defined regions of deletion overlap and the first six mutants fall, from left to right, R, Q, N, M, P, O. S and T are deletions themselves; S overlaps the R to M regions and T the Q to O regions.

15–9. Prokaryotic and phage genes appear not to have intervening sequences.

Chapter 16.   Extrachromosomal Inheritance

16–2. Although the genetic scheme predicts shell coiling perfectly, one could do experiments involving the injection of cytoplasm into eggs to test the viral hypothesis.

16–3. It follows Mendelian ratios without exception.

16–6. A sinistral Dd female produces dextral offspring. Her father would have to be DD or Dd.

16–8. $Kk \times Kk$; of sensitive exconjugants, all will produce sensitive (kk) offspring by autogamy. Of killer exconjugants, one-third (KK) will produce only killer offspring while two-thirds (Kk) will segregate both killers (KK) and sensitives (kk) through autogamy.

Chapter 17.   Population Genetics: Equilibrium, Mutation, Migration, and Genetic Drift

17–1.   a. Assuming Hardy-Weinberg equilibrium, $f(a) = 0.3$, $f(A) = 0.7$, $f(aa) = 0.09$, $f(Aa) = 0.42$, $f(AA) = 0.49$.
   b. $f(FF) = 0.33$, $f(FS) = 0.47$, $f(SS) = 0.21$, $f(F) = 0.56$, $f(S) = 0.44$. Chi-square = 1.198. The population is in Hardy-Weinberg equilibrium.

17–2. Assuming Hardy-Weinberg equilibrium, $f(t) = 0.55$, $f(T) = 0.45$.

17–3. Assuming Hardy-Weinberg equilibrium, 0.4% of persons carry the Tay-Sachs allele.

17–4. Assuming Hardy-Weinberg equilibrium, $f(I^A) = 0.1$, $f(I^B) = 0.6$, $f(I^O) = 0.3$. Chi-square $= 0$. The population is in Hardy-Weinberg equilibrium.

17–6. **a.** The allele frequencies are reversed in the two sexes. If $p = f(M)$ and $q = f(N)$, in males $p = 0.6$ and $q = 0.4$ and in females $p = 0.4$ and $q = 0.6$. Within each sex, genotypes are distributed as p^2, $2pq$, and q^2. However, we have not discussed how to statistically test two distributions (males and females) for similarity. We can test each sex for a genotype distribution on the basis of an expectation of the mean allele frequency ($p = q = 0.5$). In this case the chi-square in males is 8.0 and in females is 8.64. Hence the population is not in Hardy-Weinberg equilibrium.

b. After one generation of random mating (disregarding sampling errors), $f(MM) = 0.24$, $f(MN) = 0.52$, and $f(NN) = 0.24$. Although theoretically not yet in Hardy-Weinberg equilibrium, statistically the population (assume 200 offspring) is in equilibrium (expected proportions of 0.25, 0.50, and 0.25, chi-square $= 0.32$).

c. After a second generation of random mating, chi-square is zero.

17–7. **a.** If $p = f(a)$ and $q = f(b)$, $\overline{p} = 0.47$ and $\overline{q} = 0.53$. By using these values to calculate expecteds, for males chi-square $= 43.7$ and for females chi-square $= 24.7$. Hence both sexes are out of Hardy-Weinberg equilibrium.

17–9. $\hat{q} = 0.95$.

17–10. $m = 0.29$. Data would be needed for the allele frequency of the ancestral American white population.

17–11. The ultimate fate of the A allele is either loss or fixation. The probability of loss is 0.3. 100 generations; 2000 generations.

17–12. **a.** No; **b.** more than half.

Chapter 18.   Population Genetics: Nonrandom Mating and Natural Selection

18–1. $F_I = (1/2)^6(1 + 1/8) + (1/2)^4 + (1/2)^3$
$= 0.0176 + 0.0625 + 0.125$
$= 0.2051$

18–2. $F_I = 4(1/2)^6(1.01) = 0.063$

18–3. (See Figure 18–3.) $F_E = (1/2)^3(1.2) + (1/2)^3 = 0.275$

18–5. Fitnesses could be 1, 1/2, and 1/4 for AA, Aa, and aa, respectively (or 1, $1 - s$, and $1 - 1.5s$, where $s = 0.5$).
a. $\overline{W} = 1 - 2pqs - 1.5q^2s$
b. $\hat{q} = 0$ (stable)
c. $L = 2pqs - 1.5q^2s = 0$ at equilibrium

18–6. Fitnesses of 1, $1 - s$, and 1 for AA, Aa, and aa. Delta $- q = spq(2q - 1)/\overline{W}$, where $\overline{W} = 1 - 2spq$. $\hat{q} = 0.5$, 0, or 1. The 0.5 equilibrium is unstable.

18–8.   C. Assume $s_2 = 2s_1$; then, $\hat{q} = 1/3$. If $s_1 = 0.1$ and $s_2 = 0.2$, then the percentage of heterozygotes at equilibrium will be $2pq(100) = 2(1/3)(2/3)(100) = 44.4\%$ and $L = (2/3)s_1 = 0.067$.

18–9.   With fitnesses of $1 - s$, 1, and 1 for *AA, Aa,* and *aa,* respectively, $p_{n+1} = p(1 - sp)/(1 - sp^2)$

18–10.   See formula 18.10. $\hat{q} = 0.018$; $\hat{p} = 0.982$. The genetic load is the mutation rate, 5×10^{-5}.

18–11.   0.5; balanced lethal.

18–12.   The relative changes in the three genotypes are $30/25 = 1.2$, $60/50 = 1.2$, and $10/25 = 0.4$. We divide each by 1.2 to get fitnesses of 1, 1, and 0.33. Thus, this is selection against the homozygous recessive where $s = 0.67$. $L = 1 - sq^2$, which at equilibrium ($\hat{q} = 0$) is zero.

Glossary

A priori probability. Probability determined by the nature or geometry of an event or a situation.

A (aminoacyl) site. The site on the ribosome occupied by an aminoacyl-tRNA just prior to peptide bond formation.

Acentric fragment. A chromosome piece without a centromere.

Acrocentric. A chromosome whose centromere lies very near one end.

Active site. The part of an enzyme where the actual enzymatic function is performed.

Adaptive value. *See* fitness.

Additive model. A mechanism of quantitative inheritance in which alleles at different loci either add a fixed amount to the phenotype or add nothing.

Adenine. *See* purines.

Adjacent–1 segregation. A separation of centromeres during meiosis in a reciprocal translocation heterozygote such that unbalanced gametes are produced.

Adjacent–2 segregation. Separation of centromeres during meiosis in a translocation heterozygote such that homologous centromeres are pulled to the same pole.

Affected. Individuals in a pedigree that exhibit the specific phenotype under study.

Allele. Alternative form of a gene.

Allopatric speciation. Speciation in which the evolution of reproductive-isolating mechanisms occurs during physical separation of the populations.

Allopolyploidy. Polyploidy produced by the hybridization of two species.

Allosteric protein. A protein whose shape is changed when it binds a particular molecule. In the new shape the protein's ability to react to a second molecule is altered.

Allotype. Mutant of the nonvariant parts of immunoglobulin genes that follows the rules of simple Mendelian inheritance.

Allozygosity. Homozygosity where the two alleles are alike but unrelated.

Alternate segregation. A separation of centromeres during meiosis in a reciprocal translocation heterozygote such that balanced gametes are produced.

Aminoacyl-tRNA synthetases. Enzymes that attach amino acids to their proper tRNAs.

Amphidiploid. An organism produced by hybridization of two species followed by somatic doubling. It is an allotetraploid that appears as a normal diploid.

Anaphase. The stage of mitosis and meiosis where sister chromatids or homologous centromeres are separated by spindle fibers.

Aneuploids. Individuals or cells exhibiting aneuploidy.

Aneuploidy. The condition of a cell or of an organism that has additions or deletions of whole chromosomes from the expected, balanced number of sets.

Angiosperms. Plants whose seeds are enclosed within an ovary. Flowering plants.

Anticodon. The complementary sequence to a codon.

Antigen. A foreign substance capable of triggering an immune response in an organism.

Antimutator mutations. Mutations of DNA polymerase that decrease the overall mutation rate of a cell or of an organism.

Antiparallel strands. Strands, as in DNA, that run in opposite directions.

Ascospores. Haploid spores found in the asci of Ascomycete fungi.

Ascus. The sac in Ascomycete fungi that holds the ascospores.

Assignment test. A test that determines whether a locus belongs on a specific chromosome by the observation of the concordance of the locus and the chromosome in hybrid cell lines.

Assortative mating. The mating of individuals with similar phenotypes.

Attenuator region. A control region at the promoter end of repressible amino acid operons that exerts transcriptional control dependent on the translation of a small leader peptide gene.

Attenuator stem. *See* terminator stem.

Autogamy. Nuclear reorganization in a single *Paramecium* cell similar to the changes that occur during conjugation.

Autopolyploidy. Polyploidy in which all the chromosomes come from the same species.

Autoradiography. A technique in which radioactive molecules make their location known by exposing photographic plates.

Autosomal set. A combination consisting of one chromosome from each homologous pair in a diploid species.

Autosomes. The nonsex chromosomes.

Autotrophs. Organisms that can utilize carbon dioxide as a carbon source.

Autozygosity. Homozygosity in which the two alleles are identical by descent.

Auxotrophs. Strains of organisms that have specific nutritional requirements.

Bacillus. A rod-shaped bacterium.

Back mutation. The process that causes reversion. A change in a nucleotide pair in a mutant gene that restores the original sequence and hence the original phenotype.

Backcross. The cross of an individual with one of its parents or an organism with the same genotype as a parent.

Bacterial lawn. A continuous cover of bacteria on the surface of the growth medium.

Bacteriophages. Bacterial viruses.

Balanced lethal system. An arrangement of recessive lethals that maintains a heterozygous chromosome arrangement. Homozygotes for any lethal-bearing chromosome perish.

Balbiani rings. The larger polytene chromosomal puffs. Generally synonymous with *puffs*. See chromosome puffs.

Barr body. Heterochromatic body found in the nuclei of normal females but absent in the nuclei of normal males.

Binary fission. Simple cell division in single-celled organisms.

Binomial expansion. The terms generated when a binomial raised to a particular power is multiplied out.

Binomial theorem. The theorem that gives the terms of the expansion of a binomial raised to a particular power.

Bivalents. Structures, formed during prophase of meiosis I, consisting of the synapsed homologous chromosomes.

Blastomeres. Cells making up the blastula.

Blastopore. The embryonic opening of the future gut.

Blastula. The first developmental stage of a developing embryo.

Bottleneck. A marked reduction in size of a population that potentially leads to genetic drift.

Breakage and reunion hypothesis. A model that suggests that breakage of homologous chromatids and their rejoining account for recombination.

Bubbles. Nucleic acid configuration relating to replication in eukaryotic chromosomes or the shape of heteroduplex DNA at the site of a deletion or insertion.

Buoyant density of DNA. A measure of the density or lightness of DNA determined by the equilibrium point reached by the DNA in a density gradient.

Cancer. An informal term for a diverse class of diseases marked by abnormal cell proliferation.

Cancer-family syndromes. Pedigree patterns in which unusually large numbers of blood relatives develop certain kinds of cancers.

Cap. A sequence of methyl groups added to the 5′ end of eukaryotic mRNA.

Catabolite activator protein (CAP). A protein that when bound with cyclic AMP can bind to sites on sugar-metabolizing operons to enhance transcription of these operons.

Catabolite repression. Repression of certain sugar-metabolizing operons in favor of glucose utilization when glucose is present in the environment of the cell.

Catalyst. A substance that increases the rate of a chemical reaction without itself being permanently changed.

Cell cycle. The cycle of cell growth, replication of the genetic material, and nuclear and cytoplasmic division.

Centimorgan. One map unit.

Central dogma. The original postulate that information can only be transferred from DNA to DNA, from DNA to RNA, and from RNA to protein.

Centric fragment. A chromosome piece with a centromere.

Centrioles. Cylindrical organelles, found in eukaryotes (except in higher plants), that organize the formation of the spindle.

Centromere markers. Loci located near their centromeres.

Centromeres. Constrictions in eukaryotic chromosomes in which the kinetochores lie.

Centromeric fission. Creation of two chromosomes from one by splitting the centromere.

Chargaff's rule. Chargaff's discovery that, in the base composition of DNA, the quantity of adenine equaled the quantity of thymine and the quantity of guanine equaled the quantity of cytosine (equal purine to pyrimidine content).

Charon phages. Phage lambda derivatives used as vehicles in recombinant DNA work.

Chiasmata. X-shaped configurations seen in tetrads during the latter stages of prophase I of meiosis. They represent physical crossovers. (Singular: *chiasma*.)

Chimaeric plasmid. Hybrid, or genetically mixed, plasmids used in recombinant DNA work.

Chimeras. *See* mosaics.

Chi-square distribution. The sampling distribution of the chi-square statistic. A family of curves depending on degrees of freedom.

Chloroplast. The organelle that carries out photosynthesis and starch grain formation in plants.

Chromatids. Two identical units (sister chromatids) held together at the centromere that, at prophase of nuclear divisions, make up each chromosome. When the centromeres divide and the chromatids separate, each chromatid is then a chromosome.

Chromatin. The nucleoprotein material of the eukaryotic chromosome.

Chromomeres. Dark regions in eukaryotic chromosomes at meiosis or mitosis.

Chromosomal theory of inheritance. The theory that chromosomes are linear sequences of genes.

Chromosome. The form of the genetic material in viruses and cells. A circle of DNA in prokaryotes; a DNA or an RNA molecule in viruses; a linear nucleoprotein complex in eukaryotes.

Chromosome puffs. Diffuse, uncoiled regions in polytene chromosomes where transcription is actively taking place.

Cis. Meaning "on the near side of" and referring to geometric configurations of atoms, or mutants usually on the same chromosome.

Cis-trans complementation test. A mating test to determine whether two mutants on opposite chromosomes will complement each other. A test for allelism.

Cistron. Term coined by Benzer for the smallest unit that exhibits the *cis-trans* position effect. Synonymous with *gene* or *locus*.

ClB method. A technique devised by Muller to rapidly screen fruit flies for recessive X chromosome lethals. The *ClB* chromosome carries a recessive lethal (*l*), a dominant marker (*B*) and an inversion (crossover suppressor, *C*).

Clone. A group of cells arising from a single ancestor.

Coccus. A spherical bacterium.

Codominance. The situation in which the phenotypes independently produced by each allele as a homozygote are visible together as the phenotype in the heterozygote.

Codons. The sequences of three RNA nucleotides that specify either an amino acid or termination of translation.

Coefficient of coincidence. The percentage of observed double crossovers divided by the percentage expected.

Colicinogenic factors. *See* col plasmids.

Col plasmids. Plasmids that produce antibiotics (colicinogens) used by the host to kill other strains of bacteria.

Common ancestry. The state of two individuals when they are blood relatives. When two parents have common ancestry, their offspring will be inbred.

Competence factor. A surface protein that binds extracellular DNA and enables the cell to be transformed.

Complementarity. The correspondence of DNA bases in the double helix such that adenine in one strand is opposite thymine in the other strand and cytosine in one strand is opposite guanine in the other. This explains Chargaff's rule.

Complementation. The production of the wild-type phenotype by a cell or an organism that contains two mutant genes. If complementation occurs, the mutants are non-allelic.

Complete linkage. The state in which two loci are so close together that alleles of these loci are virtually never separated by crossing over.

Complete medium. A medium that is enriched to contain all of the growth requirements of a strain of organisms.

Component of fitness. A particular variable in the life cycle of an organism upon which selection acts.

Concordance. The amount of similarity in phenotype among individuals.

Conditional-lethal mutant. A mutant that is lethal under one condition but not lethal under another condition.

Confidence limits. A statistical term for a pair of numbers that predict, with a particular probability level, the region in which a particular parameter lies.

Conjugation. A process whereby two cells come in contact and exchange genetic material. In prokaryotes the transfer is a one-way process.

Consanguineous. Mating between blood relatives.

Conservative replication. A postulated mode of DNA replication where an intact double helix would act as a template for a new double helix.

Constitutive heterochromatin. Heterochromatin that surrounds the centromere. *See* satellite DNA.

Constitutive mutant. A mutant that is no longer under regulatory control but instead produces a fixed quantity of gene product.

Continuous replication. In DNA, uninterrupted replication allowed in the 5′ to 3′ direction by a 3′ to 5′ template.

Continuous variation. Variation measured on a continuum or distribution rather than in discrete categories (e.g., height in humans).

Copy-choice hypothesis. An incorrect hypothesis that stated that recombination resulted from the switching of the DNA replicating enzyme from one homologue to the other.

Corepressor. The metabolite that when bound to the repressor (of a repressible operon) forms a functional unit that can bind to its operator and block transcription.

Correlation coefficient. A statistic that gives a measure of how closely two variables are related.

Cotransduction. The simultaneous transduction of two or more genes.

Cot values ($cot_{1/2}$). The product of C_0, the original concentration of denatured, single-stranded DNA and t, time in seconds, giving a useful index of renaturation. $Cot_{1/2}$ values are the midpoint values on cot curves—cot values plotted against concentration of remaining single-stranded DNA—and estimate the length of unique DNA in the sample.

Coupling. Allele arrangement in which mutants are on the same chromosome and wild-type alleles on the homologue.

Covariance. A statistical value measuring the simultaneous deviations of x and y variables from their means.

Criss-cross pattern of inheritance. The phenotypic pattern of inheritance shown by traits controlled by X-linked recessive alleles in a diploid XY species.

Critical chi-square. A chi-square for a given degree of freedom and probability level to which an experimental chi-square is to be compared.

Crossbreed. Fertilization between separate individuals.

Cross-fertilization. *See* crossbreed.

Crossing over. A process in which homologous chromosomes exchange parts by a breakage-and-reunion process.

Crossovers. *See* chiasmata.

Cryptic coloration. Coloration that allows an organism to match its background and hence become less vulnerable to predation.

Cyclic AMP. A form of AMP used frequently as a second messenger in eukaryotic hormone nets and in catabolite repression in prokaryotes.

Cytokinesis. The division of a cell into two daughter cells.

Cytoplasmic inheritance. Extra-chromosomal inheritance controlled by nonnuclear genomes.

Cytosine. *See* pyrimidines.

Dauermodification. The persistance for several generations of an environmentally induced trait.

Degenerate code. A code in which several code words have the same meaning. The genetic code is degenerate because there are many instances in which different codons specify the same amino acid.

Degrees of freedom. An estimate of the number of independent categories in a particular statistical test or experiment.

Deletion chromosome. A chromosome with part deleted.

Denatured. Loss of natural configuration (of a molecule) through heat or other treatment. Denatured DNA is single stranded.

Density-gradient centrifugation. A method of separating molecular moieties dependent upon their differential sedimentation in a centrifugal gradient.

Depauperate fauna. A fauna, especially common on islands, lacking many species found in similar habitats.

Derepressed. The condition of an operon that is transcribing because repressor control has been lifted.

Development. The process of orderly change that an individual goes through in the formation of structure.

Diakinesis. The final stage of prophase I of meiosis when chiasmata terminalize.

Dicentric chromosome. A chromosome with two centromeres.

Dihybrid. An organism heterozygous at two loci.

Dimerization. The chemical union of two identical molecules.

Diploid. The state of having each chromosome in two copies per nucleus or cell.

Diplotene. The stage of prophase I of meiosis in which chromatids appear to repel each other.

Directional selection. A type of selection that removes individuals from one end of a phenotypic distribution and thus causes a shift in the distribution.

Disassortative mating. The mating of individuals with dissimilar phenotypes.

Discontinuous replication. In DNA, only interrupted replication allowed backward in 5′ to 3′ segments by a 5′ to 3′ template strand.

Discontinuous variation. Variation that falls into discrete categories.

Discrete generations. Generations that have no overlapping reproduction. All reproduction takes place between individuals in the same generation.

Dispersive replication. A postulated mode of DNA replication combining aspects of conservative and semiconservative replication.

Disruptive selection. A type of selection that removes individuals from the center of a phenotypic distribution and thus causes the distribution to become bimodal.

DNA–DNA hybridization. A technique in which, when DNA from the same or different sources is heated and then cooled, double-helix configurations will reform at homologous regions. This technique is useful for determining sequence similarities and degrees of repetitiveness among DNAs.

DNA ligase. An enzyme that closes nicks or discontinuities in one strand of double-stranded DNA by creating an ester bond between adjacent 3′OH and a 5′PO$_4$ ends on the same strand.

DNA polymerase. One of several classes of enzymes that polymerize DNA nucleotides by using single-stranded DNA as a template and that require a double-helical primer.

DNA–RNA hybridization. A technique in which, when a mixture of DNA and RNA is heated and then cooled, RNA can hybridize (form a double helix) with DNA that has a complementary nucleotide sequence.

Dominant. A trait that expresses itself even when heterozygous.

Dosage compensation. The mechanism used in species with sex chromosomes to ensure that one sex does not suffer due to the different number of sex-linked alleles in the two sexes.

Double helix. The structure of DNA that is made of two helixes rotating about the same axis.

Double reduction. The condition in polyploids in which a heterozygous individual produces homozygous gametes.

Doublesex. An allele that converts fruit fly males and females into developmental intersexes.

Dyad. A centromere with two chromatids attached.

Electrophoresis. The separation of molecular moieties by using electric current.

Elongation factors (EF-T$_s$, EF-T$_u$, EF-G). Proteins necessary for the proper elongation and translocation processes during translation at the ribosome in prokaryotes.

Empirical probability. Probability determined by observing a large number of relevant cases.

Endogenote. Bacterial host chromosome.

Endomitosis. Chromosomal replication without nuclear or cellular division that results in cells with many copies of each chromosome.

Endonucleases. Enzymes that make nicks internally in the backbone of a polynucleotide. They hydrolyze internal phosphodiester bonds.

Enriched medium. *See* complete medium.

Enzyme.   Protein catalyst.

Episomes.   Genetic particles that can either exist independently in a cell or can become integrated into the host chromosome.

Epistasis.   The masking of the action of alleles of one gene by allele combinations of another gene.

Equational division.   The second meiotic division that is equational because it does not reduce chromosome numbers.

Euchromatin.   Regions of eukaryotic chromosomes that are highly diffuse during interphase. Presumably the actively transcribing DNA of the chromosomes.

Eukaryotes.   Organisms with true nuclei.

Euploidy.   The condition of a cell or organism that has one or more complete sets of chromosomes.

Evolution.   In Darwinian terms, a gradual change in phenotypic frequencies that results in a population of individuals better adapted to survive.

Evolutionary rates.   The rate of divergence between taxa, measurable as amino acid substitutions per million years.

Excision repair.   A process whereby cells repair certain kinds of mutations by the removal of the mutated DNA strand and replacement using the good strand as a template.

Exconjugant.   Each of the two cells that separates after conjugation has taken place.

Exogenote.   DNA that a bacterial cell has incorporated through one of its sexual processes.

Exon.   A region of a gene that has intervening sequences (introns) and that is actually translated.

Exonucleases.   Enzymes that digest nucleotides from the ends of polynucleotide molecules. They hydrolyze terminal phosphodiester bonds.

Experimental design.   A branch of statistics that attempts to outline the way in which experiments should be carried out so that the data gathered will have statistical value.

Exploitation competition.   A form of competition that revolves around the superior ability to gather resources, rather than an active interaction among organisms for resources.

Expressivity.   The degree of expression of a genetically controlled trait.

Eyes.   Referring to the configuration of replicating DNA in eukaryotic chromosomes.

F_1.   *See* filial generation.

F-duction.   *See* sexduction.

F factor.   *See* fertility factor.

F-pili.   Sex pili. Hair-like projections of an F^+ or Hfr bacterium involved in anchorage during conjugation and presumably through which DNA passes.

Factorial.   The product of all integers from the specified number down to 1.

Fecundity selection.   The forces acting to cause one genotype to be more fertile than another genotype.

Feedback inhibition.   A posttranslational control mechanism in which the end product of an enzymatic pathway inhibits the activity of the first enzyme of this pathway.

Fertility factor.   The plasmid that allows a prokaryote to engage in conjugation with and pass DNA into an F^- cell.

Filial generation.   Offspring generation. F_1 is the first offspring, or filial, generation; F_2 is the second; and so on.

Fimbriae.   Fringed. Referring to the surface of bacteria with pili (hair-like projections).

First-division segregation (FDS).   The allele arrangement in spores of Ascomycetes with ordered spores that indicates no recombination between a locus and its centromere.

Fitness, *w*.   The relative reproductive success of a genotype as measured by survival, fecundity, or other life history parameters.

Fluctuation test.   An experiment by Luria and Delbrück that compared the variance in number of mutations between small cultures and subsamples of a large culture to determine the mechanism of inherited change in bacteria.

Fokker–Planck equation. An equation that describes diffusion processes and that is used by population geneticists to describe random genetic drift.

Founder effect. Genetic drift observed in a population founded by a small, nonrepresentative sample of a larger population.

Frameshift. A shift in the codon reading frame. *See* frameshift mutation.

Frameshift mutation. An addition or deletion of nucleotides that causes the codon reading frame to shift.

Frequency-dependent selection. Selection whereby a genotype is at an advantage when rare and at a disadvantage when common.

Functional alleles. Mutants that fail to complement each other in a *cis-trans* complementation test.

Fundamental number. The number of chromosome arms in a somatic cell of a particular species.

G-bands. Eukaryotic chromosomal bands produced by treatment with Giemsa stain.

β-galactosidase. The enzyme that splits lactose into glucose and galactose (coded by a gene in the *lac* operon).

β-galactoside permease. An enzyme involved in concentrating lactose in the cell (coded by a gene in the *lac* operon).

Gametic selection. The forces acting to cause differential reproductive success of one allele over another in a heterozygote.

Gametophyte. The stage of a plant life cycle that produces gametes (by mitosis). Alternates with a diploid, sporophyte generation.

Gene. Inherited determinant of the phenotype. *See* cistron, locus.

Gene conversion. In Ascomycete fungi, where a 2:2 ratio of alleles is expected after meiosis and a 3:1 ratio is sometimes observed. The mechanism of this gene conversion is explained by the Holliday model of recombination.

Gene pool. All of the alleles available among the reproductive members of a population from which gametes can be drawn.

Genetic code. The linear sequences of nucleotides that specify the amino acids during the process of translation at the ribosome.

Genetic fine structure. The structure of the gene in relation to the number and size of the smallest units of recombination and mutation.

Genetic load, L. The relative decrease in the mean fitness of a population due to the presence of genotypes that have less than the highest fitness.

Genetic polymorphism. The occurrence together in the same population of more than one allele at the same locus, with the least frequent allele occurring more frequently than can be accounted for by mutation.

Genic balance theory. The theory of Bridges that stated that the sex of a fruit fly is determined by the relative number of X chromosomes and autosomes.

Genome. The genetic complement of a prokaryote or virus. A haploid cell or gamete of a eukaryotic species.

Genotype. The genes that an organism possesses.

Germ-line theory. A theory to account for the high degree of antibody variability. The germ-line theory suggests that every B lymphocyte has all the genes for every type of immunoglobulin but only transcribes one. *See* somatic-mutation theory.

Giemsa stain. A complex of stains specific for the phosphate groups of DNA.

Gray crescent. A cortical region of the egg of frogs and some salamanders that forms just after fertilization on the side opposite sperm penetration.

Group selection. Selection for traits that would be beneficial to a population at the expense of the individual possessing the trait.

Guanine.   *See* purines.

Gynandromorphs.   Mosaic individuals having simultaneous aspects of both the male and female phenotype.

H–Y antigen.   The Histocompatibility Y-antigen, a protein found on the cell surfaces of male mammals.

Haploid.   The state of having one copy of each chromosome per nucleus or cell.

Hemizygous.   The condition of loci on the X chromosome of the heterogametic sex of a diploid species.

Heritability.   A measure of the degree to which the variance in the distribution of a phenotype is due to genetic causes.

Heterochromatin.   Chromatin that remains tightly coiled (and darkly staining) throughout the cell cycle.

Heteroduplex analysis.   Analysis in which, if double-helix DNA is formed by strands from different sources, loops and bubbles identify regions where the two DNAs differ. This heterogeneous DNA is referred to as a *heteroduplex*. Electron microscopic observation of this DNA is a useful tool in recombinant DNA work.

Heteroduplex DNA.   *See* hybrid DNA.

Heterogametic.   The sex with heteromorphic sex chromosomes and which therefore during meiosis produces different kinds of gametes in regard to the sex chromosomes.

Heterogeneous nuclear mRNA (hnRNA).   The original RNA transcripts found in eukaryotic nuclei prior to posttranscriptional modifications.

Heterokaryon.   A cell that contains two or more nuclei from different origins.

Heteromorphic chromosomes.   Chromosomes of which the members of a homologous pair are not morphologically identical (e.g., the sex chromosomes).

Heterotrophs.   Organisms that require an organic form of carbon as a carbon source.

Heterozygote.   A diploid or polyploid with different alleles at a particular locus.

Heterozygote advantage.   A selection model in which heterozygotes have the highest fitness.

Heterozygous DNA.   *See* hybrid DNA.

Hfr.   High frequency of recombination. A strain of bacteria that has incorporated an F factor into its chromosome and can then transfer the chromosome during conjugation.

Histones.   Arginine- and lysine-rich basic proteins making up a substantial portion of eukaryotic nucleoprotein.

Holoenzyme.   The complete enzyme. Usually refers to RNA polymerase when indicating the core enzyme plus the sigma factor.

Homogametic.   The sex with homomorphic sex chromosomes and which therefore only produces one kind of gamete in regard to the sex chromosomes.

Homologous chromosomes.   Members of a pair of essentially identical chromosomes that synapse during meiosis.

Homomorphic chromosomes.   Morphologically identical members of a homologous pair of chromosomes.

Homozygote.   A diploid or polyploid with identical alleles at a locus.

Hormone net.   The integration and control of the various interacting hormones in an individual.

Hormones.   Chemicals that are secreted by one type of cell and act on a second type of cell.

Hybrid DNA.   DNA whose two strands have different origins.

Hybrid.   Offspring of unlike parents.

Hybrid screening.   Radioisotope technique used to determine whether a hybrid plasmid contains a particular gene or DNA region.

Hypostatic gene.   A gene whose expression is masked by an epistatic gene.

Hypotheses, testing of.   Statistical methods for determining the probability that a data set fits a particular hypothesis about it.

Identity by descent. The state of two alleles when they are identical copies of the same ancestral allele (autozygous).

Idiogram. A photograph or diagram of the chromosomes of a cell arranged in an orderly fashion.

Idiotypic variation. Variation in the variable parts of immunoglobulin genes.

Immunity. The ability of an organism to resist infection.

Immunoglobulins. Specific proteins produced by derivatives of B lymphocytes that protect an organism from antigens.

Inbreeding. The mating of genetically related individuals.

Inbreeding coefficient, F. The probability of autozygosity.

Inbreeding depression. A depression of vigor or yield due to inbreeding.

Incestuous. A mating between blood relatives who are more closely related than the law of the land allows.

Inclusive fitness. The expansion of the concept of the fitness of a genotype to include benefits accrued to relatives of an individual, since relatives share parts of their genomes. Hence an altruistic act towards a relative may in fact be a selfish act that enhances the fitness of the genotype performing the act.

Incomplete dominance. The situation in which both alleles of the heterozygote influence the phenotype.

Independent assortment, rule of. Mendel's second rule describing the independent segregation of alleles of different loci.

Inducible system. A system, in which a coordinated group of enzymes is involved in a catabolic pathway, is inducible if the metabolite upon which it works causes transcription of the genes controlling these enzymes. These systems are primarily prokaryotic operons.

Induction. Regarding temperate phage, the process of causing a prophage to become virulent.

Industrial melanism. The darkening of moths during the recent period of industrialization in many countries.

Initiation codon. The mRNA sequence AUG, which specifies methionine, the first amino acid used in the translation process.

Initiation complex. The initiation complex of translation consisting of the 30S ribosome subunit, mRNA, N-formyl methionine tRNA, and three initiation factors.

Initiation factors (IF1, IF2, IF3). Proteins required for the proper initiation of translation.

Insertion sequences (IS). Regions of homology between host chromosomes and plasmids that allow the latter to synapse with the former and become inserted into the host chromosome by a crossover.

Inside marker. The middle locus of three linked loci.

Intercalary heterochromatin. Heterochromatin, other than centromeric heterochromatin, dispersed through eukaryotic chromosomes.

Interference competition. A form of competition that involves a fight or other active interaction among organisms.

Intergenic suppression. A mutation at a second locus that apparently restores the wild-type phenotype to a mutant at a first locus.

Interphase. The metabolically active, nondividing stage of the cell cycle.

Interrupted mating. A mapping technique that disrupts bacterial conjugation after specified time intervals.

Intersex. An organism with external sexual characteristics that have attributes of both sexes.

Intervening sequences. Sequences of DNA within a gene that are transcribed but later removed prior to translation. *See* intron.

Intragenic suppression. A second change within a mutant gene that results in an apparent restoration of the original phenotype.

Intron. A length of DNA that makes up an intervening sequence.

Inversion. The replacement of an internal section of a chromosome in the reverse orientation.

In vitro. Biological or chemical work done in the test tube (literally, "in glass") rather than in living systems.

Iojap. A locus in corn that produces variegation.

Ionizing radiation. Radiation, such as X rays, that causes atoms to release electrons and become ions.

Isochromosome. A chromosome with two genetically and morphologically identical arms.

Kappa particles. The bacteria-like particles that give a *Paramecium* the "killer" phenotype.

Karyotype. The chromosome complement of a cell.

Kinetochores. The chromosomal attachment points for the spindle fibers, located within the centromeres.

***lac* operon.** The inducible operon including three loci involved in the uptake and breakdown of lactose.

Lampbrush chromosomes. Chromosomes of amphibian oocytes having loops suggestive of a lampbrush.

Leader. The length of mRNA from the 5′ end to the initiation codon (AUG).

Leader peptide gene. A small gene within the attenuator control region of repressible amino acid operons. Translation of the gene tests the content of the amino acid whose operon is being regulated.

Leader transcript. The mRNA transcribed by the attenuator region of repressible amino acid operons. The transcript is capable of several alternate stem-loop structures dependent on the translation of a short leader peptide gene.

Leptotene. The first stage of prophase I of meiosis where chromosomes become distinct.

Level of significance. The probability value used to separate agreement or disagreement with the null hypothesis.

Linkage. The association of genes to chromosomes and the association of different loci on the same chromosome.

Linkage disequilibrium. The condition among alleles at different loci such that any allelic combination in a gamete does not occur according to the product rule of probability.

Linkage equilibrium. The condition among alleles at different loci such that any allelic combination in a gamete occurs as the product of the frequencies of each allele at its own locus.

Linkage groups. Associations of loci on the same chromosome. In a species there are as many linkage groups as there are homologous pairs of chromosomes.

Locus. The position of a gene on a chromosome. Used synonymously with *gene*. (Plural: *loci*.)

Lyon hypothesis. The hypothesis that suggests that the Barr body is an inactivated X chromosome.

Lysate. The contents released from a lysed cell.

Lysis. The breaking open of a cell by the destruction of its wall or membrane.

Lysogenic. The state of a bacterial cell that has an integrated phage in its chromosome.

Map unit. One percent recombination between two loci.

Mapping. The study of the position of genes on chromosomes.

Mapping function. The mathematical relationship between measured map distance and actual recombination frequency.

Marker. A locus whose phenotype provides

information about a chromosome or chromosome segment during genetic analysis.

Mate-killer. A phenotype of *Paramecium* induced by intracellular bacteria-like mu particles.

Maternal inheritance. Extrachromosomal inheritance controlled by non-DNA cytoplasmic substances.

Mean. The arithmetic mean, or the sum of the data values divided by the sample size.

Mean fitness of the population, $\overline{w}$. The sum of the fitnesses of the genotypes of a population weighted by their proportions; hence, a weighted mean fitness.

Meiosis. The nuclear process that results, in diploid eukaryotes, in gametes or spores with only one member of each original homologous pair of chromosomes per nucleus.

Merozygote. A partially diploid bacterial cell arising from one of the sexual processes.

Metacentric chromosome. A chromosome with a centrally located centromere.

Metafemale. A fruit fly with an X/A ratio greater than 1.0.

Metagon. An RNA necessary for the maintenance of mu particles in *Paramecium.*

Metamale. A fruit fly with an X/A ratio below 0.5.

Metaphase. The stage of mitosis or meiosis in which spindle fibers are attached to kinetochores and the chromosomes are positioned in the center of the cell.

Metaphase plate. The plane of the equator of the spindle into which chromosomes are manipulated at metaphase.

Metrical variation. *See* continuous variation.

Microtubules. Hollow cylinders made of the protein tubulin and making up, among other things, the spindle.

Mimicry. A phenomenon in which an individual of one species gains an advantage by looking like individuals of a different species.

Minimal medium. A culture medium for microorganisms that contains the minimal necessities for growth of the wild type.

Missense mutations. Mutations that change a codon for an amino acid to a codon for a different amino acid.

Mitochondrion. The eukaryotic cellular organelle in which the Krebs cycle and electron transport reactions take place.

Mitosis. The nuclear division producing two daughter nuclei identical to the original nucleus.

Mitotic apparatus. *See* spindle.

Monohybrids. Offspring of parents that differ in only one characteristic. Usually implies heterozygosity at a single locus under study.

Monosomic. A diploid cell missing a single chromosome.

Morphological species concept. *See* type-species concept.

Mosaicism. The condition of being a mosaic. *See* mosaics.

Mosaics. Individuals made up of two or more different cell lines.

mRNA. Messenger RNA. The basic function of the nucleotide sequence of mRNA is to determine the amino acid sequence in proteins.

Mu particles. Bacteria-like particles found in the cytoplasm of *Paramecium* that cause the mate-killer phenotype.

Multihybrid. An organism heterozygous at numerous loci.

Multinomial expansion. The terms generated when a multinomial raised to a particular power is multiplied out.

Mutants. Alternate alleles to the wild type. The phenotypes produced by alternate alleles.

Mutation. The process by which a gene or chromosome changes structurally and the end result of this process.

Mutational load. Genetic load, caused by mutation, that brings deleterious alleles into a population.

Mutation rate. The proportion of mutants per cell division in bacteria or single-celled organisms or the proportion of mutants per gamete in higher organisms.

Mutator mutations. Mutations of DNA polymerase that increase the overall mutation rate of a cell or of an organism.

Muton. A term coined by Benzer for the smallest mutable site within a cistron.

Natural selection. A process whereby one genotype leaves more offspring than another genotype.

Nearest-neighbor analysis. A technique of transferring radioactive atoms between adjacent nucleotides in DNA that demonstrated that the two strands of DNA run in opposite directions.

Negative interference. The phenomenon whereby a crossover in a particular region enhances the occurrence of other apparent crossovers in the same region of the chromosome.

Neo-Darwinism. The merger of classical Darwinian evolution with population genetics.

Neutral gene hypothesis. The hypothesis that suggests that most genetic variation in natural populations is not maintained by selection.

NF. *See* fundamental number.

Nickase. An enzyme that nicks one strand of double-stranded DNA during DNA replication presumably to allow torsion to be released.

Nondisjunction. The failure of a pair of homologous chromosomes to separate properly during meiosis.

Nonhistone proteins. The proteins remaining in chromatin after the histones are removed. The scaffold structure is made of nonhistone proteins.

Nonparental ditype (NPD). A spore arrangement in Ascomycetes that indicates a four-strand, double crossover between two linked loci.

Nonparentals. *See* recombinants.

Nonrecombinants. In mapping studies, offspring that have alleles arranged as in the original parents.

Nonsense codon. One of the mRNA sequences (UAA, UAG, UGA) that signals the termination of translation.

Nonsense mutations. Mutations that change a codon for an amino acid to a nonsense codon.

Normal distributions. Any of a family of bell-shaped curves defined on the basis of the mean and standard deviation.

Nuclear transplantation. The technique of placing a nucleus from one source into an enucleated cell.

Nuclease. One of several classes of enzymes that degrade nucleic acid. *See* endonucleases and exonucleases.

Nucleolus. The globular, nuclear organelle formed at the nucleolus organizer.

Nucleolus organizer. The chromosomal location of the ribosomal RNA genes around which the nucleolus forms.

Nucleoprotein. The substance of eukaryotic chromosomes consisting of proteins and nucleic acids.

Nucleosomes. Arrangements of DNA and histones forming regular spherical structures in eukaryotic chromatin.

Nucleotide. Subunits that polymerize into nucleic acids (DNA or RNA). Each nucleotide consists of a nitrogenous base, a sugar, and one or more phosphates.

Null hypothesis. The statistical hypothesis that states that there are no differences between observed and expected data.

Nullisomic. A diploid cell missing both copies of the same chromosome.

Nutritional-requirement mutants. *See* auxotrophs.

Okazaki fragments. Segments of newly replicated DNA produced during discontinuous DNA replication.

Oogenesis. The process of ovum formation in female animals.

Oogonia. Cells in females that produce primary oocytes by mitosis.

Operator. A DNA sequence that is recognized by a repressor protein or repressor-corepressor complex. When the operator is complexed with the repressor, transcription is prevented.

Operon. A sequence of adjacent genes all under the transcriptional control of the same operator.

Outbreeding. The mating of genetically unrelated individuals.

Outside marker. Loci on either side of another locus or specified region.

Ovum. Egg. The one functional product of each meiosis in female animals.

P (peptidyl) site. The site on the ribosome occupied by the peptidyl-tRNA just prior to peptide bond formation.

P_1. Parental generation.

Pachytene. The stage of prophase I of meiosis where chromatids are first distinctly visible.

Paracentric inversion. An inversion that does not include the centromere.

Paramecin. A toxin liberated by "killer" *Paramecium*.

Parameters. Measurements of attributes of a population; denoted by Greek letters.

Parapatric speciation. Speciation in which the evolution of reproductive isolating mechanisms occurs when a population enters a new niche or habitat within the range of the parent species.

Parental ditype (PD). A spore arrangement in Ascomycetes that indicates no recombination between two linked loci.

Parentals. *See* nonrecombinants.

Partial dominance. *See* incomplete dominance.

Pascal's triangle. A triangular array made up of the coefficients of the binomial expansion.

Passenger DNA. DNA incorporated into a plasmid to form a hybrid plasmid.

Path diagram. A modified pedigree showing only the direct line of descent from common ancestors.

Pedigree. A representation of the ancestry of an individual or family. A family tree.

Penetrance. The normal appearance in the phenotype of genetically controlled traits.

Peptidyl transferase. The enzyme responsible for peptide bond formation during translation at the ribosome.

Pericentric inversion. An inversion that includes the centromere.

Permissive temperature. A temperature at which temperature-sensitive mutants are normal.

Petite mutations. Mutations of yeast that produce small, anaerobic-like colonies.

Phages. *See* bacteriophages.

Phenocopy. A phenotype that is not genetically controlled but that looks like a genetically controlled phenotype.

Phenotype. The observable attributes of an organism.

Phosphodiester bond. Diester bond linking nucleotides together (between phosphoric acid and sugars) to form the nucleotide polymers DNA and RNA.

Photoreactivation. The process whereby dimerized pyrimidines (usually thymine dimers) are restored by an enzyme requiring light energy (deoxyribodipyrimidine photolyase).

Pili (fimbriae). Hair-like projections on the surface of bacteria (Latin for "hair").

Plaques. Clear area on a bacterial lawn caused by cell lysis due to viral attack.

Plasmid. A genetic particle that can exist independently in a cell's cytoplasm without the ability to integrate into the host chromosome.

Plastid. A chloroplast prior to the development of chlorophyll.

Point mutations. Mutations that are single changes in the nucleotide sequence and that consist of a replacement, addition, or deletion of a base pair.

Poky mutations. Mutations in *Neurospora* that produce a petite phenotype.

Polar bodies. The small cells (that eventually disintegrate) that are the by-products of meiosis in female animals. One functional ovum and three polar bodies result from meiosis of each primary oocyte.

Polarity. Referring either to an effect seen in only one direction from a point of origin or to the fact that linear moieties (such as a single strand of DNA) have ends that differ from each other. Polarity means directionality.

Polarity gene. A gene in mitochondrial DNA with alleles that are preferentially found in daughter mitochondria after recombination between mitochondria.

Pollen grain. The male gametophyte in higher plants.

Poly-A-tail. A sequence of adenosine nucleotides added to the 3′ end of eukaryotic mRNAs.

Polygenic inheritance. *See* quantitative inheritance.

Polynucleotide phosphorylase. An enzyme that can polymerize diphosphate nucleotides without the need for a primer. The in vivo function is probably in its reverse role, as an RNA exonuclease.

Polyploids. Organisms with whole chromosome sets greater than two.

Polysome. The configuration of several ribosomes simultaneously translating the same mRNA. Polyribosome.

Polytene chromosome. Large chromosome consisting of many chromatids formed by rounds of endomitosis followed by synapsis.

Position effect. An alteration of phenotype caused by the relative arrangement of the genetic material.

Positive interference. When the occurrence of one crossover reduces the probability that a second crossover will occur in the same region.

Postreplicative repair. A DNA repair system initiated when DNA polymerase bypasses a damaged area. Uses enzymes in the *rec* system.

Posttranscriptional modifications. The changes in eukaryotic mRNA made after transcription has been completed. These changes include additions of caps and tails and removal of introns.

Preemptor stem. A configuration of leader transcript that does not terminate transcription in attenuator-controlled amino acid operons.

Pribnow box. Relatively invariant sequence of 7 nucleotides in DNA that signal the start of transcription.

Primary oocytes. The cells that undergo the first meiotic division in female animals.

Primary spermatocytes. The cells that undergo meiosis in males.

Primary structure. Of a protein, the sequence of its polymerized amino acids.

Primer. In DNA replication, a length of double-stranded DNA that continues as a single-stranded template leaving a 3′-OH end.

Probability. The expectation of the occurrence of a particular event.

Probability theory. The conceptual framework concerned with quantification of probabilities. *See* probability.

Proband. *See* propositus.

Product rule. The rule that states that the probability of the occurrence of independent events is the product of their separate probabilities.

Progeny testing. Breeding of offspring to determine their, and their parents', genotypes.

Prokaryotes. Organisms that lack true nuclei.

Promoter. A region of DNA that signals the initiation of transcription to RNA polymerase.

Proofread. Technically, to read for the purpose of detecting errors for later correction. DNA polymerase has 3′ to 5′ exonuclease activity, which it uses during polymerization to remove nucleotides it has recently added. This is a correcting ability to remove errors in replication and is referred to as proof-reading.

Prophage. A temperate phage integrated into a host chromosome.

Prophase. The initial stage of mitosis or meiosis in which chromosomes become visible and the spindle apparatus forms.

Proplastid. Mutant plastids that do not grow and develop into chloroplasts.

Propositus (proposita). The person through whom a particular pedigree was discovered.

Prototrophs. Strains of organisms that can survive on the minimal medium.

Pseudoalleles. Alleles that are functionally but not structurally allelic.

Pseudodominance. The phenomenon in which a recessive allele shows itself in the phenotype when only one copy of the allele is present as in hemizygous alleles or alleles opposite deletions.

Punnett square. A diagrammatic representation of a particular cross used to determine the progeny of this cross.

Purines. Nitrogenous bases of which guanine and adenine are found in DNA and RNA.

Pyrimidines. Nitrogenous bases of which thymine is found in DNA, uracil in RNA, and cytosine in both.

Quantitative inheritance. The mechanism of genetic control of continuous variation.

Quantitative variation. *See* continuous variation.

Quaternary structure. Of a protein, the association of polypeptide subunits to form the final protein.

R factors. *See* R plasmids.

R plasmids. Plasmids that carry genes that control resistance to various drugs.

RAM mutants. Referring to ribosomal ambiguity (RAM). Ribosomal mutants that allow incorrect tRNAs to be incorporated into the translation process.

Random genetic drift. Changes in allelic frequency due to sampling error.

Random mating. The mating of individuals in a population such that the union of individuals with the trait under study occurs according to the product rule of probability.

Random strand analysis. Mapping studies in organisms that do not retain all the products of meiosis in a recoverable form.

Realized heritability. Heritability determined by response to selection.

Recessive. A trait that does not express itself in the heterozygous condition.

Reciprocal altruism. An apparently altruistic behavior done with the understanding that the receiver will reciprocate at some future date.

Reciprocal cross. Testing of the role of parental sex on a phenotype by repeating a particular cross with the phenotype of each sex reversed as compared to the original cross.

Reciprocal translocation. A chromosomal configuration in which the ends of two nonhomologous chromosomes are broken off and become attached to the nonhomologues.

Reciprocity. In relation to recombination, the conservation of the total amount of genetic material while allowing changes in the arrangement of alleles.

Recombinants. In mapping studies, offspring with allelic arrangements made up of combinations of the original parental arrangements.

Recombination. The nonparental arrangement of alleles in progeny that can result from either independent assortment or crossing over.

Recon. A term coined by Benzer for the smallest recombinable unit within a cistron.

***Rec* system.** Several loci controlling genes (*recA, recB, recC,* and others) involved in postreplicative DNA repair.

Reductional division. The first meiotic division that is reductional because it reduces the number of chromosomes and centromeres to half the original per daughter cell.

Regression to the mean. A phenomenon of polygenic traits in which the offspring of extremes tend toward the population mean.

Regulator gene. A gene primarily involved in control of the production of another gene's product.

Relative Darwinian fitness. *See* fitness.

Relaxed mutant. A mutant that does not exhibit the stringent response under amino acid starvation.

Release factors (RF-1, RF-2, RF-3). Proteins responsible for proper termination of translation and release of the newly synthesized polypeptide when a nonsense codon appears in the A site of the ribosome.

Renner complexes. Specific gametic chromosome combinations in *Oenothera*.

Repetitive DNA. DNA containing copies of the same nucleotide sequence.

Replica plating. A technique to rapidly transfer microorganism colonies to numerous petri plates with different media.

Replication. The process of copying.

Replicons. A replicating genetic unit including the site for the initiation of replication.

Repressible system. A system in which a coordinated group of enzymes is involved in a synthetic pathway (anabolic) if excess quantities of the end product of the pathway lead to the termination of transcription of the genes for the enzymes. These systems are primarily prokaryotic operons.

Repressor. The protein product of a regulator gene that acts to control transcription of inducible and repressible operons.

Reproductive isolating mechanisms. Environmental, behavioral, mechanical, and physiological barriers that prevent two individuals of different populations from producing viable progeny.

Reproductive success. The unit of natural selection that is measured as the relative production of offspring by a particular genotype.

Repulsion. Allele arrangement in which each homologue has mutant and wild-type alleles.

Resistance transfer factor. A plasmid that confers on its host the simultaneous resistance to several antibiotics.

Restriction endonucleases. Endonucleases that recognize certain DNA sequences and cleave that DNA. Thought to protect cells from viral infection; useful in recombinant DNA work.

Restrictive temperature. A temperature at which temperature-sensitive mutants display the mutant phenotype.

Reverse transcriptase. An enzyme that can synthesize single-stranded DNA by using RNA as a template.

Reversion. The return of a mutant to the wild type through the process of a second mutational event.

Rho. A protein that is involved in the termination of transcription and release of the transcript at the terminator sequence.

Ribosomes. Organelles at which translation takes place. Made up of two subunits consisting of RNA and proteins.

RNA phages. Phages whose genetic material is RNA. They are the simplest phages known.

RNA polymerase. The enzyme that polymerizes RNA by using DNA as a template. (Also known as *transcriptase* or *RNA transcriptase*.)

RNA replicase. A polymerase enzyme that catalyzes the self-replication of single-stranded RNA.

Robertsonian fusion. Fusion of two acrocentric chromosomes at the centromere.

Rolling circle replication. A model of DNA replication that accounts for a circular DNA molecule producing linear daughter double helixes.

rRNA.   Ribosomal RNA. RNA components of the subunits of the ribosomes.

Sampling distribution.   The distribution of frequencies with which various possible events could occur in a particular experiment or defined by a particular algebraic expression.

Satellite DNA.   Highly repetitive eukaryotic DNA primarily located around the centromeres. Satellite DNA usually has a different buoyant density than the rest of the cell's DNA.

Scaffold.   The eukaryotic chromosome structure remaining when DNA and histones have been removed; made from nonhistone proteins.

Screening technique.   A technique to determine the genotype or phenotype of an organism.

Secondary oocytes.   The cells formed by meiosis I in female animals.

Secondary spermatocytes.   The products of the first meiotic division in male animals and which undergo the second meiotic division.

Secondary structure.   Of a protein, the flat or helical configuration of the polypeptide backbone.

Second-division segregation (SDS).   The allele arrangement in spores of Ascomycetes with ordered spores that indicates a crossover between a locus and its centromere.

Segregation, rule of.   Mendel's first principle describing how genes are passed from one generation to the next.

Segregational load.   Genetic load caused when a population is segregating less fit homozygotes under heterozygote advantage.

Selection.   *See* natural selection.

Selection coefficients, *s, t*.   The sum of forces acting to prevent reproductive success of a genotype.

Selection-mutation equilibrium.   An equilibrium allele frequency resulting from the balance between selection removing an allele and mutation recreating this allele.

Selective medium.   A medium that is enriched with a particular substance to allow the growth of particular strains of organisms.

Selfed.   *See* self-fertilization.

Self-fertilization.   Fertilization in which the two gametes are from the same individual.

Semiconservative replication.   The mode by which DNA replicates. *See* template.

Semisterility.   Nonviability of a proportion of progeny.

Sex chromosomes.   Heteromorphic chromosomes whose distribution in a zygote determines the sex of the organism.

Sex-controlled traits.   Traits that appear more often in one sex than another but are neither sex linked, sex limited, or sex influenced.

Sexduction.   A process whereby a bacterium gains access to and incorporates foreign DNA brought in by a modified F factor during conjugation.

Sex-influenced traits.   Traits controlled by alleles that show a different dominance-recessiveness relationship depending on the sex of the heterozygote.

Sex-limited genes.   Autosomal genes whose phenotypes are expressed in only one sex.

Sex linked.   The inheritance pattern of loci located on the sex chromosomes (usually the X chromosome in XY species). Also refers to the loci themselves.

Sex-ratio phenotype.   A trait in *Drosophila* where females produce mostly, if not only, daughters.

Sexual selection.   The forces acting to cause one genotype to mate more frequently than another genotype.

Siblings (sibs).   Brothers and sisters.

Sigma factor.   The protein that gives promoter-recognition specificity to the RNA polymerase core enzyme.

Sister chromatids.   *See* chromatids.

Skew. A distortion of the shape of the normal distribution toward one side or the other.

Somatic doubling. A disruption of the mitotic process that produces a cell with twice the normal chromosome number.

Somatic-mutation theory. A theory to account for the high degree of antibody variability. The somatic-mutation theory suggests that mutation of a basic immunoglobulin gene accounts for all of the different types of immunoglobulins produced by B lymphocytes. *See* germ-line theory.

Spacer DNA. Regions of nontranscribed DNA between transcribed segments, as in the numerous spacer regions in the nucleolus organizer.

Speciation. A process whereby, over time, one species evolves into a different species or where one species diverges to become two or more species.

Species. A group of organisms belonging to the same species because they are capable of interbreeding to produce fertile offspring.

Sperm. The gametes of male animals.

Spermatids. The four products of meiosis in male animals. Spermatids develop into sperm.

Spermatogenesis. The process of sperm production.

Spermatogonium. A cell type in the testes of male vertebrates that gives rise to primary spermatocytes by mitosis.

Spindle. The microtubule apparatus that controls chromosome movement during mitosis and meiosis.

Spiral cleavage. The cleavage process in mollusks and some invertebrates where the spindle at mitosis is tipped in relation to the original egg axis.

Spirillum. A spiral bacterium.

Sporophyte. The stage of a plant life cycle that produces spores by meiosis and alternates with the gametophyte stage.

Stabilizing selection. A type of selection that removes individuals from both ends of a phenotype distribution and thus maintains the same mean of the distribution.

Standard deviation. The square root of the variance.

Standard error of the mean. The standard deviation divided by the square root of the sample size. It is the standard deviation of a sample of means.

Statistics. Measurements of attributes of a sample from a population; denoted by Roman letters.

Stem-loop structures. Structures formed when nucleic acid loops back on itself to form complementary double helixes (stems) topped by the loops. Lollipop-shaped structures.

Stochastic. A process with an indeterminate or random element as compared to a deterministic process that has no random element.

Stringent factor. A protein that catalyzes the formation of two unusual nucleotides during the stringent response under amino acid starvation.

Stringent response. A translational control mechanism of prokaryotes that represses tRNA and rRNA synthesis during amino acid starvation.

Structural allele. Mutant alleles that have changes at identical base pairs.

Structural genes. Nonregulatory genes.

Submetacentric chromosome. A chromosome whose centromere lies between the middle and the end, but closer to the middle.

Subtelocentric chromosome. A chromosome whose centromere lies between the middle and the end, but closer to the end.

Sum rule. The rule that states that the probability of the occurrence of one of several of a group of mutually exclusive events is the sum of the probabilities of the individual events.

Supergenes. Close physical association of several loci that usually control related aspects of the phenotype.

Suppressor gene. A gene that, when mutated, apparently restores the wild-type phenotype to a mutant of another locus.

Survival of the fittest. In evolutionary theory, survival of only those organisms best able to obtain and utilize resources (fittest). This phenomenon is the cornerstone of Darwin's theory.

Svedberg unit. A unit of sedimentation during centrifugation. Abbreviated as S, as in 50S.

Swivalase. *See* nickase.

Sympatric speciation. Speciation in which the evolution of reproductive isolating mechanisms occurs within the range and habitat of the parent species. This speciation is common in parasites.

Synapsis. The point-by-point pairing of homologous chromosomes during zygotene or in certain dipteran tissues prior to endomitosis.

Synaptinemal complex. A proteinaceous complex that mediates synapsis during zygotene and breaks down shortly thereafter.

Synteny test. A test that determines whether two loci belong to the same linkage group by observing concordance in hybrid cell lines.

Synthetic medium. A chemically defined substrate upon which microorganisms are grown.

Target theory. A theory that predicts response curves based on the number of events required to cause the phenomenon. Used to determine that point mutations are single events.

Tautomeric shift. Reversible shifts in proton position in a molecule. Bases in nucleic acids shift between keto and enol forms or between amino and imino forms.

Telocentric chromosome. A chromosome whose centromere lies at one end.

Telophase. The terminal stage of mitosis or meiosis in which chromosomes uncoil, the spindle breaks down, and cytokinesis usually occurs.

Temperate phage. A phage that can enter into lysogeny with its host.

Temperature-sensitive mutant. Mutants that are normal at a permissive temperature, but mutant at a restrictive temperature.

Template. A pattern serving as a mechanical guide. In DNA replication, each strand acts as a template for the synthesis of a new double helix.

Terminator sequence. A sequence in DNA that signals to RNA polymerase the termination of transcription.

Terminator stem. A configuration of leader transcript that signals transcription termination in attenuator-controlled amino acid operons.

Tertiary structure. Of a protein, the further folding beyond the secondary structure as well as the formation of disulfide bridges between cysteines.

Testcross. The cross of an organism with a homozygous recessive organism.

Testing of hypotheses. The determination of whether to accept or reject a proposed hypothesis based on the likelihood that the hypothesis is correct.

Tetrads. The configuration made of four chromatids first seen in pachytene. There is one tetrad—bivalent—per homologous pair of chromosomes.

Tetranucleotide hypothesis. Hypothesis, based on incorrect information, that DNA could not be the genetic material because its structure was too simple—repeating subunits containing one copy each of the four DNA nucleotides.

Tetraploids. Organisms with four whole sets of chromosomes.

Tetratype (TT). A spore arrangement in Ascomycetes that indicates a single crossover between two linked loci.

Theta structure. An intermediate structure formed during the replication of a circular DNA molecule.

β-thiogalactoside acetyltransferase. An enzyme that is involved in lactose metabolism and encoded by a gene in the *lac* operon.

Three-point cross. A cross involving three loci.

Thymine. *See* pyrimidines.

Trailer. The length of mRNA from the nonsense codon to the 3′ end (or, in prokaryotes, from a nonsense codon to the next initiation codon).

Trans. Meaning "across" and referring to geometric configurations of atoms or mutants usually on different homologous chromosomes.

Transcription. The process whereby RNA is synthesized from a DNA template.

Transduction. A process whereby a cell can gain access to and incorporate foreign DNA. The new DNA is brought in by a viral particle.

Transfer operon (tra). Sequence of loci that impart the male (F-pili producing) phenotype on a bacterium. The cell can then transfer its genes to another bacterium.

Transformation. A process whereby prokaryotes take up DNA from the environment and incorporate it into their genomes.

Transformer. An allele in fruit flies that converts chromosomal females into sterile males.

Transition mutation. A mutation in which a purine/pyrimidine base pair is replaced by a base pair in the same purine/pyrimidine relationship.

Translation. The process of protein synthesis wherein the primary structure of proteins is determined by the nucleotide sequence in RNA.

Translocase (EF-G). Elongation factor necessary for proper translocation at the ribosome during the translation process.

Translocation. A chromosomal configuration in which part of a chromosome becomes attached to a different chromosome.

Transversion. A mutation in which a purine replaces a pyrimidine or vice versa.

Trihybrid. An organism heterozygous at three loci.

Triploids. Organisms with three whole sets of chromosomes.

tRNA. Transfer RNA. Small RNA molecules that transfer amino acids to the ribosome for polymerization.

True heritability. *See* heritability.

Two-point cross. A cross involving two loci.

Two-strand double crossovers. Double crossovers that occur in only two of the four chromatids of a tetrad.

Type I error. In statistics, the rejecting of a true hypothesis.

Type II error. In statistics, the accepting of a false hypothesis.

Type-species concept. The concept that organisms that are morphologically similar belong to the same species.

Uninemic chromosome. A chromosome consisting of one double helix of DNA.

Unique DNA. A length of DNA with no repetitive nucleotide sequences.

Unusual bases. Other bases, in addition to adenine, cytosine, guanine, and uracil, found primarily in tRNAs.

Uracil. *See* pyrimidines.

Variance. The average squared deviation about the mean of a set of data.

Variegation. Patchiness. A position effect caused when particular loci are contiguous with heterochromatin.

Vehicle plasmid. A plasmid containing a piece of passenger DNA forming a hybrid plasmid, used in recombinant DNA work.

Virion. A virus particle.

Viroids. Bare RNA particles that are plant pathogens.

Wild type.   The phenotype of a particular organism as first seen in nature.

Wobble.   When the third position of an anticodon is not as closely constrained as the other positions (wobbles) and thus allows additional complementary base pairing.

X linked.   *See* sex linked.

X ray crystallography.   A photographic technique, using X rays, to determine the atomic structure of molecules that have been crystallized.

Y linked.   Inheritance pattern of loci located on the Y chromosome. Also refers to the loci themselves.

Zygotene.   The stage of prophase I of meiosis in which synapsis occurs.

Zygotic induction.   When a prophage that is passed into an F^- cell during conjugation becomes virulent.

Zygotic selection.   The forces acting to cause differential mortality of an organism at any stage in its life cycle (other than gametes).

Index

Author Index

Numbers in italics indicate full references.

Subject Index